Drugs of Abuse

METHODS IN MOLECULAR MEDICINE™

John M. Walker, Series Editor

85. **Novel Anticancer Drug Protocols,** edited by *John K. Buolamwini and Alex A. Adjei*, 2003
84. **Opioid Research:** *Methods and Protocols,* edited by *Zhizhong Z. Pan,* 2003
83. **Diabetes Mellitus:** *Methods and Protocols,* edited by *Sabire Özcan,* 2003
82. **Hemoglobin Disorders:** *Molecular Methods and Protocols,* edited by *Ronald L. Nagel,* 2003
81. **Prostate Cancer Methods and Protocols,** edited by *Pamela J. Russell, Paul Jackson, and Elizabeth A. Kingsley,* 2003
80. **Bone Research Protocols,** edited by *Stuart H. Ralston and Miep H. Helfrich,* 2003
79. **Drugs of Abuse:** *Neurological Reviews and Protocols,* edited by *John Q. Wang,* 2003
78. **Wound Healing:** *Methods and Protocols,* edited by *Luisa A. DiPietro and Aime L. Burns,* 2003
77. **Psychiatric Genetics:** *Methods and Reviews,* edited by *Marion Leboyer and Frank Bellivier,* 2003
76. **Viral Vectors for Gene Therapy:** *Methods and Protocols,* edited by *Curtis A. Machida,* 2003
75. **Lung Cancer:** *Volume 2, Diagnostic and Therapeutic Methods and Reviews,* edited by *Barbara Driscoll,* 2003
74. **Lung Cancer:** *Volume 1, Molecular Pathology Methods and Reviews,* edited by *Barbara Driscoll,* 2003
73. **E. coli:** *Shiga Toxin Methods and Protocols,* edited by *Dana Philpott and Frank Ebel,* 2003
72. **Malaria Methods and Protocols,** edited by *Denise L. Doolan,* 2002
71. *Haemophilus influenzae* **Protocols,** edited by *Mark A. Herbert, Derek W. Hood, and E. Richard Moxon,* 2002
70. **Cystic Fibrosis Methods and Protocols,** edited by *William R. Skach,* 2002
69. **Gene Therapy Protocols, 2nd ed.,** edited by *Jeffrey R. Morgan,* 2002
68. **Molecular Analysis of Cancer,** edited by *Jacqueline Boultwood and Carrie Fidler,* 2002
67. **Meningococcal Disease:** *Methods and Protocols,* edited by *Andrew J. Pollard and Martin C. J. Maiden,* 2001
66. **Meningococcal Vaccines:** *Methods and Protocols,* edited by *Andrew J. Pollard and Martin C. J. Maiden,* 2001
65. **Nonviral Vectors for Gene Therapy:** *Methods and Protocols,* edited by *Mark A. Findeis,* 2001
64. **Dendritic Cell Protocols,** edited by *Stephen P. Robinson and Andrew J. Stagg,* 2001
63. **Hematopoietic Stem Cell Protocols,** edited by *Christopher A. Klug and Craig T. Jordan,* 2002
62. **Parkinson's Disease:** *Methods and Protocols,* edited by *M. Maral Mouradian,* 2001
61. **Melanoma Techniques and Protocols:** *Molecular Diagnosis, Treatment, and Monitoring,* edited by *Brian J. Nickoloff,* 2001
60. **Interleukin Protocols,** edited by *Luke A. J. O'Neill and Andrew Bowie,* 2001
59. **Molecular Pathology of the Prions,** edited by *Harry F. Baker,* 2001
58. **Metastasis Research Protocols:** *Volume 2, Cell Behavior In Vitro and In Vivo,* edited by *Susan A. Brooks and Udo Schumacher,* 2001
57. **Metastasis Research Protocols:** *Volume 1, Analysis of Cells and Tissues,* edited by *Susan A. Brooks and Udo Schumacher,* 2001
56. **Human Airway Inflammation:** *Sampling Techniques and Analytical Protocols,* edited by *Duncan F. Rogers and Louise E. Donnelly,* 2001
55. **Hematologic Malignancies:** *Methods and Protocols,* edited by *Guy B. Faguet,* 2001
54. *Mycobacterium tuberculosis* **Protocols,** edited by *Tanya Parish and Neil G. Stoker,* 2001
53. **Renal Cancer:** *Methods and Protocols,* edited by *Jack H. Mydlo,* 2001
52. **Atherosclerosis:** *Experimental Methods and Protocols,* edited by *Angela F. Drew,* 2001
51. **Angiotensin Protocols,** edited by *Donna H. Wang,* 2001
50. **Colorectal Cancer:** *Methods and Protocols,* edited by *Steven M. Powell,* 2001
49. **Molecular Pathology Protocols,** edited by *Anthony A. Killeen,* 2001

METHODS IN MOLECULAR MEDICINE™

Drugs of Abuse

Neurological Reviews and Protocols

Edited by

John Q. Wang

School of Pharmacy, University of Missouri-Kansas City

Humana Press ☀ **Totowa, New Jersey**

© 2003 Humana Press Inc.
999 Riverview Drive, Suite 208
Totowa, New Jersey 07512
humanapress.com

All rights reserved.

No part of this book may be reproduced, stored in a retrieval system, or transmitted in any form or by any means, electronic, mechanical, photocopying, microfilming, recording, or otherwise without written permission from the Publisher. Methods in Molecular Medicine™ is a trademark of The Humana Press Inc.

All papers, comments, opinions, conclusions, or recommendations are those of the author(s), and do not necessarily reflect the views of the publisher.

Due diligence has been taken by the publishers, editors, and authors of this book to assure the accuracy of the information published and to describe generally accepted practices. The contributors herein have carefully checked to ensure that the drug selections and dosages set forth in this text are accurate and in accord with the standards accepted at the time of publication. Notwithstanding, as new research, changes in government regulations, and knowledge from clinical experience relating to drug therapy and drug reactions constantly occurs, the reader is advised to check the product information provided by the manufacturer of each drug for any change in dosages or for additional warnings and contraindications. This is of utmost importance when the recommended drug herein is a new or infrequently used drug. It is the responsibility of the treating physician to determine dosages and treatment strategies for individual patients. Further it is the responsibility of the health care provider to ascertain the Food and Drug Administration status of each drug or device used in their clinical practice. The publisher, editors, and authors are not responsible for errors or omissions or for any consequences from the application of the information presented in this book and make no warranty, express or implied, with respect to the contents in this publication.

This publication is printed on acid-free paper. ∞
ANSI Z39.48-1984 (American Standards Institute) Permanence of Paper for Printed Library Materials.

Production Editor: Robin B. Weisberg
Cover illustration: Figure 1 from Chapter 25, "Primary Striatal Neuronal Culture," by Limin Mao and John Q. Wang
Cover design by Patricia Cleary

Photocopy Authorization Policy:
Authorization to photocopy items for internal or personal use, or the internal or personal use of specific clients, is granted by Humana Press Inc., provided that the base fee of US $20.00 per copy is paid directly to the Copyright Clearance Center at 222 Rosewood Drive, Danvers, MA 01923. For those organizations that have been granted a photocopy license from the CCC, a separate system of payment has been arranged and is acceptable to Humana Press Inc. The fee code for users of the Transactional Reporting Service is: [1-58829-057-3/03 $20.00].

Printed in the United States of America. 10 9 8 7 6 5 4 3 2 1

Library of Congress Cataloging in Publication Data

Main entry under title:

Methods in molecular medicine™.

Drugs of abuse : neurological reviews and protocols / edited by John Q. Wang.
    p. ; cm. —(Methods in molecular medicine; 79)
  Includes bibliographical references and index.
  ISBN 1-58829-057-3 (alk. paper) e-ISBN 1-59259-358-5
  1. Drugs of abuse—Laboratory manuals. 2. Molecular pharmacology—Laboratory manuals.
  3. Experimental psychopharmacology—Laboratory manuals. I. Wang, John Q. II. Series.
  [DNLM: 1. Street Drugs—pharmacology. 2. Central Nervous System—drug effects. 3. Gene Expression—drug effects. 4. Neurobehavioral Manifestations—drug effects. 5. Oligonucleotide Array Sequence Analysis—methods. QV 38 D79665 2003]
  RM316.D835 2003
  615'.78'0287—dc21   2002027343

Preface

Drugs of Abuse: Neurological Reviews and Protocols is intended to provide insightful reviews of key current topics and, particularly, state-of-the-art methods for examining drug actions in their various neuroanatomical, neurochemical, neurophysiological, neuropharmacological, and molecular perspectives. The book should prove particularly useful to newcomers (graduate students and technicians) in this field, as well as to those established scientists (neuroscientists, biochemists, and molecular biologists) intending to pursue new careers or directions in the study of drugs. The book's protocols cover a wide variety of coherent methods for gathering information on quantitative changes in proteins and mRNAs at both tissue and cellular levels. Inducible gene expression in striatal neurons has been a hot topic over the last decade. Alterations in gene expression for a wide range of proteins in the striatum have been investigated in response to drug administration. Altered expression of given mRNAs and their product proteins constitutes essential molecular steps in the development of neuroplasticity related to long-term addictive properties of drugs of abuse. With the multiple labeling methods that are also described in the book, gene expression can be detected in a chemically identified cell phenotype; the expression of multiple genes of interest can be detected in a single cell simultaneously. Hundreds or thousands of gene expression products can today be detected in one experimental setup using the powerful systematic cDNA macroarray or microarray screening technology. Moreover, protocols useful in analyzing the functional roles of genes and proteins (e.g., viral-mediated gene transfer, knockout mice, and antisense strategy) are also included. Also important here is the inclusion of studies on the release kinetics of striatal dopamine, a prime brain transmitter that such psychostimulants as cocaine and amphetamine interact with, using an in vivo microdialysis or real-time voltammetry technique. This study will also expand to include the quantitative measurement of other neurotransmitters (such as acetylcholine) because increasing evidence for the role of this transmitter in the control of drug actions has emerged. The properties of drugs have also been recently linked to the activity of adult neural stem and progenitor cells in the forebrain. Therefore, a timely review and two protocol chapters describing an immunohistochemical method to examine cellular proliferation and differentiation in the adult rodent brain, along with a culture method to grow viable neural progenitors, are also provided in the book.

A further feature of *Drugs of Abuse: Neurological Reviews and Protocols* is the introduction of primary neural culture preparation for studies on intracellular signaling pathways, gene expression, and so on. These cultures provide a relatively purified and easily controlled model for the investigation of cellular events related to drug's actions. Analysis of DNA binding activity in specific sites of DNA promotor regions is now possible with an electrophoretic mobility-shift assay in the cell culture tissue, in addition to striatal tissue from living brain. It can be anticipated that the usefulness of the neural culture model will undoubtedly help expand cellular and molecular research into drugs of abuse.

The chapters in *Drugs of Abuse: Neurological Reviews and Protocols* follow the format of previous volumes in the Methods in Molecular Medicine series. All chapters have been contributed by scientists with considerable experience in the protocols covered, and each protocol has been thoroughly tried and successfully tested in the respective contributor's laboratory. In each article, a final section of Notes has proven to be particularly helpful because many of the tricks of the trade are provided there; I recommend reading them thoroughly whenever troubleshooting is necessary. Illustrative data have also been included as frequently as possible so that the reader will have an opportunity to compare well-documented data with their own results from the first run of a protocol. Good luck!

John Q. Wang

Contents

Preface v
Contributors xi

PART I: SPECIAL REVIEWS

1 The Temporal Sequence of Changes in Gene Expression
  by Drugs of Abuse
  **Peter W. Kalivas, Shigenobu Toda, M. Scott Bowers,
  David A. Baker, and M. Behnam Ghasemzadeh** 3

2 Effects of Psychomotor Stimulants on Glutamate Receptor
  Expression
  Marina E. Wolf 13

3 Adult Neural Stem/Progenitor Cells in the Forebrain: *Implications
  for Psychostimulant Dependence and Medication*
  John Q. Wang, Limin Mao, and Yuen-Sum Lau 33

4 Neuroprotective Effect of Naloxone in Inflammation-Mediated
  Dopaminergic Neurodegeneration: *Dissociation from the
  Involvement of Opioid Receptors*
  Bin Liu and Jau-Shyong Hong 43

5 Neuropharmacology and Neurotoxicity of
  3,4-Methylenedioxymethamphetamine
  Gary A. Gudelsky and Bryan K. Yamamoto 55

6 Learning and Memory Mechanisms Involved in Compulsive
  Drug Use and Relapse
  Joshua D. Berke 75

7 From Drugs of Abuse to Parkinsonism: *The MPTP Mouse
  Model of Parkinson's Disease*
  Yuen-Sum Lau and Gloria E. Meredith 103

PART II. DETECTION OF mRNA EXPRESSION IN THE STRIATUM

8 *In Situ* Hybridization with Isotopic Riboprobes for Detection of Striatal Neuropeptide mRNA Expression After Dopamine Stimulant Administration
   Yasmin L. Hurd *119*

9 Combining *In Situ* Hybridization with Retrograde Tracing and Immunohistochemistry for Phenotypic Characterization of Individual Neurons
   Catherine Le Moine *137*

10 Analysis of mRNA Expression Using Double *In Situ* Hybridization Labeling with Isotopic and Nonisotopic Probes
   John Q. Wang *153*

11 Quantification of mRNA in Neuronal Tissue by Northern Analysis
   Christine L. Konradi *161*

12 Analysis of Gene Expression in Striatal Tissue by Multiprobe RNase Protection Assay
   Neil M. Richtand *181*

13 Analysis of mRNA Expression in Striatal Tissue by Differential Display Polymerase Chain Reaction
   Joshua D. Berke *193*

14 Semiquantitative Real-Time PCR for Analysis of mRNA Levels
   Stephen J. Walker, Travis J. Worst, and Kent E. Vrana *211*

15 Application of TaqMan RT-PCR for Real-Time Semiquantitative Analysis of Gene Expression in the Striatum
   Andrew D. Medhurst and Menelas N. Pangalos *229*

16 Systematic Screening of Gene Expression Using a cDNA Macroarray
   Travis J. Worst, Willard M. Freeman, Stephen J. Walker, and Kent E. Vrana *243*

PART III. DETECTION OF PROTEIN EXPRESSION IN THE STRIATUM

17 Quantification of Protein in Brain Tissue by Western Immunoblot Analysis
   Christine L. Konradi *263*

Contents

18 Immunohistochemical and Immunocytochemical Detection
of Phosphoproteins in Striatal Neurons
Limin Mao and John Q. Wang 273

19 Analysis of Protein Expression in Brain Tissue by ELISA
Steffany A. L. Bennett and David C. S. Roberts 283

20 Dopamine Receptor Binding and Quantitative Autoradiographic
Study
Beth Levant 297

21 Analysis of DNA-Binding Activity in Neuronal Tissue
with the Electrophoretic Mobility-Shift Assay
Christine L. Konradi 315

PART IV. GENE FUNCTION ANALYSIS

22 Viral-Mediated Gene Transfer to Study the Behavioral Correlates
of CREB Function in the Nucleus Accumbens of Rats
William A. Carlezon, Jr. and Rachael L. Neve 331

23 Generating Gene Knockout Mice for Studying Mechanisms
Underlying Drug Addiction
Jianhua Zhang and Ming Xu 351

24 Antisense Approaches in Analyzing the Functional Role
of Proteins in the Central Nervous System
John Q. Wang and Limin Mao 365

PART V. STRIATAL CULTURE PREPARATION

25 Primary Striatal Neuronal Culture
Limin Mao and John Q. Wang 379

26 Primary Rat Mesencephalic Neuron–Glia, Neuron-Enriched,
Microglia-Enriched, and Astroglia-Enriched Cultures
Bin Liu and Jau-Shyong Hong 387

27 Primary Culture of Adult Neural Progenitors
Steffany A. L. Bennett and Lysanne Melanson-Drapeau 397

28 Organotypic Culture of Developing Striatum: *Pharmacological
Induction of Gene Expression*
Fu-Chin Liu 405

Part VI. Detection of Transmitter Release in the Striatum

29 Microdialysis Coupled with Electrochemical Detection: *A Way to Investigate Brain Monoamine Role in Freely Moving Animals*
   Ezio Carboni 415

30 HPLC/EC Detection and Quantification of Acetylcholine in Dialysates
   James A. Zackheim and Elizabeth D. Abercrombie 433

31 Real-Time Measurements of Phasic Changes in Extracellular Dopamine Concentration in Freely Moving Rats by Fast-Scan Cyclic Voltammetry
   Paul E. M. Phillips, Donita L. Robinson, Garret D. Stuber, Regina M. Carelli, and R. Mark Wightman 443

32 Measurement of Dopamine Uptake in Neuronal Cells and Tissues
   Yuen-Sum Lau 465

Part VII. Behavioral Assessment and Others

33 Laboratory Analysis of Behavioral Effects of Drugs of Abuse in Rodents
   Neil M. Richtand 475

34 Conditioned Place Preference: *A Simple Method for Investigating Reinforcing Properties in Laboratory Animals*
   Ezio Carboni and Cinzia Vacca 481

35 Detection of Cell Proliferation and Cell Fate in Adult CNS Using BrdU Double-Label Immunohistochemistry
   Zin Z. Khaing and Mariann Blum 499

Index 507

Contributors

ELIZABETH D. ABERCROMBIE • *Center For Molecular and Behavioral Neuroscience, Rutgers University, Newark, NJ*
DAVID A. BAKER • *Department of Physiology and Neuroscience, Medical University of South Carolina, Charleston, SC*
STEFFANY A. L. BENNETT • *Department of Biochemistry, Microbiology and Immunology, University of Ottawa, Ottawa, Canada*
JOSHUA D. BERKE • *Laboratory of Cognitive Neurobiology, Department of Psychology, Boston University, Boston, MA*
MARIANN BLUM • *Department of Pharmacology, University of Texas Health Science Center San Antonio, San Antonio, TX*
M. SCOTT BOWERS • *Department of Physiology and Neuroscience, Medical University of South Carolina, Charleston, SC*
EZIO CARBONI • *Department of Toxicology, University of Cagliari, Cagliari, Italy*
REGINA M. CARELLI • *Department of Psychology, University of North Carolina, Chapel Hill, NC*
WILLIAM A. CARLEZON, JR. • *Department of Psychiatry, Harvard Medical School, McLean Hospital, Belmont, MA*
WILLARD M. FREEMAN • *Department of Physiology and Pharmacology, Wake Forest University, School of Medicine, Winston-Salem, NC*
M. BEHNAM GHASEMZADEH • *Department of Physiology and Neuroscience, Medical University of South Carolina, Charleston, SC*
GARY A. GUDELSKY • *College of Pharmacy, University of Cincinnati, Cincinnati, OH*
JAU-SHYONG HONG • *Neuropharmacology Section, National Institute of Environmental Health Science, National Institutes of Health, Research Triangle Park, NC*
YASMIN L. HURD • *Department of Clinical Neuroscience, Psychiatry Section, Karolinska Institutet, Karolinska Hospital, Stockholm, Sweden*
PETER W. KALIVAS • *Department of Physiology and Neuroscience, Medical University of South Carolina, Charleston, SC*
ZIN Z. KHAING • *Department of Pharmacology, University of Texas Health Science Center, San Antonio, TX*
CHRISTINE L. KONRADI • *Department of Psychiatry, Harvard Medical School, McLean Hospital, Belmont, MA*
YUEN-SUM LAU • *Division of Pharmacology, University of Missouri-Kansas City, School of Pharmacy, Kansas City, MO*
CATHERINE LE MOINE • *Laboratoire d'Histologie-Embryologie CNSR UMR 5541, Université Victor Segalen Bordeaux 2, Bordeaux Cedex, France*

BETH LEVANT • *Department of Pharmacology, Toxicology and Therapeutics, University of Kansas Medical Center, Kansas City, KS*
BIN LIU • *Neuropharmacology Section, National Institute of Environmental Health Science, National Institutes of Health, Research Triangle Park, NC*
FU-CHIN LIU • *Institute of Neuroscience, National Yang-Ming University, Taipei, Taiwan*
LIMIN MAO • *Division of Pharmacology, School of Pharmacy, University of Missouri-Kansas City, Kansas City, MO*
ANDREW D. MEDHURST • *Neurodegeneration Research, GlaxoSmithKline, Harlow, UK*
LYSANNE MELANSON-DRAPEAU • *Department of Biochemistry, Microbiology and Immunology, University of Ottawa, Ottawa, Canada*
GLORIA E. MEREDITH • *Department of Cellular and Molecular Pharmacology, Chicago Medical School, North Chicago, IL*
RACHAEL L. NEVE • *Molecular Neurogenetics, Department of Psychiatry, Harvard Medical School, McLean Hospital, Belmont, MA*
MENELAS N. PANGALOS • *Neurodegeneration Research, GlaxoSmithKline, Essex, Harlow, UK*
PAUL E. M. PHILLIPS • *Department of Chemistry, University of North Carolina at Chapel Hill, NC*
NEIL M. RICHTAND • *Department of Psychiatry, Cincinnati Veterans Affairs Medical Center and University of Cincinnati College of Medicine, Cincinnati, OH*
DAVID C.S. ROBERTS • *Department of Physiology and Pharmacology, Wake Forest University School of Medicine, Winston-Salem, NC*
DONITA L. ROBINSON • *Department of Chemistry, University of North Carolina at Chapel Hill, NC*
GARRET D. STUBER • *Neuroscience Center, University of North Carolina, Chapel Hill, NC*
SHIGENOBU TODA • *Department of Physiology and Neuroscience, Medical University of South Carolina, Charleston, SC*
CINZIA VACCA • *Department of Toxicology, University of Cagliari, Cagliari, Italy*
KENT E. VRANA • *Department of Physiology and Pharmacology, Wake Forest University School of Medicine, Winston-Salem, NC*
STEPHEN J. WALKER • *Department of Physiology and Pharmacology, Wake Forest University School of Medicine, Winston-Salem, NC*
JOHN Q. WANG • *Division of Pharmacology, University of Missouri-Kansas City, School of Pharmacy, Kansas City, MO*
R. MARK WIGHTMAN • *Department of Chemistry, University of North Carolina at Chapel Hill, NC*
MARINA E. WOLF • *Department of Neuroscience, Chicago Medical School, North Chicago, IL*
TRAVIS J. WORST • *Department of Physiology and Pharmacology, Wake Forest University School of Medicine, Winston-Salem, NC*
MING XU • *Department of Cell Biology, Neurobiology and Anatomy, University of Cincinnati College of Medicine, Cincinnati, OH*

Contributors

BRYAN K. YAMAMOTO • *Department of Pharmacology, Boston University Medical School, Boston, MA*

JAMES A. ZACKHEIM • *Center For Molecular and Behavioral Neuroscience, Rutgers University, Newark, NJ*

JIANHUA ZHANG • *Department of Cell Biology, Neurobiology and Anatomy, University of Cincinnati College of Medicine, Cincinnati, OH*

I

SPECIAL REVIEWS

1

The Temporal Sequence of Changes in Gene Expression by Drugs of Abuse

Peter W. Kalivas, Shigenobu Toda, M. Scott Bowers, David A. Baker, and M. Behnam Ghasemzadeh

1. Introduction

Addiction is a complex maladaptive behavior produced by repeated exposure to rewarding stimuli (*1*). There are two primary features of addiction to all forms of natural and pharmacological stimuli. First, the rewarding stimulus associated with the addiction is a compelling motivator of behavior at the expense of behaviors leading to the acquisition of other rewarding stimuli. Thus, individuals come to orient increasing amounts of their daily activity around acquisition of the rewarding stimulus to which they are addicted. Second, there is a persistence of craving for the addictive stimulus, combined with an inability to regulate the behaviors associated with obtaining that stimulus. Thus, years after the last exposure to an addictive stimulus, reexposure to that stimulus or environmental cues associated with that stimulus will elicit behavior seeking to obtain the reward.

During the course of repeated exposures to strong motivationally relevant stimuli specific brain nuclei and circuits become engaged that mediate the addicted behavioral response. It is generally thought that different rewarding stimuli involve different brain circuits, but that regions of overlap with other motivational stimuli exist, forming a common substrate for all addictive stimuli. Studies using animal models of reward and addiction have focused on subcortical brain circuits known to be involved in drug reward, such as the dopamine projection from the ventral mesencephalon to the nucleus accumbens (*2,3*). Accordingly, molecular and electrophysiological studies of the cellular plasticity mediating the emergence of addictive behaviors have focused on the nucleus accumbens and ventral mesencephalon. However, about 5 yr ago studies emerged from both the animal literature and neuroimaging of drug addicts indicating that the expression of addicted behaviors such as sensitization and craving involved regions of the cortex and allocortex *(4–7)*. In this regard, two regions that have come to be closely associated with addiction are the amygdala and frontal cortex (including the anterior cingulate and ventral orbitofrontal cortex). In addition, the last decade of research has revealed a variety of enduring changes in gene expression produced by repeated exposure to drugs of abuse, notably psychostimulants *(3,8)*. The most long-lasting neuroadaptations that would be expected to underlie enduring behaviors associated with addiction appear to be concentrated in the nucleus accumbens and in cortical regions providing input to the nucleus accumbens, such as the prefrontal cortex. These studies are outlined and integrated with the corticostriatal circuitry postulated to be critical for the expression of behavioral characteristics of psychostimulant addiction, such as sensitization and craving.

2. Temporal and Anatomical Sequence of Changes in Gene Expression

A variety of studies using different addictive drugs, given in different dosing regimens and employing different withdrawal periods, have shown that repeated administration of addictive drugs produced short, intermediate, and enduring changes in gene expression. **Figure 1** illustrates the sequence of changes in gene expression associated with repeated cocaine administration. Five categories of cocaine-induced changes in gene expression are outlined, ranging from increases in immediate early gene (IEG) expression that diminish with repeated injections to changes in gene expression that appear only after a period of withdrawal. The data outlined in **Fig. 1** are specific for cocaine-induced changes in gene expression, and using these data as a guide certain temporal patterns of drug-induced changes in gene expression can be discerned from the extant literature. Similarly, anatomical patterns of gene expression related to various times during the chronic injection and withdrawal periods can be shown. However, there are exceptions in the anatomical discretion, and, importantly, in many brain nuclei relevant to addiction, notably the amygdala, very little data have been collected regarding changes in gene expression.

3. Rapid Response and Tolerance, Widespread in Dopamine Terminal Fields

The earliest changes in gene expression that are measurable shortly after acute drug administration occur in many brain regions, the most well studied being dopamine terminal fields such as the striatum, nucleus accumbens, and prefrontal cortex. These genes include classic IEG transcription factors such

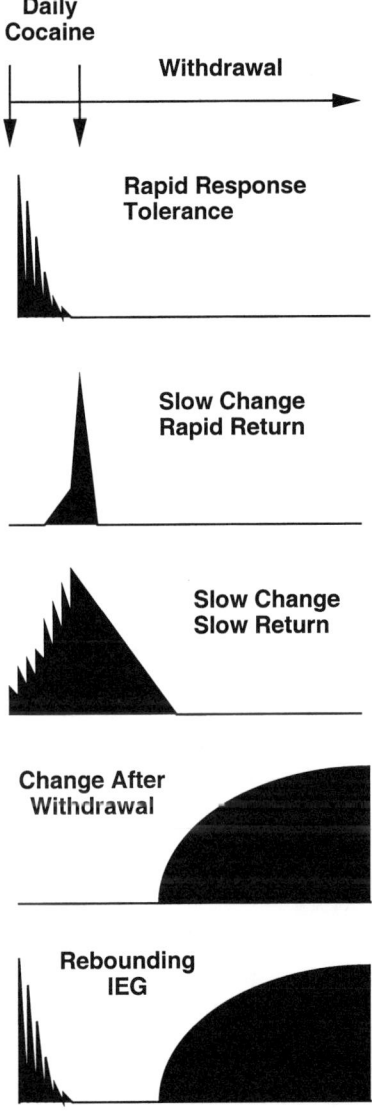

Fig. 1. Temporal pattern of changes in gene expression produced by repeated injections of cocaine.

as c-*fos* and *zif268* (*9,10*). However, both cytosolic IEGs such as homer1a and arc and extracellular IEG-like proteins such as the pentraxin narp are also induced in a number of brain regions by acute administration of cocaine (*11*). The increase in these proteins is thought to initiate changes that partly mediate the acute effects of drugs, as well as provide a background upon which subsequent, more enduring changes in gene expression can emerge. In general, the proteins encoded by IEGs have a relatively short half-life, and levels return to normal within 24 h after the injection. Moreover, with repeated administration of drug the induction produced by each injection becomes progressively less until by 1 wk of injection little or no induction is produced.

4. Slow Progressive Change and Rapid Return, Predominately in the Ventral Tegmental Area

Another category of changes in gene expression are those that gradually accumulate with repeated administration but disappear within a few days after the last injection. Interestingly, many changes in gene expression in this category are found in dopamine or nondopamine cells in the ventral tegmental area (VTA). Included are proteins encoded by genes that are directly related to dopamine transmission, such as tyrosine hydroxylase and dopamine transporters *(12–14)*. In addition, genes associated with dopamine receptor signaling such as *Giα* undergo a short-term change in expression after the last cocaine injection *(15)*. Notably, the expression of genes related to glutamate transmission such as *GluR1* and *NMDAR2* are also included in this category *(16,17)*. Taken together these changes in gene expression appear to facilitate glutamatergic activation of cells in the VTA while simultaneously diminishing the capacity of D2 dopamine autoreceptors to provide negative regulation of dopamine cell firing *(18,19)*. These changes probably contribute to known physiological alterations in dopamine cell function associated with short-term withdrawal such as increased dopamine cell firing and enhanced releasibility of dopamine, glutamate, and γ-aminobutyric acid (GABA) in the VTA *(20,23)*. In addition, the disinhibition of dopamine cells may contribute to the increased releasibility of dopamine in axon terminal fields such as the nucleus accumbens and striatum.

5. Slow Change, Slow Return, Predominately in the Striatal Dopamine Terminal Fields

This category of cocaine-induced changes in gene expression has recently received considerable attention as possible mediators of the transition from casual to addictive patterns of drug-taking *(24)*. Some of these genes are IEG-like in that they are induced by acute drug administration. However, the proteins have a relatively long half-life. As a result elevated protein levels are present for an extended period, as long as weeks after the last drug injection. The classic gene in the category is Δ-*fosB*, which has been shown to accumulate in the striatum with repeated psychostimulant exposure *(25)*. Notably, the increased expression is also associated with a redistribution of cellular expression into different striatal compartments *(26)*. In addition, changes in gene expression and protein function associated with D1 receptor signaling fall into this category, including an induction in protein kinase A, mitogen-activated protein (MAP) kinase, and phospho-cAMP response element binding protein (phospho-CREB) *(24)*. Accordingly, genes regulated by phospho-CREB, such as preprodynorphin, are also altered by repeated cocaine administration and endure for weeks after the last injection *(27,28)*. Likewise, while gene expression may not be altered, proteins regulated by protein kinase A (PKA), CdK5, or MAP kinase phosphorylation demonstrate altered function for an extended withdrawal period after the last drug injection, including sodium channels and the cystine/glutamate antiporter in the striatum *(29)*. In addition, proteins related to glutamate transmission show the slow change/slow return pattern of expression, including mGluR5, which has been recently linked to cocaine reward *(30,31)*. Also, proteins involved in other neurotransmitter systems in the striatal complex, including histidine decarboxylase and the adenosine transporter, show this temporal pattern *(30,32,33)*. These changes play a significant role in some of the enduring changes in excitability in spiny cells in the nucleus accumbens and striatum. Notably, spiny cells show more avid inhibition in response to D1 receptor stimulation and have a decreased postsynaptic response to α-amino-3-hydroxy-5-methyl-4-isoxazole propionic acid (AMPA) receptor stimulation or long-term potentiation in response to tetanic stimulation of glutamatergic afferents to the nucleus accumbens *(19,34)*.

6. Changes Only During Withdrawal, Enduring for Weeks, Predominately in the Prefrontal Cortex and Nucleus Accumbens

Members of this category of genes have undergone recent intensive study and the changes in expression generally appear after only a week or more of withdrawal from repeated drug administration. The changes are almost exclusively in the prefrontal cortex and nucleus accumbens and include a variety of gene products involved in neurotransmission, cell signaling, and glial function. However, the changes are notable in that they endure for weeks and involve a predominance of genes affecting glutamate transmission relative to dopamine transmission. Genes in this category altered by cocaine encode mGluR1, mGluR2/3, homer1bc, GluR5, A1 adenosine receptor, TrkB, BDNF, AGS3, Giα, GFAP, and vimentin. These changes in expression combine to produce a generalized decrease in signaling through group I and group II mGluR and in general serve to decrease excitability of cells in the nucleus accumbens *(30,35,36)*. In addition, the changes in glial fibrillary acidic protein (GFAP) and vimentin suggest an enduring activation of glia, which may contribute to the reduction in extracellular glutamate in the nucleus accumbens that is associated with repeated cocaine administration *(37)*.

7. Rebounding IEG

This is a category that to date contains a single gene product, nac-1 *(38,39)*. This protein has expression characteristics of the IEG class in that levels are induced by acute drug administration, and progressive tolerance to this induction occurs with repeated administration. However, similar to late expressing genes, the levels of nac-1 rise at 1 wk of withdrawal and are maintained for at least 3 wk thereafter. Experiments using viral overexpression of nac-1 and antisense oligonucleotide inhibition of protein expression reveal that nac-1 is important in the development of behavioral sensitization and in the acquisition of cocaine self-administration.

8. Anatomical Sequence of Gene Expression and the Development of Enduring Changes in Reward Circuitry

As outlined in the preceding, different brain regions demonstrate the majority of changes in gene expression in a temporal sequence. Changes to acute administration are very widespread, predominately in dopamine axon terminal fields. A large number of alterations in gene expression that exist for a relatively short duration after discontinuing repeated drug administration are found in the VTA. These changes may contribute to an increased responsiveness of dopamine cells to acute drug injection that will promote more enduring changes in gene expression in dopamine axon terminal fields such as the prefrontal cortex and nucleus accumbens. In the dopamine terminal fields the expression of proteins undergoes a transition from those that are produced during repeated drug administration and endure for a period of time after injection to changes in expression that develop later in withdrawal and endure for an extended period after the last drug injection. This temporal transition in gene expression can be seen as constituting a new baseline of cellular functioning that mediates the expression of behaviors associated with addiction, such as drug craving and sensitization. Notably, these enduring changes in expression are in the prefrontal cortex and nucleus accumbens, and the relationship between these two regions has come under increasing scrutiny as the site of primary pathology in psychostimulant addiction.

9. Conclusions

The studies reviewed in this chapter point to the possibility of a final common pathway, and possibly similar cellular neuroadaptations between drugs and stimuli that provoke craving and relapse. The extant data support a role of the projection from the prefrontal cortex to nucleus accumbens in the expression of addiction-related behaviors, such as sensitization and drug-seeking behavior, and there is abundant evidence for enduring neuroadaptations in gene expression and neuronal function in these brain regions following a bout of drug-taking. Although the studies outlined in this chapter are promising in pointing to a common point of intervention in addiction to various chemical classes of drugs, it is important to note that such a generalization based primarily on work with psychostimulants is premature, and verification will require substantially more research using other classes of drugs. Also, the temporal sequence of neuroadaptive changes during drug withdrawal points to the possible utility of targeting different pharmacotherapies at different stages of withdrawal. Thus, in early withdrawal drugs affecting dopamine transmission may be more effective, while in later withdrawal modulation of glutamate transmission may be more efficacious.

References

1. O'Brien, C. (2001) Drug addiction and drug abuse, in *The Pharmacological Basis of Therapeutics* (Hardman, J., Limbird, L., and Gilman, A. G., eds.) McGraw-Hill, New York, pp. 621–642.
2. Koshiya, K. and Kato, T. (1983) Acute changes in nigral substance P content induced by drugs acting on dopamine, muscarine and GABA receptors. *Naunyn Schmiedebergs Arch. Pharmacol.* **324,** 223–227.
3. Unterwald, E. M., Cox, B. M., Kreek, M. J., Cote, T. E., and Izenwasser, S. (1993) Chronic repeated cocaine administration alters basal and opioid-regulated adenylyl cyclase activity. *Synapse* **15,** 33–38.
4. Pierce, R. C. and Kalivas, P. W. (1997) A circuitry model of the expression of behavioral sensitization to amphetamine-like psychostimulants. *Brain Res. Rev.* **25,** 192–216.
5. Volkow, N. D. and Fowler, J. S. (2000) Addiction, a disease of compulsion and drive: involvement of the orbitofrontal cortex. *Cereb. Cortex* **10,** 318–325.
6. Grant, S., London, E. D., Newlin, D. B., et al. (1996) Activation of memory circuits during cue-elicited cocaine craving. *Proc. Natl. Acad. Sci. USA* **93,** 12040–12045.
7. Childress, A. R., Mozley, P. D., McElgin, W., Fitzgerald, J., Reivich, M., and O'Brien, C. P. (1999) Limbic activation during cue-induced cocaine craving. *Am. J. Psychiatry* **156,** 11–18.
8. White, F. J. and Kalivas, P. W. (1998) Neuroadaptations involved in amphetamine and cocaine addiction. *Drug Alcohol Depend.* **51,** 141–154.
9. Daunais, J. and McGinty, J. F. (1996) The effects of D1 or D2 dopamine receptor blockade on *zif*/268 and preprodynorphin gene expression in rat forebrain following a short-term cocaine binge. *Mol. Brain Res.* **354,** 237–248.
10. Moratalla, R., Vickers, E. A., Robertson, H. A., Cochran, B. H., and Graybiel, A. M. (1993) Coordinate expression of c-fos and jun B is induced in the rat striatum by cocaine. *J. Neurosci.* **13,** 423–433.
11. Lanahan, A. and Worley, P. (1998) Immediate-early genes and synaptic function. *Neurobiol. Learn. Mem.* **70,** 37–43.

12. Vrana, S. L., Vrana, K. E., Koves, T. R., Smith, J. E., and Dworkin, S. I. (1993) Chronic cocaine administration increases CNS tyrosine hydroxylase enzyme activity and mRNA levels and tryptophan hydroxylase enzyme activity levels. *J. Neurochem.* **61,** 2262–2268.
13. Sorg, B. A., Chen, S. Y., and Kalivas, P. W. (1993) Time course of tyrosine hydroxylase expression following behavioral sensitization to cocaine. *J. Pharmacol. Exp. Ther.* **266,** 424–430.
14. Xia, Y., Goebel, D. J., Kapatos, G., and Bannon, M. J. (1992) Quantitation of rat dopamine transporter mRNA: effects of cocaine treatment and withdrawal. *J. Neurochem.* **59,** 1179–1182.
15. Nestler, E. J., Terwilliger, R. Z., Walker, J. R., Sevarino, K. A., and Duman, R. S. (1990) Chronic cocaine treatment decreases levels of the G protein subunits $G_{i\alpha}$ and $G_{o\alpha}$ in discrete regions of rat brain. *J. Neurochem.* **55,** 1079–1082.
16. Fitzgerald, L. W., Ortiz, J., Hamedani, A. G., and Nestler, E. J. (1996) Drugs of abuse and stress increase the expression of GluR1 and NMDAR1 glutamate receptor subunits in the rat ventral tegmental area: common adaptions among cross-sensitizing agents. *J. Neurosci.* **16,** 274–282.
17. Churchill, L., Swanson, C. J., Urbina, M., and Kalivas, P. W. (1999) Repeated cocaine alters glutamate receptor subunit levels in the nucleus accumbens and ventral tegmental area of rats that develop behavioral sensitization. *J. Neurochem.* **72,** 2397–2403.
18. Wolf, M. E., White, F. J., Nassar, R., Brooderson, R. J., and Khansa, M. R. (1993) Differential development of autoreceptor subsensitivity and enhanced dopamine release during amphetamine sensitization. *J. Pharmacol. Exp. Ther.* **264,** 249–255.
19. White, F. J., Hu, X. T., Zhang, X. F., and Wolf, M. E. (1995) Repeated administration of cocaine or amphetamine alters neuronal responses to glutamate in the mesoaccumbens dopamine system. *J. Pharmacol. Exp. Ther.* **273,** 445–454.
20. White, F. J., et al. (1995) Neurophysiological alterations in the mesocorticolimbic dopamine system during repeated cocaine administration., in *The Neurobiology of Cocaine Addiction* (Hammer, R., ed.), CRC Press, Boca Raton, FL, pp. 99–120.
21. Kalivas, P. W. and Duffy, P. (1993) Time course of extracellular dopamine and behavioral sensitization to cocaine. II. Dopamine perikarya. *J. Neurosci.* **13,** 276–284.
22. Kalivas, P. W. and Duffy, P. (1998) Repeated cocaine administration alters extracellular glutamate in the ventral tegmental area. *J. Neurochem.* **70,** 1497–1502.
23. Bonci, A. and Williams, J. T. (1996) A common mechanism mediates long-term changes in synaptic transmission after chronic cocaine and morphine. *Neuron* **16,** 631–639.
24. Nestler, E. (2001) Molecular basis of long-term plasticity underlying addiction. *Nat. Rev.* **2,** 119–128.
25. Nestler, E. J., Barrot, M., and Self, D. W. (2001) DeltaFosB: a sustained molecular switch for addiction. *Proc. Natl. Acad. Sci. USA* **98,** 11042–11046.

26. Moratalla, R., Elibol, B., Vallejo, M., and Graybiel, A. M. (1996) Network-level changes in expression of inducible fos-jun proteins in the striatum during chronic cocaine treatment and withdrawal. *Neuron* **17,** 147–156.
27. Hurd, Y. L. and Herkenham, M. (1993) Molecular alterations in the neostriatum of human cocaine addicts. *Synapse* **13,** 357–369.
28. Kelz, M. B, Chen, J., Carlezon, W. A., Jr., et al. (1999) Expression of the transcription factor deltaFosB in the brain controls sensitivity to cocaine. *Nature* **401,** 272–276.
29. Zhang, X. F., Hu, X. T., and White, F. J. (1998) Whole-cell plasticity in cocaine withdrawal: reduced sodium current in nucleus accumbens neurons. *J. Neurosci.* **18,** 488–498.
30. Swanson, C. J., Baker, D. A., Carson, D., Worley, P. F., and Kalivas, P. W. (2001) Repeated cocaine administration attenuates group I metabotropic glutamate receptor-mediated glutamate release and behavioral activation: a potential role for Homer 1b/c. *J. Neurosci.* **21,** 9043–9052.
31. Chiamulera, C., Epping-Jordan, M. P., Zocchi, A., et al. (2001) Reinforcing and locomotor stimulant effects of cocaine are absent in mGluR5 null mutant mice. *Nat. Neurosci.* **4,** 873–874.
32. Kubota, Y., Ito, C., Kuramasu, A., Sato, M., and Watanabe, T. (1999) Transient increases of histamine H1 and H2 receptor mRNA levels in the rat striatum after the chronic administration of methamphetamine. *Neurosci. Lett.* **275,** 37–40.
33. Ito, C., Onodera, K., Sakurai, E., Sato, M., and Watanabe, T. (1997) Effect of cocaine on the histaminergic neuron system in the rat brain. *J. Neurochem.* **69,** 875–878.
34. Henry, D. J. and White, F. J. (1995) The persistence of behavioral sensitization to cocaine parallels enhanced inhibition of nucleus accumbens neurons. *J. Neurosci.* **15,** 6287–6299.
35. Manzoni, O., Pujalte, D., Williams, J., and Bockaert, J. (1998) Decreased presynaptic sensitivity to adenosine after cocaine withdrawal. *J. Neurosci.* **18,** 7996–8002.
36. Pierce, R., Pierce-Bancroft, A., and Prasad, B. (2000) Neurotrophin-mediated second messengers contribute to the initiation of behavioral sensitization to cocaine. *J. Neurosci.* **20,** 1–34.
37. Haile, C. N., Hiroi, N., Nestler, E. J., and Kosten, T. A. (2001) Differential behavioral responses to cocaine are associated with dynamics of mesolimbic dopamine proteins in Lewis and Fischer 344 rats. *Synapse* **41,** 179–190.
38. Cha, X. Y., Pierce, R. C., Kalivas, P. W., and Mackler, S. A. (1997) NAC-1, a rat brain mRNA, is increased in the nucleus accumbens three weeks after chronic cocaine self-administration. *J. Neurosci.* **17,** 6864–6871.
39. Mackler, S. A., Korutla, L., Cha, X. Y., Koebbe, M. J., Fournier, K. M., Bowers, M.S., and Kalivas, P. (2000) NAC-1 is a brain POZ/BTB protein that can prevent cocaine-induced sensitization in the rat. *J. Neurosci.* **20,** 6210–6217.

2

Effects of Psychomotor Stimulants on Glutamate Receptor Expression

Marina E. Wolf

1. Introduction: Addiction as a Form of Glutamate-Dependent Plasticity

It is increasingly well accepted that addiction can be viewed as a form of neuronal plasticity, even as a type of very powerful, albeit maladaptive, learning. On a behavioral level, this can be conceptualized as the transition from experimentation to compulsive drug-seeking behavior. This view of addiction has been strengthened by many recent studies demonstrating commonalities between mechanisms underlying learning and addiction. Both are associated with changes in gene expression, phosphorylation and phosphatase cascades, neurotrophin signaling, altered dendritic morphology, and activity-dependent forms of plasticity such as long-term potentiation (LTP) and long-term depression (LTD) *(1,2)*. Through these mechanisms, drugs of abuse are proposed to strengthen or weaken activity in pathways related to motivation and reward. This in turn may produce behavioral changes that drive compulsive drug-seeking behavior in addiction, including sensitization of incentive-motivational effects of drugs, enhanced ability of drug-conditioned stimuli to control behavior, and loss of inhibitory control mechanisms that normally govern reward-seeking behavior *(3,4)*.

An open question is how drugs of abuse, which initially target monoamine receptors, are able to influence mechanisms of synaptic plasticity. Glutamate is a key transmitter for synaptic plasticity, and many neuronal pathways implicated in addiction are glutamatergic *(4)*. Historically, studies of behavioral sensitization, a well-established animal model for addiction, were important in directing drug addiction research toward glutamate *(5)*. Behavioral sensitization

From: *Methods in Molecular Medicine, vol. 79: Drugs of Abuse: Neurological Reviews and Protocols*
Edited by: J. Q. Wang © Humana Press Inc., Totowa, NJ

refers to the progressive enhancement of species-specific behavioral responses that occurs during repeated drug administration and persists even after long periods of withdrawal. Although most studies have measured sensitization of locomotor activity, sensitization also occurs to the reinforcing effects of psychomotor stimulants. Behavioral sensitization is influenced by the same factors that influence addiction (stress, conditioning, and drug priming), and is accompanied by profound cellular and molecular adaptations in the neuronal circuits that are fundamentally involved in normal motivated behavior as well as addiction. Like addiction, it is extremely persistent. Robinson and Berridge *(6,7)* have argued for an incentive-sensitization view of addiction, which holds that repeated drug administration sensitizes the neuronal systems involved in drug "wanting" rather than drug "liking."

It is now acknowledged that the development of sensitization requires glutamate transmission in the midbrain, where dopamine (DA) cell bodies are located, whereas its maintenance and expression are associated with profound changes in glutamate transmission in limbic and cortical brain regions that receive dopaminergic innervation. To understand the role of glutamate transmission in sensitization, many studies have examined drug effects on glutamate transmission in these brain regions. This review focuses on cocaine and amphetamine effects on glutamate receptor expression in the midbrain (ventral tegmental area [VTA] and substantia nigra), the striatal complex (nucleus accumbens [NAc] and dorsal striatum), and the prefrontal cortex (PFC). Recent studies are emphasized, with the goal of updating a comprehensive review published 4 yr ago *(8)*.

2. Effects of Psychomotor Stimulants on Glutamate Receptor Expression in the VTA and Substantia Nigra

2.1. Role of the VTA in Behavioral Sensitization

Many lines of evidence have suggested that the development of behavioral sensitization is associated with an increase in excitatory drive to VTA DA neurons *(8)*. This provided the impetus for examining whether glutamate transmission is enhanced in the VTA during the early phase of drug withdrawal. The first evidence to support this hypothesis came from in vivo single-unit recording studies demonstrating that VTA DA neurons recorded from cocaine- or amphetamine-sensitized animals were more responsive to the excitatory effects of iontophoretic glutamate *(9)*. A subsequent study showed that increased responsiveness was selective for α-amino-3-hydroxy-5-methylisoxazole-4-propionic acid (AMPA) (there was no change in sensitivity to N-methyl-D-aspartate [NMDA] or a metabotropic glutamate receptor agonist) and transient, present 3 but not 10–14 d after discontinuing repeated drug administration *(10)*. Recently, we have shown that AMPA receptor supersensitivity can also be demonstrated in microdialysis experiments, by monitoring the ability of intra-VTA AMPA to activate VTA DA neurons and thus increase DA levels in the ipsilateral NAc. Using dual-probe microdialysis, we found that intra-VTA administration of a low dose of AMPA produced significantly greater DA efflux in the NAc of amphetamine-treated rats *(11)*. This augmented response was transient (present 3 but not 10–14 d after the last injection) and specific for AMPA, as intra-VTA NMDA administration produced a trend toward increased NAc DA levels that did not differ between groups. Thus, both microdialysis and in vivo electrophysiological data suggest an enhancement of AMPA receptor transmission onto VTA DA neurons during the early phase of drug withdrawal. An increase in glutamate receptor expression would provide a simple explanation for such findings. For this and other reasons, a number of studies have examined the effect of repeated drug administration on glutamate receptor expression in VTA. Studies on glutamate receptor expression in the substantia nigra are also considered. Although the substantia nigra has received less attention in recent years than the VTA, it exhibits similar drug-induced adaptations and is also implicated in the development of sensitization *(8)*.

2.2. Results in the VTA and Substantia Nigra

Using Western blotting, Nestler and colleagues found increased GluR1 levels in the VTA of rats killed 16–18 h after discontinuation of repeated cocaine, morphine, ethanol, or stress paradigms *(12,13)*. Increased GluR1 was not observed in the substantia nigra after repeated cocaine or morphine treatment *(12)*. The substantia nigra was not examined in stress studies *(12)*, but after repeated ethanol administration, there was a greater increase in GluR1 in the substantia nigra than in the VTA *(13)*. Repeated cocaine also increased NR1 in VTA but had no effect on GluR2, NR2A/B, or GluR6/7 *(12)*. Churchill et al. *(14)* treated rats with saline or cocaine for 7 d (15 mg/kg on d 1 and 7, 30 mg/kg on d 2–6), measuring locomotor activity after the first and last injections; those rats that showed >20% increase in locomotor activity were defined as sensitized. Then, protein levels of glutamate receptor subunits were determined by Western blotting 24 h or 3 wk after daily injections were discontinued. In agreement with results of Fitzgerald et al. *(12)*, Churchill et al. *(14)* found increased GluR1 and NR1 levels in the VTA of rats killed 1 d but not 3 wk after discontinuation of this different cocaine regimen. Interestingly, this was observed only in those cocaine-treated rats that developed sensitization. GluR2/3 was not measured after 1 d but was unaltered after 3 wk *(14)*.

In contrast, our own quantitative immunoautoradiography studies found no change in GluR1 immunoreactivity in VTA, substantia nigra, or a transitional area after 16–24 h of withdrawal from repeated amphetamine or cocaine treatment *(15)*. Importantly, this study *(15)* also failed to find a change in GluR1 immunoreactivity after 3 or 14 d of withdrawal from the same amphetamine regimen that resulted in enhanced electrophysiological *(10)* and neurochemical *(11)* responsiveness to intra-VTA AMPA at the 3-d withdrawal time. Thus, although part of the discrepancy between results from different labs may be attributable to different drug regimens, our findings suggest that increased GluR1 expression is unlikely to explain our electrophysiological or neurochemical findings of increased responsiveness to AMPA. Other possible reasons for differences between our immunoautoradiography studies and prior Western blotting studies have been discussed previously *(15)*.

Another finding relevant to this controversy is that overexpression of GluR1 in the rostral VTA using a herpes simplex virus resulted in intensification of the locomotor stimulant and rewarding properties of morphine *(16,17)*. Although this is an interesting finding, it does not necessarily imply that increased GluR1 expression is involved in the naturally occurring pathways that produce behavioral sensitization to morphine or psychomotor stimulants. A state resembling behavioral sensitization can be produced by a number of diverse experimental manipulations, all sharing the ability to produce brief but intense activation of VTA DA cells. These include repeated electrical stimulation of the VTA *(18)* or PFC *(19)*, and pharmacological disinhibition of VTA DA cells *(20)*.

In contrast to discrepant results at the protein level, all studies agree that mRNA levels for AMPA receptor subunits in the VTA are not altered during withdrawal from repeated amphetamine or cocaine. We found no change in GluR1 mRNA using reverse transcriptase-polymerase chain reaction (RT-PCR) in the VTA of rats killed 16–18 h after discontinuing repeated amphetamine or cocaine administration *(15)*. Similarly, Bardo et al. *(21)* used RNase protection assays to quantify GluR1-4 mRNA levels in the ventral mesencephalon of rats killed 30 min after the third or tenth amphetamine injection in a repeated regimen and observed no significant changes, although behavioral sensitization was demonstrated. Ghasemzadeh et al. *(22)* used RT-PCR to determine mRNA levels for GluR1-4, NR1, and mGluR5 in the VTA 3 wk after discontinuing repeated cocaine or saline injections, and found no significant changes as a result of repeated cocaine treatment, although acute cocaine challenge produced a small reduction in NR1 mRNA levels in the VTA of both naïve and sensitized rats.

As noted previously, Western blotting studies have found increased NR1 levels in the VTA of rats killed 16–24 h (but not 3 wk) after discontinuing repeated cocaine administration, suggesting that the increase is transient *(12,14)*. In contrast, using immunohistochemical methods, Loftis and Janowsky

(23) compared rats treated with repeated cocaine or saline and found significant increases in NR1 immunoreactivity in the cocaine group after 3 and 14 d of withdrawal and a trend toward an increase after 24 h. Using the same regimen as Churchill et al. *(14)*, Ghasemzadeh et al. *(22)* found no significant changes in NR1 mRNA levels in VTA using RT-PCR. We used quantitative immunoautoradiography to examine NR1 expression in VTA, substantia nigra, and a transitional area in rats killed 3 or 14 d after discontinuing repeated amphetamine administration. No changes were observed after 3 d of withdrawal, whereas NR1 immunolabeling was significantly decreased in the intermediate and caudal portions of the substantia nigra, but not in other midbrain regions, after 14 d of withdrawal *(24)*. NR1 levels in the NAc and prefrontal cortex were also decreased at this withdrawal time *(24)*. It may be relevant to note that although NMDA receptor transmission in the VTA is required for the induction of sensitization, repeated stimulation of NMDA receptors in the VTA is not sufficient to elicit sensitization *(25,26)*.

2.3. Summary: VTA and Substantia Nigra

As reviewed in **Subheading 2.1.**, both neurochemical and electrophysiological studies suggest that there is an enhancement in the responsiveness of VTA DA neurons to the excitatory effects of AMPA shortly after discontinuing repeated psychostimulant administration. An increase in AMPA receptor expression in the VTA would provide a simple explanation for these results. However, although Western blotting studies have found increased GluR1 and NR1 levels in the VTA shortly after cocaine administration is discontinued, this is not observed with immunoautoradiography following either cocaine or amphetamine administration (see *15* for discussion). More importantly, after the same drug regimens and withdrawal times that are associated with increased responsiveness of VTA DA neurons to AMPA, no changes in GluR1 are observed. Thus, although considerable evidence suggests that enhanced responsiveness of VTA DA neurons to AMPA is closely linked to the induction of sensitization, the mechanisms are likely to be more complex than a generalized increase in GluR1 expression within the VTA.

As LTP is expressed as a potentiation of AMPA receptor transmission, an alternative explanation is that sensitization is accompanied by LTP-like changes that increase the efficiency of glutamate transmission in the VTA. Although LTP appears to involve insertion of AMPA receptor subunits into synaptic sites *(27)*, there is no evidence that this is accompanied by increases in total cellular expression of AMPA receptor subunits. Supporting the involvement of LTP in the development of sensitization, a single systemic injection of cocaine to mice (sufficient to elicit behavioral sensitization) produced LTP in midbrain DA neurons *(28)*. The mechanisms responsible are probably complex.

DA-releasing stimulants could promote LTP by decreasing the opposing influence of LTD, as D2 receptor activation inhibits LTD in midbrain slices *(29,30)*. Psychostimulant-induced increases in VTA glutamate levels may also promote LTP *(31,32)*. Of course, mechanisms unrelated to LTP may also contribute to increased excitability of VTA DA neurons, including inhibition of mGluR-mediated inhibitory postsynaptic potentials (IPSPs) *(33,34)*. Finally, it should be noted that glutamate transmission in the VTA may be influenced by drug-induced alterations in other transmitter systems. Mechanisms that may contribute to sensitization-related plasticity in the VTA have been reviewed elsewhere *(2,35)*.

An interesting future direction is to study sensitization in transgenic mice with alterations in glutamate receptors or signaling pathways implicated in LTP. Chiamulera et al. *(36)* reported that mGluR5 knockout mice do not exhibit locomotor activation when injected with acute cocaine, and do not acquire cocaine self-administration. Mao et al. *(37)* found that mGluR1 knockout mice have augmented locomotor responses to amphetamine, perhaps due to impaired mobilization of inhibitory dynorphin systems that normally regulate responses to amphetamine. Using GluR1 knockout mice, Vekovischeva et al. *(38)* found that sensitization was normal when mice received repeated morphine injections in the same environment in which they were ultimately tested (context-dependent sensitization) but did not develop when the repeated treatment was given in home cages (context-independent sensitization), whereas wild-type mice developed sensitization under both conditions. Although all of these results are potentially important, it is hard to draw firm conclusions because of the possibility of altered neuronal development in glutamate receptor deficient mice.

3. Effect of Psychomotor Stimulants on Glutamate Receptor Expression in the Nucleus Accumbens and Dorsal Striatum

3.1. Role of the NAc and Striatum in Behavioral Sensitization

The NAc occupies a key position in the neural circuitry of motivation and reward. Not surprisingly, it is also critical for behavioral sensitization. While psychostimulants act in the midbrain to trigger the development of sensitization, drug actions in the NAc lead to the expression of a sensitized response. Accordingly, the VTA is associated with transient cellular adaptations during the early withdrawal period, while the NAc is the site of more persistent adaptations (*see* **refs. *10*** and ***39***). The output neurons of the NAc, medium spiny γ-aminobutyric acid (GABA) neurons, are regulated by convergent DA and glutamate inputs, although the nature of the interaction between DA and glutamate is complex and remains controversial *(40)*. Repeated psychostimulant administration leads to profound changes in both DA and glutamate transmission in the NAc *(8,39)*, and many recent studies have demonstrated that glutamate transmission in the NAc plays a critical role in drug-seeking behavior *(4)*. Therefore, many groups have examined the effects of psychostimulants on glutamate receptor expression in the NAc, as well as in the dorsal striatum. The dorsal striatum exhibits many of the same drug-induced adaptations as the NAc, although the NAc has received much more attention in recent years *(8)*.

3.2. Results in the NAc and Striatum

We have measured glutamate receptor subunit mRNA levels and immunoreactivity in rats treated for 5 d with 5 mg/kg of amphetamine or saline and perfused 3 or 14 d after the last injection. For AMPA receptor subunits, quantitative *in situ* hybridization studies showed no changes in GluR1-3 mRNA levels in the NAc after 3 d, but decreases in GluR1 and GluR2 mRNA levels were observed after 14 d *(41)*. Parallel changes were observed at the protein level using quantative immunoautoradiography *(42)*. Similarly, mRNA and protein levels for NR1 in the NAc were not altered by repeated amphetamine at the 3-d withdrawal time, but both were significantly decreased after 14 d of withdrawal *(24)*. The decreased levels of GluR1, GluR2, and NR1 subunits in amphetamine-treated rats may be functionally significant. Single-unit recording studies performed in the NAc of rats treated with the same amphetamine regimen, or a sensitizing regimen of cocaine, revealed that NAc neurons recorded from drug-treated rats were subsensitive to glutamate as compared to NAc neurons from saline-pretreated rats *(9)*. Follow-up studies showed that NAc neurons were also subsensitive to NMDA and AMPA but not a metabotropic glutamate receptor agonist (Hu and White, *unpublished observations*). However, the correspondence is not perfect. The decreases in glutamate receptor subunit expression were observed only after 14 d of withdrawal, whereas electrophysiological subsensitivity was observed after both 3 and 14 days of withdrawal. Perhaps other mechanisms account for subsensitivity at the early withdrawal time (*see* **ref. 43**). Another problem is that NAc neurons recorded from repeated cocaine treated rats also show electrophysiological subsensitivity to glutamate agonists (*see* previous discussion in this subheading), but most studies report increased glutamate receptor expression after long withdrawals from repeated cocaine administration (*see* following portions of this subheading).

Similar to our results showing no changes in glutamate receptor subunit expression in the NAc 3 d after discontinuing repeated amphetamine, Fitzgerald et al. *(12)* found no change in NAc levels of GluR1, GluR2, NR1, NR2A/B, GluR6/7, and KA-2 subunit proteins (measured by Western blotting) 16–18 h after withdrawal from repeated cocaine treatment. However, alterations are observed at later withdrawal times, and they differ from those produced by amphetamine. Churchill et al. *(14)* used Western blotting to determine protein levels of glutamate receptor subunits 24 h or 3 wk after discontinuing daily cocaine or saline injections (*see* **Subheading 2.2.** for more details). After 24 h, there were no changes in GluR1 or NMDAR1 levels in the NAc, consistent with the findings of Fitzgerald et al. *(12)*. However, after 3 wk, sensitized rats (but not cocaine-treated rats that failed to sensitize) showed a significant increase in GluR1 levels in the NAc compared to saline-treated rats. When saline-treated rats were compared to all cocaine rats (sensitized + nonsensitized), there was a trend toward increased NMDAR1 in the NAc after repeated cocaine, but this was actually more pronounced in nonsensitized rats. GluR2/3 was not changed in the NAc at either withdrawal time. Dorsal striatum was analyzed only after 3 wk of withdrawal; there were no changes in GluR1, GluR2/3, or NR1. Likewise, these subunits were unchanged in prefrontal cortex or VTA after 3 wk of withdrawal, although increases in GluR1 and NR1 were found in VTA of sensitized rats 24 h after discontinuing cocaine (*see* **Subheading 2.2.**).

Interestingly, the changes in protein levels found by Churchill et al. *(14)* were not paralleled by changes at the mRNA level. Ghasemzadeh et al. *(22)* used *in situ* hybridization histochemistry and RT-PCR to quantify glutamate receptor subunit mRNA levels 3 wk after discontinuing the same regimen of cocaine or saline injections used by Churchill et al. *(14)*. Twenty-four hours before decapitation, half the rats in each group were challenged with saline and half with cocaine. In NAc, acute cocaine decreased mRNA levels for GluR3, GluR4, and NR1, while repeated cocaine also decreased GluR3 mRNA and increased mGluR5 mRNA. The only significant effect in dorsolateral striatum was decreased NR1 mRNA after acute cocaine. The VTA and PFC were also evaluated (see **Subheadings 2.2.** and **4.2.**). Because of the complexity of the design, the reader should consult the article for an in-depth discussion of interactions between chronic cocaine treatment and acute challenge, and interesting trends that were apparent in some groups.

Scheggi et al. *(44)* used Western blotting to measure glutamate receptor subunits after administering 40 mg/kg of cocaine every other day over 14 d, testing for sensitization after 10 d of withdrawal, and killing the rats 1 wk after the test for sensitization. In NAc, significant increases in GluR1, NR1, and NR2B (but not GluR2 or NR2A) were found in sensitized rats. The changes in GluR1 and NR1 are in agreement with those reported by Churchill et al. *(14)*. In hippocampus, only the NR2B subunit was significantly elevated although there was a trend toward increased NR1 (26% increase). In the PFC, small increases (~20%) were observed for NR1 and NR2B, but these were not significant, and there was no change in GluR1. All of these changes were blocked if MK-801 was continuously infused (s.c., via osmotic minipumps) during cocaine administration, a treatment that also blocked development of sensitization, suggesting they are linked to sensitization.

Chronic cocaine treatment leads to accumulation in some NAc neurons of stable isoforms of the transcription factor ΔFosB, so Kelz et al. *(45)* used transgenic mice in which ΔFosB was induced in a subset of NAc neurons to model chronic cocaine treatment. These mice showed increased responsiveness to rewarding and locomotor-activating effects of cocaine, as well as increased expression of GluR2 in the NAc but not dorsal striatum. In a place conditioning test, rats that received intra-NAc injections of a recombinant herpes simplex virus vector encoding GluR2 spent more time in a cocaine-paired chamber than controls, while rats made to overexpress GluR1 spent less time in the cocaine-paired environment. Although this suggests that increased NAc levels of GluR2 may account for enhanced rewarding effects of cocaine in the ΔFosB-expressing mice, more work is needed to evaluate the relevance of these findings to the intact cocaine-treated animal.

NR2B is an interesting NMDAR subunit, as it is implicated in ethanol dependence *(46)* and morphine-induced conditioned place preference *(47)*. Loftis and Janowsky *(23)* measured NR2B levels using immunohistochemical methods in NAc and dorsolateral neostriatum, as well as hippocampal and cortical regions (*see* **Subheading 4.2.**). Rats were treated with 20 mg/kg of cocaine × 7 d (or saline) and killed 24 h, 72 h, or 14 d after discontinuing injections. In dorsal striatum, there were no changes after 24 or 72 h, but NR2B immunolabeling was increased after 14 d. In the NAc, NR2B was decreased in shell but not core after 24 h, no changes were present after 72 h, and there were increases in core and shell after 14 d.

Several recent studies have evaluated glutamate receptor binding after repeated cocaine. Keys and Ellison *(48)* found a decrease in [^{3}H]AMPA binding, assessed with autoradiography, in ventral striatum, and a trend in NAc, 21 d following two exposures to cocaine administered continuously for 5 d via subcutaneous pellets. Itzhak and Martin *(49)* compared NMDA receptor binding in several brain regions (striatum, amygdala, and hippocampus) in rats treated for 5 d with 15 mg/kg of cocaine (a sensitizing regimen) and mice treated for the same time with a higher dose of cocaine (35 mg/kg; a regimen that resulted in kindled seizures). No changes in NMDA receptor binding were found with the sensitizing regimen, whereas binding was elevated in all regions 3 d after the high-dose regimen was discontinued, with additional alterations occurring after the expression of kindled seizures. Szumlinski et al. *(50)* found no changes in [^{3}H]MK-801 binding in the rat striatum after a sensitizing regimen of cocaine (five daily injections of 15 mg/kg of cocaine) and 2 wk of withdrawal. Bhargava and Kumar *(51)* treated mice with a sensitizing regimen of cocaine (10 mg/kg, twice daily for 7 d). Immediately after drug treatment, [^{3}H]MK-801 binding was increased in cerebellum and spinal cord but decreased in cortex and hypothalamus. After withdrawal, binding remained decreased in cortex but other changes normalized.

Recent studies have focused on the role of metabotropic glutamate receptors in sensitization. Mao and Wang *(52)* used quantitative *in situ* hybridization histochemistry to measure mRNA levels for group I mGluRs (mGluR1 and mGluR5) in the NAc and striatum in naïve and amphetamine-sensitized rats. No changes in mGluR1 or mGluR5 mRNA levels were observed in naïve rats 3 h after acute administration of amphetamine. In contrast, 3 h after the last of five daily amphetamine injections, mGluR1 mRNA levels were increased in dorsal striatum and NAc. This effect was transient, as no changes were observed after 7, 14, or 28 d of withdrawal. A different pattern was observed for mGluR5. Levels of mRNA were decreased markedly 3 h after the final amphetamine injection, and the reduction persisted at 7-, 14-, and 28-d withdrawal times. In a rare example of concordance between amphetamine and cocaine findings, Swanson et al. *(53)* found a small but significant reduction in mGluR5 protein levels, measured by Western blotting, in the medial NAc of rats killed 3 wk after discontinuation of repeated cocaine injections. mGluR5 is postsynaptic and can negatively modulate AMPA receptor transmission. Thus, the authors suggested that cocaine-induced decreases in mGluR5 may contribute to the potentiation of AMPA receptor-mediated behavioral responses related to drug-seeking behavior that have been reported after chronic cocaine administration *(54,55)*. In the same study, repeated cocaine administration attenuated the ability of mGluR1 stimulation to decrease glutamate release and locomotor activity, but this was not accompanied by alterations in mGluR1 protein levels and may be attributable to altered expression of Homer1b/c, a scaffolding protein that regulates mGluR signaling *(53)*. Increasing evidence indicates that mGluRs play an important role in behavioral responses to psychomotor stimulants *(56)*.

A relatively unexplored question, owing primarily to technical difficulty, is whether posttranslational modification of glutamate receptors is altered after repeated drug treatment. Bibb et al. *(57)* found reduced peak amplitudes of AMPA/kainate-evoked currents in acutely dissociated striatal neurons from rats chronically treated with cocaine; other findings suggested that this was attributable to reduced PKA-dependent phosphorylation of GluR1.

3.3. Summary: NAc and Striatum

As discussed in **Subheading 3.1.**, considerable evidence implicates glutamate receptors in the striatal complex in persistent neuroadaptations associated with behavioral sensitization and drug-seeking behavior. There is some agreement that GluR1 and NR1 levels are not altered in the NAc after short withdrawals (1–3 d) from repeated cocaine or amphetamine administration. At longer withdrawal times (2–3 wk), cocaine-treated rats may show increases in GluR1, NR1, and NR2B, whereas amphetamine-treated rats show decreases in GluR1, GluR2, and NR1. There may also be persistent changes in the expression and function of group I mGluRs. The delayed onset of many of the reported changes in glutamate receptor expression is consistent with a role for the NAc in the long-term maintenance of sensitization and other drug-induced behavioral changes. However, it is difficult to reconcile opposite effects of cocaine and amphetamine on glutamate receptor expression with a role for these changes in the maintenance and expression of sensitization, as both drugs produce similar behavioral effects (augmented locomotor response) in sensitized rats. It should be kept in mind that the NAc contains heterogeneous populations of projection neurons and interneurons, and we do not know the phenotype of the neurons that experience changes in glutamate receptor subunit expression (e.g., *42,52*). Moreover, other types of drug-induced changes may contribute importantly to the excitability of NAc neurons. For example, Zhang et al. *(43)* found reduced sodium currents in NAc neurons after a short withdrawal from repeated cocaine, while Thomas et al. *(58)* found evidence for LTD in the NAc after long-term withdrawal from cocaine. In fact, growing evidence suggests that abnormal synaptic plasticity in the NAc, triggered by chronic drug treatment, leads to dysregulation of motivation- and reward-related circuits and thereby contributes to addiction *(2)*. It will be important to determine whether alterations in glutamate receptor expression contribute to the induction of altered plasticity, are involved in its expression, or represent compensatory responses to changes in the activity of glutamate-containing projections.

4. Effect of Psychomotor Stimulants on Glutamate Receptor Expression in the PFC and Other Cortical or Limbic Regions
4.1. Role of the Prefrontal Cortex in Behavioral Sensitization

The PFC is now acknowledged to play an important role in behavioral sensitization. Excitotoxic lesions of the PFC prevent the development of sensitization *(59–61)* as well as cellular changes in DA systems that are closely associated with sensitization *(61)*. The role of PFC in the expression of behavioral sensitization in response to psychostimulant challenge is more controversial. Some evidence suggests that expression of sensitization requires glutamatergic transmission between the dorsal PFC and the NAc core *(62)*. On the other hand, excitotoxic lesions of the PFC that are sufficient to prevent development of sensitization do not interfere with expression *(63,64)*. Other findings suggest that maintenance and expression of sensitization may be associated with loss of inhibitory DA tone in the PFC, leading to a loss of inhibitory control over PFC projections to subcortical regions; multiple mechanisms may contribute (e.g., **65–69**). Less has been done to examine specifically the role of glutamate transmission in the PFC in behavioral sensitization. No studies have examined the effect of intra-PFC injection of glutamate receptor antagonists and only a few microdialysis studies have assessed glutamate release in the PFC in response to stimulants *(70–73)*. Likewise, there have been relatively few studies on stimulant-induced alterations in glutamate receptor expression in the PFC as compared to the striatum and midbrain.

4.2. Results in the PFC

Using quantitative *in situ* hybridization and immunoautoradiography, we found increased GluR1 mRNA and protein levels 3 d after discontinuing repeated amphetamine administration; this effect was transient, as it was not observed in rats killed after 14 d of withdrawal *(41,42)*. This increase in GluR1 may be functionally significant, as PFC neurons recorded from amphetamine-treated rats after 3 d of withdrawal (but not 14 d of withdrawal) showed increased responsiveness to the excitatory effects of iontophoretically applied glutamate *(67)*. In studies using the same amphetamine regimen, we found a significant decrease in NR1 mRNA levels and a trend toward decreased immunolabeling after 14 d of withdrawal, but no change after 3 d *(24)*.

Cocaine also exerts complex effects on glutamate receptor subunit expression in the PFC. Churchill et al. *(14)* found no changes in PFC levels of GluR1, GluR2/3, or NMDAR1 (using Western blots) 3 wk after discontinuing a week of daily cocaine injections (see **Subheading 2.2.** for more details on this study). In another study, rats were treated with cocaine, tested for sensitization after 10 d of withdrawal, and killed 1 wk after the test for sensitization *(44;* see **Subheading 3.2.** for more details). Small increases (~20%) were observed for NR1 and NR2B in the PFC, but these were not significant, and there was no change in GluR1. Loftis and Janowsky *(23)* measured NR2B levels using immunohistochemical methods in VTA and NAc (*see* **Subheadings 2.2.** and **3.2.**), as well as dorsolateral neostriatum, the hippocampal formation (CA1, CA3, and dentate gyrus), and the cortex (medial frontal cortex, lateral frontal cortex, and parietal cortex). Rats were killed 24 h, 72 h, or 14 d after discontinuation of repeated cocaine or saline injections. There were no changes in the hippocampal formation following 24 or 72 h of withdrawal. Results in cortex depended on the region analyzed. For medial frontal cortex, there were increases at all withdrawal times. For lateral frontal cortex and parietal cortex, there was no change after 24 h, but increases after 72 h and 14 d. This study also measured neuronal nitric oxide synthase, but these results will not be discussed.

4.3. Summary: PFC and Other Cortical and Limbic Regions

Glutamate receptor expression in the PFC undergoes complex changes after drug administration is discontinued that depend on the withdrawal time and probably differ between cocaine and amphetamine, at least for NR1. In general, some results suggest that AMPA receptor subunit expression changes at early withdrawal times whereas NMDA receptor subunit expression is altered after longer withdrawals. Because relatively few studies have assessed glutamate transmission in the PFC of sensitized rats using electrophysiological or neurochemical approaches, it is difficult to assess the functional significance of observed changes. An exception is the correlation between increased responsiveness of PFC neurons to glutamate *(67)*, and increased expression of GluR1 in the PFC *(41,42)*, after short withdrawals from repeated amphetamine administration. It will be important to conduct studies on additional brain regions implicated in addiction, such as the amygdala.

5. Conclusions

It is clear that repeated administration of cocaine or amphetamine influences glutamate receptor expression in brain regions important for behavioral sensitization and addiction. However, to date, the data obtained raise more questions than they answer. One important problem is that amphetamine and cocaine produce different patterns of changes, whereas both produce behavioral sensitization. Either there are multiple ways to achieve a sensitized state, or the changes in glutamate receptor expression are not directly associated with sensitization. The picture is made more complex by different effects at different withdrawal times, different effects with different drug regimens, and lack of agreement between laboratories using similar drug regimens. Another problem is that studies of receptor expression have been conducted at the regional level, precluding identification of the types of cells exhibiting particular alterations in glutamate receptor expression after stimulant exposure. Without such information, it is hard to predict the functional effect of these alterations at the level of neuronal circuits. For example, does the increase in GluR1 expression in the PFC after repeated amphetamine occur in pyramidal neurons or interneurons, or in a subset of one of these populations? It will be important to conduct future studies in identified cells, although this is a very challenging undertaking. It will also be important to study the cellular mechanisms by which monoamine-releasing psychomotor stimulants influence the expression of glutamate receptors, as well as other aspects of glutamate neurotransmission.

References

1. Hyman, S. E. and Malenka, R. C. (2001) Addiction and the brain: the neurobiology of compulsion and its persistence. *Nat. Rev.* **2,** 695–703.
2. Wolf, M. E. (2002) Addiction and glutamate-dependent plasticity, in *Glutamate and Addiction* (Herman, B. H., Frankenheim, J., Litten, R., Sheridan, P. H., Weight, F. F., and Zukin, S. R., eds.), Humana Press, Totowa, NJ, pp. 127–142.
3. Jentsch, J. D. and Taylor, J. R. (1999) Impulsivity resulting from frontostriatal dysfunction in drug abuse: implications for the control of behavior by reward-related stimuli. *Psychopharmacology* **146,** 373–390.
4. Everitt, B. J. and Wolf, M. E. (2002) Psychomotor stimulant addiction: a neural systems perspective. *J. Neurosci.* **22,** 3312–3320.
5. Wolf, M. E. (2001) The neuroplasticity of addiction, in *Toward a Theory of Neuroplasticity* (Shaw, C. A. and McEachern, J. C., eds.), Taylor & Francis, Philadelphia, pp. 359–372.
6. Robinson, T. E. and Berridge, K. C. (1993) The neural basis of drug craving: an incentive-sensitization theory of addiction. *Brain Res. Rev.* **18,** 247–291.
7. Robinson, T. E. and Berridge, K. C. (2000) The psychology and neurobiology of addiction: an incentive-sensitization view. *Addiction* **95** *(Suppl. 2),* S91–S117.
8. Wolf, M. E. (1998) The role of excitatory amino acids in behavioral sensitization to psychomotor stimulants. *Prog. Neurobiol.* **54,** 679–720.
9. White, F. J., Hu, X.-T., Zhang, X.-F., and Wolf, M. E. (1995) Repeated administration of cocaine or amphetamine alters neuronal responses to glutamate in the mesoaccumbens dopamine system. *J. Pharmacol. Exp. Ther.* **273,** 445–454.
10. Zhang, X.-F., Hu, X.-T., White, F. J., and Wolf, M. E. (1997) Increased responsiveness of ventral tegmental area dopamine neurons to glutamate after repeated administration of cocaine or amphetamine is transient and selectively involves AMPA receptors. *J. Pharmacol. Exp. Ther.* **281,** 699–706.
11. Giorgetti, M., Hotsenpiller, G., Ward, P., Teppen, T., and Wolf, M. E. (2001) Amphetamine-induced plasticity of AMPA receptors in the ventral tegmental area: effects on extracellular levels of dopamine and glutamate in freely moving rats. *J. Neurosci..* **21,** 6362–6369.
12. Fitzgerald, L. W., Ortiz, J., Hamedani, A. G., and Nestler, E. J. (1996) Drugs of abuse and stress increase the expression of GluR1 and NMDAR1 glutamate receptor subunits in the rat ventral tegmental area: common adaptations among cross-sensitizing agents. *J. Neurosci.* **16,** 274–282.
13. Ortiz, J., Fitzgerald, L. W., Charlton, M., et al. (1995) Biochemical actions of chronic ethanol exposure in the mesolimbic dopamine system. *Synapse* **21,** 289–298.

14. Churchill, L., Swanson, C. J., Urbina, M., and Kalivas, P. W. (1999) Repeated cocaine alters glutamate receptor subunit levels in the nucleus accumbens and ventral tegmental area of rats that develop behavioral sensitization. *J. Neurochem.* **72,** 2397–2403.
15. Lu, W., Monteggia, L. M., and Wolf, M. E. (2002) Repeated administration of amphetamine or cocaine does not alter AMPA receptor subunit expression in the rat midbrain. *Neuropsychopharmacology* **26,** 1–13.
16. Carlezon, W. A., Jr., Boundy, V. A., Haile, C. N., et al. (1997) Sensitization to morphine induced by viral-mediated gene transfer. *Science* **277,** 812–814.
17. Carlezon, W. A., Jr., Haile, C. N., Coopersmith, R., et al. (2000) Distinct sites of opiate reward and aversion within the midbrain identified using a herpes simplex virus vector expressing GluR1. *J. Neurosci.* **20,** RC62.
18. Ben-Shahar, O. and Ettenberg, A. (1994) Repeated stimulation of the ventral tegmental area sensitizes the hyperlocomotor response to amphetamine. *Pharmacol. Biochem. Behav.* **48,** 1005–1009.
19. Schenk, S. and Snow, S. (1994) Sensitization to cocaine's motor activating properties produced by electrical kindling of the medial prefrontal cortex but not of the hippocampus. *Brain Res.* **659,** 17–22.
20. Steketee, J. D. and Kalivas, P. W. (1991) Sensitization to psychostimulants and stress after injection of pertussis toxin into the A10 dopamine region. *J. Pharmacol. Exp. Ther.* **259,** 916–924.
21. Bardo, M. T., Robinet, P. M., Mattingly, B. A., and Margulies, J. E. (2001) Effect of 6-hydroxydopamine or repeated amphetamine treatment on mesencephalic mRNA levels for AMPA glutamate receptor subunits in the rat. *Neurosci. Lett.* **302,** 133–136.
22. Ghasemzadeh, M. B., Nelson, L. C., Lu, X. Y., and Kalivas, P. W. (1999) Neuroadaptations in ionotropic and metabotropic glutamate receptor mRNA produced by cocaine treatment. *J. Neurochem.* **72,** 157–165.
23. Loftis, J. M. and Janowsky, A. (2000) Regulation of NMDA receptor subunits and nitric oxide synthase expression during cocaine withdrawal. *J. Neurochem.* **75,** 2040–2050.
24. Lu, W., Monteggia, L. M., and Wolf, M. E. (1999) Withdrawal from repeated amphetamine administration reduces NMDAR1 expression in the substantia nigra, nucleus accumbens and medial prefrontal cortex. *Eur. J. Neurosci.* **11,** 3167–3177.
25. Schenk, S. and Partridge, B. (1997) Effects of acute and repeated administration of *N*-methyl-D-aspartate (NMDA) into the ventral tegmental area: locomotor activating effects of NMDA and cocaine. *Brain Res.* **769,** 225–232.
26. Licata, S. C., Freeman, A. Y., Pierce-Bancroft, A. F., and Pierce, R. C. (2000) Repeated stimulation of L-type calcium channels in the rat ventral tegmental area mimics the initiation of behavioral sensitization to cocaine. *Psychopharmacology* **152,** 110–118.
27. Sheng, M. and Lee, S. H. (2001) AMPA receptor trafficking and the control of synaptic transmission. *Cell* **105,** 825–828.

28. Ungless, M. A., Whistler, J. L., Malenka, R. C., and Bonci, A. (2001) Single cocaine exposure *in vivo* induces long-term potentiation in dopamine neurons. *Nature* **411,** 583–587.
29. Jones, S., Kornblum, J. L., and Kauer, J. A. (2000) Amphetamine blocks long-term synaptic depression in the ventral tegmental area. *J. Neurosci.* **20,** 5575–5580.
30. Thomas, M. J., Malenka, R. C., and Bonci, A. (2000) Modulation of long-term depression by dopamine in the mesolimbic system. *J. Neurosci.* **20,** 5581–5586.
31. Kalivas, P. W. and Duffy, P. (1995) D1 receptors modulate glutamate transmission in the ventral tegmental area. *J. Neurosci.* **15,** 5379–5388.
32. Xue, C.-J., Ng, J. P., Li, Y., and Wolf, M. E. (1996) Acute and repeated systemic amphetamine administration: effects on extracellular glutamate, aspartate and serine levels in rat ventral tegmental area and nucleus accumbens. *J. Neurochem.* **67,** 352–363.
33. Fiorillo, C. D. and Williams, J. T. (2000) Selective inhibition by adenosine of mGluR IPSPs in dopamine neurons after cocaine treatment. *J. Neurophysiol.* **83,** 1307–1314.
34. Paladini, C. A., Fiorillo, C. D., Morikawa, H., and Williams, J. T. (2001) Amphetamine selectively blocks inhibitory glutamate transmission in dopamine neurons. *Nat. Neurosci.* **4,** 275–281.
35. Wolf, M. E. (2002) Addiction: making the connection between behavioral changes and neuronal plasticity in specific circuits. *Mol. Intervent.* **2,** 146–157.
36. Chiamulera, C., Epping-Jordan, J. P., Zocchi, A., et al. (2001) Reinforcing and locomotor stimulant effects of cocaine are absent in mGluR5 null mutant mice. *Nat. Neurosci.* **4,** 873–874.
37. Mao, L., Conquet, F., and Wang, J. Q. (2001) Augmented motor activity and reduced striatal preprodynorphin mRNA induction in response to acute amphetamine administration in metabotropic glutamate receptor 1 knockout mice. *Neuroscience* **106,** 303–312.
38. Vekovischeva, O. Y., Zamanillo, D., Echenko, O., et al. (2001) Morphine-induced dependence and sensitization are altered in mice deficient in AMPA-type glutamate receptor-A subunits. *J. Neurosci.* **21,** 4451–4459.
39. White, F. J. and Kalivas, P. W. (1998) Neuroadaptations involved in amphetamine and cocaine addiction. *Drug Alcohol. Depend.* **51,** 141–153.
40. Nicola, S. M., Surmeier, D. J., and Malenka, R. C. (2000) Dopaminergic modulation of neuronal excitability in the striatum and nucleus accumbens. *Annu. Rev. Neurosci.* **232,** 185–215.
41. Lu, W., Chen, H., Xue, C.-J., and Wolf, M. E. (1997) Repeated amphetamine administration alters the expression of mRNA for AMPA receptor subunits in rat nucleus accumbens and prefrontal cortex. *Synapse* **26,** 269–280.
42. Lu, W. and Wolf, M. E. (1999) Repeated amphetamine administration alters AMPA receptor subunit expression in rat nucleus accumbens and medial prefrontal cortex. *Synapse* **32,** 119–131.

43. Zhang, X.-F., Hu, X.-T., and White, F. J. (1998) Whole-cell plasticity in cocaine withdrawal: reduced sodium currents in nucleus accumbens neurons. *J. Neurosci.* **18,** 488–498.
44. Scheggi, S., Mangiavacchi, S., Masi, F., Gambarana, C., Tagliamonte, A., and De Montis, M. G. (2002) Dizocilpine infusion has a different effect in the development of morphine and cocaine sensitization: behavioral and neurochemical aspects. *Neuroscience* **109,** 267–274.
45. Kelz, M. B., Chen, J., Carlezon, W. A., Jr., et al. (1999) Expression of the transcription factor deltaFosB controls sensitivity to cocaine. *Nature* **16,** 272–276.
46. Narita, M., Aoki, T., and Suzuki, T. (2000) Molecular evidence for the involvement of NR2B subunit containing *N*-methyl-D-aspartate receptors in the development of morphine-induced place preference. *Neuroscience* **101,** 601–606.
47. Narita, M., Soma, M., Narita, M., Mizoguchi, H., Tseng, L. F., and Suzuki, T. (2000) Implications of the NR2B subunit-containing NMDA receptor localized in mouse limbic forebrain in ethanol dependence. *Eur. J. Pharmacol.* **401,** 191–195.
48. Keys, A. S. and Ellison, G. D. (1999) Long-term alterations in benzodiazepine, muscarinic and alpha-amino-3-hydroxy-5-methylisoxazole-4-propionic acid (AMPA) receptor density following continuous cocaine administration. *Pharmacol. Toxicol.* **85,** 144–150.
49. Itzhak, Y. and Martin, J. L. (2000) Cocaine-induced kindling is associated with elevated NMDA receptor binding in discrete mouse brain regions. *Neuropharmacology* **39,** 32–39.
50. Szumlinksi, K. K., Herrick-Davis, K., Teitler, M., Maisonneuve, I. M., and Glick, S. D. (2000) Behavioural sensitization to cocaine is dissociated from changes in striatal NMDA receptor levels. *NeuroReport* **11,** 2785–2788.
51. Bhargava, H. N. and Kumar, S. (1999) Sensitization to the locomotor stimulant effect of cocaine modifies the binding of [^{3}H]MK-801 to brain regions and spinal cord of the mouse. *Gen. Pharmacol.* **32,** 359–363.
52. Mao, L. and Wang, J. Q. (2001) Differentially altered mGluR1 and mGluR5 mRNA expression in rat caudate nucleus and nucleus accumbens in the development and expression of behavioral sensitization to repeated amphetamine administration. *Synapse* **41,** 230–240.
53. Swanson, C. J., Baker, D. A., Carson, D., Worley, P. F., and Kalivas, P. W. (2001) Repeated cocaine administration attenuates group I metabotropic glutamate receptor-mediated glutamate release and behavioral activation: a potential role for Homer. *J. Neurosci.* **21,** 9043–9052.
54. Bell, K. and Kalivas, P. W. (1996) Context-specific cross sensitization between systemic cocaine and intra-accumbens AMPA infusion in rats. *Psychopharmacology* **127,** 377–383.
55. Pierce, R. C., Bell, K., Duffy, P., and Kalivas, P. W. (1996) Repeated cocaine augments excitatory amino acid transmission in the nucleus accumbens only in rats having developed behavioral sensitization. *J. Neurosci.* **16,** 1550–1560.

56. Vezina, P. and Kim, J.-H. (1999) Metabotropic glutamate receptors and the generation of locomotor activity: interactions with midbrain dopamine. *Neurosci. Biobehav. Rev.* **23,** 577–589.
57. Bibb, J. A., Chen, J., Taylor, J. R., et al. (2001) Effects of chronic exposure to cocaine are regulated by the neuronal protein Cdk5. *Nature* **410,** 376–380.
58. Thomas, M. J., Beurrier, C., Bonci, A., and Malenka, R. C. (2001) Long-term depression in the nucleus accumbens: a neural correlate of behavioral sensitization to cocaine. *Nat. Neurosci.* **4,** 1217–1223.
59. Wolf, M. E., Dahlin, S. L., Xu, H.-T., Xue, C.-J., and White, K. (1995) Effects of lesions of prefrontal cortex, amygdala, or fornix on behavioral sensitization to amphetamine: comparison with *N*-methyl-D-aspartate antagonists. *Neuroscience* **69,** 417–439.
60. Cador, M., Bjijou, Y., Cailhol, S., and Stinus, L. (1999) D-Amphetamine-induced behavioral sensitization: implication of a glutamatergic medial prefrontal cortex–ventral tegmental area innervation. *Neuroscience* **94,** 705–721.
61. Li, Y., Hu, X.-T., Berney, T. G., et al. (1999) Both glutamate receptor antagonists and prefrontal cortex lesions prevent induction of cocaine sensitization and associated neuroadaptations. *Synapse* **34,** 169–180.
62. Pierce, R. C., Reeder, D. C., Hicks, J., Morgan, Z. R., and Kalivas, P. W. (1998) Ibotenic acid lesions of the dorsal prefrontal cortex disrupt the expression of behavioral sensitization to cocaine. *Neuroscience* **82,** 1103–1114.
63. Li, Y. and Wolf, M. E. (1997) Ibotenic acid lesions of prefrontal cortex do not prevent expression of behavioral sensitization to amphetamine. *Behav. Brain Res.* **84,** 285–289.
64. Li, Y., Wolf, M. E., and White, F. J. (1999) The expression of cocaine sensitization is not prevented by MK-801 or ibotenic acid lesions of the medial prefrontal cortex. *Behav. Brain Res.* **104,** 119–125.
65. Karler, R., Calder, L. D., Thai, D. K., and Bedingfield, J. B. (1998) The role of dopamine in the mouse frontal cortex: a new hypothesis of behavioral sensitization to amphetamine and cocaine. *Pharmacol. Biochem. Behav.* **61,** 435–443.
66. Prasad, B. M., Hochstatter, T., and Sorg, B. A. (1999) Expression of cocaine sensitization: regulation by the medial prefrontal cortex. *Neuroscience* **88,** 765–774.
67. Peterson, J. D., Wolf, M. E., and White, F. J. (2000) Altered responsiveness of medial prefrontal cortex neurons to glutamate and dopamine after withdrawal from repeated amphetamine treatment. *Synapse* **36,** 342–344.
68. Chefer, V. I., Moron, J. A., Hope, B., Rea, W., and Shippenberg, T. S. (2000) Kappa-opioid receptor activation prevents alterations in mesocortical dopamine neurotransmission that occur during abstinence from cocaine. *Neuroscience* **101,** 619–627.
69. Hedou, G., Homberg, J., Feldon, J., and Heidbreder, C. A. (2001) Expression of sensitization to amphetamine and dynamics of dopamine neurotransmission in different laminae of the rat medial prefrontal cortex. *Neuropharmacology* **40,** 366–382.

70. Stephans, S. E. and Yamamoto, B. K. (1995) Effect of repeated methamphetamine administrations on dopamine and glutamate efflux in rat prefrontal cortex. *Brain Res.* **700,** 991–906.
71. Reid, M. S., Hsu, K., Jr., and Berger, S. P. (1997) Cocaine and amphetamine preferentially stimulate glutamate release in the limbic system: studies on the involvement of dopamine. *Synapse* **27,** 95–105.
72. Del Arco, A., Martinez, R., and Mora, F. (1998) Amphetamine increases extracellular concentrations of glutamate in the prefrontal cortex of the awake rat: a microdialysis study. *Neurochem. Res.* **23,** 1153–1158.
73. Hotsenpiller, G. and Wolf, M. E. Extracellular glutamate levels in prefrontal cortex during the expression of associative responses to cocaine related stimuli. *Neuropharmacology*, in press.

3

Adult Neural Stem/Progenitor Cells in the Forebrain

Implications for Psychostimulant Dependence and Medication

John Q. Wang, Limin Mao, and Yuen-Sum Lau

1. Introduction

The question as to what exactly a stem cell is has remained contentious even after nearly three decades of debate. The prevailing view is that stem cells are cells with the capacity for unlimited or prolonged self-renewal that can produce at least one type of highly differentiated descendant. Usually, between the stem cell and its terminally differentiated progeny there is an intermediate population of committed progenitors or precursors with limited proliferative capacity and restricted differentiation potential. The term "neural stem cell" is used loosely to describe cells that (1) are derived from the nervous system; (2) have self-renewal capacity; and (3) can give rise to one specific phenotype, or more likely multiple types, of neural cells other than themselves through asymmetric cell division *(1,2)*. However, to date, it is not clear how primitive the detectable population of dividing cells in the brain is. They may represent true neural stem cells or lineage-restricted progenitor cells. Given this uncertainty, the cautious term "neural progenitor cells" is used in this chapter to describe dividing cells in the central nervous system (CNS).

Adult neural progenitor cells are the ones derived from the adult nervous system. Although it has long been thought that the neural tissue in the adult mammalian brain is entirely postmitotic, a particular surprise is the discovery of progenitor cells in unexpected brain areas, such as the subventricular zone (SVZ) and the hippocampal dentate gyrus, throughout adulthood *(1–3)*. As compared to embryonic stem cells, which tend to proliferate at high levels and spontaneously differentiate into all kinds of tissues, adult neural progenitor cells usually show low levels of cell division under normal conditions and have already made a commitment to become neural tissues. Natural proliferation and differentiation of neural progenitor cells generate either neuronal cells (neurogenesis) or glial cells (gliogenesis). Increasing evidence shows that proliferation and/or differentiation of progenitors can be altered substantially by exogenous administration of growth factors or other experimental manipulations.

Exposure to psychostimulants such as cocaine and amphetamine causes long-term mental illnesses. Brain mechanisms underlying biological actions of these stimulants are not well understood and may be related to changes in the mesolimbic and mesostriatal dopaminergic pathways. It has been suggested that drug exposure causes various cellular and molecular changes in the dopaminergic system, which lead to the development of psychoplasticity related to long-term properties of drugs of abuse. However, identification of altered neural elements responsible for psychoplasticity has not been achieved despite multidisciplinary efforts during the past few decades. Given the existence of active neural progenitor cells in several key structures of the forebrain, alteration in proliferation and/or differentiation of progenitors under dopamine-stimulated conditions might participate in the formation of psychoplasticity. This is indeed supported by the observations from recently emerging animal studies summarized in this chapter.

2. Adult Neural Progenitor Cells in the Forebrain

Adult neural progenitor cells in the SVZ and hippocampus represent the most thoroughly investigated and best characterized of such cells in the forebrain. These progenitor cells are often detected in vivo through the use of retroviruses *(4)* or thymidine autoradiography *(5)*. Recently, the thymidine analog bromodeoxyuridine (BrdU) has been used as a tracer of new DNA synthesis to label dividing cells in the CNS *(6)*. There are advantages and disadvantages to these methods *(1)*. The highest density of progenitor cells is found in the SVZ. Neuronal progenitors in the SVZ migrate tangentially (sagittally) along the rostral migratory stream into the olfactory bulb, where they differentiate into granular and periglomerular neurons *(7,8)*. In contrast, glial progenitors in the SVZ migrate radially into neighboring brain areas such as the striatum, corpus callosum, and neocortex *(9)*. Adult neural progenitors in the hippocampus are distributed throughout the medial dentate gyrus at all rostrocaudal levels *(10,11)*. They are typically observed in a thin lamina between the hilus and the granule cell layer, that is, the subgranular zone, as well as within the granule cell layer and hilus *(10,11)*. Approximately half of newborn cells in the hippocampus are believed to differentiate into neurons 3–4 wk after their birth according to their characteristic morphology of granule neurons and co-expression with the neuronal markers, such as neuron specific enolase (NSE), microtubule-associated protein-2 (MAP-2), or neuronal nuclear antigen (NeuN). A small fraction of newborn cells (~15%) adopt a glial fate as detected by their association with the astrocytic (glial fibrillary acidic protein [(GFAP]) or S100β) or oligodendrocytic markers. Newborn neurons in the adult dentate gyrus can migrate to the functional site, where they execute the programmed missions and connect appropriately into the circuitry of the hippocampus by developing synapses and axonal projections to receive and deliver signals, respectively *(12)*.

Besides the SVZ and the dentate gyrus, active adult neurogenesis and/or gliogenesis exist in other brain regions. The drug-affected area, striatum, is among those regions where cell proliferation and differentiation recently have been noticed *(13)*. After BrdU injection, cell division is consistently observed in the dorsal and ventral striatum. Divided cells are scattered throughout the area. Newborn striatal cells survive beyond 60 d, with a graduate increase in their body size and processes *(13,14)*. Although a small fraction of cells exhibit the morphological characteristics of radial glia 3 wk after birth, the vast majority of newborn cells show no obvious morphology of either projection neurons or glia. Parallel with the morphological observations, approx 10–20% of BrdU-labeled cells are immunoreactive to S100β and no BrdU cells are double labeled with NeuN even 6 wk after birth. Thus, it appears that gliogenesis, but not neurogenesis, naturally occurs in the intact striatum at a small scale, and the vast majority of newborn cells normally remain undifferentiated in this brain area. The exact primitive stage of those dividing cells in the striatum is not yet defined. However, the aforementioned study clearly demonstrates that these cells could self-renew and give rise to at least glia in the adult striatum.

3. Regulation of Adult Neurogenesis and Gliogenesis

A great deal of effort recently has been made in animal experiments to explore the regulation of adult neurogenesis/gliogenesis in the CNS by a variety of experimental manipulations. Available data show that growth factors have significant effects on the behavior of neural progenitor's both in vivo and in vitro. For example, basic fibroblast growth factor (bFGF) and epidermal growth factor (EGF) infused into the lateral ventricle of adult rats and mice profoundly increase proliferation of cells in the SVZ, but not in the dentate gyrus of the hippocampus *(15–18)*. Moreover, the two growth factors tend to influence the fate of cells, usually resulting in more glial cells and fewer neurons *(15–18)*. Increased systemic levels of bFGF by subcutaneous injection also increase cell proliferation in both the SVZ and hippocampus of neonatal and adult rats *(19)*. The effects of intraventricular bFGF are age dependent: much more increases in neurons in the neonate than those in the adult are induced following bFGF application *(19,20)*. Brain-derived neurotrophic factor (BDNF) is another mitogenic factor that increases the number of cells and probably the number of neurons in the olfactory bulb after intraventricular injection *(21)*. Like growth factors, hormones exhibit significant influences on adult neurogenesis. Glucocorticoids inhibit adult neurogenesis according to the finding that adrenalectomy increases proliferation of the progenitor population in the hippocampus, and systemic application of glucocorticoids antagonizes this influence *(22,23)*. In contrast, estrogen stimulates neurogenesis in the hippocampus of adult rats *(24)*. Besides growth factors and hormones, multiple neurotransmitter systems show their ability to modulate adult progenitor activity. Glutamatergic transmission in the CNS is the first system studied in this regard. Glutamatergic deafferentiation and pharmacological blockade of glutamate receptors (*N*-methyl-D-aspartate [NMDA]) cause an increase in all aspects of hippocampal neurogenesis *(25–27)*. In contrast, activation of NMDA receptors causes a dramatic decrease in proliferation of progenitors in the dentate gyrus *(25–27)*. Thus, glutamate appears to affect adult neurogenesis in an inhibitory fashion, as opposed to a facilitating role of serotonin in the production of new neurons via activation of the 5-hydroxytryptamine$_{1A}$ (5-HT$_{1A}$) receptor *(28)*. Running also increases hippocampal neurogenesis in adult mice. Mice housed with a running wheel show an increased number of BrdU-positive cells in the dentate gyrus *(29)*. Moreover, these mice show an increase in long-term potentiation in the dentate gyrus as compared to mice without a running wheel *(29)*. Other factors that affect adult neurogenesis include ischemia *(30)*, an enriched environment for increased social interactions and physical activity *(31)*, psychosocial stress *(27,32,33)* presumably via adrenal steroids, and bone morphogenetic protein administration *(34,35)*. Detailed mechanisms underlying the effects of the regulators described in the preceding are unclear. Further studies are needed to elucidate responsible mechanisms and interactions between those regulators.

4. Regulation of Adult Neurogenesis and Gliogenesis by Abused Substances

Studies on dopaminergic roles in the regulation of adult cytogenesis are just emerging. Two recent reports have demonstrated that the midbrain dopaminergic transmission that underlies major biological actions of psychostimulants can be a powerful regulator of adult cytogenesis. Teuchert-Noodt and co-workers found that acute treatment with methamphetamine at a high dose (25 or 50 mg/kg) suppresses dentate granule cell proliferation by 28–34% in the adult gerbil hippocampus *(36)*. Experiments carried out in this laboratory also define the amphetamine regulation of proliferation and differentiation of striatal progenitors in adult rats *(13)*. Like the effect of methamphetamine on hippocampal cytogenesis, acute administration of amphetamine (10 mg/kg) induces a rapid and transient decrease in the number of proliferating cells in the striatum, although amphetamine has no significant effect on differentiation of newborn cells. These results indicate that dopaminergic inputs control cell proliferation in striatal and hippocampal regions in an inhibitory fashion. How dopamine inhibits cell division is unclear. It is hypothesized that dopamine stimulation may prevent or reduce the synthesis or release of mitogenic factors from cells in the vicinity of progenitor cells. Alternatively, dopamine stimulation may affect cytogenesis indirectly through glutamatergic transmission. It has been shown that acute administration of amphetamine or cocaine increases glutamate release in the striatum *(37)*. The increased glutamate could then decrease cell division in the striatum as discussed in the preceding.

In contrast to decreased cell division after dopamine stimulation, dopamine depletion increases progenitor proliferation. Reduction of D_1/D_2 receptor tone with the antagonist haloperidol increases dentate granule cell proliferation in the gerbil hippocampus *(38)*. Similarly, a single or repeated injection of the neurotoxin 1-methyl-4-phenyl-1,2,3,6-tetrahydropyridine (MPTP), known to selectively damage dopaminergic terminals in the dorsal striatum and cell bodies in the substantia nigra, causes a robust proliferative response in striatal and nigral regions of adult mice *(14,39)*. Nearly all newly generated cells in the striatum, but not in the nigra, rapidly differentiate into astrocytes, whereas no neurogenesis is seen in the two affected areas even 60 d after cell birth *(14)*. Strong striatal astrogenesis after dopaminergic insult implies the participation of astroglia in dopamine repair. The unexpected lack of striatal and nigral neurogenesis after a long period of survival may be related to the extent of MPTP damage to midbrain dopaminergic cells. With the MPTP lesion model used in the aforementioned studies, a marginal loss of dopaminergic cells in the nigra is observed, and dopamine content and uptake in the striatum are rapidly recovered *(14)*. Thus, limited and transient damage to nigral cells may not raise adequate call for neurogenesis repair. It will be intriguing therefore to investigate whether a chronic MPTP model that could produce prolonged and severe loss of nigral cells or application of exogenous growth factors might induce neurogenesis, including a particular neuronal fate, such as the tyrosine hydroxylase containing dopaminergic neurons.

Given the known effects of opiates on hippocampal function, a recent attempt has been made to evaluate opiate influence on adult neurogenesis in the rat hippocampus *(40)*. Chronic morphine exposure decreases hippocampal neurogenesis by 42%. A similar effect is revealed after chronic self-administration of heroin. Because adrenalectomy and corticosterone replacement have no effect on neurogenesis, the opiate suppression is not mediated by changes in circulating levels of glucocorticoids. These findings suggest that opiates may influence hippocampal function via regulation of neurogenesis in the adult rat hippocampus.

5. Functional Roles of Adult Neural Progenitor Cells

The adult brain has long been thought to be entirely postmitotic. Hence, functional roles of adult neural progenitors in the CNS are unclear at present. It has been suggested that they are vestiges of evolution from more primitive organisms, such as fish *(41)*, in which organ and tissue self-renewal provides survival advantages in an inhospitable environment. However, along with emerging studies on this issue, some functional roles of adult neurogenesis can be speculated and tested. Under physiological conditions, neurogenesis may replace cells programmed for death with fully functional cells, even though this repopulation is considered to be very limited in adult brains. More importantly, the adult mammalian nervous system retains the capacity of adapting new demands of brain functions, such as learning, memory, and neural plasticity in response to environmental changes. It is possible that the local generation of new neuronal and non-neuronal cells in the responsible brain structures could participate in the acquisition or integration of new memories and neuroadaptation. As to region-specific roles, the SVZ is more likely a stem-cell factory conveniently located in the brain. Through proliferation, it manufactures progenitor cells infinitely and delivers them to their destinations in the whole brain *(42,43)*. As compared to the SVZ, neurogenesis in the hippocampus can directly add new granule cells in the dentate gyrus whenever a call is made for new memory or neuroadaptive formation.

Under pathophysiological conditions, inducible cytogenesis can play dual roles in a given pathophysiological process. First, cytogenesis can be provoked to process aberrant functions. For example, the neurons that are formed through normal ongoing neurogenesis do not send processes to the CA3 region of the hippocampus *(44,45)*. However, epilepsy-induced neurogenesis sends axon collaterals back onto the dentate gyrus that forms recurrent collaterals contributing to enhanced local activity for epilepsy *(44)*. Second and more significantly, cytogenesis can be stimulated to repair (rescue or compensate) for cell loss in chronic neurodegenerative diseases, such as Parkinson's disease. In this case, repopulation of missing cells by increased endogenous neurogenesis in the diseased site could be an ideal "self-repair." The newborn cells after differentiation *in situ* may partly or completely take on the exact function of the cells they replace. Further studies are needed to explore anatomical and functional "self-repair" of this kind in various neurodegenerative diseases, and results from these studies may bring about a new therapy for those diseases.

6. Implications in Psychostimulant Dependence and Medication

With limited studies, how drug exposure influences brain cytogenesis and how altered cytogenesis contributes to addictive properties of drugs of abuse can only be hypothesized at present. As described in the preceding, amphetamine exposure decreases total cell proliferation in the striatum *(13)*. Morphine, heroin, and methamphetamine decrease hippocampal neurogenesis *(36,40)*. However, alteration in proliferation of a given population of cells in the affected areas remains unidentified. It is possible that one specific phenotype of neuronal and/or non-neuronal cells is either increased or decreased in response to drug stimulation. In this case, a decreased generation of cells that normally exert an inhibitory influence on the formation of drug addiction may result in disinhibition of addictive processes. On the contrary, an increased generation of cells that are involved in the mediation of drug effects may facilitate drug addiction. In sum, drugs may develop their dependence via altering cytogenesis activity in adult brain. Future studies can be pursued to address (1) effects of drug exposure on the generation of a specific phenotype of cells in the adult forebrain, and (2) underlying mechanisms of altered neurogenesis/gliogenesis in the regulation of long-term drug actions.

The therapeutic potential of targeting neural progenitor cells for the treatment of drug addiction is obvious. Again, however, concrete suggestions will have to wait until future studies can elucidate changes in cytogenesis in response to drug exposure and how those changes regulate drug actions. In the future, it is expected that pharmacological agents can be developed to inhibit or facilitate the new generation of specific phenotype of cells, depending on the cell function in drug addiction. Reprogrammed endogenous cell birth through exogenously administered agents could then prevent drug dependence and addiction.

In summary, recent convincing evidence has shown a profound influence of drug exposure on neurogenesis/gliogenesis in the affected brain areas in adult animals. Altered cytogenesis may participate in the modulation of addictive properties of drugs. More studies are needed to define the underlying mechanisms of the drug effects and contribution of neural progenitors to drug actions in order to develop an effective therapy for drug addiction by targeting neural stem cells.

Acknowledgments

The authors' studies described in this chapter were supported by grants from the NIH (DA10355 and MH61469) and an UMRB grant from the University of Missouri.

References

1. Gage, F. H. (2000) Mammalian neural stem cells. *Science* **287**, 1433–1438.
2. McKay, R. (1997) Stem cells in the central nervous system. *Science* **276**, 66–71.
3. Smart, I. (1961) The subependymal layer of the mouse brain and its cell production as shown by radioautography after thymidine-H^3 injection. *J. Comp. Neurol.* **116**, 325–338.
4. Price, J. and Thurlow, L. (1988) Cell lineage in the rat cerebral cortex: a study using retroviral-mediated gene transfer. *Development* **104**, 473–482.
5. Altman, J. and Das, G. D. (1965) Autoradiographic and histological evidence of postnatal hippocampal neurogenesis in rats. *J. Comp. Neurol.* **124**, 319–335.
6. Altman, J. and Das, G. D. (1967) Postnatal neurogenesis in the guinea pig. *Nature* **214**, 1098–1101.
7. Lois, C. and Alvarez-Buylla, A. (1993) Proliferating subventricular zone cells in the adult mammalian forebrain can differentiate into neurons and glia. *Proc. Natl. Acad. Sci. USA* **90**, 2074–2077.
8. Luskin, M. D. (1993) Restricted proliferation and migration of postnatally generated neurons derived from the forebrain subventricular zone. *Neuron* **11**, 173–189.
9. Levision, S. W., Chuang, C., Abramson, B. J., and Goldman, J. E. (1993) The migrational patterns and developmental fates of glial precursors in the rat subventricular zone are temporally regulated. *Development* **119**, 611–622.
10. Gage, F. H., Kempermann, G., Palmer, T. D., Peterson, D. A., and Jasodhara, R. (1998) Multipotent progenitor cells in the adult dentate gyrus. *J. Neurobiol.* **36**, 249–266.
11. Cameron, H. A., Woolley, C. S., McEwen, B. S., and Gould, E. (1993) Differentiation of newly born neurons and glia in the dentate gyrus of the adult rat. *Neuroscience* **56**, 337–344.
12. Markakis, E. A. and Gage, F. H. (1999) Adult-generated neurons in the dentate gyrus send axonal projections to field CA3 and are surrounded by synaptic vesicles. *J. Comp. Neurol.* **406**, 449–460.
13. Mao, L. and Wang, J. Q. (2001) Gliogenesis in the striatum of the adult rat: alteration in neural progenitor population after psychostimulant exposure. *Dev. Brain Res.* **130**, 41–51.
14. Mao, L., Lau, Y. S., Petroske, E., and Wang, J. Q. (2001) Profound astrogenesis in the striatum of adult mice following nigrostriatal dopaminergic lesion by repeated MPTP administration. *Dev. Brain Res.* **131**, 57–65.
15. Calzà, L., Giardino, L., Pozza, M., Bettelli, C., Micera, A., and Aloe, L. (1998) Proliferation and phenotype regulation in the subventricular zone during experi-

mental allergic encephalomyelitis: in vivo evidence of a role for nerve growth factor. *Proc. Natl. Acad. Sci. USA* **95**, 3209–3214.
16. Craig, C. G., Tropepe, V., Morshead, C. M., Reynolds, B. A., Weiss, S., and van der Kooy, D. (1996) In vivo growth factor expansion of endogenous subependymal neural precursor cell populations in the adult mouse brain. *J. Neurosci.* **16**, 2649–2658.
17. Gritti, A., Frolichsthal-Schoeller, P., Galli, R., et al. (1999) Epidermal and fibroblast growth factors behave as mitogenic regulators for a single multipotent stem cell-like population from the subventricular region of the adult mouse forebrain. *J. Neurosci.* **19**, 3287–3297.
18. Kuhn, H. G., Winkler, J., Kempermann, G., Thal, L. J., and Gage, F. H. (1997) Epidermal growth factor and fibroblast growth factor-2 have different effects on neural progenitors in the adult rat brain. *J. Neurosci.* **17**, 5820–5829.
19. Wagner, J. P., Black, I. B., and DiCicco-Bloom, E. (1999) Stimulation of neonatal and adult brain neurogenesis by subcutaneous injection of basic fibroblast growth factor. *J. Neurosci.* **19**, 6006–6016.
20. Tao, Y., Black, I. B., and DiCicco-Bloom, E. (1996) Neurogenesis in neonatal rat brain is regulated by peripheral injection of basic fibroblast growth factor (bFGF). *J. Comp. Neurol.* **376**, 653–663.
21. Zigova, T., Pencea, V., Wiegand, S. J., and Luskin, M. B. (1998) Intraventricular administration of BDNF increases the number of newly generated neurons in the adult olfactory bulb. *Mol. Cell. Neurosci.* **11**, 234–245.
22. Gould, E., Cameron, H. A., Daniels, D. C., Woolley, C. S., and McEwen, B. S. (1992) Adrenal hormones suppress cell division in the adult rat dentate gyrus. *J. Neurosci.* **12**, 3642–3650.
23. Cameron, H. A. and Gould, E. (1994) Adult neurogenesis is regulated by adrenal steroids in the dentate gyrus. *Neuroscience* **61**, 203–209.
24. Tanapat, P., Hastings, N. B., Reeves, A. J., and Gould, E. (1999) Estrogen stimulates a transient increase in the number of new neurons in the dentate gyrus of the adult female rat. *J. Neurosci.* **19**, 5792–5801.
25. Cameron, H. A., McEwen, B. S., and Gould, E. (1995) Regulation of adult neurogenesis by excitatory input and NMDA receptor activation in the dentate gyrus. *J. Neurosci.* **15**, 4687–4692.
26. Gould, E., Cameron, H. A., and McEwen, B. S. (1994) Blockade of NMDA receptors increases cell death and birth in the developing rat dentate gyrus. *J. Comp. Neurol.* **340**, 551–565.
27. Gould, E., McEwen, B. S., Tanapat, P., Galea, L. A. M., and Fuchs, E. (1997) Neurogenesis in the dentate gyrus of the adult tree shrew is regulated by psychosocial stress and NMDA receptor activation. *J. Neurosci.* **17**, 2492–2498.
28. Gould, E. (1999) Serotonin and hippocampal neurogenesis. *Neuropsychopharmacology* **21**, 46S–51S.
29. Praag, H., Christie, van B. R., Sejnowski, T. J., and Gage, F. H. (1999) Running enhances neurogenesis, learning, and long-term potentiation in mice. *Proc. Natl. Acad. Sci. USA* **96**, 13427–13431.

30. Liu, J., Solway, K., Messing, R. O., and Sharp, F. R. (1998) Increased neurogenesis in the dentate gyrus after transient global ischemia in gerbils. *J. Neurosci.* **18,** 7768–7778.
31. Kempermann, G., Kuhn, H. G., and Gage, F. H. (1998) Experience-induced neurogenesis in the senescent dentate gyrus. *J. Neurosci.* **18,** 3206–3212.
32. Gould, E. and Tanapat, P. (1999) Stress and hippocampal neurogenesis. *Biol. Psychiatry* **46,** 1472–1479.
33. Gould, E., Tanapat, P., McEwen, B. S., Flugge, E., and Fuchs, E. (1998) Proliferation of granule cell precursors in the dentate gyrus of adult monkeys is diminished by stress. *Proc. Natl. Acad. Sci. USA* **95,** 3168–3171.
34. Li, W., Cogswell, C. A., and LoTurco, J. J. (1998) Neuronal differentiation of precursors in the neocortical ventricular zone is triggered by BMP. *J. Neurosci.* **18,** 8853–8862.
35. Mabie, P. C., Mehler, M. F., and Kessler, J. A. (1999) Multiple roles of bone morphogenetic protein signaling in the regulation of cortical cell number and phenotype. *J. Neurosci.* **19,** 7077–7088.
36. Teuchert-Noodt, G., Dawirs, P. R., and Hildebrandt, K. (2000) Adult treatment with methamphetamine transiently decreases dentate granule cell proliferation in the gerbil hippocampus. *J. Neural. Transm.* **107,** 133–143.
37. Wang, J. Q. and McGinty, J. F. (1999) Glutamate/dopamine interactions mediate the effects of psychostimulant drugs. *Addict. Biol.* **4,** 141–150.
38. Dawirs, P. R., Hildebrandt, K., and Teuchert-Noodt, G. (1998) Adult treatment with haloperidol increases dentate granule cell proliferation in the gerbil hippocampus. *J. Neural. Transm.* **105,** 317–327.
39. Kay, J. N. and Blum, M. (2000) Differential response of ventral midbrain and striatal progenitor cells to lesions of the nigrostriatal dopaminergic projection. *Dev. Neurosci.* **22,** 56–67.
40. Eisch, A. J., Barrot, M., Schad, C. A., Self, D. W., and Nestler, E. J. (2000) Opiates inhibit neurogenesis in the adult rat hippocampus. *Proc. Natl. Acad. Sci. USA* **97,** 7579–7584.
41. Alvarado, A. S. and Newmark, P. A. (1998) The use of planarians to dissect the molecular basis of metazoan regeneration. *Wound Repair Regen.* **6,** 413–420.
42. Gould, E., Reeves, A. J., Graziano, M. S. A., and Gross, C. G. (1999) Neurogenesis in the neocortex of adult primates. *Science* **286,** 548.
43. Lois, C. and Alvarez-Buylla, A. (1994) Long-distance neuronal migration in the adult mammalian brain. *Science* **264,** 1145–1148.
44. Markakis, E. and Gage, F. H. (1999) Adult-generated neurons in the dentate gyrus send axonal projections to field CA3 and are surrounded by synaptic vesicles. *J. Comp. Neurol.* **406,** 449–460.
45. Stanfield, B. B. and Trice, J. E. (1988) Evidence that granule cells generated in the dentate gyrus of adult rats extend axonal projections. *Exp. Brain Res.* **72,** 399–406.

4

Neuroprotective Effect of Naloxone in Inflammation-Mediated Dopaminergic Neurodegeneration

Dissociation from the Involvement of Opioid Receptors

Bin Liu and Jau-Shyong Hong

1. The Opioid System

Historically, the nociceptive/analgesic effect of naturally occurring opiates such as morphine has long been recognized by humans. Advances in research in the last several decades have revealed the existence of the so-called endogenous opioid peptides, can be divided into three classes: dynorphins, enkephalins, and β-endorphins. Contrary to the initial understanding, in addition to the cells of the central nervous system, those of peripheral tissues such as cardiac myocytes and heart tissues also express opioid peptides *(1–3)*. The wide distribution of opioid peptides throughout the body underscores their involvement in a variety of cellular activities including pain regulation, respiration, immune responses, and ion channel activity *(4)* as well as possibly pathophysiological conditions such as asthma, alcoholism, and eating disorders *(5–7)*.

Endogenous opioid peptides are synthesized as biologically inactive precursor polypeptides termed preprodynorphin, preproenkephalin, and preproopiomelanocortin, the last being a precursor of β-endorphins *(8)*. Precise processing by the action of highly regulated proteolytic enzymes converts the inert polypeptides into active fragments of varying lengths and bioactivity. It is now well known that endogenous opioid peptides exert their bioactivity through binding to cell surface receptors. Intensive research by means of ligand binding and molecular cloning studies in the last 40 yr has identified at least three classes of classic opioid receptors: δ, κ, and μ *(9,10)*. They are all G-protein-linked receptors with seven transmembrane domains and multiple potential phosphorylation sites in the terminal sequences. Pharmacological studies have further suggested the existence of subgroups within the three types of opioid receptors. However, molecular cloning has yet to confirm this. Nevertheless, various endogenous opioid peptides exhibit differential affinity toward the three types of opioid receptors. For example, methionine-enkephalin and leucine-enkephalin preferentially bind δ-opioid receptor and dynorphins show some selectivity toward κ-opioid receptor, while β-endorphin has a slightly higher affinity for the μ-opioid receptor than the other two types of receptors *(9,11)*. The dissociation constants of endogenously opioid peptides for opioid receptors are in the range of 10^{-9} M. Naturally occurring opiates such as morphine elicit physiologic responses by mimicking endogenous opioid peptides in binding to opioid receptors.

2. Naloxone

Naloxone is a synthetic and nonselective antagonist against all three classic opioid receptors. Structurally, it closely resembles the naturally occurring opiate morphine (**Fig. 1**). As with morphine, the binding of naloxone to opioid receptors is stereospecific, such that (–)-naloxone (also called L-naloxone for the levorotatory configuration of the asymmetric carbon) binds the μ-opioid receptor with the same dissociation constant as that for morphine ($\sim 2 \times 10^{-9}$ M). The affinity of (+)-naloxone (also called *R*-naloxone), however, is three to four orders of magnitude less than that for (–)-naloxone *(12,13)*. Therefore, ever since their synthesis and characterization, the naloxone enantiomers have been one of the most powerful tools in the investigation of the involvement of opioid receptors in various systems.

Over the last 30 yr, the ever increasing realization of the involvement of opioid systems in a wide variety of physiological as well as pathophysiological conditions, beyond the initially described roles in the nociceptive/analgesic systems, has certainly prompted an intensive screening of opioid receptor antagonists for potential therapeutic purposes. Because of its potent antagonistic activity, ease of crossing the blood–brain barrier, and relatively low systemic toxicity, (–)-naloxone has been tested for beneficial effects in a variety of experimental disease models. Mechanistically, the efficacy in the experimental treatment of conditions such as opiate dependence is certainly related to its activity as an opioid receptor antagonist *(14)*, whereas in the treatment of eating disorders *(15)* and alcoholism *(16)*, the opioid system most likely plays a role.

However, the underlying mechanisms of action for the experimental treatment of traumatic brain and spinal cord injuries *(17)*, myocardial and cerebral stroke *(18–22)*, and sepsis *(23)* are far from clear and are definitely not

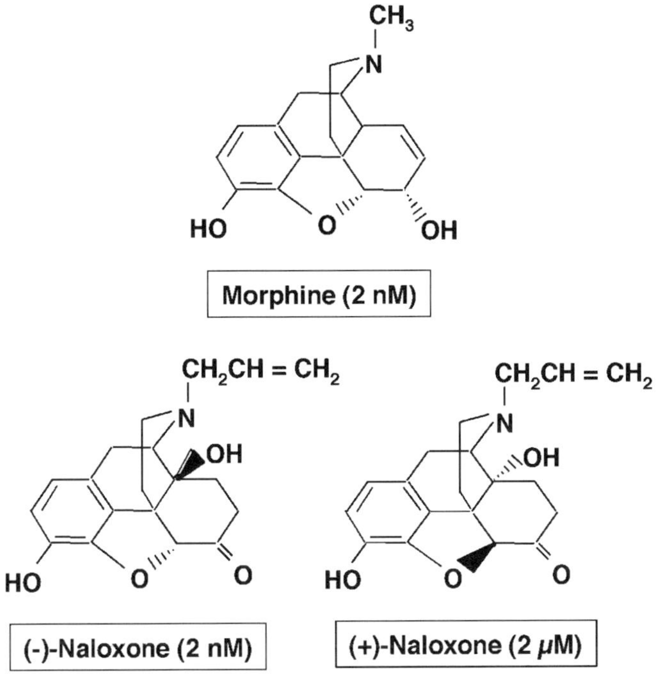

Fig. 1. Structures of morphine and naloxone stereoisomers. Numbers in parentheses indicate the approximate affinity to μ-type opioid receptor.

exclusively related to the opioid system. The notion of nonopioid actions of naloxone are discussed further toward the end of this chapter. Closer examination, at the molecular and cellular levels, of those disease conditions seems to suggest the existence of a common theme: the involvement of immune activities, in particular, the inflammatory responses.

3. Role of the Inflammatory Process in Ischemia, Brain and Spinal Cord Injuries, and Sepsis

Recently, increasing evidence has implicated the involvement of the inflammatory process in the pathogenesis of a variety of diseases. For example, in ischemic brain injury, the inflammatory response can be induced by the initial phase of neuronal death triggered by excitotoxicity. Cytotoxic factors produced as a result of the inflammatory response are thought to exacerbate the neuronal damage in the later phase of disease progression *(24–26)*. Similarly, in the case of myocardial tissue injury, a cell death induced inflammatory response is also suspected of playing a role in the final outcome of the disease *(27,28)*. An equally interesting issue is the realization of an inflammatory response induced by damage to the peripheral and central nervous systems. Again, it is thought that the initial injury to nervous tissues would trigger a local inflammatory response that contributes to secondary and lasting injury with persistent pain as one of the outcomes *(29,30)*. In sepsis, inflammation in response to the initial endotoxin challenge and subsequent damage to various organs are main components of the factors leading to the eventual multiorgan failure *(31)*.

4. Role of Inflammation in Neurodegenerative Disorders

Previously, the brain has been considered an immune privileged environment partly owing to the existence of the brain–blood barrier. However, inflammatory responses in the brain have been increasingly associated with pathogenesis of several degenerative neurological disorders including Parkinson's disease (PD), Alzheimer's disease, AIDS dementia, and amyotrophic lateral sclerosis (ALS, *32–34*).

Two types of glial cells, namely microglia and astrocytes, are the primary players of the inflammatory process in the brain *(35,36)*. Under normal conditions, microglia serve a function of immune surveillance. Astrocytes, on the other hand, act to maintain ionic homeostasis, buffer the action of neurotransmitters, and secrete nerve growth factors. In response to immunological stimuli or injuries in the brain, glia, especially microglia, readily become activated. Traditionally, injury- and/or neuronal death-induced glial activation have been called reactive gliosis with the term reactive microgliosis specifically referring to the activation of microglia. Perhaps in analogy to certain components of the immune response observed in the peripheral system, activated astrocytes and microglia which are either resident or, as some have speculated, recruited to the injury site play a key role in tissue repair through phagocytosis and secretion of various trophic factors. However, activated glia, especially microglia, produce a wide array of proinflammatory factors including cytokines such as tumor necrosis factor-α (TNF-α), interleukin-1β (IL-1β), IL-6, chemokines, free radicals such as nitric oxide (NO) and superoxide, and fatty acid metabolites such as eicosanoids *(33,36–39)*. Overproduction of a large number of these factors by glia contributes to inflammation and additional neuronal death in the brain and has been considered to be closely related to the pathogenesis of AD, PD, AIDS dementia, and ALS. This notion is supported by several lines of evidence. First, postmortem analysis of brains of patients with the aforementioned degenerative neurological disorders revealed the occurrence of microglial and astroglial activation *(32–34,40)*. Second, production of free radicals (NO, superoxide), cytokines (TNF-α, IL-1β), and eicosanoids has been observed frequently in cultures of glia and/or microglia following stimulation with amyloid peptides, HIV coat protein gp120, and the inflammagen bacterial endotoxin lipopolysaccharide (LPS) *(38,41–44)*. Third, addition of some of these factors, often in various combinations, to neuron/glia cultures has been shown to induce neuronal death *(45–47)*. Fourth, inhibition of the production of cytokines such as TNF-α and IL-1β, free radicals such as NO and reactive oxygen intermediates, and eicosanoids such as prostaglandins affords neuroprotection *(42,48–52)*. Interestingly, neurons may not be solely the innocent victims of inflammation-mediated degeneration. Interactions between neurons and glial cells via molecules such as the neural cell adhesion molecules may actually serve to suppress the immune response of glial cells, and loss of the "neuronal suppression" may be part of the mechanism of action underlying reactive gliosis *(53,54)*.

5. Microglial Activation: Part of the Etiology of PD?

The hallmark of PD is the progressive degeneration of the nigrostriatal dopaminergic system involving the loss of dopaminergic neurons in the substantia nigra and their fibers in the striatum. Sufficient damage to the dopaminergic pathways, over time, eventually leads to disorders in movement regulation. It has now been recognized that microglial activation is involved in the neurodegenerative process of PD *(55–58)*. Furthermore, epidemiological studies appear to suggest that microglial activation, as a consequence of exposure to infectious agents and environmental toxins and occurrence of early-life traumatic brain injuries, may play a role in the early stage of the pathogenetic process of PD *(59–64)*. Some of the important clues in favor of the hypothesis that microglial activation will result in dopaminergic neurodegeneration have come from experiments with neuron-glia cultures stimulated with the bacterial endotoxin LPS. Indeed, LPS neurotoxicity requires the presence of glia, and activation of glia, especially of microglia, leads to the degeneration of dopaminergic neurons *(65)*. In addition, intranigral injection of LPS activates microglia and induces the degeneration of nigral dopaminergic neurons *(66–68)*. Microglial activation occurs as early as 6 h after the infusion of LPS into rat brains, and significant production of cytokines (TNF-α, IL-1β) and free radicals (NO and superoxide) can be detected 2–12 hr after stimulation of neuron-glia cultures with LPS *(67,69,70)*. In contrast, significant degeneration of dopaminergic neurons in vitro and in vivo was not observed until after the occurrence of significant microglial activation, indicating that microglial activation and production of neurotoxic factors precede dopaminergic neurodegeneration. These results demonstrate that, at least in rodents, inflammagen (LPS)-induced microglial activation is capable of causing the degeneration of nigral dopaminergic neurons.

6. Neuroprotective Effect of Naloxone Stereoisomers on Dopaminergic Neurons in the Inflammation-Related Model of PD

The establishment of both in vitro and in vivo inflammation-related models of PD has enabled the search for and study of the mechanism of action responsible for a variety of neuroprotective agents. Of particular interest is the neuroprotective effect of the naloxone stereoisomers. In the in vitro neuron-glia culture system, 1 µ*M* (–)-naloxone afforded significant protection of dopaminergic neurons against LPS-induced degeneration. Interestingly, (+)-naloxone, which lacks opioid receptor binding activity, was equally effective *(69)*. The neuroprotective effect of naloxone was most likely unrelated to the opioid system because both compounds effectively inhibited the activation of microglia and their production of NO, TNF-α, and especially the superoxide free radical *(69,82;* **Fig. 2**). The in vitro observations were confirmed by in vivo studies where systemic administration of naloxone with an osmotic minipump reduced the loss of nigral dopaminergic neurons induced by LPS injection *(67,68)*. Again, both (–)-naloxone and (+)-naloxone were equally effective *(67)*. Interestingly, over the last several decades, several groups have described nonopioid effects of (+)-naloxone in various systems *(71–74)*. In addition, Simpkins and associates have reported that both naloxone stereoisomers are capable of suppressing the chemoattractant-induced activation of human neutrophils *(75,76)*. Therefore, it is possible that naloxone, regardless of its stereoconfigurations, is an effective modulator of immune cell activity. The appreciation of nonopioid and immune modulatory activity of naloxone and, at the same time, the increasing awareness of the role of inflammation in many disease conditions, may help redefine the mechanism of action for the observed efficacy of naloxone in the experimental treatment of a variety of pathological conditions. For example, whether the beneficial effects observed for naloxone in the treatment of spinal cord and traumatic brain injuries and myocardiac/cerebral stroke are related to any potential inhibitory effect on the secondary inflammatory response occurred following the initial phase of cell death remains to be examined *(17–19)*. Interestingly, in the experimental treatment of Alzheimer's disease, naloxone was found to be much more promising in younger patients *(77,78)* than in more advanced cases *(79,80)*. It would be of great interest to determine whether these results imply that reduction by naloxone of the inflammatory response at an earlier stage of the progression of the disease will have a more favorable final outcome. Similarly, the beneficial effect of naloxone in the experimental treatment of bacterial sepsis may also be related to its negative modulatory effect on the immune cells, although it may be

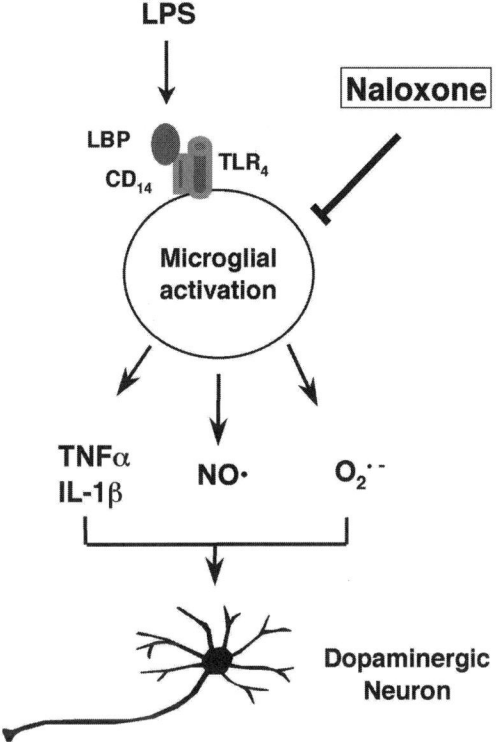

Fig. 2. Proposed mechanism of action for the neuroprotective effect of naloxone. By inhibiting the LPS-stimulated release of neurotoxic factors such as cytokines TNF-α and IL-1β and free radicals such as NO and superoxide, naloxone affords protection to dopaminergic neurons against inflammation-mediated degeneration.

true that inhibition of neutrophil activation alone may not be sufficient to reverse the course of the catastrophic cascade of the multiorgan failures *(23,81)*.

In animal studies and clinical trials, (–)-naloxone has a relatively large safety margin. (+)-Naloxone, devoid of opioid activity, would be a better choice as a candidate agent for potential use in the treatment of neurodegenerative diseases such as PD and inflammation-related disorders in general.

References

1. Smith, A. P. and Lee, N. M. (1988) Pharmacology of dynorphin. *Annu. Rev. Pharmacol. Toxicol.* **28,** 123–140.
2. Herz, A. (1993) *Opioids*, Vol. 1. Springer-Verlag, Berlin.
3. Barron, B. A. (2000) Cardiac opioids. *Proc. Soc. Exp. Biol. Med.* **224,** 1–7.

4. Roy, S. and Loh, H. (1996) Effects of opioids on the immune system. *Neurochem. Res.* **21,** 1375–1386.
5. Groneberg, D. A. and Fischer, A. (2001) Endogenous opioids as mediators of asthma. *Pulm. Pharmacol. Ther.* **14,** 383–389.
6. Gianoulakis, C. (2001) Influence of the endogenous opioid system on high alcohol consumption and genetic predisposition to alcoholism. *J. Psychiatry Neurosci.* **26,** 304–318.
7. Volpicelli, J. R. (2001) Alcohol abuse and alcoholism: an overview. *J. Clin. Psychiatry* **20,** 4–10.
8. Kieffer, B. L. (1995) Recent advances in molecular recognition and signal transduction of active peptides: receptors for opioid peptides. *Cell. Mol. Neurobiol.* **15,** 615–635.
9. Minami, M. and Satoh, M. (1995) Molecular biology of the opioid receptors: structures, functions and distributions. *Neurosci. Res.* **23,** 121–145.
10. Jordan, B. A., Cvejic, S., and Devi, L. A. (2000) Opioids and their complicated receptor complexes. *Neuropsychopharmacology* **23,** S5–S18.
11. Knapp, R. J., Malatynska, E., Collins, N., et al. (1995) Molecular biology and pharmacology of cloned opioid receptors. *FASEB J.* **9,** 516–525.
12. Iijima, I., Minamikawa, J-I., Jacobsen, A. E., Brossi, A., and Rice, K. E. (1978) Studies in the (+)-morphinin series-5. Synthesis and biological properties of (+)-naloxone. *J. Med. Chem.* **21,** 398–400.
13. Marcoli, M., Ricevuti, G., Mazzone, A., Pasotti, D., Lecchini, S., and Frigo, G. M. (1989) A stereoselective blockade by naloxone of opioid and non-opioid-induced granulocyte activation. *Int. J. Immunopharmacol.* **11,** 57–61.
14. Ward, J., Hall, W., and Mattick, R. P. (1999) Role of maintenance treatment in opioid dependence. *Lancet* **353,** 221–226.
15. de Zwaan, M. and Mitchell, J. E. (1992) Opiate antagonists and eating behavior in humans: a review. *J Clin. Pharmacol.* **32,** 1060–1072.
16. Anton, R. F. (2001) Pharmacologic approaches to the management of alcoholism. *J. Clin. Psychiatry* **20,** 11–17.
17. Seidl, E. C. (2000) Promising pharmacological agents in the management of acute spinal cord injury. *Pharm. Pract. Manag.* **20,** 21–27.
18. Hosobuchi, T., Baskin, D. S., and Woo, S. K. (1982) Reversal of induced ischemic neurologic deficit in gerbil by the opiate antagonist naloxone. *Science* **215,** 69–71.
19. Fallis, R. J., Fisher, M., and Lobo, R. A. (1983) A double blind trial of naloxone in the treatment of stroke. *Stroke* **15,** 627–629.
20. Holaday, J. W. and Macolm, D. S. (1986) Endogenous opioids and their antagonist in endotoxic shock, in *9th International Congress of Infectious and Parasitic Diseases: Bacterial Infection, Antibacterial Chemotherapy* (Marget, W., Lang, W., and Gabler-Sanberger, E., eds.), pp. 425–435, Munich.
21. Chen, C., Cheng, J. F. C., Liao, S. L., Chen, W. Y., Lin, N. N., and Kuo, J. S. (2000) Effects of naloxone on lactate, pyruvate metabolism and antioxidant enzyme activity in rat cerebral ischemia/reperfusion. *Neurosci. Lett.* **287,** 113–116.

22. Chen, C. J., Liao, S. L., Chen, W. Y., Hong, J. S., and Kuo, J. S. (2001) Cerebral ischemia/reperfusion injury in rat brain: effects of naloxone. *NeuroReport* **12,** 1245–1249.
23. Napolitano, L. M. (2000) Naloxone therapy in shock: the controversy continues. *Crit. Care Med.* **28,** 887–888.
24. Stoll, G., Jander, S., and Schroeter, M. (1998) Inflammation and glial responses in ischemic brain lesions. *Prog. Neurobiol.* **56,** 149–171.
25. Dirnagl, U., Iadecola, C., and Moskowitz, M. A. (1999) Pathobiology of ischaemic stroke: an integrated view. *Trends Neurosci.* **22,** 391–397.
26. del Zoppo, G., Ginis, I., Hallenbeck, J. M., Iadecola, C., Wang, X., and Feuerstein, G. Z. (2000) Inflammation and stroke: putative role for cytokines, adhesion molecules and iNOS in brain response to ischemia. *Brain Pathol.* **10,** 95–112.
27. Vallance, P., Collier, J., and Bhagat, K. (1997) Infection, inflammation, and infarction: does acute endothelial dysfunction provide a link? *Lancet* **349,** 1391–1392.
28. Mehta, J. L. and Li, D. Y. (1999) Inflammation in ischemic heart disease: response to tissue injury or a pathogenetic villain? *Cardiovasc. Res.* **43,** 291–299.
29. Lu, J., Ashwell, K. W., and Waite, P. (2000) Advances in secondary spinal cord injury: role of apoptosis. *Spine* **25,** 1859–1866.
30. DeLeo, J. A. and Yezierski, R. P. (2001) The role of neuroinflammation and neuroimmune activation in persistent pain. *Pain* **90,** 1–6.
31. Tjardes, T. and Neugebauer, E. (2002) Sepsis research in the next millennium: concentrate on the software rather than the hardware. *Shock* **17,** 1–8.
32. McGeer, P. L., Itagaki, S., Boyes, B. E., and McGeer, E. G. (1988) Reactive microglia are positive for HLA-DR in the substantia nigra of Parkinson's and Alzheimer's disease brains. *Neurology* **38,** 1285–1291.
33. Dickson, D. W., Lee, S. C., Mattiace, L. A., Yen, S. H., and Brosnan, C. (1993) Microglia and cytokines in neurological disease, with special reference to AIDS and Alzheimer's disease. *Glia* **7,** 75–83.
34. Raine, C. S. (1994) Multiple sclerosis: immune system molecule expression in the central nervous system. *J. Neuropathol. Exp. Neurol.* **53,** 328–337.
35. Kreutzberg, G. W. (1996) Microglia: a sensor for pathological events in the CNS. *Trends Neurosci.* **19,** 312–318.
36. Aloisi, F. (1999) The role of microglia and astrocytes in CNS immune surveillance and immunopathology. *Adv. Exp. Med. Biol.* **468,** 123–133.
37. Banati, R. B., Gehrmann, J., Schubert, P., and Kreutzberg, G. W. (1993) Cytotoxicity of microglia. *Glia* **7,** 111–118.
38. Minghetti, L. and Levi, G. (1998) Microglia as effector cells in brain damage and repair: focus on prostanoids and nitric oxide. *Prog. Neurobiol.* **54,** 99–125.
39. Streit, W. J., Conde, J. R., and Harrison, J. K. (2001) Chemokines and Alzheimer's disease. *Neurobiol. Aging* **22,** 909–913.
40. Gonzalez-Scarano, F. and Baltuch, G. (1999) Microglia as mediators of inflammatory and degenerative diseases. *Annu. Rev. Neurosci.* **22,** 219–240.
41. Boje, K. M. and Arora, P. K. (1992) Microglial-produced nitric oxide and reactive nitrogen oxides mediate neuronal cell death. *Brain Res.* **587,** 250–256.

42. Chao, C. C., Hu, S., Ehrlich, L., and Peterson, P. K. (1992) Interleukin-1 and tumor necrosis factor-alpha synergistically mediate neurotoxicity: involvement of nitric oxide and of *N*-methyl-D-aspartate receptors. *Brain Behav. Immun.* **9,** 355–365
43. Dawson, V. L. (1995) Nitric oxide: role in neurotoxicity. *Clin. Exp. Pharmacol. Physiol.* **22,** 305–308.
44. Kong, L. Y., Wilson, B. C., McMillian, M. K., Bing, G., Hudson, P. M., and Hong, J. S. (1996) The effects of the HIV-1 envelope protein gp120 on the production of nitric oxide and proinflammatory cytokines in mixed glial cell cultures. *Cell Immunol.* **172,** 77–83.
45. Chao, C. C., Hu, S., Molitor, T. W., Shaskan, E. G., and Peterson, P. K. (1995) Activated microglia mediate neuronal cell injury via a nitric oxide mechanism. *J. Immunol.* **149,** 2736–2741.
46. Jeohn, G-H., Kong, L-Y., Wilson, B., Hudson, P., and Hong, J-S. (1998) Synergistic neurotoxic effects of combined treatments with cytokines in murine primary mixed neuron/glia cultures. *J. Neuroimmunol.* **85,** 1–10.
47. McGuire, S. O., Ling, Z. D., Lipton, J. W., Sortwell, C. E., Collier, T. J., and Carvey, P. M. (2001) Tumor necrosis factor alpha is toxic to embryonic mesencephalic dopamine neurons. *Exp. Neurol.* **169,** 219–230.
48. Kong, L. Y., Maderdrut, J. L., Jeohn, G. H., and Hong, J. S. (1999) Reduction of lipopolysaccharide-induced neurotoxicity in mixed cortical neuron/glia cultures by femtomolar concentrations of pituitary adenylate cyclase-activating polypeptide. *Neuroscience* **91,** 493–500.
49. Kong, L. Y., Jeohn, G. H., Hudson, P. M., Du, L., Liu, B., and Hong, J. S. (2000) Reduction of lipopolysaccharide-induced neurotoxicity in mouse mixed cortical neuron/glia cultures by ultralow concentrations of dynorphins. *J. Biomed. Sci.* **7,** 241–247.
50. Jeohn, G. H., Wilson, B., Wetsel, W. C., and Hong, J. S. (2000) The indolocarbazole Go6976 protects neurons from lipopolysaccharide/interferon-gamma-induced cytotoxicity in murine neuron/glia co-cultures. *Brain Res. Mol. Brain Res.* **79,** 32–44.
51. Liu, B., Qin, L., Yang, S-N., Wilson, B. C., Liu, Y., and Hong, J. S. (2001) Femtomolar concentrations of dynorphins protect rat mesencephalic dopaminergic neurons against inflammatory damage. *J. Pharmacol. Exp. Ther.* **298,** 1133–1141.
52. Araki, E., Forster, C., Dubinsky, J. M., Ross, M. E., and Iadecola, C. (2001) Cyclo-oxygenase-2 inhibitor ns-398 protects neuronal cultures from lipopolysaccharide-induced neurotoxicity. *Stroke* **32,** 2370–2375.
53. Chang, R. C., Hudson, P., Wilson, B., et al. (2000) Immune modulatory effects of neural cell adhesion molecules on lipopolysaccharide-induced nitric oxide production by cultured glia. *Brain Res. Mol. Brain Res.* **81,** 197–201.
54. Chang, R. C., Chen, W., Hudson, P., Wilson, B., Han, D. S., and Hong, J. S. (2001) Neurons reduce glial responses to lipopolysaccharide (LPS) and prevent injury of microglial cells from over-activation by LPS. *J. Neurochem.* **76,** 1042–1049.

55. Vawter, M. P., Dillon-Carter, O., Tourtellotte, W. W., Carvey, P., and Freed, W. J. (1996) TGFbeta1 and TGFbeta2 concentrations are elevated in Parkinson's disease in ventricular cerebrospinal fluid. *Exp. Neurol.* **142,** 313–322.
56. Cassarino, D. S., Fall, C. P., Swerdlow, R. H., et al. (1997) Elevated reactive oxygen species and antioxidant enzyme activities in animal and cellular models of Parkinson's disease. *Biochim. Biophys. Acta* **1362,** 77–86.
57. Banati, R. B., Daniel, S. E., and Blunt, S. B. (1998) Glial pathology but absence of apoptotic nigral neurons in long-standing Parkinson's disease. *Mov. Disord.* **13,** 221–227.
58. Hunot, S., Dugas, N., Faucheux, B., et al. (1999) Fc epsilonRII/CD23 is expressed in Parkinson's disease and induces, in vitro, production of nitric oxide and tumor necrosis factor-alpha in glial cells. *J. Neurosci.* **19,** 3440–3447.
59. Ravenholt, R. T. and Foege, W. H. (1982) 1918 influenza, encephalitis lethargica, parkinsonism. *Lancet* **8303,** 860–864.
60. Factor, S. A., Sanchez-Ramos, J., and Weiner, W. J. (1988) Trauma as an etiology of parkinsonism: a historical review of the concept. *Mov. Disord.* **3,** 30–36.
61. Mattock, C., Marmot, M., and Stern, G. (1998) Could Parkinson's disease follow intrauterine influenza? a speculative hypothesis. *J. Neurol. Neurosurg. Psychiatry* **51,** 753–756.
62. Williams, D. B., Annegers, J. F., Kokmen, E., O'Brien, P. C., and Kurland, L. T. (1991) Brain injury and neurologic sequelae: a cohort study of dementia, parkinsonism, and amyotrophic lateral sclerosis. *Neurology* **41,** 1554–1557.
63. Casals, J., Elizan, T. S., and Yahr, W. H. (1998) Postencephalitic parkinsonism—a review *J. Neural. Transm.* **105,** 645–676.
64. Plassman, B. L., Havlik, R. J., Steffens, D. C., et al. (2000) Documented head injury in early adulthood and risk of Alzheimer's disease and other dementias. *Neurology* **55,** 1158–1166.
65. Bronstein, D. M., Perez-Otano, I., Sun, V., et al. (1995) Glia-dependent neurotoxicity and neuroprotection in mesencephalic cultures. *Brain Res.* **704,** 112–116.
66. Castano, A., Herrera, A. J., Cano, J., and Machado, A. (1998) Lipopolysaccharide intranigral injection induces inflammatory reaction and damage in nigrastriatal dopaminergic system. *J. Neurochem.* **70,** 1584–1592.
67. Liu, B., Jiang, J. W., Wilson, B. C., et al. (2000) Systemic infusion of naloxone reduces degeneration of rat substantia nigral dopaminergic neurons induced by intranigral injection of lipopolysaccharide. *J. Pharmacol. Exp. Ther.* **295,** 125–132.
68. Lu, X., Bing, G., and Hagg, T. (2000) Naloxone prevents microglia-induced degeneration of dopaminergic substantia nigra neurons in adult rats. *Neuroscience* **97,** 285–291.
69. Liu, B., Du, L., and Hong, J. S. (2000) Naloxone protects rat dopaminergic neurons against inflammatory damage through inhibition of microglia activation and superoxide generation. *J. Pharmacol. Exp. Ther.* **293,** 607–617.

70. Kim, W. G., Mohney, R. P., Wilson, B., Jeohn, G. H., Liu, B., and Hong, J. S. (2000) Regional difference in susceptibility to lipopolysaccharide-induced neurotoxicity in the rat brain: role of microglia. *J. Neurosci.* **20,** 6309–6316.
71. Dunwiddie, T. V., Perez-Reyes, E., Rice, K. C., and Palmer, M. R. (1982) Stereoselectivity of opiate antagonists in rat hippocampus and neocortex: responses to (+) and (–) isomers of naloxone. *Neuroscience* **7,** 1691–1702.
72. Kim, J. P., Goldberg, M. P., and Choi, D. W. (1987) High concentrations of naloxone attenuate *N*-methyl-D-aspartate receptor-mediated neurotoxicity. *Eur. J. Pharmacol.* **138,** 133–136.
73. Chatterjie, N., Alexander, G. J., Sechzer, J. A., and Lieberman, K. W. (1996) Prevention of cocaine-induced hyperactivity by a naloxone isomer with no opiate antagonist activity. *Neurochem. Res.* **21,** 691–693.
74. Chatterjie, N., Sechzer, J. A., Lieberman, K. W., and Alexander, G. J. (1998) Dextro-naloxone counteracts amphetamine-induced hyperactivity. *Pharmacol. Biochem. Behav.* **59,** 271–274.
75. Simpkins, C. O., Ives, N., Tate, E., and Johnson, M. (1985) Naloxone inhibits superoxide release from human neutrophils. *Life Sci.* **37,** 1381–1386.
76. Simpkins, C. O., Alailima, S. T., and Tate, E. A. (1986) Inhibition by naloxone of neutrophil superoxide release: a potentially useful antiinflammatory effect. *Circ. Shock* **20,** 181–191.
77. Tariot, P. N., Upadhyaya, A., Sunderland, T., et al. (1999) Physiologic and neuroendocrine responses to intravenous naloxone in subjects with Alzheimer's disease and age-matched controls. *Biol. Psychiatry* **46,** 412–419.
78. Tariot, P. N., Gross, M., Sunderland, T., Weingartner, H., Murphy, D. L., and Cohen, R. M. (1988) High-dose naloxone in older normal subjects: implications for Alzheimer's disease. *J. Am. Geriatr. Soc.* **36,** 681–686.
79. Serby, M., Resnick, R., Jordan, B., Adler, J., Corwin, J., and Rotrosen, J. P. (1986) Naltrexone and Alzheimer's disease. *Prog. Neuropsychopharmacol. Biol. Psychiatry.* **10,** 587–590.
80. Henderson, V. W., Roberts, E., Wimer, C., et al. (1989) Multicenter trial of naloxone in Alzheimer's disease. *Ann. Neurol.* **25,** 404–406.
81. Holaday, J. W. and Faden, A. I. (1978) Naloxone reversal of endotoxin hypotension suggests role of endorphins in shock. *Nature* **275,** 450–451.
82. Chang, R. C., Rota, C., Glover, R. E., Mason, R. P., and Hong, J. S. (2000) A novel effect of an opioid receptor antagonist, naloxone, on the production of reactive oxygen species by microglia: a study by electron paramagnetic resonance spectroscopy. *Brain Res.* **854,** 224–229.

5

Neuropharmacology and Neurotoxicity of 3,4-Methylenedioxymethamphetamine

Gary A. Gudelsky and Bryan K. Yamamoto

With the continuing and increasing popularity of the amphetamine analog 3,4-methylenedioxymethamphetamine (MDMA, Ecstasy) as a drug of abuse, concern also has increased regarding the long-term psychological and neurochemical effects of this drug. The acute psychological effects of MDMA include a mild sense of euphoria and sense of well being and increased ability to interact with others *(1)* which contribute to the drug's popularity. Recent studies in laboratory animals and humans indicate that repeated exposure to MDMA elicits long-term changes in neurochemistry and behavior that are viewed as resulting from a selective neurotoxicity of 5-hydroxytryptamine (5-HT)-containing axon terminals.

1. Acute Neurochemical and Physiological Effects of MDMA

MDMA exerts a pronounced stimulatory effect on the release of 5-HT in the brain, as evidenced under in vitro conditions from rat brain slices and synaptosomes *(2,3)*, as well as in vivo *(4,5)*. The MDMA-induced release of 5-HT is most likely a carrier-mediated process that involves the interaction of MDMA with the 5-HT uptake site *(5,6)*.

It also is well documented that MDMA increases dopamine release in vitro *(3)* and in vivo *(7,8)*. The MDMA-induced release of dopamine involves both carrier-mediated and impulse-dependent processes *(9,10)*. The contribution of impulse-dependent processes to MDMA-induced dopamine release appears to involve 5-HT neuronal systems, inasmuch as the magnitude of MDMA-induced dopamine release is modulated by 5-HT$_2$ receptors *(10–13)*.

From: *Methods in Molecular Medicine, vol. 79: Drugs of Abuse: Neurological Reviews and Protocols*
Edited by: J. Q. Wang © Humana Press Inc., Totowa, NJ

Recently, MDMA also has been shown to increase the release of acetylcholine in striatal slices and in the cortex and hippocampus in vivo *(14–16)*. Although the MDMA-induced release of acetylcholine in the striatum involves histaminergic mechanisms *(14)*, the neurotransmitter interactions underlying the ability of MDMA to enhance cholinergic function in the cortex and hippocampus are unknown at present.

The excessive release of dopamine and 5-HT induced by MDMA is thought to mediate many of the physiological and behavioral effects of the drug. MDMA produces a 5-HT_2 receptor-dependent increase in the serum concentrations of the hormones prolactin and corticosterone *(17)*. MDMA also produces a marked increase in core body temperature, and evidence supports the view that activation of 5-HT_2 receptors contributes to this response *(17,18)*. Motor function also is enhanced by MDMA. A characteristic 5-HT behavioral syndrome is produced by MDMA *(19,20)*, and low doses of MDMA induce hyperlocomotion that is mediated, in part, by activation of $5\text{-HT}_{1B/1D}$ receptors *(21)*.

Recently, the acute effects of MDMA on social function in rats have been investigated. Morley and McGregor *(22)* report that MDMA can elicit both anxiogenic and anxiolytic effects that are dependent on the test situation employed. MDMA elicits anxiogenic effects in the elevated plus maze and the emergence test. Anxiolytic effects of acute MDMA administration are evident in the reduction of aggressive behavior and increase in the duration of social interaction *(22)*.

2. Long-Term Effects of MDMA: 5-HT Neurotoxicity

The single or repeated administration of MDMA consistently has been shown to result in long-term reductions in (1) 5-HT concentrations in multiple brain regions of the rat and nonhuman primate *(23,24)*, (2) the density of 5-HT uptake sites *(25)*, (3) the activity of tryptophan hydroxylase *(26)*, and (4) the density of fine axons of 5-HT neurons *(27)*. The monoamine-depleting effect of MDMA is selective for 5-HT. Dopamine and norepinephrine concentrations are unaffected by MDMA; an exception is the mouse, in which MDMA depletes dopamine *(28,29)*. In nonhuman primates and humans exposed to MDMA, data from positron emission tomography and single photon emission computed tomography studies are indicative of MDMA-induced reductions in the density of 5-HT transporters *(30,31)*.

The effect of MDMA on brain concentrations of 5-HT is biphasic and can be divided into early and late or long-lasting phases. An early, reversible phase of 5-HT depletion occurs within 3–6 h after its administration, after which 5-HT concentrations return to normal values *(23)*. A long-lasting depletion of 5-HT occurs 2–3 d after drug treatment, and this depletion of 5-HT is evident in most brain regions containing 5-HT terminals *(32)*. There is only a partial recovery, at best, to control concentrations of 5-HT after depletion produced by MDMA. In fact, 5-HT concentrations remain depleted in most brain regions up to 1 yr following MDMA administration *(32,33)*. The fact that 5-HT cell bodies are spared by MDMA may allow for the regeneration of "pruned" 5-HT axons. Indeed, the regeneration of 5-HT axon terminals has been reported following a "neurotoxic" regimen of MDMA, although the pattern of reinnervation is abnormal *(34)*.

On the basis of the persistence of MDMA effects on 5-HT terminals and the accompanying functional changes (*see* **Subheading 7.**), there is a general concensus that MDMA induces selective 5-HT neurotoxicity, that is, degeneration of 5-HT axon terminals. Nevertheless, there is a lack of consistent histopathological or cytochemical changes that usually accompany neurotoxicity in MDMA-treated rats. Specifically, MDMA induces little increase in silver staining for degenerating neurons *(35,36)*, and there is little induction of glial fibrillary acidic protein *(37)*. However, consistent with the view that MDMA produces neurotoxicity, presumably within 5-HT axon terminals, is the finding that MDMA treatment results in a reduction of anterograde axonal transport of labeled material to forebrain regions containing 5-HT innervation *(38)*. In addition, MDMA promotes the cleavage of the cytoskeletal protein tau in the hippocampus *(39)*. These findings support the conclusion that MDMA produces structural brain damage in the rodent brain that accompanies the long-term depletion of brain 5-HT.

3. Role of Dopamine in MDMA-Induced Neurotoxicity

The excessive release of dopamine elicited by MDMA is proposed to contribute, in part, to the 5-HT neurotoxicity produced by this drug. A correlation exists between the extent of dopamine release and extent of long-term 5-HT depletion induced by MDMA and structurally related compounds *(8)*. Moreover, attenuation of the MDMA-induced release of dopamine by the lesioning of neurons *(40,41)*, treatment with dopamine uptake inhibitors *(42,43)*, or inhibition of dopamine synthesis *(26)* affords protection against MDMA-induced 5-HT neurotoxicity. Furthermore, elimination of dopamine transporters in the striatum with the use of antisense oligonucleotides also prevents the MDMA-induced depletion of striatal 5-HT *(44)*. Conversely, facilitation of dopamine release with 3,4-dihydroxy-L-phenylalanine (L-DOPA) or 5-HT$_2$ agonists results in an augmentation of the long-term depletion of 5-HT induced by MDMA *(12,45)*. The ontogeny of MDMA-induced 5-HT neurotoxicity also appears to be dependent on the ability of MDMA to release dopamine *(41)*.

The contribution of excessive extracellular dopamine to the process of MDMA neurotoxicity is predicated on the fact that dopamine itself is cytotoxic *(46,47)*. Along these lines, dopamine, present in excessive synaptic concentrations, may be taken up by an activated 5-HT transporter to initiate dopamine-dependent oxidative processes within the 5-HT terminal *(48)*.

4. Role of 5-HT and Its Transporter in MDMA-Induced Neurotoxicity

Although a sustained activation of the dopamine transporter and increase in the release of dopamine appear necessary for the expression of MDMA-induced neurotoxicity, it is not sufficient to explain the pattern of neurotoxicity exhibited selectively by MDMA and not by other amphetamines. The acute action of MDMA on 5-HT terminals themselves appears to contribute to the selective depletion of 5-HT produced by this agent. Amphetamine, which increases dopamine release but has little effect on 5-HT release, or 5-methoxy-6-methyl-2-aminoindane (MMAI), which selectively enhances 5-HT release, given alone do not produce 5-HT toxicity *(49)*. However, the coadministration of MMAI and amphetamine does produce 5-HT depletion. Thus, a concomitant increase in the extracellular concentration of dopamine and activation of the 5-HT transporter appear necessary for 5-HT neurotoxicity.

It can be envisioned that activation of the 5-HT transporter by MDMA renders 5-HT terminals vulnerable to further initiators of toxicity. Indeed, MDMA-induced 5-HT neurotoxicity is prevented in animals in which the 5-HT transporter is inhibited by fluoxetine *(50,51)*. Actions of MDMA on the 5-HT transporter may be responsible for the generation of reactive oxygen species, i.e., free radicals, or for an increase in intracellular calcium concentrations *(52)*, both of which could contribute to the process of toxicity.

The direct administration of MDMA into the brain does not result in 5-HT neurotoxicity *(53–56)*. This has led to the hypothesis that a neurotoxc metabolite of MDMA is formed peripherally. 5-(Glutathion-S-yl)-α-methyldopamine is one reactive metabolite of MDMA that has been proposed to mediate its toxicity *(57,58)*. MDMA undergoes demethylenation to form N-methyl-α-methyldopamine. This catechol may undergo further oxidation to form a reactive quinone that reacts readily with glutathione (GSH) to form thioether conjugates of α-methyldopamine. GSH and N-acetylcysteine conjugates of α-methyldopamine produce selective 5-HT neurotoxicity when injected directly into the brain *(57)*. Moreover, inhibition of the breakdown of thioether conjugates of α-methyldopamine augments MDMA-induced 5-HT neurotoxicity, whereas the administration of GSH to reduce the entry of these thioether conjugates into the brain diminishes MDMA-induced neurotoxicity *(50,59)*.

Thus, MDMA neurotoxicity may result from the peripheral formation and subsequent actions of thioether conjugates of N-methyl-α-methyldopamine. The mechanism by which these reactive metabolites of MDMA induce neurotoxicity is unclear but may involve an interaction of the metabolite with the 5-HT transporter and the redox cycling of these compounds to generate reactive oxygen species.

5. Role of Hyperthermia in MDMA-Induced Neurotoxicity

Considerable attention has been given to the role of environmental and/or core body temperature in the long-term neurotoxic effects of MDMA on 5-HT terminals. Maintenance of rats at a cool ambient temperature attenuates the MDMA-induced depletion of brain 5-HT *(60,61)*. In addition, the magnitude of the thermal response to MDMA is correlated with the extent of MDMA-induced 5-HT neurotoxicity *(62)*. Importantly, many drugs (e.g., dizolcipine, ketanserin, haloperidol) that afford protection against MDMA-induced 5-HT neurotoxicity attenuate MDMA-induced hyperthermia *(63,64)*. Thus, it is not possible to distinguish between the roles of specific neurotransmitter receptors and diminished hyperthermia in the neuroprotective actions of these drugs.

Although hyperthermia may contribute to the extent of MDMA toxicity, it is neither necessary nor sufficient for the expression of this toxicity. Thus, MDMA is capable of producing 5-HT neurotoxicity in the absence of drug-induced hyperthermia *(63)*, as well as under conditions in which MDMA by itself evokes hypothermia *(62)*. In addition, although MDMA is capable of producing hyperthermia in rats of postnatal age 21 d, there is no long-term depletion of brain 5-HT produced by MDMA in rats at this age *(41)*.

6. Role of Oxidative and Bioenergetic Stress in MDMA Neurotoxicity

Increasing evidence suggests that MDMA-induced 5-HT neurotoxicity results from increased free radical formation and the subsequent induction of a state of oxidative stress. Support for a free radical hypothesis of MDMA toxicity is based on findings that (1) MDMA increases the formation of free radicals, (2) MDMA produces cellular damage that often accompanies free radical formation, and (3) free radical scavengers and/or antioxidants attenuate MDMA-induced 5-HT neurotoxicity.

Although free radicals are short-lived reactive species, in the presence salicylic acid, a stable adduct, that is, 2,3-dihydroxybenzoic acid (DHBA), is formed that can be quantified analytically. MDMA produces a rapid and sustained increase in the extracellular concentration of 2,3-DHBA in brain *(43,51,*

65,66). The MDMA-induced generation of hydroxyl radicals appears to be dependent on the activation of both the dopamine and 5-HT transporter *(43,51)*.

There are several potential sources for free radicals generated by MDMA. Dopamine may undergo enzymatic or nonenzymatic oxidation to form superoxide radical and hydrogen peroxide. The importance of superoxide radicals in MDMA neurotoxicity is evident in transgenic mice that overexpress superoxide dismutase, which is responsible for the degradation of superoxide radicals. These transgenic mice are resistant to MDMA-induced neurotoxicity *(67)*. Alternatively, metabolites of MDMA may serve as sources of free radicals. Quinone thioethers described in the preceding section have the capability of redox cycling to produce reactive oxygen species.

Reactive species that contribute to MDMA toxicity may not be limited to oxygen-based radicals but also may include reactive nitrogen species, for example, nitric oxide and peroxynitrite. Inhibitors of nitric oxide synthase (NOS) provide protection against MDMA-induced dopamine depletion in the mouse *(68)*, as well as 5-HT depletion in the rat *(69,70)*. However, it has not been possible to ascribe the neuroprotective effects of these drugs specifically to the inhibition of NOS, inasmuch as NOS inhibitors, for the most part, markedly attenuate MDMA-induced hyperthermia *(69)*. However, the NOS inhibitor S-methyl-L-thiocitrulline attenuates MDMA-induced dopamine toxicity in the mouse without modifying MDMA-induced hyperthermia *(68)*. S-Methyl-L-citrulline also attenuates the long-term depletion of 5-HT, as well as dopamine, in the striatum of the rat following the intrastriatal administration of MDMA and malonate (Gudelsky, *unpublished observations*).

The exact mechanism whereby MDMA increases the formation of reactive nitrogen species is unknown, but it is unlikely to involve an increase in the release of glutamate *(71)* or glutamate receptor stimulation *(68)*. Conceivably, treatment with MDMA may produce an increase in intracellular calcium through impairment of cellular energetics *(61)*, disruption of mitochondrial function *(72)*, or activation of protein kinase C *(52)* that ultimately results in the activation of NOS. Regardless of the nature of the reactive oxygen or nitrogen species, the importance of the 5-HT transporter itself in the generation of reactive oxygen or nitrogen species is underscored by the finding that MDMA-induced free radical formation is absent in rats treated with fluoxetine *(51)* or in rats in which 5-HT terminals have been disrupted by fenfluramine *(65)*.

The administration of MDMA also results in cellular damage or changes consistent with the induction of oxidative stress. MDMA increases the formation of malondialdehyde-related substances that are indicative of free radical-induced lipid peroxidation *(73,74)*. MDMA also increases the formation of nitro-tyrosine residues (Yamamoto, *unpublished observations*) that is consistent with nitric oxide- or peroxynitrite-induced protein nitration. Finally, a neurotoxic regimen of MDMA depletes brain concentrations of the endogenous antioxidants vitamin E and ascorbic acid *(75)*.

The contribution of free radical-induced damage in the process of MDMA neurotoxicity is inferred from the findings that the administration of antioxidants *(50,59,75)* or spin trap agents *(76,77)* prevents the MDMA-induced depletion of brain 5-HT. Furthermore, as discussed previously, overexpression of superoxide dismutase renders transgenic mice resistant to MDMA-induced dopamine toxicity *(67)*.

In addition to a role of oxidative stress in MDMA-induced neurotoxicity, alterations in energy metabolism also may contribute to the process of neurotoxicty induced by psychostimulant drugs. Methamphetamine reduces brain concentrations of ATP *(78)* and increases the extracellular concentration of lactate *(79)*. In addition, the administration of energy substrates attenuates dopamine neurotoxicity elicited by methamphetamine *(79,80)*. These findings suggest that psychostimulants may acutely impair mitochondrial function. Indeed, methamphetamine and MDMA acutely inhibit the activity of the mitochondrial enzyme cytochrome oxidase *(72)*. Furthermore, the combined administration of methamphetamine and malonate, a complex II inhibitor of mitochondrial function, synergize to deplete striatal dopamine concentrations *(81,82)*. The intrastriatal administration of malonate and MDMA, neither of which alone depletes tissue 5-HT concentrations, together produces significant reductions in striatal 5-HT and dopamine concentrations *(55)*. These data suggest a role for bioenergetic stress in the long-term effects of MDMA on 5-HT, and possibly dopamine, nerve terminals.

MDMA, as well as methamphetamine and parachloroamphetamine, appear to disrupt cellular energetics by further promoting glycogenolysis *(61,83)*. MDMA and methamphetamine produce a transient decrease in brain glycogen concentrations that may involve the 5-HT$_2$ receptor-dependent activation of glycogen phosphorylase *(84)*. MDMA also produces a sustained elevation in the extracellular concentration of glucose in the brain *(61)*. The increased extracellular concentration of glucose may be indicative of increaesd regional cerebral blood flow and glucose utilization. This ability of MDMA to promote glycogenolysis is associated with the hyperthermia produced by MDMA *(61)*. Thus, the involvement of hyperthermia in the process by which MDMA depletes energy stores may be the same mechanism through which hyperthermia exacerbates MDMA-induced 5-HT neurotoxicity.

7. Functional Consequences of MDMA Neurotoxicity

Although the long-term effects of MDMA on 5-HT axon terminals have been well documented, the potential functional consequences associated with MDMA-induced 5-HT depletion have been investigated only recently. It is known that the administration of a 5-HT depleting regimen of MDMA results in diminished behavioral, thermal, neurochemical, and neuroendocrinological responses to a subsequent administration of MDMA or other 5-HT releasing drugs *(20,85–87)*. These diminished responses appear to be due to a reduced amount of evoked 5-HT release *(20,88)* in animals treated previously with a neurotoxic regimen of MDMA.

However, alterations in 5-HT receptor sensitivity induced by high-dose MDMA administration also may underlie abnormal responses to 5-HT agonists or releasing agents *(89)*. The repeated administration of MDMA increases the density of 5-HT$_{1A}$ receptors in the frontal cortex and augments the hypothermic response produced by the 5-HT$_{1A}$ agonist 8-hydroxy-2-(di-n-propylamino) tetralin hydrobromide (8-OH-DPAT) *(90)*. Prior MDMA treatment also results in an increased serum prolactin response to fenfluramine *(87)*.

In addition to impaired or abnormal responses evoked by pharmacological agents in MDMA-treated rats, numerous physiological responses also are rendered abnormal by prior, repeated MDMA administration. For example, rats treated with a 5-HT depleting regimen of MDMA exhibit reduced diurnal and nocturnal locomotor activity *(91)*. Furthermore, thermoregulation in rats is disrupted by the repeated administration of MDMA. This is evidenced by exaggerated hyperthermia in response to an elevated environmental temperature *(92,93)*. Neurotransmitter responses to stress also are altered in rats given MDMA repeatedly. Restraint stress results in increased extracellular concentrations of 5-HT and dopamine in the frontal cortex and hippocampus; these responses are diminished in MDMA-treated rats *(94)*.

The repeated administration of MDMA also disrupts cycles of sleep/wakefulness in the rat. According to Jones *(95)*, 5-HT facilitates the transition from slow-wave sleep to REM sleep and inhibits wakefulness. Therefore, it follows that a depletion of 5-HT may increase wakefulness. Consistent with this hypothesis, **Fig. 1** shows that rats pretreated with neurotoxic doses of MDMA exhibit increased bout lengths of wakefulness and increased overall time spent in wakefulness but do not exhibit any significant changes in REM or slow wave sleep (**Fig. 2**).

Behaviors related to anxiety and learning and memory also are abnormal in MDMA-treated animals, although some of these data are conflicting. Morley et al. *(96)* report that rats given a neurotoxic regimen of MDMA exhibit greater anxiety-related behaviors, as assessed in an elevated plus maze and social interaction and emergence tests. In contrast, Mechan et al. *(97)* conclude that MDMA-treated rats display a reduction in anxiety-related behaviors.

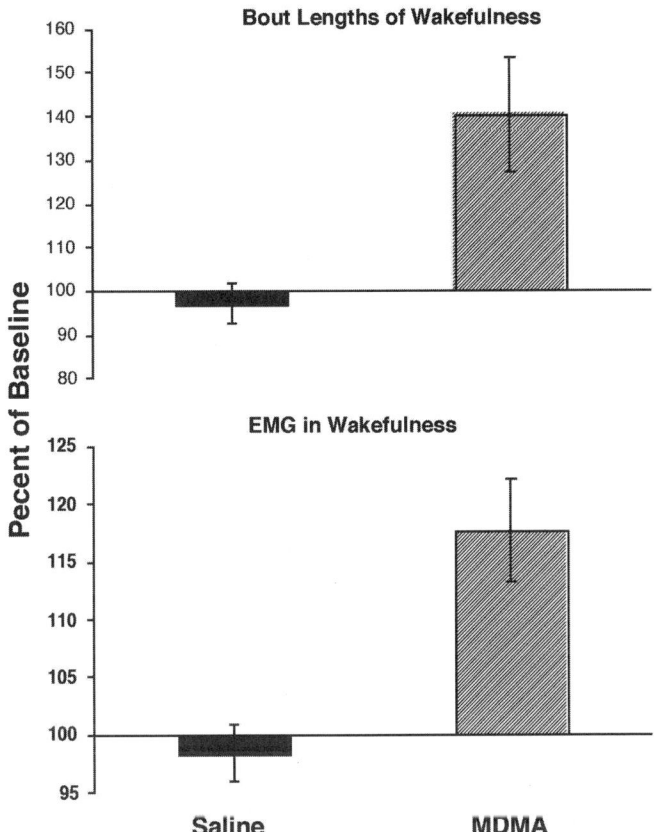

Fig. 1. Baseline EEG and neck EMG recordings were taken for 2 d prior to injections of saline or MDMA (10 mg/kg, i.p. every 2 h for a total of four injections). The data presented are those from recordings obtained 7 d after MDMA treatment and are expressed as percent of the baseline recordings taken prior to MDMA treatment.

Differences in the strain of rat used in these studies may account for differences in behavior.

Importantly, reduction in social interaction in MDMA-treated rats may not be related to depletion of brain 5-HT. Social interaction is reduced in rats treated with a moderate dosage regimen of MDMA that does not produce 5-HT neurotoxicity *(96)* and in rats treated on postnatal d 39 *(98)*, at which time rats are insensitive to 5-HT neurotoxicity induced by MDMA *(41,62)*. Thus,

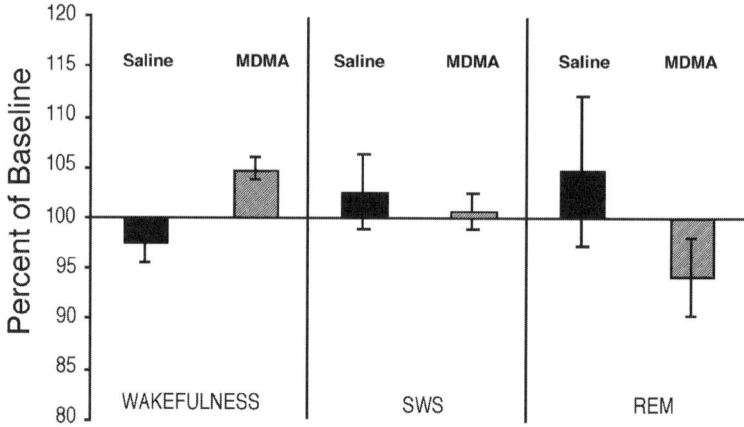

Fig. 2. Treatment of rats and recordings are the same as described in the legend for Fig. 1. The duration of time spent in wakefulness, slow-wave sleep (SWS), and REM sleep (REM) is plotted as a percentage of values obtained prior to drug treatment.

MDMA may induce long-term alterations in behavior that are unrelated to the long-term depletion of brain 5-HT.

Treatment of adult or neonatal rats with MDMA also results in impairments in learning and memory, and these effects also may be unrelated to long-term deficits in brain 5-HT. In adult rats, MDMA treatment results in deficits in object recognition *(96)*. The administration of MDMA on postnatal d 11–20 results in deficits in tests of sequential learning and spatial learning and memory when rats are tested as adults, and these deficits are evident in the absence of significant depletions in brain 5-HT *(99)*.

8. Summary

The existing data indicate that MDMA produces long-term deficits in markers of 5-HT axon terminals in the rodent brain. Increased cleavage of the cytoskeletal protein tau, impairment of axonal transport, and functional consequences associated with a 5-HT depleting regimen of MDMA support the view that MDMA induces structural brain damage, that is, axonal degeneration. A confluence of oxidative stress and bioenergetic stress induced by MDMA is hypothesized to underlie the process of MDMA neurotoxicity (**Fig. 3**). The actions of MDMA on the 5-HT transporter to promote free radical formation and/or intracellular calcium may synergize with MDMA-induced disturbances in cellular energetics and hyperthermia to effect selective toxicity to 5-HT axon terminals.

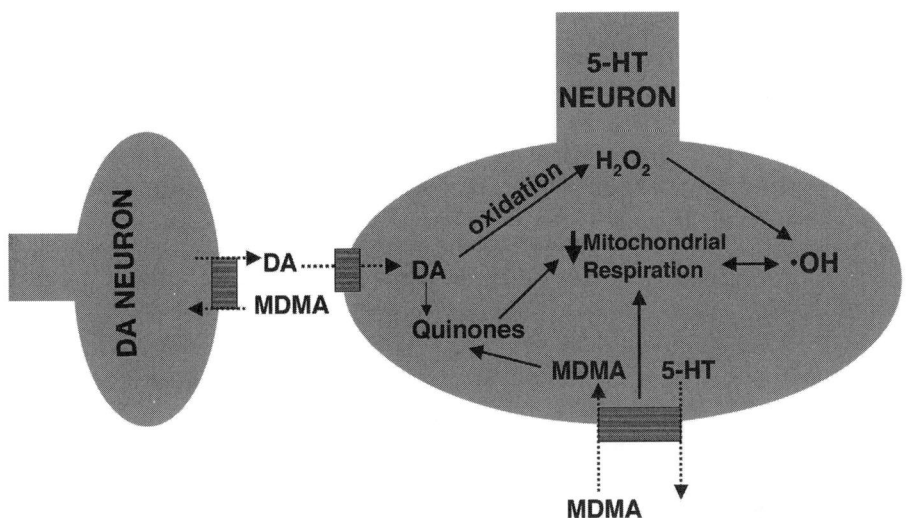

Fig. 3. Hypothetical model of the mechanism of MDMA-induced 5-HT neurotoxicity. The convergence of oxidative stress and energetic stress in the mechanism of MDMA neurotoxicity is depicted in which (1) MDMA (or a toxic metabolite)-induced activation of the 5-HT transporter facilitates free radical formation and energy depletion which act in concert to (2) promote mitochondrial impairment and (3) further generate free radicals and the cellular toxicity associated with these processes.

Acknowledgments

This work was supported by USPHS Grants DA07427 and DA07606.

References

1. Parrott, A. C. (2002) Recreational Ecstasy/MDMA, the serotonin syndrome, and serotonergic neurotoxicity. *Pharmacol. Biochem. Behav.* **71,** 837–844.
2. Nichols, D. E., Lloyd, D. H., Hoffman, A. J., Nichols, M. D., and Yim, G. K. (1982) Effects of certain hallucinogenic amphetamine analogues on the release of [^{3}H]serotonin from rat brain synaptoxomes. *J. Med. Chem.* **25,** 530–535.
3. Johnson, M. P., Hoffman, A. J., and Nichols, D. E. (1986) Effects of the enantiomers of MDA, MDMA and related analogues on [^{3}H]serotonin and [^{3}H]dopamine release from superfused rat brain slices. *Eur. J. Pharmacol.* **132,** 269–276.
4. Gough, B., Ali, S. F., Slikker, W., Jr., and Holson, R. R. (1991) Acute effects of 3,4-methylenedioxymethamphetamine (MDMA) on monoamines in rat caudate. *Pharmacol. Biochem. Behav.* **39,** 619–623.
5. Gudelsky, G. A. and Nash, J. F. (1996) Carrier-mediated release of serotonin by 3,4-methylenedioxymethamphetamine: implications for serotonin–dopamine interactions. *J. Neurochem.* **66,** 243–249.

6. Rudnick, G. and Wall, S. C. (1992) The molecular mechanism of "ecstasy" [3,4-methylenedioxymethamphetamine (MDMA)]: serotonin transporters are targets for MDMA-induced serotonin release. *Proc. Natl. Acad. Sci. USA* **89,** 1817–1821.
7. Yamamoto, B. K. and Spanos, L. J. (1988) The acute effects of methylenedioxymethamphetamine on dopamine release in the awake-behaving rat. *Eur. J. Pharmacol.* **148,** 195–203.
8. Nash, J. F. and Nichols, D. E. (1991) Microdialysis studies on 3,4-methylenedioxyamphetamine and structurally related analogues. *Eur J. Pharmacol.* **200,** 53–58.
9. Nash, J. F. and Brodkin, J. (1991) Microdialysis studies on 3,4-methylenedioxymethamphetamine-induced dopamine release: effect of dopamine uptake inhibitors. *J. Pharmacol. Exp. Ther.* **259,** 820–825.
10. Yamamoto, B. K., Nash, J. F., and Gudelsky, G. A. (1995) Modulation of methylenedioxymethamphetamine-induced striatal dopamine release by the interaction between serotonin and γ-aminobutyric acid in the substantia nigra. *J. Pharmacol. Exp. Ther.* **273,** 1063–1070.
11. Nash, J. F. (1990) Ketanserin pretreatment attenuates MDMA-induced dopamine release in the striatum as measured by in vivo microdialysis. *Life Sci.* **47,** 2401–2408.
12. Gudelsky, G. A., Yamamoto, B. K., and Nash, J. F. (1994) Potentiation of 3,4-methylenedioxymethamphetamine-induced dopamine release and serotonin neurotoxicity by 5-HT$_2$ agonists. *Eur. J. Pharmacol.* **264,** 325–330.
13. Schmidt, C. J., Sullivan, C. K., and Fadayel, G. M. (1994) Blockade of striatal 5-hydroxytryptamine$_2$ receptors reduces the increase in extracellular concentrations of dopamine produced by the amphetamine analogue 3,4-methylenedioxymethamphetamine. *J. Neurochem.* **62,** 1382–1389.
14. Fisher, H. S., Zerning, G., Schatz, D. S., Humpel, C., and Saria, A. (2000) MDMA (Ecstasy) enhances basal acetylcholine release in brain slices of the rat striatum. *Eur. J. Neurosci.* **12,** 1385–1390.
15. Acquas, E., Marrocu, P., Pisanu, A., et al. (2001) Intravenous administration of ecstasy (3,4-methylenedioxymethamphetamine) enhances cortical and striatal acetylcholine release in vivo. *Eur. J. Pharmacol.* **418,** 207–211.
16. Nair, S. and Gudelsky, G. A. (2002) MDMA enhances acetylcholine release in the prefrontal cortex and hippocampus of the rat. *Soc. Neurosci. Abstr.* No. 809.17.
17. Nash, J. F., Jr., Meltzer, H. Y., and Gudelsky, G. A. (1988) Elevation of serum prolactin and corticosterone concentrations in the rat after the administration of 3,4-methylenedioxymethamphetamine. *J. Pharmacol. Exp. Ther.* **245,** 873–879.
18. Schmidt, C. J., Black, C. K., Abbate, G. M., and Taylor V. L. (1990) Methylenedioxymethamphetamine-induced hyperthermia and neurotoxicity are independently mediated by 5-HT$_2$ receptors. *Brain Res.* **529,** 85–90.
19. Spanos, L. J. and Yamamoto, B. K. (1989) Acute and subchronic effects of methylenedioxymethamphetamine [(±)MDMA] on locomotion and serotonin syndrome behavior in the rat. *Pharmacol. Biochem. Behav.* **32,** 835–840.

20. Shankaran, M. and Gudelsky, G. A. (1999) A neruotoxic regimen of MDMA suppresses behavioral, thermal and neurochemical responses to subsequent MDMA administration. *Psychopharmacology* **147,** 66–72.
21. McCreary, A. C., Bankson, M. G., and Cunningham, K. A. (1999) Pharmacological studies of the acute and chronic effects of (+)-3,4-methylenedioxymethamphetamine on locomotor activity: role of 5-hydroxytryptamine$_{1A}$ and 5-hydroxytryptamine$_{1B/1D}$ receptors. *J. Pharmacol. Exp. Ther.* **290,** 965–973.
22. Morley, K. C. and McGregor, I. S. (2000) (±)-3,4-Methylenedioxymethamphetamine (MDMA, 'Ecstasy') increases social interaction in rats. *Eur. J. Pharmacol.* **408,** 41–49.
23. Schmidt, C. J. (1987) Neurotoxicity of the psychedelic amphetamine, methylenedioxymethamphetamine. *J. Pharmacol. Exp. Ther.* **240,** 1–7.
24. Ricaurte, G. A. Forno, L., Wilson, M. A., et al. (1988) (+)-3,4-Methylenedioxymethamphetamine selectively damages central serotonergic neurons in nonhuman primates. *JAMA* **260,** 51–55.
25. Battaglia, G., Yeh, S. Y., O'Hearn, E., Molliver, M. E., Kuhar, M., and De Souza, E. B. (1987) 3,4-Methylenedioxymethamphetamine and 3,4-methylenedioxyamphetamine destroy serotonin terminals in rat brain: quantification of neurodegeneration by measurement of [^{3}H]paroxetine-labeled serotonin uptake sites. *J. Pharmacol. Exp. Ther.* **242,** 911–921.
26. Stone, D. M., Stahl, D., Hanson, G. R., and Gibb, J. W. (1986) The effects of 3,4-methylenedioxymethamphetamine (MDMA) on monoaminergic systems in rat brain. *Eur. J. Pharmacol.* **128,** 41–49.
27. O'Hearn, E., Battaglia, G., De Souza, E. B., Kuhar, M. J., and Molliver, M. E. (1988) Methylenedioxyamphetamine (MDA) and methylenedioxy-methamphetamine (MDMA) cause selective ablation of serotonergic axon terminals in forebrain: immunocytochemical evidence for neurotoxicity. *J. Neurosci.* **8,** 2788–2803.
28. Logan, B. J., Laverty, R., Sanderson, W. D., and Yee, Y. B. (1988) Differences between rats and mice in MDMA (methylenedioxy-methylamphetamine) neurotoxicity. *Eur. J. Pharmacol.* **152,** 227–234.
29. Miller, D. B. and O'Callaghan, J. P. (1994) Environment, drug and stress-induced alterations in body temperature affect the neurotoxicity of substituted amphetamines in the C57Bl/6J mouse. *J. Pharmacol. Exp. Ther.* **270,** 752–760.
30. McCann, U., Szabo, Z., Scheffel, U., Dannais, R., and Ricaurte, G. A. (1998) Positron emission tomographic evidence of toxic effect of MDMA(Ecstasy) on brain serotonin neurons in human beings. *Lancet* **352,** 1433–1437.
31. Reneman, L., Lavalaye, J., Schmand, B., et al. (2001) Cortical serotonin transport density and verbal memory in individuals who stopped using 3,4-methylenedioxymethamphetamine (MDMA or Ecstasy). *Arch. Gen. Psychiatry* **58,** 901–906.
32. Sabol, K. E., Lew, R., Richards, J. B., Vosmer, G. L., and Seiden, L. S. (1996) Methylenedioxymethamphetamine-induced serotonin deficits are followed by partial recovery over a 51-week period. Part I: Synaptosomal uptake and tissue concentrations. *J. Pharmacol. Exp. Ther.* **276,** 846–854.

33. Lew, R., Sabol, K. E., Chou, C., Vosmer, G. L., Richards, J., and Seiden, L. S. (1996) Methylenedioxymethamphetamine-induced serotonin deficits are followed by partial recovery over a 52-week period. Part II: Radioligand binding and autoradiography studies. *J. Pharmacol. Exp. Ther.* **276,** 855–865.
34. Fischer, C., Hatzidimitriou, G., Wios, J., Katz, J., and Ricaurte, G. (1995) Reorganization of ascending 5-HT axon projections in animals previously exposed to the recreational drug (±)3,4-methylenedioxymethamphetamine (MDMA, "Ecstasy"). *J. Neurosci.* **15,** 5476–5485.
35. Commins, D. L., Vosmer, G., Virius, R., Woolverton, W. L., Schuster, C. R., and Seiden, L. S. (1987) Biochemical and histological evidence that methylenedioxymethamphetamine (MDMA) is toxic to neurons in the rat brain. *J. Pharmacol. Exp. Ther.* **241,** 338–344.
36. Scallet, A. C., Lipe, G. W., Ali, S. F., Holson, R. R., Frith, C. H., and Slikker, W., Jr. (1988) Neuropathological evaluation by combined immunohistochemistry and degeneration-specific methods: application of methylenedioxymethamphetamine. *Neurotoxicology* **9,** 529–538.
37. O'Callaghan, J. P. and Miller, D. B. (1993) Quantification of reactive gliosis as an approach to neurotoxicity, in *Assessing Neurotoxicity of Drugs of Abuse* (Erinoff, L., ed.), NIDA Monograph 136, pp. 188–212.
38. Callahan, B. T., Cord, B. J., and Ricaurte, G. A. (2001) Long-term impairment of anterograde axonal transport along fiber projections originating in the rostral raphe nuclei after treatment with fenfluramine or methylenedioxymethamphetamine. *Synapse* **40,** 113–121.
39. Gudelsky, G. A., Wallace, T. L., Vorhees, C. V., and Zemlan, F. P. (2001) Methamphetamine enhances the cleavage of the cytoskeletal protein tau in the striatum and hippocampus of the rat. *Soc. Neurosci. Abstr.* 358.6.
40. Schmidt, C. J., Black, C. K., Abbate, G. M., and Taylor, V. L. (1990) Antagonism of the neurotoxicity due to a single administration of methylenedioxymethamphetamine. *Eur. J. Pharmaocol.* **181,** 59–70.
41. Aguirre, N., Barrionuevo, M., Lasheras, B., and Del Río, J. (1998) The role of dopaminergic systems in the perinatal sensitivity to 3,4-methylenedioxymethamphetamine-induced neurotoxicity in rats. *J. Pharmacol. Exp. Ther.* **286,** 1159–1165.
42. Stone, D. M., Johnson, M., Hanson, G. R., and Gibb, J. W. (1988) Role of endogenous dopamine in the central serotonergic deficits induced by 3,4-methylenedioxymethamphetamine. *J. Pharmacol. Exp. Ther.* **247,** 79–87.
43. Shankaran, M., Yamamoto, B. K., and Gudelsky, G. (1999) Mazindol attenuates the 3,4-methylenedioxymethamphetamine-induced formation of hydroxyl radicals and long-term depletion of serotonin in the striatum. *J. Neurochem.* **72,** 2516–2522.
44. Kanthasamy, A. and Nichols, D. E. (2002) Unilateral injection of dopamine transporter antisense oligonucleotide into the substantia nigra protects against MDMA-induced serotonergic deficits in the ipsilateral striatum. *Soc. Neurosci. Abstr.,* No. 445.18.

45. Schmidt, C. J., Taylor, V. L., Abbate, G. M., and Nieduzak, T. R. (1991) 5-HT$_2$ antagonists stereoselectively prevent the neurotoxicity of 3,4-methylenedioxymethamphetamine by blocking the acute stimulation of dopamine synthesis: reversal by L-dopa. *J. Pharmacol. Exp. Ther.* **256,** 230–235.
46. Michel, P. P. and Hefti, F. (1990) Toxicity of 6-hydroxydopamine and dopamine for dopaminergic neurons in culture. *J. Neurosci. Res.* **26,** 428–435.
47. Filloux, F. and Townsend, J. J. (1993) Pre- and postsynaptic neurotoxic effects of dopamine demonstrated by intrastriatal injection. *Exp. Neurol.* **119,** 79–88.
48. Sprague, J. E., Everman, S. L., and Nichols, D. E. (1998) An integrated hypothesis for the serotonergic axonal loss induced by 3,4-methylene-dioxymethamphetamine. *Neurotoxicology* **19,** 427–442.
49. Johnson, M. P. and Nichols, D. E. (1991) Combined administration of a nonneurotoxic 3,4-methylenedioxymethamphetamine analogue with amphetamine produces serotonin neurotoxicity in rats. *Neuropharmacology* **30,** 819–822.
50. Schmidt, C. J. and Kehne, J. H. (1990) Neurotoxicity of MDMA: neurochemical effects. *Ann. NY Acad. Sci.* **600,** 665–680.
51. Shankaran, M., Yamamoto, B. K., and Gudelsky, G. A. (1999) Involvement of the serotonin transporter in the formation of hydroxyl radicals induced by 3,4-methylenedioxymethamphetamine. *Eur. J. Pharmacol.* **385,** 103–110.
52. Kramer, H. K., Poblete, J. C., and Azmitia, E. C. (1997) Activation of protein kinase C (PKC) by 3,4-methylenedioxymethamphetamine (MDMA) occurs through the stimulation of serotonin receptors and transporter. *Neuropsychopharmacology* **17,** 117–129.
53. Molliver, M. E., O'Hearn, E., Battaglia, G., and DeSouza, E. B. (1986) Direct intracerebral administration of MDA and MDMA does not produce serotonin neurotoxicity. *Soc. Neurosci. Abstr.* **12,** 1234.
54. Paris, J. M. and Cunningham, K. A. (1991) Lack of serotonin neurotoxicity after intraraphe microinjection of (+)-3,4-methylenedioxymethamphetamine (MDMA). *Brain Res. Bull.* **28,** 115–119.
55. Nixdorf, W. L., Burrows, K. B., Gudelsky, G. A., and Yamamoto, B. K. (2001) Enhancement of 3.4-methylenedioxymethamphetamine neurotoxicity by the energy inhibitor malonate. *J. Neurochem.* **77,** 647–654.
56. Esteban, B., O'Shea, E., and Camarero, J. (2001) 3,4-Methylenedioxymethamphetamine induces monoamine release, but not toxicity, when administered centrally at a concentration occurring following a peripherally injected neurotoxic dose. *Psychopharmacology* **154,** 251–260.
57. Miller, R. T., Lau, S. S., and Monks, T. J. (1997) 2,4-bis-(Glutathion-S-yl)-α-methyldopamine, a putative metabolite of 3,4-methylenedioxy-methamphetamine, decreases brain serotonin concentrations. *Eur. J. Pharmacol.* **323,** 173–180.
58. Bai, F., Jones, D. C., Lau, S. S., and Monks, T. J. (2001) Serotonergic neurotoxicity of 3,4-(±)-methylenedioxyamphetamine and 3,4-(±)-methylendioxymethamphetamine (Ecstasy) is potentiated by inhibition of γ-glutamyl transpeptidase. *Chem. Res. Toxicol.* **14,** 863–870.

59. Gudelsky, G. A. (1996) Effect of ascorbate and cysteine on the 3-4-methylenedioxymethamphetamine-induced depletion of brain serotonin. *J. Neural Transm.* **103,** 1397–1404.
60. Malberg, J. E. and Seiden, L. S. (1998) Small changes in ambient temperature cause large changes in 3,4-methylenedioxy-methamphetamine (MDMA)-induced serotonin neurotoxicity and core body temperature in the rat. *J. Neurosci.* **18,** 5086–5094.
61. Darvesh, A. S., Shankaran, M., and Gudelsky, G. (2002) 3,4-methylenedioxymethamphetamine produces glycogenolysis and increases the extracellular concentration of glucose in the rat brain. *J. Pharmacol. Exp. Ther.* **300,** 138–144.
62. Broening, H. W., Bowyer, J. F., and Slikker, W. Jr. (1995) Age-dependent sensitivity of rats to the long-term effects of the serotonergic neurotoxicant (±)-3,4-methylenedioxymethamphetamine (MDMA) correlates with the magnitude of the MDMA-induced thermal response. *J. Pharmacol. Exp. Ther.* **275,** 325–333.
63. Farfel, G. M. and Seiden, L. S. (1995) Role of hyperthermia in the mechanism of protection against serotonergic toxicity. I. Experiments using 3,4-methylenedioxymethamphetamine, dixolcipine, CGS 19755 and NBQX. *J. Pharmacol. Exp. Ther.* **272,** 860–867.
64. Malberg, J. E., Sabol, K. E., and Seiden, L. S. (1996) Co-administration of MDMA with drugs that protect against MDMA neurotoxicity produces different effects on body temperature in the rat. *J. Pharamcol. Exp. Ther.* **278,** 258–267.
65. Colado, M. I., O'Shea, E., Granados, R., Murray, T. K., and Green, A. R. (1997) *In vivo* evidence for free radical involvement in the degeneration of rat brain 5-HT following administration of MDMA ('ecstasy') and *p*-chloramphetamine but not the degeneration following fenfluramine. *Br. J. Pharmacol.* **121,** 889–900.
66. Colado, M. I., O'Shea, E., Granados, R., Esteban, B., Martin, A. B., and Green, A. R. (1999) Studies on the role of dopamine in the degeneration of 5-HT nerve endings in the brain of Dark Agouti rats following 3,4-methylenedioxymethamphetamine (MDMA or ecstasy) administration. *Br. J. Pharmacol.* **126,** 911–924.
67. Cadet, J. L., Ladenheim, B, Baum, I., Carlson, E., and Epstein, C. (1994) CuZn-superoxide dismutase (CuZnSOD) transgenic mice show resistance to the lethal effects of methylenedioxyamphetamine (MDA) and of methylenedioxymethamphetamine (MDMA). *Brain Res.* **655,** 259–262.
68. Colado, M. I., Camarero, J., Mechan, A. O., et al. (2001) A study of the mechanisms involved in the neurotoxic action of 3,4-methylenedioxymethamphetamine (MDMA, 'ecstasy') on dopamine neurones in mouse brain. *Br. J. Pharmacol.* **134,** 1711–1723.
69. Taraska, T. and Finnegan, K. T. (1997) Nitric oxide and the neurotoxic effects of methamphetamine and 3,4-methylenedioxymethamphetamine. *J. Pharmacol. Exp. Ther.* **280,** 941–947.
70. Zheng, Y. and Laverty, R. (1998) Role of brain nitric oxide in (±)3,4-methylenedioxymethamphetamine (MDMA)-induced neurotoxicity in rats. *Brain Res.* **795,** 257–263.

71. Nash, J. F. and Yamamoto, B. K. (1992) Methamphetamine neurotoxicity and striatal glutamate release: comparison to 3,4-methylenedioxymethamphetamine. *Brain Res.* **581,** 237–243.
72. Burrows, K. B., Gudelsky, G. A., and Yamamoto, B. K. (2000) Rapid and transient inhibition of mitochondrial function following methamphetamine or 3,4-methylenedioxymethamphetamine administration. *Eur. J. Pharmacol.* **398,** 11–18.
73. Sprague, J. E. and Nichols, D. E. (1995) The monoamine oxidase-B inhibitor L-deprenyl protects against 3,4-methylenedioxy-methamphetamine-induced lipid peroxidation and long-term serotonergic deficits. *J. Pharmacol. Exp. Ther.* **273,** 667–673.
74. Colado, M. I., O'Shea, E., Ados, R., Misra, A., Murray, T. K., and Green, A. R. (1997) A study of the neurotoxic effect of MDMA ('ecstasy') on 5-HT neurones in the brains of mothers and neonates following administration of the drug during pregnancy. *Br. J. Pharmacol.* **121,** 827–833.
75. Shankaran, M., Yamamoto, B. K., and Gudelsky, G. A. (2001) Ascorbic acid prevents 3,4-methylenedioxymethamphetamine (MDMA)-induced hydroxyl radical formation and the behavioral and neurochemical consequences of the depletion of brain 5-HT. *Synapse* **40,** 55–64.
76. Colado, M. I. and Green, A. R. (1995) The spin trap reagent α-phenyl-*N*-tert-butyl nitrone prevents 'ecstasy'-induced neurodegeneration of 5-hydroxytryptamine neurons. *Eur. J. Pharmacol.* **280,** 343–346.
77. Yeh, S. Y. (1999) *N-tert*-Butyl-alpha-phenylnitrone protects against 3,4-methylenedioxymethamphetamine-induced depletion of serotonin in rats. *Synapse* **31,** 169–177.
78. Chan, P., Di Monte, D., Luo, J. J., DeLaney, L. E., Irwin, I., and Langston, J. W. (1994) Rapid ATP loss caused by methamphetamine in the mouse striatum: relationship between energy impairment and dopaminergic neuroxicity. *J. Neurochem.* **62,** 2484–2487.
79. Stephans, S. E., Whittingham, T. S., Douglas, A. J., Lust, W. D., and Yamamoto, B. K. (1998) Substrates of energy metabolism attenuate methamphetamine-induced neurotoxicity in striatum. *J. Neurochem.* **71,** 613–621.
80. Huang, N. K., Wan, F. J., Tseng, C. J., and Tung, C. S. (1997) Nicotinamide attenuates methamphetamine-induced striatal dopamine depletion in rats. *NeuroReport* **8,** 1883–1885.
81. Burrows, K. B., Nixdorf, W. L., and Yamamoto, B. K. (2000) Central administration of methamphetamine synergizes with metabolic inhibition to deplete striatal monoamines. *J. Pharmacol. Exp. Ther.* **292,** 853–860.
82. Albers, D. S., Zeevalk, G., and Sonsalla, P. K. (1996) Damage to dopaminergic nerve terminals in mice by combined treatment of intrastriatal malonate with systemic methamphetamine or MPTP. *Brain Res.* **718,** 217–220.
83. Huether, G., Zhou, D., and Rüther, E. (1997) Causes and consequences of the loss of serotonergic presynapses elicited by the consumption of 3,4-methylenedioxymethamphetamine (MDMA, "ecstasy") and its congeners. *J. Neural Transm.* **104,** 771–794.

84. Poblete, J. C. and Azmitia, E. C. (1995) Activation of glycogen phosphorylase by serotonin and 3,4-methylenedioxymethamphetamine in astroglial-rich primary cultures: involvement of the 5-HT$_{2A}$ receptor. *Brain Res.* **680,** 9–15.
85. Series, H. G., Masurier, M., Gartside, S., Franklin, M., and Sharp, T. (1995) Behavioral and neuroendocrine responses to *d*-fenfluramine in rats treated with neurotoxic amphetamines. *J. Psychopharmacol.* **9,** 214–219.
86. Bauman, M. H., Ayestas, M., and Rothman, R. B. (1998) Functional consequences of central serotonin depletion produced by repeated fenfluramine administration to rats. *J. Neurosci.* **18,** 9069–9077.
87. Poland, R. E., Lutchmansingh, P., McCracken, J. T., et al. (1997) Abnormal ACTH and prolactin responses to fenfluramine in rats exposed to single and multiple doses of MDMA. *Psychopharmacology* **131,** 411–419.
88. Gartside, S., McQuade, R., and Sharp, T. (1996) Effects of repeated administration of 3,4-methylenedioxymethamphetamine on 5-hydroxytryptamine neuronal activity and release in the rat brain in vivo. *J. Pharmacol. Exp. Ther.* **279,** 277–283.
89. Aguirre, N., Frechilla, D., García-Osta, A., Lasheras, B., and Del Río, J. (1997) Differential regulation by methylenedioxymethamphetamine of 5-hydroxytryptamine$_{1A}$ receptor density and mRNA expression in rat hippocampus, frontal cortex, and brainstem: the role of corticosteroids. *J. Neurochem.* **68,** 1099–1105.
90. Aguirre, N., Ballaz, S., Lasheras, B., and Del Río, J. (1998) MDMA ('Ecstasy') enhances 5-HT$_{1A}$ receptor density and 8-OH-DPAT-induced hypothermia: blockade by drugs preventing 5-hydroxytryptamine depletion. *Eur. J. Pharmacol.* **346,** 181–188.
91. Wallace, T. L., Gudelsky, G., and Vorhees, C. V. (2001) Alterations in diurnal and nocturnal locomotor activity in rats treated with a monoamine-depleting regimen of methamphetamine or 3,4-methylenedioxymethamphetamine. *Psychopharmacology* **153,** 321–326.
92. Dafters, R. I. and Lynch, E. (1998) Persistent loss of thermoregulation in the rat induced by 3,4-methylenedioxymethamphetamine (MDMA or "Ecstasy") but not by fenfluramine. *Psychopharmacology* **138,** 207–212.
93. Mechan, A. O., O'Shea, E., Elliott, J. M., Colado, M. I., and Green, A. R. (2001) A neurotoxic dose of 3,4-methylenedioxymethamphetamine (MDMA; ecstasy) to rats results in a long term defect in thermoregulation. *Psychopharmacology* **155,** 413–418.
94. Matuszewich, L., Filon, M. E., Finn, D. A., and Yamamoto, B. K. (2002) Altered forebrain neurotransmitter responses to immobilization stress following 3,4-methylenedioxymethamphetamine. *Neuroscience* **110,** 41–48.
95. Jones, B. E. (2000) Basic mechanisms of sleep-wake state, in *Principles and Practice of Sleep Medicine*, 3rd ed. (Kryger, M. H., Roth, T., and Dement, W. C., eds.), W. B. Saunders, Philadelphia, pp. 134–154.
96. Morley, K. C., Gallate, J. E., Hunt, G. E., Mallet, P. E., and McGregor, I. S. (2001) Increased anxiety and impaired memory in rats 3 months after administration

of 3,4-methylenedioxymethamphetamine ("Ecstasy"). *Eur. J. Pharmacol.* **433,** 91–99.
97. Mechan, A., Moran, P., Elliott, J. M., Young, A., Joseph, M. H., and Green, A. R. (2002) A study of the effect of a single neurotoxic dose of 3,4-methylenedioxymethamphetamine (MDMA; ecstasy) on the subsequent long-term behaviour of rats in the plus maze and open field. *Psychopharmacology* **159,** 167–175.
98. Fone, K. C. F., Beckett, S. R. G., Topham, I. A., Swettenham, J., Ball, M., and Maddocks, L. (2002) Long-term changes in social interaction and reward following repeated MDMA administration to adolescent rats without accompanying serotonergic neurotoxicity. *Psychopharmacology* **159,** 437–444.
99. Broening, H. W., Morford, L. L., Inman-Wood, S. L., Fukumura, M., and Vorhees, C. V. (2001) 3,4-Methylenedioxymethamphetamine (Ecstasy)-induced learning and memory impairments depend on the age of exposure during early development. *J. Neurosci.* **21,** 3228–3235.

6

Learning and Memory Mechanisms Involved in Compulsive Drug Use and Relapse

Joshua D. Berke

1. Introduction

Chronic drug abuse is a complex behavioral and social phenomenon, that stems from a diverse set of underlying neural mechanisms. However, two defining features of drug addiction make it especially difficult to treat. First, addiction is *compulsive*—individuals often continue or resume drug use despite a conscious, stated wish to quit. Second, it is *persistent*—relapse to active drug use can occur despite years of abstinence. This chapter discusses evidence that the inappropriate engagement of neural mechanisms involved in normal associative learning is responsible for these key behavioral aspects of drug addiction. Central to understanding these mechanisms are the distinctions between reward and reinforcement, between goal-directed and automatized behavior, and between information-dense patterns of synaptic plasticity and information-poor neuronal adaptations. This chapter is intended to complement a previous review *(1)* by elaborating on some particular forms of associative learning that may be central to drug addiction, although limitations of space prevent a full consideration of the many roles that learning has been suggested to play in drug addiction *(2–5)*.

2. Why Associative Learning?

Drug addiction is generally defined as compulsive use of drugs despite adverse consequences. It typically features periods of ongoing drug use, in which drugs are actively sought and taken, and periods of drug abstinence, terminated by relapse. The proclivity to relapse is central to addiction—"It is not so difficult to get individuals to stop using a drug; the problem is to keep them from restarting" *(6)*. Relapse can occur despite months or even years of abstinence (long after the end of any withdrawal symptoms), and is very often provoked by specific cues or contexts—meeting an acquaintance with whom the addict previously used drugs, or perhaps passing a location where drugs were obtained. Any explanation of addiction will therefore need to account for the specificity, persistence, and compulsive nature of drug-taking and relapse.

Humans have particular brain mechanisms for allowing specific acquired behaviors to be persistently expressed, and we call this associative learning. Associative mechanisms are therefore excellent candidates for both the specificity and the persistence of relapse liability. However, most learning models of addiction have had difficulty providing a compelling model of compulsion—the apparent *loss of control* over drug-taking. (The meaning of "loss of control" is obviously critical, and is discussed further in **Subheading 7.**) Individuals frequently face situations of conflicting goals and desires, and need to make decisions about which course of action to pursue. If addicts are able to state clearly to others and to themselves a resolute intention not to restart drug use, why should they not make the appropriate choice when faced with drug-related cues? Although people may *use* drugs for a host of reasons, including social pressures, novelty-seeking, pleasure-seeking, relief of withdrawal symptoms, and other forms of self-medication, drug *addiction* is essentially a failure of adaptive decision-making.

Some theories have suggested that associative learning has a necessary but not sufficient role, and that the dysregulation of some additional process underlies compulsion. For example, Robinson and Berridge argued that drugs could produce sensitization of mechanisms underlying "incentive motivation"; cues associated with drug use would then evoke excessive "wanting" of drugs *(7)*. Another, complementary proposal is that dysregulation of homeostatic mechanisms result in altered reward "set points" *(8)* and that these act in conjunction with sensitization processes to produce failed self-regulation of drug-taking.

The major difficulty with such ideas is in accounting for specificity and persistence in drug addiction. For example, if addicts have generally sensitized incentive motivation, why don't they show uncontrollable wanting of food or sex, in addition to drugs? Although there is some evidence that such broad cross-sensitization might occur in some animal models *(9,10)*, it does not appear to be a prominent feature of human addiction. Robinson and Berridge suggested that "The expression of this sensitized system is focused expressly on stimuli that have become associated with its excessive activation, so drugs and drug-associated stimuli become irresistibly attractive ('wanted')" *(7)*. Yet they give no account of how or why such "focusing" should take place. Further, if one allows that there is a mechanism that focuses wanting on specific stimuli (i.e., a form of associative learning), it is not obvious that invoking a general sensitization process is necessary. General changes in responsiveness can certainly occur in some circumstances, including withdrawal. In the case of cocaine, for example, heavy use is typically followed by a period of "decreased activation, anxiety, lack of motivation, and boredom, with markedly diminished intensity of normal pleasurable experiences (anhedonia)" *(11)*. Yet this period is transient, lasting days to weeks at most—far too short to account for the persistent proclivity toward relapse.

The remainder of this chapter argues that particular forms of associative learning are sufficient to account for the compulsive aspects of addiction, as well as its specificity and persistence. Although nonassociative mechanisms can dynamically interact with learning processes to sculpt patterns of behavior, it is suggested that the hijacking of normal mechanisms allowing for acquisition and "unplanned" performance of learned habits is largely responsible for the failure of addicts to make timely, adaptive decisions.

3. Multiple Memories in Multiple Systems

To describe how drug-associated learning might produce compulsive behaviors, it is necessary to have some overall conception of how different forms of learning occur and how they are normally orchestrated in the service of adaptive behavior. It is now well established that there are multiple "memory systems" in the mammalian brain (reviewed in *12,13*). These memory systems are functionally semi-independent, and often concerned with distinct aspects of the animal's experience. Although the learning/memory literature is too vast to review adequately here, one fundamental conceptual distinction is essential to the present discussion. This distinction has been expressed in various ways; on the one side we have "declarative," "explicit," "cognitive," flexible, conscious memories, on the other side lie "procedural," "implicit," inflexible, unconscious "habits." The central point is that the performance of some learned actions involves conscious manipulation of knowledge and/or expectations, whereas other behaviors are performed more automatically, without the need for conscious attention or care. This conceptual divide in part reflects distinct historical strands of psychological theorizing—emphasizing stimulus–stimulus (S–S) vs stimulus–response (S–R) associations (for review *see* **ref. *14***).

Mishkin proposed that these conceptual distinctions had a neuroanatomical foundation, with medial temporal lobe structures such as the hippocampus being critical for declarative/episodic memories and corticostriatal circuits playing a key role in habits *(15)*. Since then, a substantial body of evidence has emerged to support this hypothesis, including data from humans *(16,17)* and nonhuman primates *(18)*. Some of the clearest demonstrations come from the work of White, Packard, and colleagues, who trained rats in a set of radial maze tasks. In the "win-shift" task animals received food rewards for choosing maze arms that they had not recently visited; in the "win-stay" task they were rewarded for choosing an illuminated arm, with their recent history of spatial positions being less important. Lesions *(19,20)* and local drug infusions *(21)* demonstrated that the hippocampus is important for acquisition and performance of the win-shift task (and not win-stay), whereas the (dorsal) striatum is closely involved in the win-stay task (and not win-shift). The task components are similar in each case (maze, rewards, etc.) but are combined under different contingencies. The involvement of distinct neural circuits thus appears to reflect their differing "rules of operation" *(22)* or information processing "styles" *(13)*. Among other roles, the hippocampus is essential for the ability to keep track of the when, where, and what of particular recent experiences (episodic memory; reviewed in *23,24*). In contrast, the dorsal striatum appears to be critical for the progressive acquisition of fixed response tendencies (habits)—in this case, an approach response to a light stimulus.

In some task situations the relationship between the biologically meaningful events and other task elements is ambiguous, and "correct" performance can rely on either of two or more strategies. This has been classically demonstrated in a four-arm radial maze ("plus maze"), with an animal consistently placed in one arm ("South") and allowed to find food consistently in another ("East") *(25)*. The animal learns to retrieve the food rapidly; the central experimental question is what exactly is learned—a fixed response habit (e.g., "turn right!") or a more cognitive use of spatial information (e.g., "food is in East—go there!"). A probe test—in this case, starting from the North arm—can help to resolve this, as the two theories make opposite predictions about the animal's choice (West if turning right, East if using spatial information). Many experiments on this "place vs response" question *(26)* established that both strategies are learned in parallel. Which strategy predominates depends on a variety of factors such as availability of local and distal cues, and also on the amount of training received by the animal. Early in training rats tend to make use of a hippocampus-dependent spatial/cognitive strategy, but with many repetitions the fixed response "habit," dependent on the dorsal striatum, comes to dominate behavior *(27)*. In a sense, the strategies implemented by different neural circuits "compete" for behavioral output. Acquisition of the striatum-dependent win-stay task is considerably accelerated in animals with a hippocampal lesion *(19)*; in fact, hippocampus-damaged animals tend to act generally as if they were easily dominated by stimulus–response habits *(28,29)*. Inactivation of the hippocampus early in plus-maze training leads rats to use the nascent striatum-dependent habit, while inactivation of the dorsal striatum late in training reveals that the hippocampus-dependent spatial strategy is still available, even if normally unused *(30)*. The normal temporal transition from a cognitive to a habitual mode of behavior is also clear in our own human behavior. For example, driving a car along a complex route to a new workplace requires considerable attention and a deliberate choice at each junction. With extended practice, the task becomes automatic, performed at a low level of awareness—the driver can simultaneously engage in an unrelated conversation, and no delibrate choices need to be made.

A "habit" can be either a very "simple" response to a stimulus (e.g., a lever-press) or a more complex, temporally extended action (as in the driving example). In either case the component muscle group firing sequence, or sequence of actions, becomes "chunked" *(31)* and thereafter performed in a "closed-loop" fashion—smooth, effortless, without the need for online monitoring. Having the ability to learn habits is useful because it allows behaviors that have worked well in the past, over and over again, to be performed without the need to expend time or effort in deciding the best course of action. That is, habits allow evaluative decision-making to be skipped.

Avoiding the costs of decision-making does, however, produce an obvious problem if circumstances are liable to change. While acting in a habitual mode other, potentially more beneficial actions will not even be properly considered. To reduce this danger, habits are normally acquired slowly, thereby hopefully reflecting stable regularities of the world (invariances; *22*). An automatic program may even be engaged inappropriately unless care is taken—such "actions-not-as-planned" are common in human behavior *(32)*. For example, many people have had the experience of finding themselves driving to work when another location was the originally intended goal. This reflects the "stimulus–response" nature of habits—no *outcome* is directly encoded as part of the learned association *(33)*. In one recent experiment, rats were trained for varying amounts of time in the win-stay task, using food rewards *(34)*. The rats were then placed in another environment and the same food was paired with a nausea-inducing injection (i.e., conditioned taste aversion) to "devalue" that food. When subsequently returned to the win-stay task, rats with little pretraining were slow to run on the task, as if they were less motivated than before. In contrast, those rats that had received extensive pretraining ran swiftly from the start to the food, despite the fact that after attaining the food, they declined to eat it.

Such examples help to illustrate the critical difference between "reward" and (positive) "reinforcement" (discussed in **ref. 35**). Usually, these are linked processes—if an action produces consequences that are acutely evaluated as pleasant (rewarding), the animal is persistently more likely to subsequently perform the same action (i.e., it is reinforced). But the subsequent performance may be either because the animal has that same reward "in mind" as an intended goal, or simply because of a strengthened stimulus–response association, without any explicit expectation. The ability of reinforcers to strengthen habits directly has been termed the "enhancing function" of reinforcers *(36)*.

4. Prefrontal Cortex and Flexible Control Over Learning and Performance

Flexible behavior requires that habits normally be suppressed when inappropriate, as part of an integrated, adaptive decision-making process. This is the domain of the prefrontal cortex (PFC), which, in summary, makes use of a wide range of current and previously stored information to select and initiate actions predicted to serve long-term goals *(37–40)*. This functional role is evident from human patients with damage to regions of PFC *(41)*. Although the exact deficits depend on the specific regions affected, and can be subtle *(42)*, such individuals generally have difficulty with planning, especially long-term planning. Their behavior tends to be stimulus bound—that is, they react to features of their environment as they occur (e.g., "utilization behavior"; *43*) rather than spontaneously initiating proactive action plans. Once engaged in an activity or a cognitive set, they often have difficulty switching to another task or mindset. Animal lesion studies generally give analogous results—for example, difficulty in switching from one task variant to another *(44,45)*.

Such abilities of the PFC reflect and depend on its anatomical relationships. The PFC receives information from medial temporal lobe "memory" structures such as hippocampus and related cortical areas *(12,46,47)*; such "working-with-memory" is likely central to the temporal organization of behavior and the prediction of the consequences of competing behavioral options *(48–50)*. The selection of actions and the suppression of inappropriate responses is thought to rely on cortex–basal ganglia circuits *(51,52)*. All cortical regions, including PFC subregions, send projections to the striatum, which in turn projects to other basal ganglia regions, thence to thalamus and back to cortex in semi-segregated "loop" circuits; lesions to striatal regions produce behavioral deficits similar to lesioning the cortical area that projects to that striatal region *(53,54)*. The complex anatomy of these latter circuits is not reviewed in detail here *(55–57)*; while fascinating computational theories of how such anatomical arrangements subserve action selection and initiation are being developed *(58,59)* they remain rather speculative. It is clear, however, that the PFC sits at the top of a control hierarchy, being able to either delegate performance of actions to lower levels or to monitor them and assume control as necessary. Deliberate actions and conscious thoughts rely more on PFC projections to "ventral" (and more rostral and medial) parts of striatum, whereas more automatic behavior can be subserved by circuits involving more posterior cortical projections to "dorsal" (and caudal and lateral) parts of striatum. These circuits are involved in both performance of actions and also plasticity of those actions; in this way, habit learning is believed to involve altering information processing in cortex–"dorsal" striatal projections *(35)*. However, the dichotomy between deliberate goal-directed actions and more automatized habits is not a rigid distinction; there is likely a corresponding continuum between cortex–basal ganglia circuits involved in each mode of behavior.

5. Dopamine and Reinforcement Learning

Within striatal circuits the neuromodulator dopamine is considered by many to act as a learning signal, controlling neural plasticity *(60)*. One of the main lines of evidence has come from recordings of midbrain dopamine neurons whose axons project to wide regions of forebrain *(61)*—including the striatum, which has a very high density of dopamine receptors. In macaque monkeys such cells often show brief (~0.1 s) increases in firing rate, particularly to unexpected rewards, or to unexpected cues that signal upcoming rewards (reviewed in *62*). The ability of dopamine cells to ignore fully expected rewards and shift firing to cues that predict rewards has a strong resemblance to "error signals" in certain formal learning theories *(63)* and certain computational models of "reinforcement learning"*(64)*. Both classes of model refer to situations in which the extent of associative learning is controlled by a simple signal that provides feedback on the overall success or failure of an action or expectation. Because the signal does not provide detailed information on the exact nature of any errors committed, it is sometimes described as a "critic" rather than a "teacher" *(65)*.

Natural changes in dopamine cell firing rate are typically subtle—an extra spike or two over a tonic rate of a few per second—and often clearly visible only in cumulative records of many trials *(66)*. Such modest changes in dopamine cell firing rate may potentially lead to substantial changes in striatal dopamine release *(67)*. However, increased activity of dopamine cells may also produce no increase in dopamine release in terminal regions. One recent study found that direct electrical stimulation of this cell population would produce increased release of dopamine in ventral striatum, but only if the animal was not expecting the stimulation *(68)*. This implies an exquisite local control over rapid changes in dopamine release that may have as important a part in dopaminergic control of reinforcement learning as altered firing of midbrain dopamine cells. Both mechanisms are likely responsible, in part, for behavioral observations that "the more expected a reinforcer, the less effective it is" *(69)*. Evaluative and anticipatory circuits involving amygdala, hippocampus, and PFC can control dopamine release as part of the normal top-down, cognitive control over learning processes *(70,71)*.

However, dopamine affects not just plasticity of circuits through striatum, but also expression of already established striatum-dependent behaviors. The ability to respond to salient stimuli, whether internally or externally generated, relies on the presence of striatal dopamine; complete loss of striatal dopamine causes severe Parkinson's disease in humans and experimental animals. Conversely, events that are generally "arousing"—including rewards, stressors, and novelty—can provoke increases in dopamine release lasting minutes or more (*see* references in **ref. 72**). Dopamine thus clearly has an important role in psychomotor activation and attentional processes. The relationship between slow and fast shifts in dopamine release remains poorly understood—despite a long-standing recognition of the importance of distinguishing phasic and tonic dopamine release *(73)*.

6. Addictive Drugs and Learning Processes

Striatal dopamine is also elevated by drugs of abuse. Indeed, although such drugs have a wide variety of effects in numerous parts of the brain, the shared ability to enhance striatal dopamine release remains the best candidate for a key common action (*see* references in **ref. 1**). The mechanisms by which drugs produce altered striatal dopamine release are varied—for example, nicotine likely achieves this both through actions on midbrain dopamine neurons *(74)* and through local modulation of dopamine release in striatum *(75)*. By altering release of dopamine throughout the striatum, drugs of abuse can alter both acute information-processing and long-lasting plasticity across a wide range of cortex–basal ganglia circuits. If striatal dopamine can act as a reinforcement learning signal, then it makes sense that self-administration of drugs that elevate dopamine is reinforced. In fact, multiple kinds of learning contribute to any given drug's reinforcing properties *(76,77)*. Besides affecting a range of cortex–basal ganglia circuits, abused drugs can also affect neuromodulators in other brain systems important for motivated behavior, such as hippocampus and amygdala *(78)*.

Dopaminergic modulation of cortex–ventral striatal circuits appears to be important for "rewarding" sensations. The deliberate desire to repeat such experiences doubtless underlies much human drug use. However, some drugs such as nicotine produce mild, if any, pleasant sensations during initial use, and yet can be highly addictive. The ability of drugs to affect other aspects of reinforcement may therefore be of greater importance to addiction than their rewarding effects. For example, Viaud and White *(79)* found that local amphetamine injections directly into striatal regions that receive visual cortical inputs enhance learning of a visually conditioned response, while injections into regions concerned with olfactory information enhance learning of a response to an olfactory cue. Such enhancing effects are dopamine dependent *(80)*, and occur even when injections are given shortly *after* the training session. In other words, dopamine-releasing drugs can affect *consolidation (81)* of striatum-dependent learning. Note that in those experiments and others *(82)*, the animals were injected by the experimenter, and the action whose learning was enhanced was not related to drug-taking. Drug-induced enhancement of learning is not dependent on drug-produced affective states, or on having drugs as a goal object of behavior. Rather, any very recently performed action that may still have an active representation in striatum can be affected *(1,60)*. When the action performed is drug self-administration, it is drug-taking behavior that will be consolidated (the cellular and molecular mechanisms for this are discussed in **Subheading 8**).

7. Exaggerated Habit Learning and Compromised Executive Control

As described in the preceding, reinforcement learning is normally under cognitive control; both evaluation and expectation are key determining factors gating striatal dopamine release. This is a large part of the reason for the normal connection beween reward (i.e., a positively evaluated result) and other processes involved in positive reinforcement (such as consolidation of actions). It also forms a natural brake on normal habit learning. Actions that produce fully expected consequences do not generally undergo further plasticity, and so asymptotically stabilize. Abused drugs that directly enhance striatal dopamine release can foil such natural constraints *(1,83)*. The behavioral features of human drug addiction are consistent with increasingly strong "habit" learning. Actions involved in drug intake generally become increasingly stereotyped ("ritualized"), automatic, and stimulus-bound *(84)*, and there is an overall diminution of behavioral repertoire *(85)*. The inability to regulate or terminate drug-taking despite adverse long-term consequences indicates a compromised capacity for flexible, adaptive decision-making, and is thus highly suggestive of inadequate PFC control over behavior.

Could exaggerated stimulus–response-like habits account for the *compulsive* aspects of drug addiction? To address this it is worth very briefly distinguishing between three different possible senses of "compulsive behavior" in addiction. The one most directly corresponding to the everyday sense of compulsion is that addicts are bystanders at the very moment of drug-taking, conscious that they are doing something that they do not want to do but helpless to prevent it. Such radical dissociations of control may occur under unusual neurological circumstances but this does not appear to be the standard problem in drug addiction.

A second sense of compulsion might be that drugs are excessively "wanted." One might imagine that the process of PFC-arbitrated evaluation/competition between competing behavioral responses remains fairly normal in drug addiction, and the real problem is that drug-taking is simply valued (i.e., wanted) more than many other options *(86)*. This is certainly true for many drug users—probably the great majority—and is also perhaps closer to the more causal sense of "addiction" often used to refer to nondrug situations—"I'm addicted to chocolate," and so on. However, it is unclear that "compulsion" is really an appropriate way to refer to this phenomenon. Rather than demonstrating loss of control, doing what one wants is surely the very essence of control—which is not to say that this is not a serious problem in the overall phenomenon of drug abuse.

In the third sense of compulsion, drug-taking (and especially relapse) involves not so much a failure to execute the intended decision, but rather a failure to make a timely, deliberate, evaluative decision at all. Self-reports of relapse are often highly suggestive of this—e.g., "going on automatic pilot" *(4)*, "everything that's on my mind just kind of disappears" *(87)*. Rather than indicating a failure to rein in overwhelming desires, such examples are consistent with failure of other aspects of executive function—attention to one's own actions *(88)* and evaluation of their likely consequences. Relapse is seen as compulsive because, even though the addict may have had an explicit, clear goal of avoiding drug intake, this goal was not "in mind" at the critical time and so did not direct behavior. Acting in this more habitual mode is perhaps more accurately described as "omission of control" rather than "loss of control."

What then to make of "craving"—subjective reports of excessive desire for drugs? It is important to note that even in our regular daily experience strong feelings of wanting tend especially to arise in situations of conflict—for example, when a desired object is visible but not yet available, or when short-term desires clash with long-term goals. When there are no barriers to action we tend to simply act *(89)*. As noted by Tiffany *(90)*, conscious drug cravings may similarly reflect cognitive processes evoked when automatized drug use patterns are impeded in some way, either by external factors or by a deliberate effort to avoid drug intake. Craving is often stimulus triggered, and it has been noted that craving is greatly diminished or absent in settings in which drugs are understood to be unavailable, such as hospitalization *(91)*. Conversely, in drug-associated settings (e.g., a bar) thoughts of drug use and/or preparatory actions (e.g., reaching into pocket for a cigarette) may be frequent and intrusive, with cravings reflecting the effort involved in continuously suppressing them.

Cravings may be indications of cognitive conflict rather than the "cause" of relapse—which very often happens on just those occasions when no such conflict occurred.

Acting deliberately vs habitually is not an all-or-nothing distinction. An accurate account of drug abuse will likely incorporate multiple levels of habits, motivations, and cognitive control. Rather than relying on a wholly distinctive pathophysiology, addiction can be seen as an extension or exaggeration of our normal human difficulties with controlling our behavior to conform with long-term personal or societal goals.

Viewed perhaps simplistically, the failure of executive control might arise either from specific, strong habits that are intrinsically "hard" to overcome, or from a more general deficit in executive function. In fact, both may be important, particularly in certain subpopulations. Some neuropsychological tests of drug abusers have found deficits in PFC functions *(92,93)*. General problems with self-control may account for why substance abuse has higher prevalence in subjects with comorbid additional psychiatric conditions. Naturally occuring individual variation in aspects of executive control such as impulsivity *(94,95)* may also contribute to the poorly understood issue of why some people are more vulnerable to drug addiction than others. But it is also important to understand that an abstinent former drug abuser trying to avoid relapse may be effectively bombarded by drug-related cues and contexts unless a drastic break is made with former friends, locations, and so forth. Even a normal, functional mechanism for inhibitory/supervisory control that was 99% effective would not prevent relapse within a few months if faced with daily challenges.

Lumping most PFC processes together under the label of "executive control" is obviously a crude approximation. Other areas may also contribute to executive processes, and within frontal cortex distinct subregions can subserve distinct aspects of executive functioning (reviewed in *96,97*). It is not clear what neural processes are involved in normal "predecisions" to bring into working memory the relevant features of competing behavioral options and their likely outcomes. One can speculate that, for example, communication between hippocampus and prefrontal cortical regions may provide essential information about whether the current "episode" of experience is entirely routine or requires more on-line monitoring. Emotional "alert" cues may also be important for this process of engaging adaptive decision-making *(98)*. This and other aspects of evaluative processing may rely especially on orbitofrontal cortex *(99,100)*, while response selection may be more dependent on dorsolateral PFC *(101)*. The continuing development and application of sophisticated neuropsychological tests in conjunction with neuroimaging studies should further disaggregate and redefine processes such as disinhibition and attention to action *(102)*.

8. Habit Learning and Synaptic Plasticity

I turn now to the cellular and molecular mechanisms underlying dopamine-modulated learning in cortex–basal ganglia circuits (*see* **ref. *1***). Although the anatomical complexities of such circuits indicate multiple potential loci for plasticity, the cortical projections to striatum have received particular attention. Single striatal neurons sample information from a wide array of cortical cells *(103)*, and it is tempting to consider that plasticity of these synaptic connections might readily serve as a "switchboard" linking complex stimulus representations to behavioral responses. In this view, sets of corticostriatal synapses that are "successful," in the *local* sense of participating in driving firing of striatal neurons, become temporarily eligible for persistent strengthening; if the action in which they are participating is successful in the *global* sense of producing a reinforcement signal, then such strengthening occurs. Relative to more complex issues of motivation and goal-directed behavior, the control of habit learning by dopamine or abused drugs thus seems intuitively (if perhaps deceptively) straightforward to conceptualize in neurobiological terms.

Long-term potentiation (LTP) of the strength of corticostriatal connections can readily occur in vivo *(104,105)* and stimulation of dopamine neurons recently has been shown to modulate corticostriatal synaptic strength in direct correlation with behavioral reinforcement *(106)*. Although LTP can involve a host of cellular and subcellular mechanisms, especially in early phases *(107)*, the persistent substrate for much synaptic plasticity is generally believed to be structural changes in synaptic connectivity patterns *(108–110)*. It is likely that the persistent effects of addictive drugs on habit learning are also ultimately manifested through altered structural patterns of synaptic connectivity *(1)*. Repeated doses of amphetamine, cocaine, morphine, and nicotine can all provoke dendritic growth and synaptic change in rat ventral striatum and PFC *(111–114)*; conversely, removal of striatal dopamine causes broad decreases in measures of synaptic connectivity *(115–118)*.

In this framework the two major intracellular-level questions thus become: (1) What is the basis of synaptic eligibility traces? and (2) How does dopamine act to consolidate the strength of eligible synapses? In the case of hippocampal CA3→CA1 plasticity, it has been proposed that there are certain molecules that "tag" synapses at which coactivation of pre- and postsynaptic neurons has led to calcium entry through *N*-methyl-D-aspartate (NMDA) receptors *(119)*; the identity of such molecules is unknown but is the subject of active investigation. Persistent synaptic plasticity involves signaling to the nucleus, and transient changes in gene expression. A wide array of addictive drugs cause transient gene changes in striatum, which are dependent on coactivation of dopamine D1 receptors and glutamate NMDA receptors (*see* references in **ref. *1***). Just as occurs with many other forms of learning *(120,121)*, signaling pathways involving cAMP, protein kinase A, and the transcription factor cAMP-response element binding protein (CREB) are believed to be of central importance for this process, along with other transcription factors that participate through complex feedback mechanisms *(122)*. The resulting changes in gene expression appear to involve both increases in overall transcription of certain genes *(123–125)* and also altered patterns of neuronal splicing *(126,127)*. The newly produced transcripts include mRNAs such as *arc (128)*, which are translocated to dendrites and may serve as a cell-wide signal interacting with local synaptic tags.

Into this overall scheme some cautionary notes should be added. First, although dopamine or abused drugs can affect LTP in multiple brain pathways *(129–132)*, it is essential to consider carefully the informational capacity of particular loci of synaptic change. Projections from the PFC to midbrain dopamine cells can display LTP in response to cocaine *(133)*, and stimulation of nicotinic acetylcholine receptors can also affect LTP in this synaptic population *(134)*. However, although the timing of dopamine cell firing may be precisely controlled and convey useful information, the resulting release of dopamine across wide terminal fields in forebrain is highly unlikely to have detailed content beyond a general reinforcement and/or alerting signal. LTP of projections to dopamine cells is therefore not a likely substrate of specific learned addictive behaviors—although it might have an important role in adjusting thresholds for subsequent associative learning in the forebrain.

Second, although LTP is a model mechanism for altered patterns of synaptic connectivity underlying drug addiction, whether LTP has the persistence to account for addictive memory formation has come into question *(135)* as part of the overall question of whether "LTP = memory" *(136,137)*. The potential persistence of dopamine-modulated corticostriatal LTP in particular remains unsettled; it is worth noting that the corticostriatal enhancement produced by direct stimulation of dopaminergic cells *(106)* faded within a few hours. It has become ever more clear that even within the hippocampal CA1 region there are multiple forms of LTP involving distinct signaling pathways *(107)*. For example, manipulations of CREB/ATF transcriptions factors that diminish the LTP induced by combined dopamine D1 agonist and mild electrical stimulation do not disrupt LTP induced by purely electrical stimulation patterns *(138)*. Persistent modification of synaptic connectivity is thus not a unitary phenomenon, and our understanding of which particular molecular mechanisms are the enduring substrate of learned habits is far from complete. Nonetheless, altered information-rich patterns of communication between neurons remain the best current model for associative learning in general, and specific drug-related habits in particular.

Third, to draw the appropriate conclusions from molecular studies of drug abuse mechanisms, one must keep in mind that CREB-mediated transcription is involved in far more than synaptic plasticity. The ability of drugs to affect gene expression also plays a key part in compensatory adaptations to abnormal levels of dopamine stimulation. The best known of these adaptations is CREB-mediated induction of the neuropeptide dynorphin *(139)*. Dynorphin can be released from striatal neurons and act on presynaptic dopamine terminals via κ-opioid receptors, to reduce tonic dopamine release *(140,141)*. Increased dynorphin, and hence reduced dopamine levels, may be responsible for the psychomotor depression following human binge use of cocaine *(142)*. Similar behavioral results can be produced in experimental animals by artificially prolonged increases in CREB activity *(143,144)*; such manipulations may therefore be a better model of processes underlying drug withdrawal states than drug reward or drug addiction.

9. Addictive Drugs and Dopamine Dynamics

It is clearly important that learning be as specific as possible, and it is therefore hardly surprising that there is a tight feedback control of dopamine signaling. This occurs at every stage from receptors *(145)* to gene expression *(125)*. The ability of abused drugs to bypass or subvert the normal control over dopamine release can result in unusually intense and prolonged homeostatic adaptations. A general principle of such changes is that they are "information-poor"—they are not specific to certain cues or actions. Besides dynorphin induction, another likely example is a broad change in the voltage-dependent conductances of striatal neurons ("whole-cell plasticity") *(146)*. An understanding of such processes, and how they affect dopamine neurotransmission at multiple time scales, will be essential for understanding the idiosyncratic features of each abused drug. Whereas processes such as receptor desensitization can occur very rapidly *(75)*, other compensatory adaptations have slower onsets and can last for weeks or perhaps even months *(147)*.

Considering the dynamics of these opposing processes provides a partial explanation for observations that more rapid routes of drug administration (e.g., intravenous) are more addictive than slower routes (e.g., oral). In addition to higher peak brain drug concentrations reached for a given amount of drug, the faster routes provide quicker and more transient increases in dopamine release *(148,149)*. This in turn provokes less induction of slower, homeostatic changes that would blunt the effects of subsequent drug doses *(150)*. Similar considerations underlie observations that dopaminergic drugs with brief time courses of action cause behavioral changes different from those that persist longer in the brain, and that intermittent injections of drugs have consequences different from closely spaced or continuous infusions. A classic example of this latter effect, as discussed by Post *(151)*, is that intermittent psychostimulant injections cause behavioral sensitization while closely spaced injections produce tolerance. A schematic illustration of why this is the case is given in **Fig. 1**.

A drug or neuromodulator by itself does not contain rich information, and yet drug-related behaviors can be highly specific. As indicated previously, such specificity derives from the patterns of information being processed around the time of drug administration. Such information represents the cues and context that happen to be present, together with either the specific action of drug self-administration or whatever else the animal happens to be doing at the time of passive receipt of the drug. Once a specific behavior has been reinforced, altered release of dopamine is no longer required for performance of that behavior. For example, specific learned cues can evoke drug-reinforced lever pressing in monkeys without increasing striatal dopamine release *(152)*. However, if drugs are acutely received as well, the acute striatal dopamine release will potentiate performance of previously learned actions, while at the same time further reinforcing such actions for future occasions. The unfortunate potential for positive feedback of learning is clear, and is likely responsible for much sensitization and also the development of stereotypies—in which natural behaviors such as sniffing in rats become pathologically narrowed, fragmented, and intensified *(153)*. A similar process in human cortical–basal ganglia circuits may be responsible for psychosis in long-term psychostimulant abusers *(154,155)*.

10. Conclusions

This review addressed whether addiction reflects changes in the brain that are specific to particular patterns of altered information processing (i.e., associative learning) or a more general dysregulation of brain processes such as motivation, "reward," or inhibitory control over actions. If addiction indeed arises from inappropriate associative learning, this has important implications for molecular studies of drug abuse. In particular, it does not make sense to assume that persistent behavioral changes must necessarily be paralleled by persistent changes in overall levels or modification states of particular proteins within given brain regions. Rather, the key lies in the way information is represented and processed, and this in turn arises from altered patterns of communication between neurons. When large doses of drugs are repeatedly given, gross molecular changes can certainly be observed and can certainly affect behavior—but generally in a broad manner that more reflects compensatory adaptations to drugs than drug addiction per se.

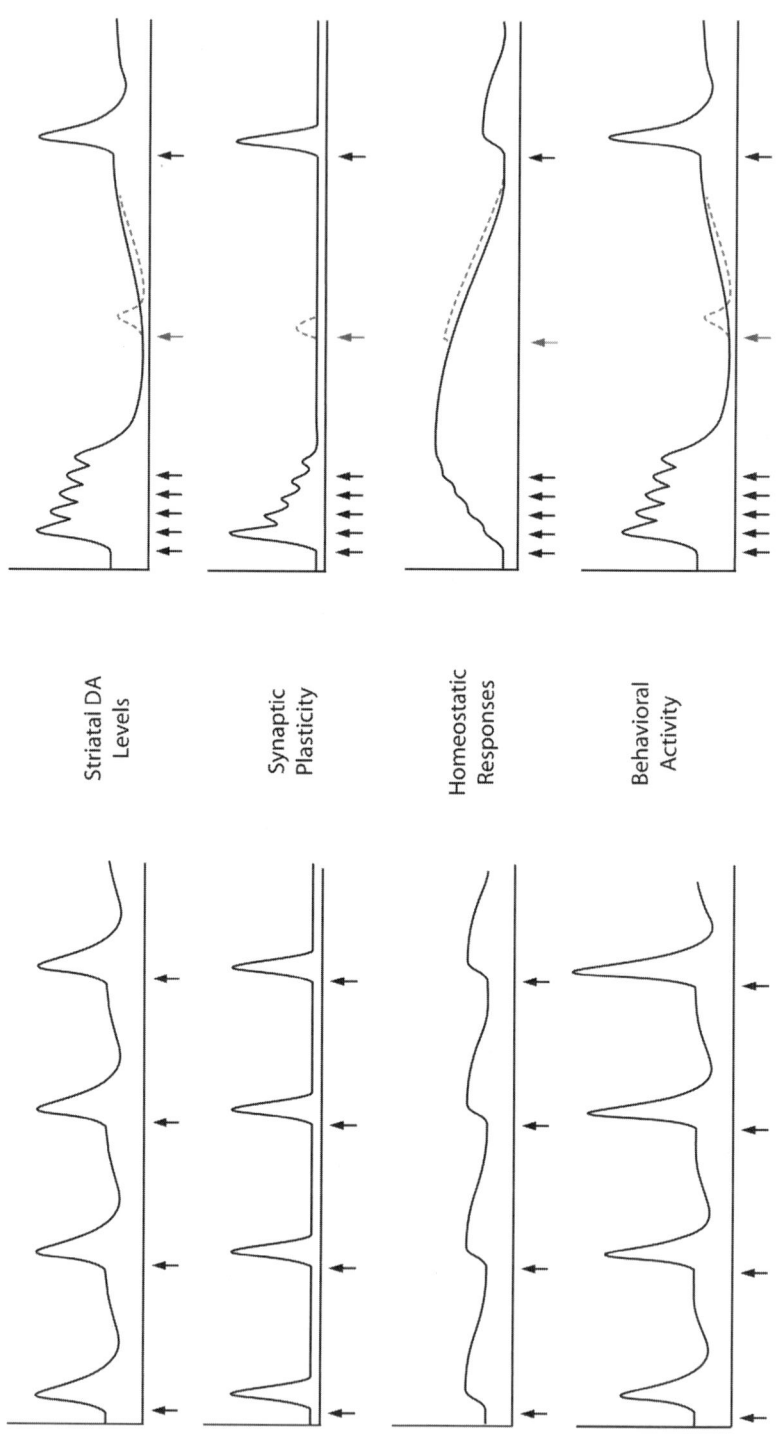

More specifically, it was argued that the ability of addictive drugs to evoke exaggerated habit learning in cortical–basal ganglia circuits is likely responsible for the persistent and compulsive aspects of addiction. Compared to other contemporary accounts of the role of learning in addiction *(5,83)*, this view thus places more emphasis on the reinforcing, rather than rewarding, actions of abused drugs. Maladaptive behavior is seen largely as the result of acting in a habitual mode in which adaptive decision-making is not properly engaged. So far, animal models of drug self-adminstration have provided fascinating information about the neural substrates of motivated behavior, which is highly relevant to the *general* phenomenon of drug abuse. The challenge they face is to reveal more about the processes of control and loss of control over drug intake that characterize addiction. This is particularly true for experiments examining "reinstatement" of drug taking following extinction, which is often erroneously referred to as "relapse" *(156)*.

Drug addiction was defined here and elsewhere as a persistent tendency to lose control over drug intake (despite negative consequences). It is worth reemphasizing that the great majority of drug use is not well described as addiction in this sense (*see also* **ref. 86**). Much of the seeking out of drugs that characterizes active drug abuse is indicative of planned, goal-directed behavior. For example, observations that increases in cigarette taxes cause substantial declines in smoking rates imply that, for the majority of smokers, smoking competes with (and can lose to) other options in an evaluative decision-making process. To view drug addiction as the serious psychiatric condition that it is may involve accepting that not all heavy or even antisocial drug use falls into this category.

Nor do I mean to imply that all the drug user needs are to ensure that evaluative decision-making is consistently engaged, and all will be well. Sometimes

---

Fig. 1. *(previous page)* Simplified, qualitative model illustrating dynamic interactions between associative sensitization and homeostatic adaptations in response to repeated injections of an addictive drug such as amphetamine. On the **left**, widely spaced injections *(arrows)* cause transient elevation of striatal dopamine, which in turn produces brief periods of synaptic plasticity without evoking large compensatory homeostatic responses. Synaptic plasticity causes consolidation of the behaviors acutely provoked by the drug, producing enhanced behavioral responses to subsequent drug injections (context-dependent sensitization; *bottom*). On **right**, closely spaced injections of the same drug provoke large homeostatic responses, which produce tolerance to a challenge injection if given during the subsequent withdrawal period *(shaded arrow)*. This withdrawal period is time limited, and if enough time elapses, subsequent challenge injections are not blunted *(solid arrow)*.

we delude ourselves about the consequences of our actions, and sometimes we deliberately choose options that we know to have negative long-term consequences (*see 157* for an interesting discussion of deliberation gone awry). Self-delusion and weakness of resolve are among a diversity of factors contributing to maladaptive decision-making across individuals, and remain poorly characterized at a neurobiological level. Rather than viewing drug abuse en masse as either a personal failing or a "brain disease" *(158)*, the extent to which we can describe and distinguish such factors will be the extent to which we can strike the right balance of sanctions, education, and compassionate treatment.

Acknowledgments

I thank Drs. Jill McGaughy, Howard Eichenbaum, and Anna Grzymała-Busse for their helpful comments on an earlier version of the manuscript.

References

1. Berke, J. D. and Hyman, S. E. (2000) Addiction, dopamine, and the molecular mechanisms of memory. *Neuron* **25,** 515–532.
2. Wikler, A. (1971) Some implications of conditioning theory for problems of drug abuse. *Behav. Sci.* **16,** 92–97.
3. Stewart, J. (1992) Neurobiology of conditioning to drugs of abuse. *Ann. NY Acad. Sci.* **654,** 335–346.
4. O'Brien, C. P., Childress, A. R., McLellan, A. T., and Ehrman, R. (1992) A learning model of addiction. *Res. Publ. Assoc. Res. Nerv. Ment. Dis.* **70,** 157–177.
5. Everitt, B. J., Parkinson, J. A., Olmstead, M. C., Arroyo, M., Robledo, P., and Robbins, T. W. (1999) Associative processes in addiction and reward. The role of amygdala–ventral striatal subsystems. *Ann. NY Acad. Sci.* **877,** 412–438.
6. O'Brien, C. P., Childress, A. R., Ehrman, R., and Robbins, S. J. (1998) Conditioning factors in drug abuse: can they explain compulsion? *J. Psychopharmacol.* **12,** 15–22.
7. Robinson, T. E. and Berridge, K. C. (1993) The neural basis of drug craving: an incentive-sensitization theory of addiction. *Brain Res. Brain Res. Rev.* **18,** 247–291.
8. Koob, G. F. and Le Moal, M. (1997) Drug abuse: hedonic homeostatic dysregulation. *Science* **278,** 52–58.
9. Fiorino, D. F. and Phillips, A. G. (1999) Facilitation of sexual behavior and enhanced dopamine efflux in the nucleus accumbens of male rats after D-amphetamine-induced behavioral sensitization. *J. Neurosci.* **19,** 456–463.
10. Wyvell, C. L. and Berridge, K. C. (2001) Incentive sensitization by previous amphetamine exposure: increased cue-triggered "wanting" for sucrose reward. *J. Neurosci.* **21,** 7831–7840.
11. Gawin, F. H. (1991) Cocaine addiction: psychology and neurophysiology. *Science* **251,** 1580–1586.

12. Eichenbaum, H. and Cohen, N. J. (2001) *From Conditioning to Conscious Recollection: Memory Systems of the Brain.* Oxford University Press, New York.
13. White, N. M. and McDonald, R. J. (2002) Multiple parallel memory systems in the brain of the rat. *Neurobiol. Learn. Mem.* **77,** 125–184.
14. Packard, M. G. (2001) On the neurobiology of multiple memory systems: Tolman versus Hull, system interactions and the emotion–memory link. *Cogn. Proc.* **2,** 3–24.
15. Mishkin, M., Malamut, B., and Bachevalier, J. (1984) Memories and habits: two neural systems, in *Neurobiology of Learning and Memory* (Lynch, G., McGaugh, J. L., and Weinberger, N. M., eds.), Guildford, New York, pp. 65–77.
16. Knowlton, B. J., Mangels, J. A., and Squire, L. R. (1996) A neostriatal habit learning system in humans. *Science* **273,** 1399–1402.
17. Poldrack, R. A., Clark, J., Pare-Blagoev, E. J., et al. (2001) Interactive memory systems in the human brain. *Nature* **414,** 546–550.
18. Fernandez-Ruiz, J., Wang, J., Aigner, T. G., and Mishkin, M. (2001) Visual habit formation in monkeys with neurotoxic lesions of the ventrocaudal neostriatum. *Proc. Natl. Acad. Sci. USA* **98,** 4196–4201.
19. Packard, M. G., Hirsh, R., and White, N. M. (1989) Differential effects of fornix and caudate nucleus lesions on two radial maze tasks: evidence for multiple memory systems. *J. Neurosci.* **9,** 1465–1472.
20. McDonald, R. J. and White, N. M. (1993) A triple dissociation of memory systems: hippocampus, amygdala, and dorsal striatum. *Behav. Neurosci.* **107,** 3–22.
21. Packard, M. G. and White, N. M. (1991) Dissociation of hippocampus and caudate nucleus memory systems by posttraining intracerebral injection of dopamine agonists. *Behav. Neurosci.* **105,** 295–306.
22. Sherry, D. S. and Schacter, D. L. (1987) The evolution of multiple memory systems. *Psychol. Rev.* **94,** 439–454.
23. Eichenbaum, H. (2000) A cortical–hippocampal system for declarative memory. *Nat. Rev. Neurosci.* **1,** 41–50.
24. Suzuki, W. A. and Clayton, N. S. (2000) The hippocampus and memory: a comparative and ethological perspective. *Curr. Opin. Neurobiol.* **10,** 768–773.
25. Tolman, E. C., Ritchie, B. F., and Kalish, D. (1946) Studies in spatial learning. II. Place learning versus response learning. *J. Exp. Psychol.* **36,** 221–229.
26. Restle, F. (1957) Discrimination of cues in mazes: a resolution of the "place-vs.-response" question. *Psychol. Rev.* **64,** 217–228.
27. Packard, M. G. (1999) Glutamate infused posttraining into the hippocampus or caudate–putamen differentially strengthens place and response learning. *Proc. Natl. Acad. Sci. USA* **96,** 12881–12886.
28. Hirsh, R. (1974) The hippocampus and contextual retrieval of information from memory: a theory. *Behav. Biol.* **12,** 421–444.
29. Devenport, L. D. and Holloway, F. A. (1980) The rat's resistance to superstition: role of the hippocampus. *J. Comp. Physiol. Psychol.* **94,** 691–705.

30. Packard, M. G. and McGaugh, J. L. (1996) Inactivation of hippocampus or caudate nucleus with lidocaine differentially affects expression of place and response learning. *Neurobiol. Learn. Mem.* **65,** 65–72.
31. Graybiel, A. M. (1998) The basal ganglia and chunking of action repertoires. *Neurobiol. Learn. Mem.* **70,** 119–136.
32. Reason, J. (1979) Actions not as planned: the price of automatization, in *Aspects of Consciousness*, Vol. 1: *Psychological Issues* (Underwood, G. and Stevens, R., eds.), Academic Press, New York.
33. Balleine, B. W. and Dickinson, A. (1998) Goal-directed instrumental action: contingency and incentive learning and their cortical substrates. *Neuropharmacology* **37,** 407–419.
34. Sage, J. R. and Knowlton, B. J. (2000) Effects of US devaluation on win-stay and win-shift radial maze performance in rats. *Behav. Neurosci.* **114,** 295–306.
35. White, N. M. (1989) Reward or reinforcement: what's the difference? *Neurosci. Biobehav. Rev.* **13,** 181–186.
36. White, N. M. and Milner, P. M. (1992) The psychobiology of reinforcers. *Annu. Rev. Psychol.* **43,** 443–471.
37. Goldman-Rakic, P. S. (1987) Circuitry of primate prefrontal cortex and regulation of behavior by representational memory, in *Handbook of Physiology*, Vol. 5 (Plum, F., ed.), American Physiological Society, Bethesda, MD, pp. 373–417.
38. Winocur, G. and Eskes, G. (1998) Prefrontal cortex and caudate nucleus in conditional associative learning: dissociated effects of selective brain lesions in rats. *Behav. Neurosci.* **112,** 89–101.
39. Passingham, R. E. (1993) *The Frontal Lobes and Voluntary Action*, Vol. 21. Oxford University Press, New York.
40. Fuster, J. M. (2001) The prefrontal cortex—an update: time is of the essence. *Neuron* **30,** 319–333.
41. Stuss, D. T. and Levine, B. (2002) Adult clinical neuropsychology: lessons from studies of the frontal lobes. *Annu. Rev. Psychol.* **53,** 401–433.
42. Damasio, A. R., Tranel, D., and Damasio, H. (1990) Individuals with sociopathic behavior caused by frontal damage fail to respond autonomically to social stimuli. *Behav. Brain Res.* **41,** 81–94.
43. Lhermitte, F., Pillon, B., and Serdaru, M. (1986) Human autonomy and the frontal lobes. Part I: Imitation and utilization behavior: a neuropsychological study of 75 patients. *Ann. Neurol.* **19,** 326–334.
44. Dias, R., Robbins, T. W., and Roberts, A. C. (1996) Dissociation in prefrontal cortex of affective and attentional shifts. *Nature* **380,** 69–72.
45. Ragozzino, M. E., Detrick, S., and Kesner, R. P. (1999) Involvement of the prelimbic–infralimbic areas of the rodent prefrontal cortex in behavioral flexibility for place and response learning. *J. Neurosci.* **19,** 4585–4594.
46. Hasegawa, I., Hayashi, T., and Miyashita, Y. (1999) Memory retrieval under the control of the prefrontal cortex. *Ann. Med.* **31,** 380–387.

47. Laroche, S., Davis, S., and Jay, T. M. (2000) Plasticity at hippocampal to prefrontal cortex synapses: dual roles in working memory and consolidation. *Hippocampus* **10**, 438–446.
48. Winocur, G. and Moscovitch, M. (1999) Anterograde and retrograde amnesia after lesions to frontal cortex in rats. *J. Neurosci.* **19**, 9611–9617.
49. Floresco, S. B., Seamans, J. K., and Phillips, A. G. (1997) Selective roles for hippocampal, prefrontal cortical, and ventral striatal circuits in radial-arm maze tasks with or without a delay. *J. Neurosci.* **17**, 1880–1890.
50. Berke, J. D. and Eichenbaum, H. B. (2001) Drug addiction and the hippocampus. *Science* **294**, 1235.
51. Mink, J. W. (1996) The basal ganglia: focused selection and inhibition of competing motor programs. *Prog. Neurobiol.* **50**, 381–425.
52. Wise, S. P., Murray, E. A., and Gerfen, C. R. (1996) The frontal cortex–basal ganglia system in primates. *Crit. Rev. Neurobiol.* **10**, 317–356.
53. Cummings, J. L. (1993) Frontal–subcortical circuits and human behavior. *Arch. Neurol.* **50**, 873–880.
54. Rolls, E. T. (1999) The brain and emotion. Oxford University Press, Oxford.
55. Gerfen, C. R. and Wilson, C. J. (1996) The basal ganglia, in *Handbook of Chemical Neuroanatomy*, Vol. 12: *Integrated Systems of the CNS*, Part III: (Swanson, L. W., Bjorklund, A., and Hökfelt, T., eds.), Elsevier, Amsterdam.
56. Haber, S. N., Fudge, J. L., and McFarland, N. R. (2000) Striatonigrostriatal pathways in primates form an ascending spiral from the shell to the dorsolateral striatum. *J. Neurosci.* **20**, 2369–2382.
57. Joel, D. and Weiner, I. (2000) The connections of the dopaminergic system with the striatum in rats and primates: an analysis with respect to the functional and compartmental organization of the striatum. *Neuroscience* **96**, 451–474.
58. Bar-Gad, I. and Bergman, H. (2001) Stepping out of the box: information processing in the neural networks of the basal ganglia. *Curr. Opin. Neurobiol.* **11**, 689–695.
59. Redgrave, P., Prescott, T. J., and Gurney, K. (1999) The basal ganglia: a vertebrate solution to the selection problem? *Neuroscience* **89**, 1009–1023.
60. Wickens, J. (1990) Striatal dopamine in motor activation and reward-mediated learning: steps towards a unifying model. *J. Neural Transm. Gen. Sect.* **80**, 9–31.
61. Fallon, J. H. and Loughlin., S. E. (1995) Substantia nigra, in *The Rat Nervous System* (Paxinos, G., ed.), Academic Press, New York, pp. 215–238.
62. Schultz, W. (1998) Predictive reward signal of dopamine neurons. *J. Neurophysiol.* **80**, 1–27.
63. Waelti, P., Dickinson, A., and Schultz, W. (2001) Dopamine responses comply with basic assumptions of formal learning theory. *Nature* **412**, 43–48.
64. Sutton, R. S. and Barto, A. G. (1998) *Reinforcement Learning: An Introduction*. The MIT Press, Cambridge, MA.

65. Pennartz, C. M. (1996) The ascending neuromodulatory systems in learning by reinforcement: comparing computational conjectures with experimental findings. *Brain Res. Brain Res. Rev.* **21,** 219–245.
66. Schultz, W., Apicella, P., and Ljungberg, T. (1993) Responses of monkey dopamine neurons to reward and conditioned stimuli during successive steps of learning a delayed response task. *J. Neurosci.* **13,** 900–913.
67. Gonon, F. (1997) Prolonged and extrasynaptic excitatory action of dopamine mediated by D1 receptors in the rat striatum *in vivo. J. Neurosci.* **17,** 5972–5978.
68. Garris, P. A., Kilpatrick, M., Bunin, M. A., Michael, D., Walker, Q. D., and Wightman, R. M. (1999) Dissociation of dopamine release in the nucleus accumbens from intracranial self-stimulation. *Nature* **398,** 67–69.
69. Holland, P. C. (1993) Cognitive aspects of classical conditioning. *Curr. Opin. Neurobiol.* **3,** 230–236.
70. Legault, M., Rompre, P. P., and Wise, R. A. (2000) Chemical stimulation of the ventral hippocampus elevates nucleus accumbens dopamine by activating dopaminergic neurons of the ventral tegmental area. *J. Neurosci.* **20,** 1635–1642.
71. Gallagher, M. (2000) The amygdala and associative learning, in *The Amygdala: A Functional Analysis*, 2nd edit. (Aggleton, J. P., ed.), Oxford University Press, Oxford, pp. 311–329.
72. Salamone, J. D. (1996) The behavioral neurochemistry of motivation: methodological and conceptual issues in studies of the dynamic activity of nucleus accumbens dopamine. *J. Neurosci. Methods* **64,** 137–149.
73. Grace, A. A. (1991) Phasic versus tonic dopamine release and the modulation of dopamine system responsivity: a hypothesis for the etiology of schizophrenia. *Neuroscience* **41,** 1–24.
74. Pidoplichko, V. I., DeBiasi, M., Williams, J. T., and Dani, J. A. (1997) Nicotine activates and desensitizes midbrain dopamine neurons. *Nature* **390,** 401–404.
75. Zhou, F. M., Liang, Y., and Dani, J. A. (2001) Endogenous nicotinic cholinergic activity regulates dopamine release in the striatum. *Nat. Neurosci,* **4,** 1224–1229.
76. White, N. M. (1996) Addictive drugs as reinforcers: multiple partial actions on memory systems. *Addiction* **91,** 921–949.
77. Everitt, B. J. Cardinal., R. N. Hall, J., Parkinson, J. A., and Robbins, T. W. (2000) Differential involvement of amygdala subsystems in appetitive conditioning and drug addiction, in *The Amygdala: A Functional Analysis*, 2nd edit. (Aggleton, J. P., ed.), Oxford University Press, Oxford, pp. 353–390.
78. Rosenkranz, J. A. and Grace, A. A. (2002) Dopamine-mediated modulation of odour-evoked amygdala potentials during pavlovian conditioning. *Nature* **417,** 282–287.
79. Viaud, M. D. and White, N. M. (1989) Dissociation of visual and olfactory conditioning in the neostriatum of rats. *Behav. Brain Res.* **32,** 31–42.
80. White, N. M. (1988) Effect of nigrostriatal dopamine depletion on the post-training, memory-improving action of amphetamine. *Life Sci.* **43,** 7–12.
81. McGaugh, J. L. (2000) Memory—a century of consolidation. *Science* **287,** 248–251.

82. Packard, M. G., Cahill, L., and McGaugh, J. L. (1994) Amygdala modulation of hippocampal-dependent and caudate nucleus-dependent memory processes. *Proc. Natl. Acad. Sci. USA* **91,** 8477–8481.
83. Di Chiara, G. (1998) A motivational learning hypothesis of the role of mesolimbic dopamine in compulsive drug use. *J. Psychopharmacol.* **12,** 54–67.
84. Tiffany, S. T. and Carter, B. L. (1998) Is craving the source of compulsive drug use? *J. Psychopharmacol.* **12,** 23–30.
85. Koob, G. F., Rocio, M., Carrera, A., et al. (1998) Substance dependence as a compulsive behavior. *J. Psychopharmacol.* **12,** 39–48.
86. Bigelow, G. E., Brooner, R. K., and Silverman, K. (1998) Competing motivations: drug reinforcement vs non-drug reinforcement. *J. Psychopharmacol.* **12,** 8–14.
87. Hyman, S. E. (2001) A 28-year-old man addicted to cocaine. *JAMA* **286,** 2586–2594.
88. Passingham, R. E. (1998) Attention to action, in *The Prefrontal Cortex* (Roberts, A. C., Robbins, T. W., and Weiskrantz, L., eds.), Oxford University Press, Oxford, pp. 131–143.
89. James, W. (1890) *The Principles of Psychology.* Dover, New York (1950 edition).
90. Tiffany, S. T. (1990) A cognitive model of drug urges and drug-use behavior: role of automatic and nonautomatic processes. *Psychol. Rev.* **97,** 147–168.
91. Gawin, F. H. and Khalsa-Denison, M. E. (1996) Is craving mood-driven or self-propelled? Sensitization and "street" stimulant addiction. *NIDA Res. Monogr.* **163,** 224–250.
92. Rogers, R. D., Everitt, B. J., Baldacchino, A., et al. (1999) Dissociable deficits in the decision-making cognition of chronic amphetamine abusers, opiate abusers, patients with focal damage to prefrontal cortex, and tryptophan-depleted normal volunteers: evidence for monoaminergic mechanisms. *Neuropsychopharmacology* **20,** 322–339.
93. Bechara, A., Dolan, S., Denburg, N., Hindes, A., Anderson, S. W., and Nathan, P. E. (2001) Decision-making deficits, linked to a dysfunctional ventromedial prefrontal cortex, revealed in alcohol and stimulant abusers. *Neuropsychologia* **39,** 376–389.
94. Jentsch, J. D. and Taylor, J. R. (1999) Impulsivity resulting from frontostriatal dysfunction in drug abuse: implications for the control of behavior by reward-related stimuli. *Psychopharmacology (Berl.)* **146,** 373–390.
95. Evenden, J. L. (1999) Varieties of impulsivity. *Psychopharmacology (Berl.),* **146,** 348–361.
96. Robbins, T. W. (1996) Dissociating executive functions of the prefrontal cortex. *Philos. Trans. R. Soc. Lond. B Biol. Sci.* **351,** 1463–1470; discussion 1470–1471.
97. Shallice, T. and Burgess, P. (1996) The domain of supervisory processes and temporal organization of behaviour. *Philos. Trans. R. Soc. Lond. B Biol. Sci.* **351,** 1405–1411; discussion 1411–1412.
98. Bechara, A., Damasio, H., and Damasio, A. R. (2000) Emotion, decision making and the orbitofrontal cortex. *Cereb. Cortex* **10,** 295–307.

99. Schoenbaum, G., Chiba, A. A., and Gallagher, M. (1998) Orbitofrontal cortex and basolateral amygdala encode expected outcomes during learning. *Nat. Neurosci.* **1,** 155–159.
100. Rolls, E. T. (2000) The orbitofrontal cortex and reward. *Cereb. Cortex* **10,** 284–294.
101. Rowe, J. B., Toni, I., Josephs, O., Frackowiak, R. S., and Passingham, R. E. (2000) The prefrontal cortex: response selection or maintenance within working memory? *Science* **288,** 1656–1660.
102. Rahman, S., Sahakian, B. J., Cardinal, R. N., Rogers, R. D., and Robbins, T. W. (2001) Decision making and neuropsychiatry. *Trends Cogn. Sci.* **5,** 271–277.
103. Kincaid, A. E., Zheng, T., and Wilson, C. J. (1998) Connectivity and convergence of single corticostriatal axons. *J. Neurosci.* **18,** 4722–4731.
104. Charpier, S. and Deniau, J. M. (1997) In vivo activity-dependent plasticity at cortico–striatal connections: evidence for physiological long-term potentiation. *Proc. Natl. Acad. Sci. USA* **94,** 7036–7040.
105. Charpier, S., Mahon, S., and Deniau, J. M. (1999) In vivo induction of striatal long-term potentiation by low-frequency stimulation of the cerebral cortex. *Neuroscience* **91,** 1209–1222.
106. Reynolds, J. N., Hyland, B. I., and Wickens, J. R. (2001) A cellular mechanism of reward-related learning. *Nature* **413,** 67–70.
107. Sanes, J. R. and Lichtman, J. W. (1999) Can molecules explain long-term potentiation? *Nat. Neurosci.* **2,** 597–604.
108. Bailey, C. H., Bartsch, D., and Kandel, E. R. (1996) Toward a molecular definition of long-term memory storage. *Proc. Natl. Acad. Sci. USA* **93,** 13445–13452.
109. Engert, F. and Bonhoeffer, T. (1999) Dendritic spine changes associated with hippocampal long-term synaptic plasticity. *Nature* **399,** 66–70.
110. Bozdagi, O., Shan, W., Tanaka, H., Benson, D. L., and Huntley, G. W. (2000) Increasing numbers of synaptic puncta during late-phase LTP: N-cadherin is synthesized, recruited to synaptic sites, and required for potentiation. *Neuron* **28,** 245–259.
111. Robinson, T. E. and Kolb, B. (1997) Persistent structural modifications in nucleus accumbens and prefrontal cortex neurons produced by previous experience with amphetamine. *J. Neurosci.* **17,** 8491–8497.
112. Robinson, T. E. and Kolb, B. (1999) Alterations in the morphology of dendrites and dendritic spines in the nucleus accumbens and prefrontal cortex following repeated treatment with amphetamine or cocaine. *Eur. J. Neurosci.* **11,** 1598–1604.
113. Robinson, T. E. and Kolb, B. (1999) Morphine alters the structure of neurons in the nucleus accumbens and neocortex of rats. *Synapse* **33,** 160–162.
114. Brown, R. W. and Kolb, B. (2001) Nicotine sensitization increases dendritic length and spine density in the nucleus accumbens and cingulate cortex. *Brain Res.* **899,** 94–100.
115. Ingham, C. A., Hood, S. H., and Arbuthnott, G. W. (1989) Spine density on neostriatal neurones changes with 6-hydroxydopamine lesions and with age. *Brain Res.* **503,** 334–348.

116. Ingham, C. A., Hood, S. H., van Maldegem, B., Weenink, A., and Arbuthnott, G. W. (1993) Morphological changes in the rat neostriatum after unilateral 6-hydroxydopamine injections into the nigrostriatal pathway. *Exp. Brain Res.* **93**, 17–27.
117. Ingham, C. A., Hood, S. H., Taggart, P., and Arbuthnott, G. W. (1998) Plasticity of synapses in the rat neostriatum after unilateral lesion of the nigrostriatal dopaminergic pathway. *J. Neurosci.* **18**, 4732–4743.
118. Meredith, G. E., Ypma, P., and Zahm, D. S. (1995) Effects of dopamine depletion on the morphology of medium spiny neurons in the shell and core of the rat nucleus accumbens. *J. Neurosci.* **15**, 3808–3820.
119. Frey, U. and Morris, R. G. (1998) Synaptic tagging: implications for late maintenance of hippocampal long-term potentiation. *Trends Neurosci.* **21**, 181–188.
120. Milner, B., Squire, L. R., and Kandel, E. R. (1998) Cognitive neuroscience and the study of memory. *Neuron* **20**, 445–468.
121. Silva, A. J., Kogan, J. H., Frankland, P. W., and Kida, S. (1998) CREB and memory. *Annu. Rev. Neurosci.* **21**, 127–148.
122. Sanyal, S., Sandstrom, D. J., Hoeffer, C. A., and Ramaswami, M. (2002) AP-1 functions upstream of CREB to control synaptic plasticity in Drosophila. *Nature* **416**, 870–874.
123. Graybiel, A. M., Moratalla, R., and Robertson, H. A. (1990) Amphetamine and cocaine induce drug-specific activation of the c-fos gene in striosome-matrix compartments and limbic subdivisions of the striatum. *Proc. Natl. Acad. Sci. USA* **87**, 6912–6916.
124. Cole, A. J., Bhat, R. V., Patt, C., Worley, P. F., and Baraban, J. M. (1992) D1 dopamine receptor activation of multiple transcription factor genes in rat striatum. *J. Neurochem.* **58**, 1420–1426.
125. Berke, J. D., Paletzki, R. F., Aronson, G. J., Hyman, S. E., and Gerfen, C. R. (1998) A complex program of striatal gene expression induced by dopaminergic stimulation. *J. Neurosci.* **18**, 5301–5310.
126. Berke, J. D., Sgambato, V., Zhu, P. P., Lavoie, B., Vincent, M., Krause, M., and Hyman, S. E. (2001) Dopamine and glutamate induce distinct striatal splice forms of Ania-6, an RNA polymerase II-associated cyclin. *Neuron* **32**, 277–287.
127. Bottai, D., Guzowski, J. F., Schwarz, M. K., et al. (2002) Synaptic activity-induced conversion of intronic to exonic sequence in Homer 1 immediate early gene expression. *J. Neurosci.* **22**, 167–175.
128. Fosnaugh, J. S., Bhat, R. V., Yamagata, K., Worley, P. F., and Baraban, J. M. (1995) Activation of arc, a putative "effector" immediate early gene, by cocaine in rat brain. *J. Neurochem.* **64**, 2377–2380.
129. Huang, Y. Y. and Kandel, E. R. (1995) D1/D5 receptor agonists induce a protein synthesis-dependent late potentiation in the CA1 region of the hippocampus. *Proc. Natl. Acad. Sci. USA* **92**, 2446–2450.
130. Otmakhova, N. A. and Lisman, J. E. (1998) D1/D5 dopamine receptors inhibit depotentiation at CA1 synapses via cAMP-dependent mechanism. *J. Neurosci.* **18**, 1270–1279.

131. Kulla, A. and Manahan-Vaughan, D. (2000) Depotentiation in the dentate gyrus of freely moving rats is modulated by D1/D5 dopamine receptors. *Cereb. Cortex* **10**, 614–620.
132. Gurden, H., Takita, M., and Jay, T. M. (2000) Essential role of D1 but not D2 receptors in the NMDA receptor-dependent long-term potentiation at hippocampal–prefrontal cortex synapses in vivo. *J. Neurosci.* **20**, RC106.
133. Ungless, M. A., Whistler, J. L., Malenka, R. C., and Bonci, A. (2001) Single cocaine exposure in vivo induces long-term potentiation in dopamine neurons. *Nature* **411**, 583–587.
134. Mansvelder, H. D. and McGehee, D. S. (2000) Long-term potentiation of excitatory inputs to brain reward areas by nicotine. *Neuron* **27**, 349–357.
135. Nestler, E. J. (2001) Molecular neurobiology of addiction. *Am. J. Addict.* **10**, 201–217.
136. Stevens, C. F. (1998) A million dollar question: does LTP = memory? *Neuron* **20**, 1–2.
137. Martin, S. J., Grimwood, P. D., and Morris, R. G. (2000) Synaptic plasticity and memory: an evaluation of the hypothesis. *Annu. Rev. Neurosci.* **23**, 649–711.
138. Pittenger, C., Huang, Y. Y., Paletzki, R. F., et al. (2002) Reversible inhibition of CREB/ATF transcription factors in region CA1 of the dorsal hippocampus disrupts hippocampus-dependent spatial memory. *Neuron* **34**, 447–462.
139. Cole, R. L., Konradi, C., Douglass, J., and Hyman, S. E. (1995) Neuronal adaptation to amphetamine and dopamine: molecular mechanisms of prodynorphin gene regulation in rat striatum. *Neuron* **14**, 813–823.
140. Shippenberg, T. S. and Rea, W. (1997) Sensitization to the behavioral effects of cocaine: modulation by dynorphin and kappa-opioid receptor agonists. *Pharmacol. Biochem. Behav.* **57**, 449–455.
141. Steiner, H. and Gerfen, C. R. (1998) Role of dynorphin and enkephalin in the regulation of striatal output pathways and behavior. *Exp. Brain Res.* **123**, 60–76.
142. Hurd, Y. L. and Herkenham, M. (1993) Molecular alterations in the neostriatum of human cocaine addicts. *Synapse* **13**, 357–369.
143. Carlezon, W. A., Jr., Thome, J., Olson, V. G., et al. (1998) Regulation of cocaine reward by CREB. *Science* **282**, 2272–2275.
144. Pliakas, A. M., Carlson, R. R., Neve, R. L., Konradi, C., Nestler, E. J., and Carlezon, W. A., Jr. (2001) Altered responsiveness to cocaine and increased immobility in the forced swim test associated with elevated cAMP response element-binding protein expression in nucleus accumbens. *J. Neurosci.* **21**, 7397–7403.
145. Dumartin, B., Caille, I., Gonon, F., and Bloch, B. (1998) Internalization of D1 dopamine receptor in striatal neurons in vivo as evidence of activation by dopamine agonists. *J. Neurosci.* **18**, 1650–1661.
146. Zhang, X. F., Hu, X. T., and White, F. J. (1998) Whole-cell plasticity in cocaine withdrawal: reduced sodium currents in nucleus accumbens neurons. *J. Neurosci.* **18**, 488–498.

147. Volkow, N. D., Fowler, J. S., and Wang, G. J. (1999) Imaging studies on the role of dopamine in cocaine reinforcement and addiction in humans. *J. Psychopharmacol.* **13,** 337–345.
148. Di Chiara, G. and Imperato, A. (1988) Drugs abused by humans preferentially increase synaptic dopamine concentrations in the mesolimbic system of freely moving rats. *Proc. Natl. Acad. Sci. USA* **85,** 5274–5278.
149. Pontieri, F. E., Tanda, G., and Di Chiara, G. (1995) Intravenous cocaine, morphine, and amphetamine preferentially increase extracellular dopamine in the "shell" as compared with the "core" of the rat nucleus accumbens. *Proc. Natl. Acad. Sci. USA* **92,** 12304–12308.
150. Di Ciano, P., Blaha, C. D., and Phillips, A. G. (2002) Inhibition of dopamine efflux in the rat nucleus accumbens during abstinence after free access to d-amphetamine. *Behav. Brain Res.* **128,** 1–12.
151. Post, R. M. (1980) Intermittent versus continuous stimulation: effect of time interval on the development of sensitization or tolerance. *Life Sci.* **26,** 1275–1282.
152. Bradberry, C. W., Barrett-Larimore, R. L., Jatlow, P., and Rubino, S. R. (2000) Impact of self-administered cocaine and cocaine cues on extracellular dopamine in mesolimbic and sensorimotor striatum in rhesus monkeys. *J. Neurosci.* **20,** 3874–3883.
153. Robbins, T. W., Mittleman, G., O'Brien, J., and Winn, P. (1990) The neuropsychological significance of stereotypy induced by stimulant drugs, in *Neurobiology of Stereotyped Behavior* (Cooper, S. J. and Dourish, C. T., eds.), Clarendon Press, Oxford, pp. 25–63.
154. Satel, S. L., Southwick, S. M., and Gawin, F. H. (1991) Clinical features of cocaine-induced paranoia. *Am. J. Psychiatry* **148,** 495–498.
155. Sato, M., Chen, C. C., Akiyama, K., and Otsuki, S. (1983) Acute exacerbation of paranoid psychotic state after long-term abstinence in patients with previous methamphetamine psychosis. *Biol. Psychiatry* **18,** 429–440.
156. Vorel, S. R., Liu, X., Hayes, R. J., Spector, J. A., and Gardner, E. L. (2001) Relapse to cocaine-seeking after hippocampal theta burst stimulation. *Science* **292,** 1175–1178.
157. Dennett, D. C. (1984) *Elbow Room: The Varieties of Free Will Worth Wanting.* Oxford University Press, New York.
158. Leshner, A. I. (1997) Addiction is a brain disease, and it matters. *Science* **278,** 45–47.

7

From Drugs of Abuse to Parkinsonism

The MPTP Mouse Model of Parkinson's Disease

Yuen-Sum Lau and Gloria E. Meredith

1. Introduction

It began with a single case of drug abuse in Maryland *(1)*, followed by four reported cases in California *(2)* in which young heroin addicts self-injected homemade "synthetic heroin" analogs contaminated with an impure chemical byproduct, 1-methyl-4-phenyl-1,2,3,6-tetrahydroperidine (MPTP), and consequently they developed severe Parkinson-like syndrome. These cases were considered extremely unusual, as Parkinson's disease (PD) is a slow, progressive, neurodegenerative disorder normally affecting older patients, with the age at onset for the majority during the 60s. It was quickly determined that MPTP is a potent neurotoxic agent that selectively destroys the central dopaminergic neurons, creates a deficit in dopamine transmission, and results in neurological symptoms indistinguishable from those of classical PD. Idiopathic PD patients and MPTP-intoxicated individuals all exhibit the cardinal signs of bradykinesia, rigidity, resting tremor, and gait disturbance *(3,4)*. With such an important discovery, MPTP has been widely used in various types of in vitro and in vivo models for elucidating possible pathophysiological mechanisms and for exploring new therapeutic and neuroprotective approaches hoping to slow the disease process and/or reverse the debilitating symptomology of the disease. A recent survey of Medlines showed that publications involving MPTP research have steadily increased since 1983 (**Fig. 1**).

2. Neurologic Findings in PD

Although PD was first described in 1817, very little is known about its etiology. On a neuroanatomical basis, it is clear that dopaminergic neurons

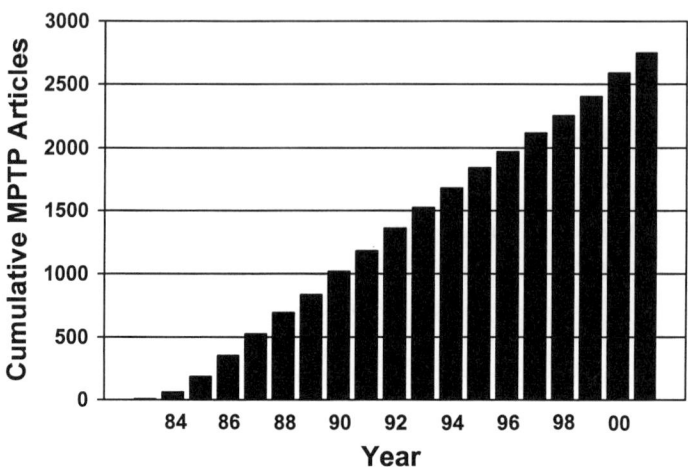

Fig. 1. Cumulative number of publications on MPTP studies since 1983. *Source:* Medlines.

within the substantia nigra are lost at a relatively accelerated rate compared to the normal aging process *(5,6)*. In human PD, clinical symptoms do not fully develop until at least 70–80% of the striatal nerve terminals and 50–60% of the substantia nigra pars compacta cells are permanently lost *(7–11)*. As a result of neuronal cell death and terminal loss, the neurochemical synthesis and release of the key transmitter, dopamine (DA), in the basal ganglia are dysfunctional and insufficient. Current drug therapy with 3,4-dihydroxy-L-phenylalanine (L-DOPA) or DA agonists only corrects the symptoms temporarily, but does not alter the course of the disease or protect neurons from degeneration due to unknown causes.

Pathologically, PD is characteristically accompanied by widespread formation of Lewy bodies and dystrophic neurites detected in the degenerated nigrostriatal dopaminergic and other cortical/subcortical neurons postmortem *(12,13)*. Lewy bodies found in the substantia nigra and locus ceruleus of PD are typically described as concentric, intracytoplasmic inclusions consisting of a dense granular core and surrounded by a halo of radiating filaments *(13)*. The Lewy bodies exhibit abnormally phosphorylated neurofilaments with an accumulation of proteins (such as ubiquitin and α-synuclein) and lipids *(14–18)*. The major clinical and neuropathological features of human PD are summarized in **Table 1**.

3. Research Models of PD

To establish a valid experimental model that closely resembles PD by keeping the cardinal symptomatology of human PD in mind (**Table 1**), the

Table 1
Major Clinical and Neuropathological Features of Human PD

Major types of symptomatology	Characteristic features in human PD
Motor deficits	Rigidity, bradykinesia, resting tremor, gait disturbance
Progressive nigral cell death	Gradual loss of TH-immunoreactive, dopaminergic neurons in the substantia nigra pars compacta
Depletion of DA transmission	Reduced level of striatal terminal DA, its metabolites and uptake transporter; clinically responsive to DA prodrug, L-DOPA
Detection of postmortem Lewy bodies	Concentric, intracytoplasmic inclusion bodies with a dense core, a peripheral halo, and radiating neurofilaments; immunoreactive to proteins such as α-synuclein and ubiquitin, and contains lipids

model should exhibit as many of the phenotypic features of the disease as possible. These features should include (1) persistent depletion of close to 80% of the striatal DA and its metabolites, (2) pronounced reduction of striatal sites for DA uptake, (3) significant (near 50%) loss of substantia nigral cells, (4) marked deficit in the animal's motor performance, and (5) formation and accumulation of inclusion bodies in nigral neurons.

There are several existing experimental models of PD, which have been developed and characterized for specific purposes and used in studies that examine certain symptomatology of PD or for determining the underlying mechanisms. These models include the unilateral 6-hydroxydopamine lesion rat produced by the chemical-induced degeneration of DA neurons *(19,20)* and Parkinsonism induced pharmacologically either by depleting the stored DA from nerve terminals (e.g., reserpine) or by blocking the postsynaptic DA receptors using neuroleptic drugs (e.g., haloperidol, phenothiazines) *(21)*. Other models have been reported by using insecticides/pesticides (e.g., rotenone) *(22,23)* or herbicides (e.g., paraquat) *(24,25)*. Transgenic models have also been generated in the fruit fly, *Drosophila melanogaster (26)*, and the mouse *(27)*. These transgenic models overexpress human α-synuclein, a constituent of Lewy bodies.

As mentioned earlier, MPTP neurotoxicity in humans is irreversible and the resulting clinical and biochemical profiles closely resemble those of the idiopathic PD *(3,4)*. This popularly used neurotoxin has been tested in many species of animals and showed strikingly different results. Although similar symptoms can also be replicated in nonhuman primates *(28)*, their sensitivity to the MPTP insult is variable *(29)*. More significantly, the MPTP-induced parkinsonian symptoms in primates tend to reverse spontaneously over time *(30)*. While classical Lewy bodies are not detected in humans *(31)* or nonhuman primates *(32)* intoxicated with MPTP, α-synuclein aggregations are found in the substantia nigra of MPTP-treated baboons *(33)*. Even though the use of nonhuman primate MPTP model may be invaluable for preclinical evaluation of new therapeutic approaches, it is costly and not practical for investigating the basic questions about neurological and pathological mechanisms of this disease, as it requires a large number of animals and is difficult to justify for large-scale experimental use.

4. The Mouse Model of PD

The induction of Parkinsonism by MPTP in some rodents has generated a large volume of neurochemical, pharmacological, and anatomical findings. Rats are generally resistant to MPTP neurotoxicity *(34–36)* but mice, such as the C57BL/6 strain, are susceptible *(34,35,37)*. In the present review, our main focus is on the mouse MPTP model.

4.1. The Acute and Subacute Mouse MPTP Model

To induce a robust depletion of striatal DA in mice, large doses of MPTP and frequent injections are required. In mouse studies, MPTP is commonly administered either by an acute or subacute regimen. The acute MPTP studies generally follow the model that was initially examined by Sonsalla and Heikkila *(38)*, who injected mice four times with 20 mg/kg of MPTP at 1- or 2-hr intervals within a day. With this acute model, it was indicated that most mice do not survive after three or more injections *(39,40)*. In our experience, the high animal mortality caused by short-interval, multiple-dose MPTP injections is a combined result of hypothermia, catalepsy-like immobility, absence of food and water intake, and peripheral organ toxicity. Although this multiple-dose, acute MPTP model can effectively cause rapid and drastic depletion of striatal DA levels, because of its unpredictable and extremely high fatality, the use of this animal model is quite limited.

The subacute model was originally developed by Heikkila et al. *(34)*, in which mice were typically injected with a single daily dose of MPTP at 30 mg/kg for 5–10 d. Most published results are based on studies conducted within a few days to a month after MPTP treatment is terminated. In the subacute model, MPTP produces decrements in the striatal level of DA and its metabolites along with a reduction in striatal synaptosomal DA uptake *(34,41,42)*. However, when

Table 2
Major Features of Subacute Mouse MPTP Model of PD

Major types of symptomatology	Features in subacute MPTP model
Motor deficits	Initial increase followed by decrease of motor activity. The activity returns to normal within a week. No long-term motor deficit
Progressive nigral cell death	Initial decrease, but later returning to normal of TH-immunoreactivity in the substantia nigra. No long-term loss of nigral cells
Depletion of DA transmission	Reduced level of striatal terminal DA, its metabolites and uptake; these levels return to normal after extended survival period.
Detection of postmortem Lewy bodies	No evidence of detecting the formation of α-synuclein- and ubiquitin-containing inclusion bodies

In this model, C57/Bl mouse is typically treated with a single daily dose of MPTP at 30 mg/kg for 5–10 d. Most studies are carried out within a few days to a month following the treatment.

survival times are extended, the neurotoxic effects of MPTP in mice are reversed gradually *(40–43)*. A loss of tyrosine hydroxylase (TH)-immunoreactive cells in the substantia nigra has been shown in some studies *(34,44,45)* but not in others *(41,42)*. Behaviorally, the motor response to subacute MPTP is quite variable. In general, MPTP causes a transient hyperactivity immediately after injection, followed by some decrease in activity, but the activity returns to normal after 5–7 d *(46)*. Despite the evidence for DA reductions, animals treated subacutely with MPTP do not exhibit long-term motor abnormalities *(29,40)*. No evidence has indicated that the subacute MPTP treatment results in the formation of neuronal inclusion bodies containing α-synuclein or ubiquitin. The characteristic features of the subacute mouse MPTP model of PD are listed in **Table 2**.

Other attempts to improve the mouse model include the use of more potent neurotoxic analogs of MPTP *(47–49)* and the use of older mice *(50–53)*. Although the magnitude of striatal DA depletion can be increased under these circumstances, progressive and persistent DA loss over a long survival period is not well established.

The mechanisms of MPTP neurotoxicity generated from either in vitro or in vivo subacute studies suggest that after systemic administration, MPTP is readily absorbed and crosses the blood–brain barrier. The toxin is converted to an active toxic metabolite, 1-methyl-4-phenylpyridinium (MPP$^+$), by the monoamine oxidase type B enzyme located in the astroglial cells *(54)*. The produced MPP$^+$ is taken up into the nigrostriatal DA nerve terminals selectively by the DA uptake transporter (DAT) *(55)*. The elevated cytoplasmic MPP$^+$ level may in turn cause the release and accumulation of the excitatory amino acid transmitter glutamate and Ca^{2+} within the afflicted neurons, resulting in an inhibition of complex I of the mitochondrial electron transport chain system and a production of free radicals *(56,57)*. Impairment of respiration and oxidative damage of the mitochondria may lead to the depletion of cellular ATP and eventual neuronal death *(58)*. Subacute MPTP injections and MPP$^+$ production may induce apoptosis in dopaminergic neurons *(59)* and that could serve to initiate neuronal degeneration. On the other hand, subacute MPTP treatment causes endogenous reactive gliogenesis from striatal progenitor cells in response to neuronal injury *(60,61)*. Nevertheless, these mechanistic pathways are established under acute or subacute conditions; whether they are associated with the slow, progressive development of neurodegeneration in PD is not understood.

Subacute MPTP treatment in mice produces only a transient neurotoxic insult, which reverses spontaneously and has no long-term neurodegenerative consequences, and thus may not be relevant to the progressive nature of human PD. Therefore, it would be extremely valuable to develop a long-term animal model that closely mimics the phenotypes and neuropathology exhibited by PD patients. Availability of such an animal model would allow further testing of mechanisms underlying the disease process and might lead to the discovery of novel, neuroprotective measures for slowing or arresting the progressive deterioration of motor performance in PD patients.

4.2. The Chronic Mouse MPTP/Probenecid Model

Pharmacokinetic consideration could be one of the underlying reasons why acute and subacute MPTP injections do not produce a sustained neurological insult in laboratory animals. It has been established that MPTP in rodents, following its peripheral administration, is rapidly excreted through the kidney *(62)*. After reaching the central nervous system (CNS), this toxin and its active metabolite, MPP$^+$, are quickly cleared from the brain *(63)*. Hence, investigators tend to intensify MPTP neurotoxicity by giving high doses and/or shortening the time period between successive toxin injections (*see* the preceding discussion on acute and subacute models). Other approaches have also been adopted to enhance MPTP neurotoxicity in mice by using agents that inhibit the central clearance of MPP$^+$ and/or the renal excretion of MPTP. When MPTP is coadministered with diethyldithiocarbamate *(64,65)* or acetaldehyde *(66)*, which is reported to reduce the clearance of MPP$^+$ from the brain, DA depletion in mice is enhanced or prolonged. However, investigations into other characteristics of PD in these models have not been extensively examined.

Alternatively, we have used probenecid as an adjuvant. Probenecid is a well-known drug that inhibits renal tubular secretion of certain compounds and raises their plasma levels by promoting renal reabsorption *(67)*. In addition to its renal action, probenecid also acts on the CNS by inhibiting the clearance of some neurochemicals from choroid plexus and parenchymal cells into plasma, and thus elevates their levels in the brain *(68)*. Two seminal findings were generated from our initial study *(43)*. First, we showed that a single dose of injected MPTP will achieve a peak DA depletion for several days before reversing to normal, which clearly indicates that giving another dose within a short period of time to boost its neurotoxicity may not be necessary or beneficial, but will increase the risk of animal death. Second, we discovered that probenecid effectively inhibits renal excretion of MPTP and potentiates its neurotoxicity by reaching a maximal depletion of striatal DA lasting 4–5 d after a single treatment with the combined agents.

We developed a chronic mouse MPTP/probenecid PD model by administering 10 doses of MPTP plus probenecid over a slow time course. For a period of 5 wk, we treat C57BL/6 mice, twice a week (3.5 d apart), first with 250 mg/kg of probenecid (made in dimethyl sulfoxide [DMSO], 30-µL injection volume using a 50-µL Hamilton syringe with a disposable 26G$_{3/8}$ Tuberculin needle) intraperitoneally, and approx 30 min later with MPTP (MPTP HCl dissolved in saline) subcutaneously.

We have tested two doses of MPTP at 15 mg/kg and 25 mg/kg (as MPTP HCl salt form), respectively. With probenecid, MPTP at 15 mg/kg produces about 60% DA depletion without recovery for at least 6 mo *(43)*, and at 25 mg/kg, it causes a nearly 80% loss of DA 6 mo later *(40)*. The latter dose appears to be the maximally tolerated dose when given with probenecid. All the treated animals survive under this regimen. Either increasing the dose of MPTP higher than 25 mg/kg with the same dose of probenecid in this chronic protocol, or injecting the animals with MPTP at 25 mg/kg plus probenecid on a daily basis results in significant death of animals (*unpublished results*).

In this chronic MPTP/probenecid mouse model of PD (10 doses of 250 mg/kg of probenecid and 25 mg/kg of MPTP HCl over 5 wk) *(40)*, we observe an immediate and robust (>90%) loss of striatal DA, its metabolite 3,4-dihydroxyphenylacetic acid (DOPAC), and uptake. The reduction in DA level persists throughout a survival period of 6 mo. The TH immunoreactivity is downregulated soon after the treatment, but it gradually recovers mainly in the fibers, not in nigral cells *(40)*. The functionality of these recovered TH-positive fibers is not clear. It is not likely that they produce sufficient DA, as its

Table 3
Major Features of Chronic Mouse MPTP/Probenecid Model of PD

Major types of symptomatology	Features in chronic MPTP/probenecid model
Motor deficits	Initial motor deficit (rotarod performance) is detected at 3 wk and deteriorated further 6 mo after treatment.
Progressive nigral cell death	Decrease of TH-immunoreactivity in the substantia nigra, associated with significant loss of substantia nigra neurons
Depletion of DA transmission	Persistent reduction in the levels of striatal terminal DA, its metabolites and uptake
Detection of postmortem Lewy bodies	Murine inclusion bodies immunoreactive to α-synuclein and ubiquitin; contain lipofuscin and lysosomal structures. Morphologically, these inclusion bodies resemble those detected in the cortex of PD.

In this model, the C57/Bl mouse is treated twice a week (3.5 d apart) with a dose of MPTP HCl at 25 mg/kg, s.c. and probenecid at 250 mg/kg, i.p. for 5 wk (a total of 10 treatments). The studies are carried out up to 6 mo following the treatment *(40,69)*.

striatal levels remain low. The loss of DA neurons in the substantia nigra 6 mo after treatment is estimated to be at least 60%. The first sign of motor deficit in chronic MPTP/probenecid treated mice as evidenced by the rotarod performance test is detected at 3 wk and persists for at least 6 mo after treatment *(40)*.

Pathologically, we further detect an increasing number of abnormal inclusion bodies in the cytoplasm of nigral dopaminergic and cortical neurons beginning at 3 wk after chronic MPTP/probenecid treatment *(69)*. These inclusions are granular and filamentous in appearance, and are immunopositive to α-synuclein and ubiquitin, resembling those Lewy body structures found in the cortex of human PD. At the ultrastructural level, we further confirm that these inclusions contain a dense and granular core similar to that of the classical Lewy bodies. In addition, numerous lobulated, secondary lysosomes filled with lipofuscin are observed in the cytoplasm of α-synuclein-immunoreactive neurons, and the density of these neurons is significantly reduced. The major characteristic features of this chronic MPTP/probenecid mouse model of PD are summarized in **Table 3**.

5. Conclusion

We have shown that the chronic MPTP/probenecid mouse shares many key features of human PD. This model is an improvement over the conventional acute and subacute models and is an important and a potentially useful model for studying disease progression, mechanisms of nigrostriatal neurodegeneration, and neuroprotective strategies in PD.

Acknowledgments

We thank the following colleagues who have contributed to the development of the chronic MPTP/probenecid mouse model of PD: Shanon Callen, Richard Callison, James Crampton, Elizabeth Petroske, Christopher Runice, Karen Santa Cruz, Susan Totterdell, Karen Trobough, and John Wilson. Our research on PD is supported by the Health Future Foundation, National Parkinson Foundation, University of Missouri and National Institute of Neurological Diseases and Stroke (NS41799).

References

1. Davis, G. C., Williams, A. C., Markey, S. P., et al. (1979) Chronic Parkinsonism secondary to intravenous injection of meperidine analogues. *Psychiatr. Res.* **1,** 249–254.
2. Langston, J. W., Ballard, P., Tetrud, J. W., and Irwin, I. (1983) Chronic Parkinsonism in humans due to a product of meperidine-analog synthesis. *Science* **219,** 979–980.
3. Ballard, P. A., Tetrud, J. W., and Langston, J. W. (1985) Permanent human parkinsonism due to 1-methyl-4-phenyl-1,2,3,6-tetrahydropyridine (MPTP): seven cases. *Neurology* **35,** 949–956.
4. Burns, R. S., LeWitt, P. A., Ebert, M. H., Pakkenberg, H., and Kopin, I. J. (1985) The clinical syndrome of striatal dopamine deficiency. Parkinsonism induced by 1-methyl-4-phenyl-1,2,3,6-tetrahydropyridine (MPTP). *N. Engl. J. Med.* **312,** 1418–1421.
5. Scherman, D., Desnos, C., Darchen, F., Pollak, P., Javoy-Agid, F., and Agid, Y. (1989) Striatal dopamine deficiency in Parkinson's disease: role of aging. *Ann. Neurol.* **26,** 551–557.
6. McGeer, P. L., Itagaki, S., Akiyama, H., and McGeer, E. G. (1989) Comparison of neuronal loss in Parkinson's disease and aging, in *Parkinsonism and Aging* (Calne, D. B., Comi, G., Crippa, D., Horowski, R., and Trabucchi, M., eds.), Raven Press, New York, pp. 25–34.
7. Bernheimer, H., Birkmayer, W., Hornykiewicz, O., Jellinger, K., and Seitelberger, F. (1973) Brain dopamine and the syndromes of Parkinson and Huntington. Clinical, morphological and neurochemical correlations. *J. Neurol. Sci.* **20,** 415–455.
8. Kish, S. J., Shannak, K., and Hornykiewicz, O. (1988) Uneven pattern of dopamine loss in the striatum of patients with idiopathic Parkinson's disease. Pathophysiologic and clinical implications. *N. Engl. J. Med.* **318,** 876–880.

9. Agid, Y. (1991) Parkinson's disease: pathophysiology. *Lancet* **337,** 1321–1324.
10. Fearnley, J. M., and Lees, A. J. (1991) Ageing and Parkinson's disease: substantia nigra regional selectivity. *Brain* **114,** 2283–2301.
11. Frost, J. J., Rosier, A. J., Reich, S. G., et al. (1993) Positron emission tomographic imaging of the dopamine transporter with 11C-WIN 35,428 reveals marked declines in mild Parkinson's disease. *Ann. Neurol.* **34,** 423–431.
12. Braak, H., Braak, E., Yilmazer, D., de Vos, R. A., Jansen, E. N., and Bohl, J. (1996) Pattern of brain destruction in Parkinson's and Alzheimer's diseases. *J. Neural Transm.* **103,** 455–490.
13. Forno, L. S. (1996) Neuropathology of Parkinson's disease. *J. Neuropathol. Exp. Neurol.* **55,** 259–272.
14. Kuzuhara, S., Mori, H., Izumiyama, N., Yoshimura, M., and Ihara, Y. (1988) Lewy bodies are ubiquitinated. A light and electron microscopic immunocytochemical study. *Acta Neuropathol. (Berl.)* **75,** 345–353.
15. Bancher, C., Lassmann, H., Budka, H., et al. (1989) An antigenic profile of Lewy bodies: immunocytochemical indication for protein phosphorylation and ubiquitination. *J. Neuropathol. Exp. Neurol.* **48,** 81–93.
16. Spillantini, M. G., Crowther, R. A., Jakes, R., Hasegawa, M., and Goedert, M. (1998) Alpha-synuclein in filamentous inclusions of Lewy bodies from Parkinson's disease and dementia with Lewy bodies. *Proc. Natl. Acad. Sci. USA* **95,** 6469–6473.
17. Gai, W. P., Yuan, H. X., Li, X. Q., Power, J. T., Blumbergs, P. C., and Jensen, P. H. (2000) In situ and in vitro study of colocalization and segregation of alpha-synuclein, ubiquitin, and lipids in Lewy bodies. *Exp. Neurol.* **166,** 324–333.
18. Shimura, H., Schlossmacher, M. G., Hattori, N., et al. (2001) Ubiquitination of a new form of alpha-synuclein by parkin from human brain: implications for Parkinson's disease. *Science* **293,** 263–269.
19. Ungerstedt, U. (1976) 6-Hydroxydopamine-induced degeneration of the nigrostriatal dopamine pathway: the turning syndrome. *Pharmacol. Ther.* **2,** 37–40.
20. Schwarting, R. K. and Huston, J. P. (1996) The unilateral 6-hydroxydopamine lesion model in behavioral brain research. Analysis of functional deficits, recovery and treatments. *Prog. Neurobiol.* **50,** 275–331.
21. Montastruc, J. L., Llau, M. E., Rascol, O., and Senard, J. M. (1994) Drug-induced parkinsonism: a review. *Fund. Clin. Pharmacol.* **8,** 293–306.
22. Ferrante, R. J., Schulz, J. B., Kowall, N. W., and Beal, M. F. (1997) Systemic administration of rotenone produces selective damage in the striatum and globus pallidus, but not in the substantia nigra. *Brain Res.* **753,** 157–162.
23. Betarbet, R., Sherer, T. B., MacKenzie, G., Garcia-Osuna, M., Panov, A. V., and Greenamyre, J. T. (2000) Chronic systemic pesticide exposure reproduces features of Parkinson's disease. *Nat. Neurosci.* **3,** 1301–1306.
24. Corasaniti, M. T., Strongoli, M. C., Rotiroti, D., Bagetta, G., and Nistico, G. (1998) Paraquat: a useful tool for the in vivo study of mechanisms of neuronal cell death. *Pharmacol. Toxicol.* **83,** 1–7.
25. Thiruchelvam, M., Richfield, E. K., Baggs, R. B., Tank, A. W., and Cory-Slechta, D. A. (2000) The nigrostriatal dopaminergic system as a preferential target of

repeated exposures to combined paraquat and maneb: implications for Parkinson's disease. *J. Neurosci.* **20,** 9207–9214.
26. Feany, M. B. and Bender, W. W. (2000) A *Drosophila* model of Parkinson's disease. *Nature* **404,** 394–398.
27. Masliah, E., Rockenstein, E., Veinbergs, I., et al. (2000) Dopaminergic loss and inclusion body formation in alpha-synuclein mice: implications for neurodegenerative disorders. *Science* **287,** 1265–1269.
28. Burns, R. S., Chiueh, C. C., Markey, S. P., Ebert, M. H., Jacobowitz, D. M., and Kopin, I. J. (1983) A primate model of parkinsonism: selective destruction of dopaminergic neurons in the pars compacta of the substantia nigra by *N*-methyl-4-phenyl-1,2,3,6-tetrahydropyridine. *Proc. Natl. Acad. Sci. USA* **80,** 4546–4550.
29. Gerlach, M. and Riederer, P. (1996) Animal models of Parkinson's disease: an empirical comparison with the phenomenology of the disease in man. *J. Neural Transm.* **103,** 987–1041.
30. Eidelberg, E., Brooks, B. A., Morgan, W. W., Walden, J. G., and Kokemoor, R. H. (1986) Variability and functional recovery in the *N*-methyl-4-phenyl-1,2,3,6-tetrahydropyridine model of parkinsonism in monkeys. *Neuroscience* **18,** 817–822.
31. Langston, J. W., Forno, L. S., Tetrud, J., Reeves, A. G., Kaplan, J. A., and Karluk, D. (1999) Evidence of active nerve cell degeneration in the substantia nigra of humans years after 1-methyl-4-phenyl-1,2,3,6-tetrahydropyridine exposure. *Ann. Neurol.* **46,** 598–605.
32. Forno, L. S., DeLanney, L. E., Irwin, I., and Langston, J. W. (1993) Similarities and differences between MPTP-induced parkinsonsim and Parkinson's disease. Neuropathologic considerations. *Adv. Neurol.* **60,** 600–608.
33. Kowall, N. W., Hantraye, P., Brouillet, E., Beal, M. F., McKee, A. C., and Ferrante, R. J. (2000) MPTP induces alpha-synuclein aggregation in the substantia nigra of baboons. *NeuroReport* **11,** 211–213.
34. Heikkila, R. E., Hess, A., and Duvoisin, R. C. (1984) Dopaminergic neurotoxicity of 1-methyl-4-phenyl-1,2,5,6-tetrahydropyridine in mice. *Science* **224,** 1451–1453.
35. Lau, Y.-S. and Fung, Y. K. (1986) Pharmacological effects of 1-methyl-4-phenyl-1,2,3,6-tetrahydropyridine (MPTP) on striatal dopamine receptor system. *Brain Res.* **369,** 311–315.
36. Zuddas, A., Fascetti, F., Corsini, G. U., and Piccardi, M. P. (1994) In brown Norway rats, MPP$^+$ is accumulated in the nigrostriatal dopaminergic terminals but it is not neurotoxic: a model of natural resistance to MPTP toxicity. *Exp. Neurol.* **127,** 54–61.
37. Hamre, K., Tharp, R., Poon, K., Xiong, X., and Smeyne, R. J. (1999) Differential strain susceptibility following 1-methyl-4-phenyl-1,2,3,6-tetrahydropyridine (MPTP) administration acts in an autosomal dominant fashion: quantitative analysis in seven strains of *Mus musculus*. *Brain Res.* **828,** 91–103.
38. Sonsalla, P. K. and Heikkila, R. E. (1986) The influence of dose and dosing interval on MPTP-induced dopaminergic neurotoxicity in mice. *Eur. J. Pharmacol.* **129,** 339–345.

39. Walters, T. L., Irwin, I., Delfani, K., Langston, J. W., and Janson, A. M. (1999) Diethyldithiocarbamate causes nigral cell loss and dopamine depletion with nontoxic doses of MPTP. *Exp. Neurol.* **56**, 62–70.
40. Petroske, E., Meredith, G. E., Callen, S., Totterdell, S., and Lau, Y.-S. (2001) Mouse model of parkinsonism: a comparison between subacute MPTP and chronic MPTP/probenecid treatment. *Neuroscience* **106**, 589–601.
41. Hallman, H., Lange, J., Olson, L., Stromberg, I., and Jonsson, G. (1985) Neurochemical and histochemical characterization of neurotoxic effects of 1-methyl-4-phenyl-1,2,3,6-tetrahydropyridine on brain catecholamine neurones in the mouse. *J. Neurochem.* **44**, 117–127.
42. Ricaurte, G. A., Langston, J. W., Delanney, L. E., Irwin, I., Peroutka, S. J., and Forno, L. S. (1986) Fate of nigrostriatal neurons in young mature mice given 1-methyl-4-phenyl-1,2,3,6-tetrahydropyridine: a neurochemical and morphological reassessment. *Brain Res.* **376**, 117–124.
43. Lau, Y.-S., Trobough, K. L., Crampton, J. M., and Wilson, J. A. (1990) Effects of probenecid on striatal dopamine depletion in acute and long-term 1-methyl-4-phenyl-1,2,3,6-tetrahydropyridine (MPTP)-treated mice. *Gen. Pharmacol.* **21**, 181–187.
44. Seniuk, N. A., Tatton, W. G., and Greenwood, C. E. (1990) Dose-dependent destruction of the coeruleus-cortical and nigral-striatal projections by MPTP. *Brain Res.* **527**, 7–20.
45. Bezard, E., Dovero, S., Bioulac, B., and Gross, C. E. (1997) Kinetics of nigral degeneration in a chronic model of MPTP-treated mice. *Neurosci. Lett.* **234**, 47–50.
46. Markey, S. P. and Schmuff, N. R. (1986) The pharmacology of the parkinsonian syndrome producing neurotoxin MPTP (1-methyl-4-phenyl-1,2,3,6-tetrahydropyridine) and structurally related compounds. *Med. Res. Rev.* **6**, 389–429.
47. Fuller, R. W., Robertson, D. W., and Hemrick-Luecke, S. K. (1987) Comparison of the effects of two 1-methyl-4-phenyl-1,2,3,6-tetrahydropyridine analogs, 1-methyl-4-(2-thienyl)-1,2,3,6-tetrahydropyridine and 1-methyl-4-(3-thienyl)-1,2,3,6-tetrahydropyridine, on monoamine oxidase in vitro and on dopamine in mouse brain. *J. Pharmacol. Exp. Ther.* **240**, 415–420.
48. Youngster, S. K., Sonsalla, P. K., and Heikkila, R. E. (1987) Evaluation of the biological activity of several analogs of the dopaminergic neurotoxin 1-methyl-4-phenyl-1,2,3,6-tetrahydropyridine. *J. Neurochem.* **48**, 929–934.
49. Ikeda, H., Markey, C. J., and Markey, S. P. (1992) Search for neurotoxins structurally related to 1-methyl-4-phenylpyridine (MPP^+) in the pathogenesis of Parkinson's disease. *Brain Res.* **575**, 285–298.
50. Jarvis, M. F. and Wagner, G. C. (1985) Age-dependent effects of 1-methyl-4-phenyl-1,2,5,6-tetrahydropyridine (MPTP). *Neuropharmacology* **24**, 581–583.
51. Ricaurte, G. A., Irwin, I., Forno, L. S., DeLanney, L. E., Langston, E., and Langston, J. W. (1987) Aging and 1-methyl-4-phenyl-1,2,3,6-tetrahydropyridine-induced degeneration of dopaminergic neurons in the substantia nigra. *Brain Res.* **403**, 43–51.

52. Saitoh, T., Niijima, K., and Mizuno, Y. (1987) Long-term effect of 1-methyl-4-phenyl-1,2,3,6-tetrahydropyridine (MPTP) on striatal dopamine content in young and mature mice. *J. Neurol. Sci.* **77,** 229–235.
53. Ali, S. F., David, S. N., and Newport, G. D. (1993) Age-related susceptibility to MPTP-induced neurotoxicity in mice. *Neurotoxicology* **14,** 29–34.
54. Singer, T. P., Castagnoli, Jr., N., Ramsay, R. R., Trevor, A. J. (1987) Biochemical events in the development of parkinsonism induced by 1-methyl-4-phenyl-1,2,3,6-tetrahydropyridine. *J. Neurochem.* **49,** 1–8.
55. Gainetdinov, R. R., Fumagalli, F., Jones, S. R., and Caron, M. G. (1997) Dopamine transporter is required for in vivo MPTP neurotoxicity: evidence from mice lacking the transporter. *J. Neurochem.* **69,** 1322–1325.
56. Leist, M., Volbracht, C., Fava, E., and Nicotera, P. (1998) 1-Methyl-4-phenyl-pyridinium induces autocrine excitotoxicity, protease activation, and neuronal apoptosis. *Mol. Pharmacol.* **54,** 789–801.
57. Cleeter, M. W., Cooper, J. M., and Schapira, A. H. (1992) Irreversible inhibition of mitochondrial complex I by 1-methyl-4-phenylpyridinium: evidence for free radical involvement. *J. Neurochem.* **58,** 786–789.
58. Chan, P., DeLanney, L. E., Irwin, I., Langston, J. W., and Di Monte, D. (1991) Rapid ATP loss caused by 1-methyl-4-phenyl-1,2,3,6-tetrahydropyridine in mouse brain. *J. Neurochem.* **57,** 348–351.
59. Tatton, N. A. and Kish, S. J. (1997) In situ detection of apoptotic nuclei in the substantia nigra compacta of 1-methyl-4-phenyl-1,2,3,6-tetrahydropyridine-treated mice using terminal deoxynucleotidyl transferase labelling and acridine orange staining. *Neuroscience* **77,** 1037–1048.
60. Kay, J. N. and Blum, M. (2000) Differential response of ventral midbrain and striatal progenitor cells to lesions of the nigrostriatal dopaminergic projection. *Dev. Neurosci.* **22,** 56–67.
61. Mao, L., Lau, Y.-S., Petroske, E., and Wang, J. Q. (2001) Profound astrogenesis in the striatum of adult mice following nigrostriatal dopaminergic lesion by repeated MPTP administration. *Brain Res. Dev. Brain Res.* **131,** 57–65.
62. Lau, Y.-S., Crampton, J. M., and Wilson, J. A. (1988) Urinary excretion of MPTP and its primary metabolites in mice. *Life Sci.* **43,** 1459–1464.
63. Johannessen, J. N., Chiueh, C. C., Burns, R. S., and Markey, S. P. (1985) Differences in the metabolism of MPTP in the rodent and primate parallel differences in sensitivity to its neurotoxic effects. *Life Sci.* **36,** 219–224.
64. Corsini, G. U., Pintus, S., Chiueh, C. C., Weiss, J. F., and Kopin, I. J. (1985) 1-Methyl-4-phenyl-1,2,3,6-tetrahydropyridine (MPTP) neurotoxicity in mice is enhanced by pretreatment with diethyldithiocarbamate. *Eur. J. Pharmacol.* **119,** 127–128.
65. Irwin, I., Wu, E. Y., DeLanney, L. E., Trevor, A., and Langston, J. W. (1987) The effect of diethyldithiocarbamate on the biodisposition of MPTP: an explanation for enhanced neurotoxicity. *Eur. J. Pharmacol.* **141,** 209–217.
66. Zuddas, A., Corsini, G. U., Schinelli, S., Barker, J. L., Kopin, I. J., and di Porzio, U. (1989) Acetaldehyde directly enhances MPP$^+$ neurotoxicity and delays its elimination from the striatum. *Brain Res.* **501,** 11–22.

67. Rennick, B. R. (1972) Renal excretion of drugs: tubular transport and metabolism. *Annu. Rev. Pharmacol.* **12,** 141–156.
68. Miller, T. B. and Ross, C. R. (1976) Transport of organic cations and anions by choroid plexus. *J. Pharmacol. Exp. Ther.* **196,** 771–777.
69. Meredith, G. E., Totterdell, S., Petroske, E., Santa Cruz, K., Callison, R. C., and Lau, Y.-S. (2002) Lysosomal malfunction accompanies alpha-synuclein aggregation in a progressive mouse model of Parkinson's disease. *Brain Res.* **956,** 156–165.

II

DETECTION OF mRNA EXPRESSION IN THE STRIATUM

8

In Situ Hybridization with Isotopic Riboprobes for Detection of Striatal Neuropeptide mRNA Expression After Dopamine Stimulant Administration

Yasmin L. Hurd

1. Introduction

Stimulant drugs such as cocaine, amphetamine, and methylphenidate induce their primary pharmacological actions through elevation of dopamine levels in the brain. The most dense innervation of dopamine nerve terminals in the central nervous system (CNS) is found within the striatum (caudate nucleus, putamen, and nucleus accumbens) *(1)*. This brain region is central to the wide range of actions of psychostimulant drugs on motor behavior, cognition, motivation, and reward and is intricately linked with mesocorticolimbic structures, also innervated by dopamine, such as the amygdaloid complex and prefrontal cortex. The neuroanatomical organization and regulation of the striatal dopaminergic system as well as its relevance to motor function and drug reinforcement have been well studied. Elevation of striatal dopamine levels as a consequence of psychostimulant drug administration leads to activation of distinct dopamine receptor subtypes (D_1–D_3) that are differentially expressed within distinct striatal neuronal populations. The predominant striatal cells are medium spiny projection neurons (70–80%) and medium aspiny interneurons (20–25%) that all contain the inhibitory neurotransmitter γ-aminobutyric acid (GABA), but contain different neuropeptides. A major focus in regard to stimulant effects in the striatum has been directed toward the medium spiny neurons that not only have distinct neuropeptidergic content, but also discrete efferent anatomical connectivity. Medium spiny neurons in the dorsal striatum that contain the opioid peptide dynorphin and the tachykinin substance P express primarily dopamine D_1 receptors and constitute the "direct" striatonigral pathway *(2,3)*. In contrast, neurons containing the opioid neuropeptide enkephalin predominantly express dopamine D_2 receptors and make up the "indirect" striatopallidal circuitry *(3,4)*. A similar organization is present in the nucleus accumbens ventral striatal region, with dynorphin primarily expressed in neurons with D_1 and D_3 dopamine receptors *(5,6)*. Thus, elevation of dopamine levels following administration of stimulant drugs will lead to differential activation of dopamine receptors and subsequently modulation of distinct striatal neuropeptide output circuits.

There are different methodologies by which striatal neuropeptidergic neurons can be studied in relation to stimulant drug use. Alteration of gene expression is an important mechanism through which long-term effects are maintained in the brain and the compulsive, repeated use of addictive drugs implies impairments at the level of gene expression. Measurements of mRNA levels can be achieved through a variety of methodologies, but the technique of *in situ* hybridization histochemistry (ISHH) has proven the most useful to detect mRNA expression levels with intact anatomical integrity. Thus it is possible by the ISHH technique to identify the discrete neuronal populations that express the preprodynorphin (striatonigral) and preproenkephalin (striatopallidal) mRNAs. Using such methodology, selective alterations of the opioid neuropeptide genes (increased preprodynorphin and decreased preproenkephalin) were revealed in the striatum of human cocaine users *(7)*. A similar up-regulation of the striatal preprodynorphin gene has also been documented in a number of experimental rat models following administration of cocaine *(8–12)* and amphetamines *(13,14)* that has generated much attention on the dynorphin/D_1/striatonigral neurons in the CNS actions of stimulant drugs. The following details the ISHH procedure that can be used to visualize the relative expression levels of the striatal neuropeptide mRNAs following cocaine administration in the rat and human brain.

2. Materials

The following subheadings are organized as to the materials needed for (1) brain tissue preparation and sectioning, (2) brain section pretreatment, (3) probe preparation, (4) hybridization, (5) post-hybridization, and (6) autoradiography. This outline corresponds to that provided in subsequent subheadings.

2.1. Brain Tissue Preparation and Sectioning

1. 2-Methylbutane (isopentane; Sigma).
2. Tissuetek OCT Compound (Histolab).
3. Superfrost microscope glass (Brain Research Laboratories).
4. SuperFrost Plus microscope glass (Menzel-Gläser).

2.2. Brain Section Pretreatment

1. Diethyl pyrocarbonate (DEPC; Sigma) distilled water: 0.1% in distilled water, mix well and let stand overnight in a fume hood; autoclave for 2 h.
2. 10X Phosphate-buffered saline (PBS), pH 7.4 (Life Technologies, Invitrogen).
3. 4% Paraformaldehyde in 1X PBS, pH 7.4.
4. Proteinase K (Sigma).
5. Triethanolamine HCl (TEA; Sigma).
6. Acetic anhydride (Sigma).
7. Chloroform Normapur Analytical (Sigma).
8. Ethanol (Merk Eurolab).

2.3. Probe Preparation

1. [^{35}S-UTP (^{33}P; PerkinElmer Life Sciences or Amersham Biosciences).
2. 2.5 mM ΔUTP (equal volumes, e.g., 5 μL, of 10 mM ATP, CTP, and GTP (Life Technologies).
3. RNA polymerase (e.g., SP6, T7, or T3; Life Technologies).
4. 1 U/μL of DNase I (Life Technologies).
5. 40 U/μL of RNase inhibitor (Life Technologies, Invitrogen).
6. Spin column (MicroSpin S-200 HR Columns; Amersham Biosciences).
7. DL-Dithiothreitol (DTT; Sigma).

2.4. Hybridization

1. 20X Standard sodium citrate buffer (SSC; Life Technologies; 1X = 3 M sodium chloride, 0.3 M sodium citrate, pH 7.0).
2. Hybridization buffer: 1 mg/mL of sheared salmon testes DNA (Sigma), 500 μg/mL of yeast tRNA (Sigma), 2X Denhardt's solution (Sigma), 20% dextran sulfate (Amersham Biosciences), and 8X SSC.
3. Formamide (Sigma).
4. DTT and DEPC distilled water.

2.5. Post-hybridization

1. RNase A (Ribonuclease A; Sigma).
2. RNase A buffer: 0.5 M NaCl, 0.04 M Tris-HCl, pH 8.0, and 1 mM EDTA.
3. Ammonium acetate (Sigma).
4. SSC, DTT, and ethanol.

2.6. Autoradiography

1. Kodak Biomax MR film (Amersham Biosciences).
2. ^{14}C polymer standards (American Radiolabeled Chemicals Inc.).
3. D-19 Developer (Kodak).
4. Rapid Fix (Kodak).
5. Acetic acid (Sigma).

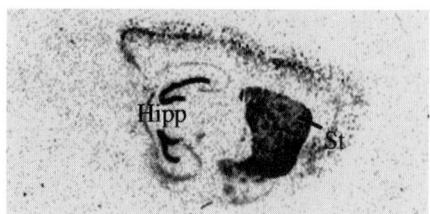 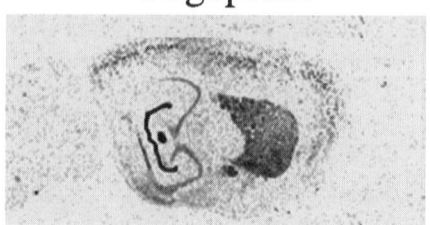

Fig. 1. Rat brain sections (sagittal orientation) processed with a riboprobe and olionucleotide antisense probe against the preprodynorphin mRNA; same film exposure time. The preprodynorphin hybridization signals are highly abundant in the striatum (St) and hippocampus (Hipp).

6. NBT 2 Emulsion (Kodak).
7. Ammonium acetate (Sigma).
8. Cresyl violet (Sigma).
9. Ethanol.
10. Xylene (Sigma).
11. Pertex mounting media (Histolab).

3. Methods

ISHH is typically performed with either synthetic oligonucleotide (DNA; normally 40 bases) or ribonucleotide (RNA; normally 100–1000 bases) probes on freshly frozen brain tissue. The experimental procedures described in the following subheadings detail the methodology for ISHH using ribonucleotide probes (riboprobes), as such protocols normally result in higher hybridization signals with greater signal-to-noise ratio. Although the shorter oligonucleotide probes should have higher tissue penetrance than the riboprobes, the resulting hybridization signal is still much lower than with the riboprobe (see **Fig. 1**), most likely due to the higher amount of radiolabeled nucleotides per hybridized mRNA molecule obtained with the riboprobes. In addition, the high stability of the RNA–RNA (as compared to RNA–DNA and DNA–DNA) duplexes allow for increased stringency conditions (e.g., increased temperature and low salt concentration) to be used during the hybridization and post-hybridization washes which result in much lower nonspecific background signals. Moreover, the use of riboprobes has thus far proven better than use of oligonucleotides for the ISHH studies of the human brain, which normally have higher background signals (due in part to the longer postmortem intervals than animal studies). Nevertheless, oligonucleotides are still useful, especially for highly abundant mRNAs *(15)*. There are a number of ISHH riboprobe procedures that have been published in regard to visualizing mRNA expression levels in brain sections, but the basic hybridization procedures described in the following have been based on riboprobe preparations described originally by Young *(15,16)* and Whitfield et al. *(17)*.

3.1. Brain Tissue Preparation and Sectioning

Immediately dissect out the animal brain specimens that will be used for the ISHH analysis (*see* **Note 1**). Freeze the brains by immersion into dry ice-cooled (–32°C to –40°C) isopentane for 30 s (*see* **Note 2**). Store the brains at –30 to –80°C (preferably with desiccant) in a sealed container if the brains will not be immediately sectioned that day. For human studies, the brain specimens should be cut into blocks of tissue approx 1.5 cm thick of the brain areas (e.g., striatum) of interest. Freeze the brain blocks by immersion into dry ice cooled (–32°C to –40°C) isopentane for 1 min and store frozen (with dessicant) at 80°C until use.

In preparation for cutting the brain sections, allow the brain to acclimate to the cyrostat temperature (–18 to –20°C) for at least 1 h. Mount the brain onto the cryostat pedestal on dry ice with embedding matrix in the orientation at which the brain will be sectioned (e.g., coronal). Cut the sections (12–20 μm thick; *see* **Note 3**), mount onto superfrost microscope glass slide (*see* **Note 4**), dry for approx 1 min on a slide warmer plate (30°C), and place the slide into a slide box maintained at least –18°C. For the rat brain, serial sections can be taken at only one level of the nucleus accumbens (e.g., 1.7 posterior to Bregma *[18]* or throughout the rostrocaudal extent of the striatum; sections from four to six different striatal levels can be placed on the same microscope slide). Store the section-mounted glass slides at –30 to – 80°C with desiccant until the ISHH experiment is performed (*see* **Note 5**).

3.2. Brain Section Pretreatment

All solutions (including the ethanol) should be made with autoclaved DEPC-treated distilled water. All containers should also be treated with DEPC for all the prehybridization steps. The prehybridization conditions should be carried out at room temperature, preferably under a fume hood (*see* **Notes 6 and 7**).

1. Place slides in a slide holder and bring to room temperature (tissue retention to the slides during the subsequent hybridization procedures can be improved by incubating the slides at 37°C for about 10 min).
2. Fix the tissue by immersing the slides in 4% paraformaldehyde for 5 min, and rinse slides twice in 1X PBS.
3. Acetylation of the tissue: Rinse the slides in 0.1 *M* TEA–0.9% saline, pH 8.0. During this rinse make a fresh solution of 0.25% acetic anhydride in 0.1 *M* TEA–0.9% saline, pH 8.0. Incubate the sections in this solution for 10 min and rinse subsequently in 2X SSC.

4. Dehydrate the sections in a series of escalating ethanols: 70% (1 min), 80% (1 min), 95% (2 min), and 100% (1 min), and delipidate in chloroform for 5 min.
5. Rehydrate slightly in 100% and 95% ethanol for 1 min each, and air-dry by standing the slides vertically in a slide holder.

The sections can be used immediately for hybridization or stored frozen with desiccant (see **Note 8**).

3.3. Probe Preparation

^{35}S-labeled riboprobes can be synthesized from (1) cDNA inserts into transcription plasmid vectors (e.g., PGEM4Z; the transcription vector has to be linearized prior to probe labeling with suitable restriction enzymes) or (2) cDNA templates obtained from polymerase chain reaction (PCR) with 5′ extension containing a promoter sequence (T7 and T3 are generally best). For example, good results have been obtained from riboprobes derived from the rat preprodynorphin synthesized from the cDNA fragment of the gene (bases 466–1101) *(19)* and for the human preprodynorphin, a 1.2 kb of exon 4 or a 478 fragment (bases 11–488) *(20)*. The following procedure is used for high specific activity ^{35}S riboprobe labeling (see **Note 9**).

1. Thaw the following components on ice (see **Note 10**) and add together in a sterile eppendorf tube at room temperature: 6 µL of 5X transcription buffer, 3 µL of 100 m*M* DTT, 5.5 µL of 2.5 m*M* ΔUTP, 1 µL of 1µg/µL linearized plasmid vector or PCR product, 12 µL of [^{35}S]UTP (12.5 µCi/µL) (see **Notes 11** and **12**), 0.6 µL of RNase inhibitor (40 U/µL), 1 µL of RNA polymerase (e.g., T3 or T7), and 0.9 mL of DEPC H$_2$O.
2. Incubate for 60 min at 37°C.
3. Add DEPC H$_2$O to 50 µL.
4. Add 0.5 µL of RNase inhibitor (40 U/µL) and 1 µL of DNase (1 U/µg).
5. Incubate 37°C for 10 min.
6. Bring the volume up to the capacity of the spin column with DEPC H$_2$O.
7. Column purify according to the manufacturer's protocol and add 1.0 µL of 5 *M* DTT (fresh).
8. Count 1 µL of probe on a scintillation counter.

3.4. Hybridization

1. Calculate the total amount of probe and hybridization buffer needed for the experiment (see **Notes 13** and **14**). The amount of buffer should relate to the area to be covered (i.e., the coverslip area). Typically, a volume of 0.1 µL/mm^2 is used. Thus, for a coverslip of 24 × 40 mm, a total hybridization solution volume of approx 96 µL is place over the slide. The probe concentration generally depends on the level of the mRNA expression and a concentration curve can be carried out to determine the optimal probe concentration. For many moderate to highly

expressed transcripts such as preprodynorphin, a concentration of 2×10^3 cpm-labeled probe/mm^2 is used.

$$\text{(desired cpm/slide)} / \text{(actual cpm)} \times \text{(no. of slides)} = \mu\text{L of probe}$$

$$(\mu\text{L/slide}) \times \text{(no. of slides)} = \text{total hybridization solution}$$

$$\text{formamide volume} = 50\% \text{ hybridization solution}$$

$$5\ M \text{ DTT volume} = 4\% \text{ of the total hybridization buffer}$$

$$\text{(total hybridization buffer)} - (\mu\text{L of probe}) - \text{formamide} - \text{DTT} = \mu\text{L of hybridization buffer}$$

2. Heat the probe and 2X hybridization buffer separately for 10 min at 90°C.
3. Cool immediately on ice for 5 min.
4. Prepare the heat-denaturated hybridization solution in a fresh tube: add formamide, DTT, hybridization buffer, and finally the labeled probe (*see* **Note 15**).
5. Mix the hybridization solution thoroughly and let stand for a few minutes so that any bubbles can rise to the surface.
6. Apply the hybridization solution to the tissue.
7. Slowly coverslip so that the solution covers all the tissue.
8. Place the slides in an incubation tray with filter paper saturated with 4× SSC–50% formamide in the bottom (the slides should not be in direct contact with the filter paper) and a cap filled with the same SSC–formamide solution to maintain humidity during incubation. Cover the tray and seal. Place the trays in an incubator at 55–65°C overnight (approx 16 h; *see* **Note 16**).

3.5. Post-Hybridization

No DEPC-treated solutions or dishes are needed for the post-hybridization procedure (*see* **Note 17**). Fill a glass dish container with 2X SSC containing 1 mM DTT at room temperature. Remove the incubation tray from the incubator. Carry out all post-hybridization washes under a fume hood. Allow the coverslip to slide off and place the slides in the container with the 2X SSC–DTT solution. If there is resistance, then dip the slide in a separate container with 2X SSC–1 mM DTT solution and let the coverslip float off.

1. Rinse the slides for 10 min at 37°C in RNase A buffer.
2. Incubate the slides with 40 µg/mL of RNase A in RNase A buffer for 30 min at 37°C (*see* **Note 18**).
3. Process through a series of SSC washes gradually desalting (increasing stringency) at room temperature: 2X SSC–1 mM DTT (5 min), 2X SSC–1 mM DTT (5 min), 1X SSC–1 mM DTT (10 min), and 0.5X SSC–1 mM DTT (10 min). If increased stringency is required, the following wash can be included at this step: 0.5X SSC–50% formamide–1 mM DTT (60 min at 48°C), 0.1X SSC–1 mM DTT (60 min at 53°C), and 0.1X SSC–1 mM DTT (1 min at room temperature).

4. Rinse in a series of ethanol washes with the ammonium acetate (*see* **Note 19**): 50% EtOH–300 mM NH$_4$Ac (1 min), 70% EtOH–300 mM NH$_4$Ac (1 min), 90% EtOH–300 mM NH$_4$Ac (1 min), 95% EtOH–300 mM NH$_4$Ac (1 min), and 100% EtOH (1 min).
5. Air-dry the slides at room temperature by standing them upright in a slide holder rack.

3.6. Autoradiography

3.6.1. Film Autoradiography

Place slides in film cassettes and appose to Biomax MR film for days to weeks (depending on the intensity of the hybridization signal such that under optimal conditions the signal is robust, but not saturated; for example, approx 7 and 14 d for abundantly expressed mRNAs such as the preprodynorphin in the rat and human striatum, respectively). A standard (rat brain pastes with varied known concentrations of ^{35}S activity or polymer microscale standard composed of known radioactivity values should also be placed in the cassette. Develop the films in dark room (under safelight conditions) for 5 min in Kodak D-19 at room temperature. Rinse subsequently in short-stop (1.5% acetic acid in deionized water), fix for 2.5 min in Kodak Rapid Fix without hardener, and wash for at least 20 min under running tap water. Air-dry the films.

3.6.2. Microscopic Autoradiography

For microscopic analysis of the autoradiographic signal, the hybridization sections should be coated in Kodak NTB-2 nuclear track emulsion (diluted 50:50 with dH$_2$O) in a dark room.

3.6.2.1. Emulsion Slide Dipping

1. Set water bath at 42°C.
2. Making 50 mL of the nuclear track emulsion: Fill a plastic vial (e.g., 50-mL Falcon) or elongated glass jar with 25 mL of 600 mM ammonium acetate. Add (using a plastic spoon) the emulsion into the vial until the fluid is displaced to 50 mL.
3. Melt the emulsion by placing the vial in the water bath (*see* **Note 20**).
4. Let it stand for at least 30 min, but stir (slowly) periodically with a plastic spoon or clean blank microscope slide to disperse any bubbles.
5. Using a clean glass slide, check that there are no bubbles. When all bubbles are eliminated, then the experimental slides can be dipped.
6. Dip slides slowly, wipe the emulsion off the back of the slide with a Kimwipe, and place the slide vertically in a test tube rack.
7. Let the slides dry at room temperature for at least 2 h.
8. Transfer slides to slide boxes, seal well (place slide box in black bag that is tightly wrapped), and store at 4°C or –20°C (to reduce high background of intense signals).

Control

Cocaine user

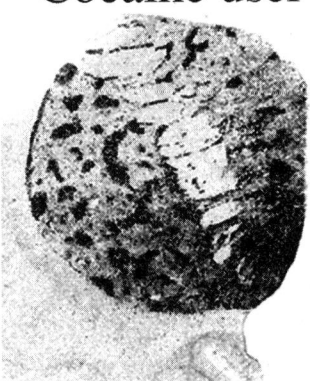

Fig. 2. Film autoradiograms showing the preprodynorphin mRNA expression in the striatum of human controls and cocaine users. Note the heterogeneous expression pattern of the hybridization signal with higher expression in the patch compartment and higher mRNA expression levels in association with cocaine use.

3.6.2.2. DEVELOPING THE EMULSION-COATED SLIDES

Develop slides after approx three to four times the film exposure period that resulted in a strong signal.

1. Let the slide boxes warm in room temperature while still sealed in the darkroom.
2. Place the slides into stainless steel racks.
3. Chill the developing solution on ice to 16°C.
4. Remove the developing solutions from the ice and process the slides while the solution is at this temperature.
5. Develop the slides in D-19 for 2 min. Gently agitate rack every 30 s.
6. Transfer to distilled water for 15 s.
7. Transfer to Rapid Fix without hardener for 2 min; agitate every 30 s.
8. Rinse slides gently under running tap water for 20 min.
9. Counterstain: Cresyl violet for 2 min (5 min for human) followed by sequential immersion in 50%, 75%, 90%, and 100% ethanol and xylene, and coverslip with Pertex mounting medium.

3.6.3. Film Quantification

The ISHH technique cannot be used to determine the absolute amount of the mRNA levels. Only information regarding the relative expression levels within neuronal populations can be obtained from the film autoradiographic signals (*see* **Fig. 2**) and the cellular analysis of the silver grains (*see* **Fig. 3**). Film images can be captured for analysis by either a video camera (in conjunc-

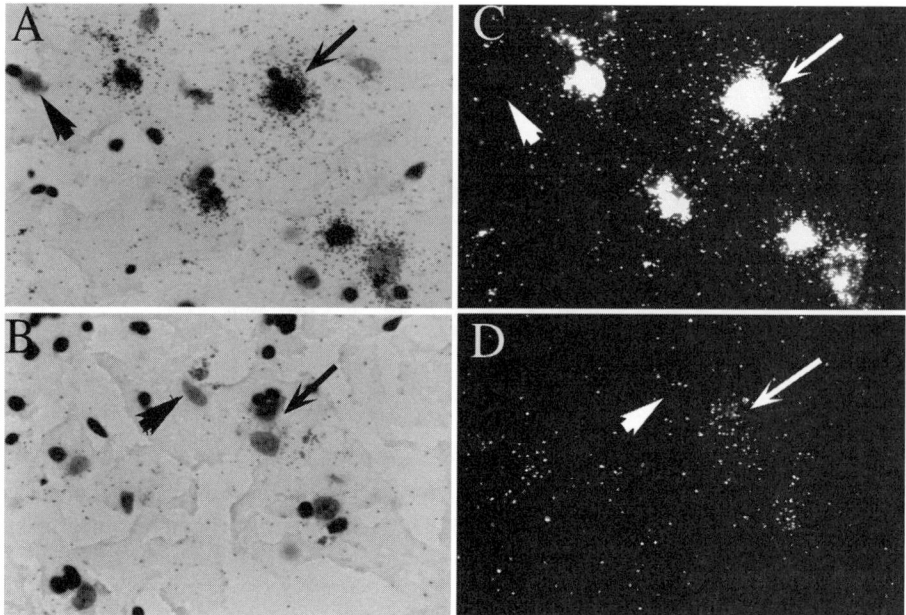

Fig. 3. High-magnification images of the human striatum taken under bright-field (**A**, **C**) and episcopic (**B**, **D**) illumination. Panels **A** and **B** show cells within the patch compartment and panels **C** and **D** show cells within the matrix compartment. White grain clusters (episcopic illumination) identify the preprodynorphin mRNA expressing cells. Note the intense expression of the preprodynorphin mRNA in cells within the patch compartment as compared to the matrix. *Filled arrows* point to examples of positively labeled cells and *arrowheads* to nonlabeled cells.

tion with a light board) or flatbed scanner. The scanners have proven more reliable than light boards for capturing the film images. Normally a scanning resolution of 300–400 dots per inch is suitable for even rat brain images.

Scan in film images with settings such that the pixel values are within the normal gray scale (0–255). Save the file as TIFF file images. The TIFF images can be opened in various image programs to determine the optical density light transmittance levels. Freeware computer programs, for example, NIH IMAGE by Wayne Rasband (http://rsb.info.nih.gov/nih-image), can be downloaded to Mac computers. There is a corresponding free PC version of the IMAGE program, Scion Image, available through Scion Corporation (http://www.scioncorp.com). The MCID (Imaging Research, Ontario, Canada) computer analysis system is also used by many research autoradiography groups. Delineate the region of interest using the computer mouse and measure the density. The measurements can be copied to other programs for data handling and statistical analysis. The best measurable hybridization signals should be above background, but not saturated (values over 200 can lead to problems; film reexposure is then necessary to bring values into correct transmittance range). The standards should also be measured and used to make a standard curve, converting the light transmittance to dpm/mg values (normally Rodbard curve in IMAGE provides the best curve fit for the values). The standard values used to generate the standard curve should only cover the range (just over the highest number) of the experimental values. Subtract background (e.g., white matter tract) from the measurements taken in the brain area of interest.

3.6.4. Cellular Analysis

Counting silver grains can help to determine the cellular localization of the film autoradiographic signals and to assess whether alterations revealed by the film analysis are due to a general increase (or decrease) of the signal in all cells expressing the transcript of interest, or whether there is a differential response within discrete cell populations. Silver grains can be manually counted over individual neurons within a specified area under a microscope, or computer-assisted grain counting can be carried out where the relative grain density visualized in the microscope are converted to a digital pixel image *(21)*.

3.7. Controls

The specificity of the antisense riboprobe (complementary sequence to the mRNA transcript) of interest should be assessed with controls such as:

1. The use of a sense riboprobe (same sequence as the mRNA transcript), which should normally result in a homogeneous signal similar to background levels (*see* **Fig. 4**).
2. The use of a different antisense riboprobe of the same size and approximate guanosine/cytosine (GC) content, which should result in a heterogeneous signal that shows a distinct anatomical pattern to the antisense riboprobe of interest.
3. The use of other riboprobes targeted against the same transcript, but at different nonoverlapping regions of the RNA, which should result in the same hybridization pattern.

4. Notes

4.1. Brain Tissue Preparation and Sectioning

1. Gloves should be used at all time to decrease RNase contamination and for protection against the radioisotope.
2. Over-freezing (< −45°C isopentane) of the brain specimens can cause tissue cracks that can make it difficult to get good brain sections when cutting in the cryostat.

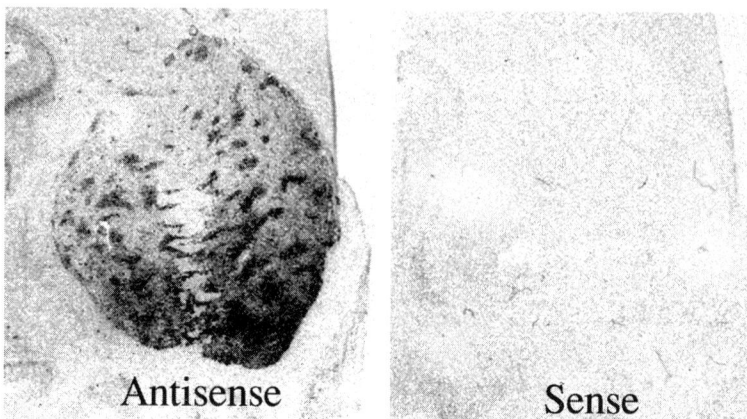

Fig. 4. *In situ* hybridization signal obtained using an antisense and sense riboprobe against the human preprodynorphin gene. Note the absence of specific hybridization signal with the use of the sense control probe.

3. Whole hemisphere human brain cryosections can be cut even up to 100 μm thick with good hybridization results *(22)*. However, these large sections are not appropriate for discrete microscopic evaluation of the hybridization signal.
4. Superfrost and poly-L-lysine coated microscope slides are better than gelatin-subbed slides for very high stringency ISHH conditions *(23)*. However, the gelatin-subbed slides can work perfectly well under normal hybridization conditions and is easily made in the laboratory *(17)*. Poly-L-lysine-coated slides can also be prepared in the lab, but it has been more cost effective and more reliable to buy precleaned coated slides that are immediately ready for use. Both poly-L-lysine and Superfrost slides have proven very reliable, but Superfrost slides are also normally available in large sizes and thus very useful for processing the human brain sections that often require high stringency conditions *(24)*.
5. Slide-mounted brain sections can be stored frozen for a long period of time. We have successfully used brain sections frozen up to even 8 yr and have obtained good levels of hybridization signals. The hybridization signal appear to be most susceptible to repeated freeze–thawing. Thus, slides should always be sorted in the cold (e.g., in the cryostat or on dry ice). Storage with desiccant also helps to prevent moisture buildup.

4.2. Brain Section Pretreatment

6. Prefixation of the tissue is important to preserve the tissue morphology, prevent further degradation of the RNA, and enhance penetration of the probe. The prehybridization step is also equally critical for decreasing background signal as other phases of the ISHH procedure. For example, improper autoclaving of the DEPC (especially if stored in plastic bottles) and impure chloroform can lead

to high background levels. Acetylation of the tissue is important for reducing background signals (perhaps due to reducing the electrostatic binding of the probe to basic [positively charged] amino groups on the brain section). It is critical that the acetic anhydride is fresh and added to the TEA buffer just prior to use.

7. Prefixation of tissue with Proteinase K, which can improve penetrance of the probe, has not proven to enhance the signal in the freshly frozen brain sections processed under the current fixation procedure. However, when testing new probes for which the hybridization signal is weak, experiments can be carried out to determine whether or not the Proteinase K pretreatment (5–10 mg/L of Proteinase K in 100 mM Tris, pH 8; 50 mM EDTA; 30 min at 37°C) can improve the signal. Note that increasing the Proteinase K concentration will damage the tissue, so test experiments should be carried out to determine the best concentration for optimized hybridization signal with minimal tissue damage.
8. Pretreated brain sections can be stored (at least –30°C) for a number of days (brain specimens have even been used up to 6 mo) prior to the hybridization procedure without significant loss of mRNA expression levels.

4.3. Probe Preparation

9. The size of the labeled probe can be determined on an acrylamide gel. High specific labeling can result in smaller bands on the gels in addition to the major band at the size of the cDNA insert. The synthesis of full-length riboprobes is obtained mainly with low specific labeling in which unlabeled UTP is added during the probe labeling. Riboprobes >1000 nucleotides can be used, but alkaline hydroxylase (especially for those >1500 bases) is recommended after the probe labeling to ensure adequate penetrance of the resulting shorter fragments.
10. It is always best to use freshly prepared labeled probes. However, labeled probes (stored at –30°C) can be used a few days later (generally a week), but some probe degradation can occur, resulting in increased background. When possible, all reagents and chemicals including the isotopes should be stored frozen in aliquots (to prevent freeze–thawing problems and to decrease the possibility of contamination).
11. Although ^{35}S is the most common isotope used for riboprobe labeling, ^{33}P has proven very effective for visualizing low expressing transcript (as well as those with normal expression levels) with minimal background signal. For example, opioid receptor mRNAs are normally much lower than the opioid neuropeptide genes and the use of ^{33}P has proven much more effective for visualization of the receptor mRNAs *(25)*. However, the half-life of ^{33}P is only 25 d, so this isotope is not suitable when cellular visualization of very low expressing transcripts that require long exposure of the emulsion-dipped slides.
12. The synthesis of CTP-labeled riboprobes works equally well for ISHH experiments, and CTP- and UTP-labeled probes can even be combined when attempting to increase the hybridization signal, for example, very low expressing transcripts. However, increasing the concentration of the hybridization probes can also increase background nonspecific signal, so it is important to determine the

probe concentration at which high saturated specific signals is obtained with low background noise.

4.4. Hybridization

13. A critical consideration of ISHH is stringency, which refers to the condition favoring dissociation of nucleic acid duplexes; specific RNA–RNA duplexes will withstand high stringency conditions as compared to nonspecific hybrids. Increased stringency can be obtained, for example, by increased hybridization temperature, increased formamide concentration, and decreased salt concentration. Most probes hybridize specifically to their target mRNAs at a T_m (melting temperature) of − 25°C. Increasing the hybridization temperature can increase tissue damage; thus it is recommended to mount tissue on Superfrost or poly-L-lysine glass slides. Inclusion of formamide (destabilizes nucleic acid duplexes, thus lowering the effective T_m) allows the hybridization to be carried out at lower temperatures. Only fresh formamide should be used for the hybridization solution. Moreover, uneven hybridization signal can be present if the SSC–formamide solution in the hybridization tray is not fresh.
14. Denhardt's solution and dextran sulfate are important components of the hybridization buffer. Denhardt's solution, which consists of Ficoll, bovine serum albumin, and polyvinylpyrrolidone, helps to decrease nonspecific binding to proteins and nucleic acids. The anionic polymer dextran sulfate greatly enhances the apparent rate of hybridization, thus leading to higher hybridization signals. Only DEPC-treated distilled water should be used in the hybridization buffer.
15. The addition of DTT is important for ^{35}S-labeled riboprobes, as it helps to prevent formation of the disulfide bond of the ^{35}S to uracil. Although ^{33}P does not have a sulfide bond, we have found that the inclusion of DTT is still beneficial for decreasing background signal. DTT should be weighed under a fume hood and as such it is best to make aliquots of the DTT and store in a refrigerator, adding DEPC-treated distilled water only when needed for each experiment.
16. If there are many hybridization trays because of high processing of slides, then immediately put each tray into the incubator after it is filled with the coverslipped slides before completing the next set of trays. This will avoid hybridization at low stringency (room temperature) for the sections in trays that were completed first.

4.5. Post-Hybridization

17. It is important to have the post-hybridization solutions at the proper temperature before beginning the post-hybridization washes.
18. The RNase A treatment is important because it will digest single-stranded RNA, essentially removing non- or poorly hybridized labeled probe, leaving the double-stranded hybridized pairs intact. Sometimes RNase T1 (1 U/mL) can also be added to the RNase A post-hybridization incubation when the signal-to-noise ratio for low expressing transcripts is still not optimal. The RNase T1 digests portions of the single-stranded RNA that are not digested by RNase A.

All containers exposed to the RNase enzymes should be kept separate from those used during all other hybridization procedures or treated with 0.1 M NaOH to inactivate the RNase enzymes. Treat the glassware by letting it stand for a few minutes with 0.1 M NaOH and rinse thoroughly.
19. The inclusion of ammonium acetate in the ethanol washes protects against denaturation of the *in situ* hybrids, which is particularly important for brain sections in which high magnification analysis will be performed.

4.6. Autoradiography

20. Aliquots of the nuclear emulsion–ammonium acetate solution should be made to avoid repeated freeze–thawing of the solution. It is critical that the emulsion is devoid of air bubbles prior to dipping the experimental slides. Once dipped, the slides should be allowed to dry very slowly (only at room temperature), as rapid drying can distort the emulsion layer.

Acknowledgments

I am grateful to the past and present members of my laboratory who have contributed to the development of the ISHH procedure. I thank Pernilla Fagergren for critical review of the manuscript and Barbro Berthelsson for technical support in carrying out the ISHH experiments. This work was supported by the National Institutes of Health (DA08914 and DA12030), and the Swedish Research Council (11252).

References

1. Björklund, A. and Lindvall, O. (1984) Dopamine-containing systems in the CNS, in *Handbook of Chemical Neuroanatomy*, Vol. 2: *Classical Transmitters in the CNS*, Part I (Björklund, A. and Hökfelt, T., eds.), Elsevier, Amsterdam, pp. 55–122.
2. Le Moine, C., Normand, E., and Bloch, B. (1991) Phenotypical characterization of the rat striatal neurons expressing the D1 dopamine receptor gene. *Proc. Natl. Acad. Sci. USA* **88,** 4205–4209.
3. Gerfen, C. R., Enber, T. M., Susel, Z., et al. (1990) D1 and D2 dopamine receptor regulated gene expression of striatonigral and striatopallidal neurons. *Science* **250,** 1429–1432.
4. Le Moine, C., Normand, E., Guitteny, A. F., Fouque, B., Teoule, R., and Bloch, B. (1990) Dopamine receptor gene expression by enkephalin neurons in rat forebrain. *Proc. Natl. Acad. Sci. USA* **87,** 230–234.
5. Curran, E. J. and Watson, S. J. (1995) Dopamine receptor mRNA expression patterns by opioid peptide cells in the nucleus accumbens of the rat: a double in situ hybridization study. *J. Comp. Neurol.* **361,** 57–76.
6. Le Moine, C. and Bloch, B. (1996) Expression of the D3 dopamine receptor in peptidergic neurons of the nucleus accumbens: comparison with the D1 and D2 dopamine receptors. *Neuroscience* **73,** 131–143.

7. Hurd, Y. L. and Herkenham, M. (1993) Molecular alterations in the neostriatum of human cocaine addicts. *Synapse* **13,** 357–369.
8. Hurd, Y. L., Brown, E., Finlay, J., Fibiger, H. C., and Gerfen, C. (1992) Cocaine self-administration differentially alters mRNA expression of striatal peptides. *Mol. Brain Res* **13,** 165–170.
9. Daunais, J. B., Roberts, D. C. S., and McGinty, J. F. (1993) Cocaine self-administration increases prodynorphin, but not c-fos, mRNA in rat striatum. *NeuroReport* **4,** 543–546.
10. Spangler, R., Unterwald, E. M., and Kreek, M. J. (1993) 'Binge' cocaine administration induces a sustained increase of prodynorphin mRNA in rat caudate-putamen. *Mol. Brain Res.* **19,** 323–327.
11. Steiner, H. and Gerfen, C. (1993) Cocaine-induced c-fos messenger RNA is inversely related to dynorphin expression in the striatum. *J. Neurosci.* **13,** 5066–5081.
12. Mathieu-Kia, A. M. and Besson, M. J. (1998) Repeated administration of cocaine, nicotine and ethanol: effects on preprodynorphin, preprotachykinin A and preproenkephalin mRNA expression in the dorsal and the ventral striatum of the rat. *Mol. Brain Res.* **54,** 141–151.
13. Wang, J. Q. and McGinty, J. F. (1995) Dose-dependent alteration in zif/268 and preprodynorphin mRNA expression induced by amphetamine or methamphetamine in rat forebrain. *J. Pharmacol. Exp. Ther.* **273,** 909–917.
14. Bronstein, D. M., Pennypacker, K. R., Lee, H., and Hong, J. S. (1996) Methamphetamine-induced changes in AP-1 binding and dynorphin in the striatum: correlated, not causally related events? *Biol Signals* **5,** 317–333.
15. Young, W. S., III (1989) In situ hybridization histochemical detection of neuropeptide mRNA using DNA and RNA probes. *Methods Enzymol.* **168,** 702–761.
16. Young, W. S., III (1990) In situ hybridization histochemistry, in *Handbook of Chemical Neuroanatomy*, Vol. 8, (Björklund, A., Hökfelt, T., Wouterlod, F., and Van Den Pol, A. N., eds.), Elsevier, Amsterdam, pp. 481–511.
17. Whitfield, H. J., Brady, L. S., Smith, M. A., Mamalaki, E., Fox, R. J., and Herkenham, M. (1990) Optimization of cRNA probe in situ hybridization methodology for localization of glucoroticoid receptor mRNA in rat brain: a detailed protocol. *Cell. Mol. Neurobiol.* **10,** 145–157.
18. Paxinos, G. and Watson, C. (1986) *The Rat Brain in Stereotaxic Coordinates*, 2nd edit. Academic Press, New York.
19. Civelli, O., Douglas, J., Goldstein, A., and Herbert, E. (1985) Sequence and expression of the rat prodynorphin gene. *Proc. Natl. Acad. Sci. USA* **82,** 4291–4295.
20. Horikawa, S., Takai, T., Toyosato, M., et al. (1983) Isolation and structural organization of the human preproenkephalin B gene. *Nature* **306,** 611–614.
21. Chesselet, M.-F. and Weiss-Wunder, L. T. (1994) Quantification of in situ hybridization histochemistry, in *In Situ Hybridization in Neurobiology* (Eberwine, J. H., Valention, K. L., and Barchas, J. D., eds.), Oxford University Press, New York, pp. 114–123.

22. Hurd, Y. L. (1996) Differential messenger RNA expression of prodynorphin and proenkephalin in the human brain. *Neuroscience* **72,** 767–783.
23. Wilcox, J. N. (1993) Fundamental principles of in situ hybridization. *J. Histochem. Cytochem.* **41,** 1725–1733.
24. Österlund, M. K., Keller, E., and Hurd, Y. L. (2000) The human amygdaloid complex is characterized by high expression of the estrogen receptor a mRNA. *Neuroscience* **95,** 333–342.
25. Peckys, D. and Landwehrmeyer, G. B. (1999) Expression of mu, kappa, and delta opioid receptor messenger RNA in the human CNS: a 33P in situ hybridization study. *Neuroscience* **88,** 1093–1135.

9

Combining *In Situ* Hybridization with Retrograde Tracing and Immunohistochemistry for Phenotypic Characterization of Individual Neurons

Catherine Le Moine

1. Introduction

In situ hybridization, using radioactive and nonradioactive probes, is at present, a widespread technique that has been developed and improved over the past 15 yr, and has become an extremely powerful tool for the study of cellular interactions through gene expression and gene regulation. Progress in sensitivity, the wide range of available markers, and the development of image analysis systems for quantification, especially at the cellular level, have made this approach particularly useful for the investigation of the cellular interactions and regulations of many heterocellular systems, especially the central nervous system *(1–4)*.

If the distribution of an mRNA is a first and necessary step for the anatomical and cellular analysis of its expression, the possibility of a phenotypical characterization of neurons expressing a given gene by using double labelings is one of the main reasons for interest in *in situ* hybridization. Various strategies have been developed, not only for double *in situ* hybridization, but also for coupling *in situ* hybridization techniques with retrograde tracing and immunohistochemistry *(5–8)*. These strategies have been very useful from various aspects, particularly the identification of neuronal populations expressing many neuroreceptors or expressing neuronal markers following specific stimulations *(9–13)*. This chapter describes different possibilities for the coupling of *in situ* hybridization with retrograde tracing and immunohistochemistry, as well as the troubleshootings inherent in these couplings. Some

From: *Methods in Molecular Medicine, vol. 79: Drugs of Abuse: Neurological Reviews and Protocols*
Edited by: J. Q. Wang © Humana Press Inc., Totowa, NJ

protocols are illustrated with examples for the detection of dopamine receptor subtypes or Fos expression in identified neurons of the striatum and the cortex.

2. Materials

1. Cryostat sections of 1–4% paraformaldehyde (PFA)-perfused rat brain for coupling retrograde tracing with *in situ* hybridization (10–15 μm thick). PFA is prepared in 0.1 M phosphate buffer, pH 7.2.
2. Vibratome sections (30–50 μm) or free-floating cryostat sections (12–15 μm) from 1–4% PFA-perfused rat brain.
3. FluoroGold (Interchim) for retrograde tracing.
4. ^{35}S-labeled nucleotides (PerkinElmer Life Sciences).
5. Probes: Synthetic oligonucleotides, cDNA clones, or cRNA.
6. Materials needed for standard *in situ* hybridization protocols using radioactive probes including kits for probe labeling (tailing for oligonucleotides, nick translation for cDNA, in vitro transcription for cRNA), RNase-free glassware, diethyl pyrocarbonate water (DEPC)-treated buffers, humidified incubator and water bath, 8X saline sodium citrate (SSC—1.2 M sodium chloride, 0.12 M trisodium citrate), absolute ethanol.
7. Hybridization buffer for oligonucleotide and cDNA probes: 2X SSC, 50% formamide, 500 μg/mL of salmon sperm DNA, 1% Denhardt's solution, 5% sarcosyl, 250 μg/mL of yeast tRNA, 200 mM dithiothreitol (DTT), and 20 mM NaH$_2$PO$_4$.
8. Hybridization buffer for cRNA probes: 20 mM Tris-HCl, pH 7.0, 1 mM EDTA; 300 mM NaCl, 50% formamide, 10% dextran sulfate, 1% Denhardt's solution, 250 μg/mL of yeast tRNA, 100 μg/mL of salmon sperm DNA, 100 mM DTT, 0.1% sodium dodecyl sulfate (SDS), and 0.1% sodium thiosulfate.
9. RNase A (10 mg/mL aliquots) for the post-hybridization with cRNA probes.
10. X-ray films (Kodak BIOMAX MR) and photographic emulsion (Ilford K5).
11. Developer and fixative for autoradiography, and material for staining (toluidine blue or Mayer's hemalun eosin).
12. Materials needed for immunohistochemistry: Rabbit polyclonal anti-Fos antibody (Santa Cruz Biotechnology), biotinylated sheep anti-rabbit secondary antibody (Amersham), avidin–biotin complex (Vector ABC Elite), standard phosphate-buffered saline (PBS) or Tris-buffered saline (TBS), Triton X-100, diaminobenzidine (DAB), and normal goat serum (NGS).
13. Microscope fitted out with epipolarization under fluorescence.

3. Methods

This section first outlines the various protocols that can be used for *in situ* hybridization, and then describes different procedures for coupling *in situ* hybridization with retrograde tracing and immunohistochemistry.

3.1. Tissue Preparation

For the coupling of *in situ* hybridization with retrograde tracing or immunohistochemistry, the protocols described here need to be performed on fixed tissue.

1. Perfuse rat brains intracardiacally with 30 mL of 0.9% NaCl, then with 250 mL of 1–4% PFA in 0.1 M phosphate buffer.
2. Proceed for cryostat or vibratome sections as described in the following corresponding methodological sections.

3.2. Labeling of the Probes

All the probes are labeled using standard protocols as originally described *(10,14)*. Oligonucleotide probes are labeled with [^{35}S]dATP by "tailing," that is, addition of nucleotides at the 3′ end. Nick translation is used to label cDNA probes by incorporation of the same radiolabeled deoxynucleotides. cRNA probes are synthesized by in vitro transcription from cDNA clones using [^{35}S]UTP from 50 ng of linearized plasmid and with the appropriate RNA polymerase (SP6, T7, or T3). For the choice of the probes and radiolabeled nucleotides, *see* **Note 1**.

3.3. In Situ *Hybridization*

Whatever the various protocols used, *in situ* hybridization is performed with the three usual steps of pretreatment, hybridization of the probe, and post-hybridization washes, with some specific steps depending on the type of probe used.

3.3.1. Pretreatment

1. Thaw the slides for 10 min at room temperature.
2. Wash twice for 30 min in 4X SSC–0.1% Denhardt's solution (only for oligonucleotides and cDNAs).
3. Wash the slides twice for 10 min in 4X SSC, then incubate for 10 min in 4X SSC–1.33% triethanolamine–0.25% acetic anhydride, pH 8.0. This step is useful to decrease the background by acetylation of free radicals.
4. Rinse twice 5 min in 4X SSC.
5. Dehydrate in graded ethanol (70%, 80%, 90%, 100%) and air-dry.

3.3.2. Hybridization

1. Microcentrifuge the labeled probe (usually stored in ethanol–acetate) for 1 h at maximum speed.
2. Remove and check the supernatant, and dry the pellet under vaccum.

3. Resuspend the pellet in 50 µL of 1X TE for oligonucleotides and cDNAs and 24 µL of 1X TE– 50 mM DTT for cRNA probes.
4. For oligonucleotide and cDNA probes, dilute the probes into 35–50 µL of the appropriate hybridization buffer to a concentration of 30 pg/µL (for cDNA probes, approx 150,000 cpm/slide) or 3 pg/µL (for oligonucleotide probes, approx 100,000–300,000 cpm/slide), and hybridize overnight at 40°C in a humidified chamber.
5. For cRNA probes, dilute the probe into 50 µL of the appropriate hybridization buffer to a concentration of 1–5 ng/slide ($0.5 \times 10^6 - 2 \times 10^6$ cpm/slide), and hybridize overnight at 55°C in a humidified chamber.

3.3.3. Post-Hybridization Washes for cDNA and Oligonucleotide Probes

1. Wash the slides briefly in cold 4X SSC.
2. Wash twice for 30 min in 1X SSC at room temperature under agitation.
3. Wash twice for 30 min in 1X SSC at 40°C under agitation.
4. For oligonucleotide probes, add an additional wash step of 30 min (twice) in 0.1X SSC at 40°C under agitation.
5. Dehydrate twice in 100% ethanol and air-dry.

3.3.4. Post-Hybridization Washes for cRNA Probes

1. Wash the slides twice 5 min in 4X SSC under agitation.
2. Wash for 15 min at 37°C in RNase A (20 µg/mL in 0.5 M NaCl–10 mM Tris-HCl, pH 7.0/0.25 mM EDTA).
3. Wash the slides again in 2X SSC (5 min, twice), 1X SSC (5 min), 0.5X SSC (5 min) at room temperature, and in 0.1X SSC at 65°C (30 min, twice) under agitation.
4. Place the slides at room temperature in 0.1× SSC to cool.
5. Dehydrate in graded ethanol (70%, 80%, 90%, 100%) and air-dry.

3.4. Coupling In Situ Hybridization with Retrograde Tracing

Coupling *in situ* hybridization with retrograde tracing is usually straightforward when using fluorescent tracers (e.g., FluoroGold) that can be visualized under the microscope without any additional step. In such a case, there is no interference in the modalities of detection of the different markers. Thus, as described in the following protocol, it appears to be the easiest way to perform coupling of the two techniques (e.g., *5,9,15*).

1. Retrograde tracer (FluoroGold, 4% in saline) is either pressure injected or iontophoretically injected (e.g., 8 µA, 10 s on/10 s off, over 20 min according to **ref. *15***) at the site of interest (*see* **Notes 2** and **3**).
2. Allow the animals to survive 5–7 d (time may vary according to specific conditions) before perfusion with 1–4% PFA in 0.1 M phosphate buffer (*see* **Note 4**).

… Phenotypic Double Labeling

3. Post-fix the brains in the same fixative for 1 h, then cryoprotect in 0.1 M phosphate buffer containing 15–30% sucrose, and freeze in isopentane at approx −40°C or into vapors of liquid nitrogen.
4. Cut and collect cryostat sections (10–15 µm) on gelatin-coated slides, and store at −80°C until used.
5. Perform *in situ* hybridization with radioactive probes (*see* **Subheading 3.3.**) with the appropriate probes (cDNAs, oligonucleotides, cRNA probes) (*see* **Note 4**).
6. Dry the sections, then dip into Ilford K5 emulsion (diluted 1:3 in 1X SSC).
7. Keep in the dark for up to 3 mo (according to the level of the mRNA to be detected), develop, and mount without counterstaining.
8. Neurons retrogradely labeled with FluoroGold are identified on a fluorescence microscope with an excitation filter (L420–Y455). Double labeling with *in situ* hybridization (silver grains) are analyzed by combining dark-field with epifluorescence and by varying the intensity of the transmitted light in dark-field (*see* **Fig. 1**).

3.5. Coupling In Situ Hybridization with Immunohistochemistry

The most commonly used method to couple *in situ* hybridization with immunohistochemistry is to perform *in situ* hybridization with radioactive probes on free-floating cryostat sections followed by the immunodetection and then the development of the autoradiographic signal (e.g., *8,16–18*). Various alternative experimental conditions may be used depending on the type and abundance of the markers to be detected, and some are also described here (*see* **Note 5**).

Two types of protocols are described here. One is based on the classical use of radioactive probes followed by immunohistochemistry on cryostat free-floating sections. This protocol has been used to detect dopamine receptor mRNAs in subpopulations of cortical interneurons together with the cytoplasmic proteins, calbindin and parvalbumin *(17)*. The other starts with the immunohistochemical detection of Fos proteins followed by *in situ* hybridization with oligonucleotide probes on vibratome sections, and is illustrated for the detection of Fos and neuropeptide mRNAs in the striatum to identify activated neurons after amphetamine *(13)*.

3.5.1. Combining In Situ Hybridization with Radioactive Probes and Immunohistochemistry on Cryostat Free-Floating Sections

1. Perfuse adult male rats through the heart with 250 mL of 4% PFA for 15 min (*see* **Note 6**).
2. Dissect out the brains and post-fix 1 h in the same fixative, cryoprotect in 15% sucrose–0.1 M phosphate buffer, and freeze over vapors of liquid nitrogen.
3. Cut 15–20-µm sections on a cryostat and collect them as free-floating in 4X SSC.

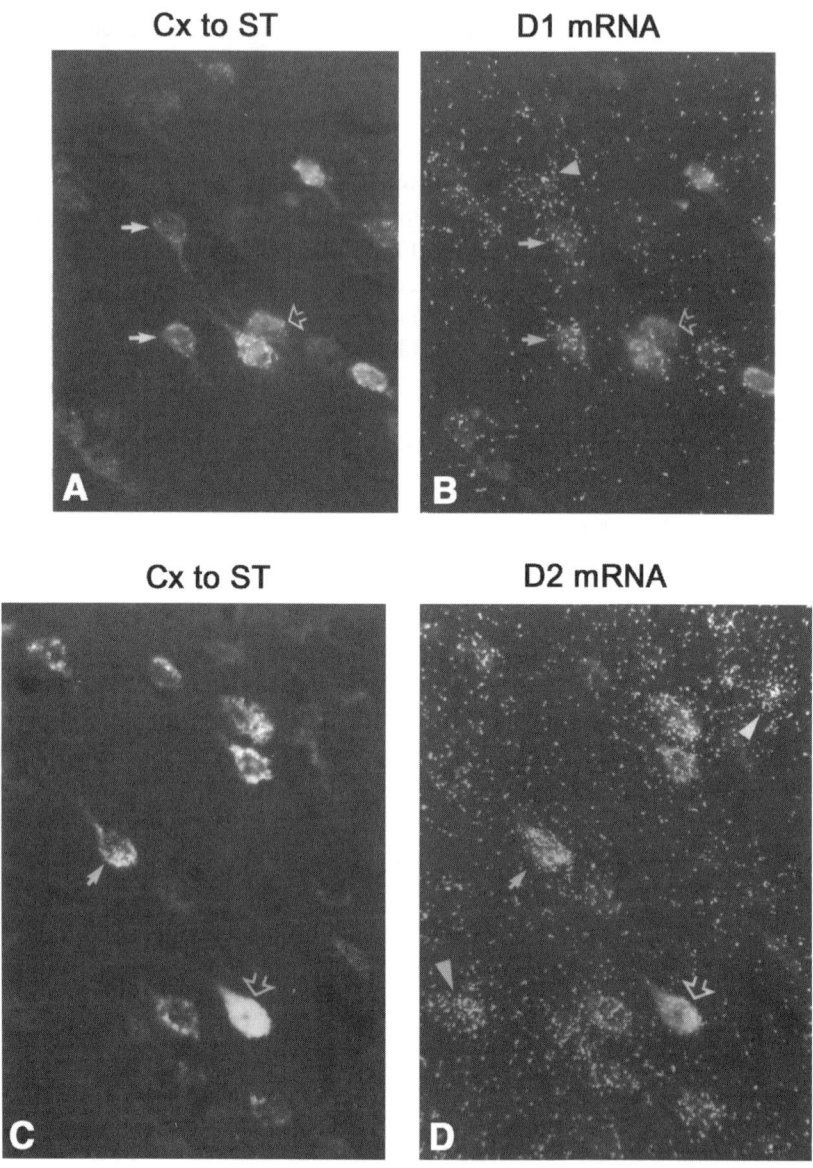

Fig. 1. Coupling *in situ* hybridization with retrograde tracing. Double labeling with FluoroGold and D1 or D2 dopamine receptor cRNA probes in the prefrontal cortex following FluoroGold (4% in saline) injection in the striatum. (**A, C**) FluoroGold-labeled neurons in the layer V of the prefrontal cortex. (**B, C**) show double-labeled neurons visualized by combining dark-field (silver grains) with epifluorescence as in **A** and **C** for FluoroGold. *White arrows* point to neurons containing both FluoroGold and D1 or D2 mRNAs in **B** and **D**, respectively. The *open arrows* point to neurons

4. Perform *in situ* hybridization immediately in a small Petri dish with ^{35}S-labeled probes (*see* **Note 7**).
5. Pretreat the sections as free-floating as described in **Subheading 3.3.1.** depending on the probes used but without dehydration.
6. After acetylation, rinse three times in 4X SSC.
7. Hybridize overnight at 55°C for cRNA probes and at 40°C for cDNA and oligonucleotide probes. Amounts of probes and volumes of hybridization buffer (500 µL–1 mL depending on the size of the Petri dish) need to be adapted according to the appropriate probe concentrations as described in **Subheading 3.3.2.**
8. For post-hybridization, wash the sections according to **Subheadings 3.3.3.** and **3.3.4.** depending on the type of probes.
9. Transfer the sections in 0.1X SSC at room temperature to cool, rinse three times in PBS, and proceed directly to immunohistochemistry.
10. Incubate the sections 24–72 h at 4°C with the primary antibody diluted in PBS–0.3% Triton. This step may be slightly modified according to specific conditions for the use of specific antibodies.
11. Reveal immunoreactivity with the streptavidin–biotin method: Biotinylated secondary antibody (1:200 in PBS for 1–2 h at room temperature) and avidin–biotin complex (1:200 in PBS for 1–2 h at room temperature).
12. Develop the signal in 0.05% DAB–0.01% hydrogen peroxide in TBS, pH 7.6, for 5 min (*see* **Note 8**).
13. Wash in TBS, then in distilled water.
14. Mount the sections on slides and allow to dry.
15. Autoradiographic *in situ* hybridization signal is then revealed by dipping the slides into Ilford K5 emulsion (diluted 1:3 in 1X SSC).
16. Expose for 2 up to 14 wk (according to mRNA levels), develop, and mount in Eukitt without counterstaining, or with slight counterstaining.
17. Data obtained by using such a protocol are illustrated in **Fig. 2**.

3.5.2. Combining Immunohistochemistry and In Situ Hybridization on Vibratome Sections

This protocol is based on the immunohistochemical detection of Fos proteins followed by *in situ* hybridization with oligonucleotide probes as originally

---

Fig. 1. *(continued)* labeled only with FluoroGold and the *arrowheads* to neurons containing D1 or D2 mRNAs without being retrogradely labeled. Please note that there is no interference or nonspecific labeling with this technique. Cx, cortex; ST, striatum. (Reprinted from Gaspar et al. D1 and D2 receptor gene expression in the rat frontal cortex: cellular localization in different classes of efferent neurons. *Eur. J. Neurosci.* (1995) **7,** 1050–1063, with kind permission from Blackwell Science.)

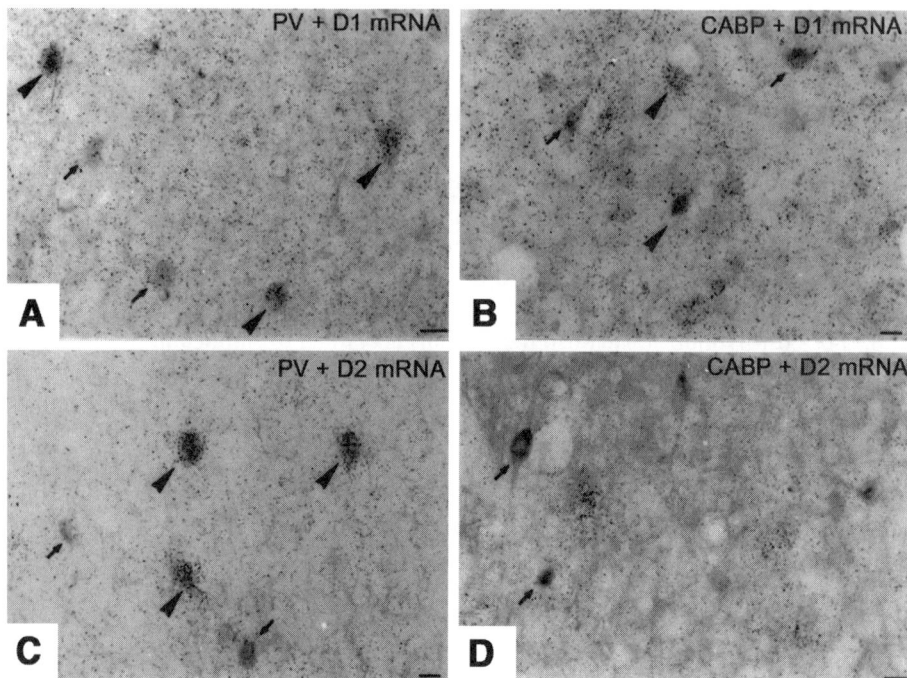

Fig. 2. Coupling *in situ* hybridization with immunohistochemistry. Free-floating cryostat sections treated by *in situ* hybridization to detect D1 and D2 dopamine receptor mRNAs with cRNA probes, followed by the immunodetection of calbindin (CABP) and parvalbumin (PV) in the cortex. The primary antibodies were incubated overnight at room temperature: monoclonal antisera to PV (1:5000 in PBS–0.3% Triton) and CABP (1:30,000 in PBS–0.3% Triton) (from Swant., Switzerland). CABP- and PV-positive neurons appear as dark staining and mRNAs as silver grains. Colocalization of parvalbumin is frequently observed both with D1 receptor mRNA **(A)** and D2 receptor mRNA **(C)**. A few CABP-immunoreactive neurons are observed, also containing D1 mRNA **(B)**, and almost none with D2 mRNA **(D)**. *Arrowheads* point to double-labeled neurons, and *small arrows* to neurons containing only PV or CABP immunoreactivity. (Reprinted from Le Moine and Gaspar. Subpopulations of cortical GABAergic interneurons differ by their expression of D1 and D2 dopamine receptor subtypes. *Mol. Brain Res.* (1998) **58,** 231–236, with kind permission of Elsevier Science.)

described *(12)* to detect striatal neuropeptide mRNAs to identify activated neurons after acute or chronic amphetamine *(13)*.

1. Perfuse adult male rats with 250 mL of 2–4% PFA for 15 min (*see* **Note 9**).
2. Dissect out the brains and post-fix overnight in the same fixative.

Phenotypic Double Labeling

3. Cut 30-μm sections on a vibratome and collect them as free-floating in PBS (*see* **Note 7**).
4. Perform the immunodetection of Fos by first incubating overnight at 4°C with a polyclonal anti-Fos antibody raised in rabbit (1:4,000 in PBS–0.3% Triton).
5. Reveal immunoreactivity with the streptavidin–biotin method: Biotinylated sheep anti-rabbit secondary antibody (1:200 in PBS for 1.5 h at room temperature) and avidin–biotin complex (1:200 in PBS for 1.5 h at room temperature).
6. Develop the signal in 0.05% DAB–0.01% hydrogen peroxide in TBS, pH 7.6, for 5 min (*see* **Note 8**).
7. Wash in TBS, and mount the sections on gelatin-coated slides and allow to dry.
8. Pretreat the slides and hybridize overnight at 40°C in a humidified chamber as described in **Subheading 3.3.** for oligonucleotide probes.
9. Wash the slides briefly in cold 4X SSC, then twice for 30 min in 1X SSC at room temperature, twice for 30 min in 1X SSC at 40°C, and finally 30 min (twice) in 0.1X SSC at 40°C under agitation.
10. Dehydrate twice in 100% ethanol and air-dry.
11. Autoradiographic *in situ* hybridization signal is then revealed by dipping the slides into Ilford K5 emulsion (diluted 1:3 in 1X SSC).
12. Expose for 3 wk, develop, and mount in Eukitt with or without counterstaining.
13. Double-labeled neurons are visualized as illustrated in **Fig. 3**.

3.5.3. Alternative Protocols

Other protocols may be used depending on the type of coupling expected, with only minor experimental modifications.

1. *In situ* hybridization with radioactive probes on cryostat sections on slides followed by immunohistochemistry. In such a case, the immunohistochemical signal can be revealed either before or after dipping in emulsion *(19–21)* (*see* **Note 10**).
2. *In situ* hybridization with nonradioactive probes (digoxigenin- or biotin-labeled) on cryostat sections on slides or free-floating sections followed by immunohistochemistry *(11,22)* (*see* **Note 11**).
3. Coupling *in situ* hybridization and immunohistochemistry on adjacent sections to avoid technical incompatibilities between the two detection procedures *(27–29)* (*see* **Note 12**).

4. Notes

1. For *in situ* hybridization, the choice of the probe depends on the mRNA levels to be detected, but also on the availability of these probes. Note that the best sensitivity is obtained with cRNAs and oligonucleotides. Different nucleotides may be used to label these probes, mainly ^{35}S and ^{33}P which have both good resolution and sensitivity. ^{3}H-labeled probes might be used for highly expressed mRNAs; nevertheless the sensitivity is not appropriate.

Fos IHC + Enk mRNA   **Fos IHC + SP mRNA**

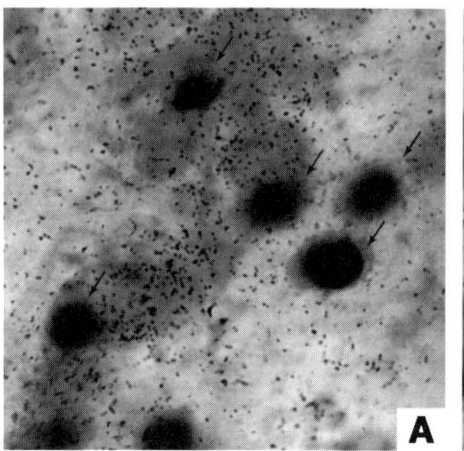 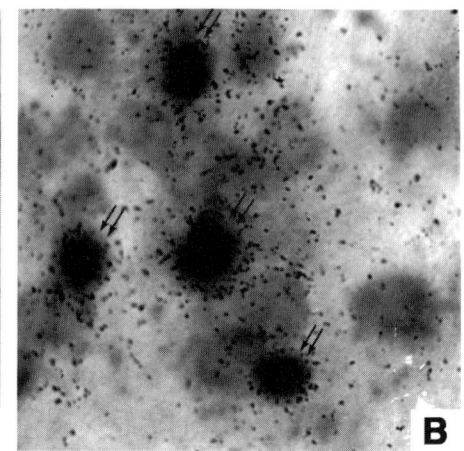

Fig. 3. Coupling *in situ* hybridization with immunohistochemistry. Free-floating vibratome sections treated for the immunohistochemical detection of Fos proteins, followed by *in situ* hybridization on slides to detect enkephalin (**A**) and substance P (**B**) mRNAs with oligonucleotide probes in the striatum. Fos-positive nuclei appear with dark staining and mRNAs as silver grains. The double labelings show that most of the Fos-positive neurons *(arrows)* are enkephalin negative (**A**) but substance P positive (**B**, *double arrows*) after acute amphetamine treatment. (Reprinted from Jaber et al. Acute and chronic amphetamine treatments differently regulate neuropeptide messenger RNA levels and Fos immunoreactivity in rat striatal neurons. *Neuroscience* (1995) **65**, 1041–1050 with kind permission from Dr. Mohamed Jaber and Elsevier Science.)

2. Coupling *in situ* hybridization with retrograde tracing is quite straightforward when using fluorescent tracers because there is no technical interference in the detection of the different markers. Indeed these fluorescent tracers may be visualized directly after fixation of the tissue without any further immunohistochemical steps. The protocol described in this chapter is based on the use of FluoroGold *(15)* as also described by others *(5,9)*. Some authors have reported the use of other fluorescent tracers such as Rhodamine–latex microspheres *(23)* or DiL *(24)*.
3. When using retrograde tracers that need to be visualized after an additional step using antibodies or silver enhancement, for example *(6,16,25,26)*, the technical problems encountered—if there are some—will be similar to those described in **Note 5** for the coupling with immunohistochemistry.
4. Whatever the tracer, the protocol described here may be performed with cRNA, cDNAs, or oligonucleotide probes, with a fixation allowing good sensitivity and appropriate preservation of the tracer. Note that a fixation with a concentration of 2% PFA may be a good compromise.

5. For coupling *in situ* hybridization with the immunohistochemical detection of an antigen, several strategies theoretically can be considered. Nevertheless these strategies have to take into account the compatibility of the experimental conditions, and there is no standard protocol. From a general point of view, including for combined *in situ* hybridization, double labeling approaches classically decrease the sensitivity of detection for both techniques. Thus the use of cRNA probes for *in situ* hybridization is recommended for low mRNA expression from an *in situ* hybridization aspect. However the stringency conditions that are necessary for the use of these probes (high temperatures, presence of detergents, RNase A treatment, etc.) may alter the histological preservation of the tissue and consequently the quality and intensity of the immunohistochemistry. The two methods have also to be compatible in terms of tissue fixation. Whatever the type of probes used, no or low fixation gives the best sensitivity for *in situ* hybridization. On the other hand, the best conditions for immunohistochemical detection depend to a large extent on the type of antigen to be detected (neuropeptides, receptors, enzymes, etc.) but strong fixation is sometimes required for an optimal immunohistochemical signal. Thus each experimental procedure of coupling will have to be designed depending on the type of markers to be visualized, and the choice for a given protocol most of the time will rely on a "compromising" procedure allowing a satisfying visualization of the mRNA together with the antigen.
6. When using cryostat free-floating sections, 4% PFA fixation is more appropriate to limit damaging of the 15–20-µm sections during the processing of the various hybridization steps
7. Cryostat free floating or vibratome sections may be also stored in cryoprotectant (30% sucrose–10% glycerol in PBS) at –80°C before use.
8. When using the peroxidase method for immunohistochemical detection, it is recommended not to use nickel to amplify the signal obtained with DAB to avoid further nonspecific radioactive hybridization signal.
9. For vibratome sections, the thickness has to be around 30 µm for a good cellular resolution; thus brains need to be fixed at least with 2% PFA to allow an homogeneous sectioning with this thickness.
10. When using cryostat sections on slides, which can be appropriate to visualize highly expressed mRNA or antigens (e.g., for neuropeptides), the immunohistochemical signal can be revealed before or after emulsion. In the first case, one needs to check for the nonspecific accumulation of silver grains. In the latter case, the development of the signal throught the emulsion will decrease the intensity of the immunohistochemical signal *(20,21)*.
11. Nonradioactive *in situ* hybridization using biotin- or digoxigenin-labeled nucleotides is classically revealed by an alkaline phosphatase-conjugated complex and gives a purple staining after incubation in nitroblue tetrazolium/5-bromo-4-chloro-3-indolyl phosphate (NBT/BCIP) substrates. Combined immunohistochemistry with the peroxidase–anti-peroxidase method will give a brown staining after incubation with DAB, which can be differentiated easily from the

purple one if (1) the global intensity of the signal is not too strong and/or (2) the DAB signal is not amplified by nickel.
12. An alternate possibility is to couple *in situ* hybridization and immunohistochemistry on very thin sections, allowing one to analyze the same neurons on adjacent sections *(27–29)*. This has the advantage of minimizing technical incompatibilities between the various detection procedures. Nevertheless this approach also has several disadvantages: (1) the necessity of working on slides with very thin sections (3–4 µm cryostat or 1–2 µm semithin sections) with a lowest sensitivity for the detection of medium to low expressed antigens or mRNAs; (2) the difficulty of detecting the same neurons on two adjacent sections especially for small and medium sized neurons (except on semithin sections); (3) the longer period of time needed to analyze the data with the necessity of using two microscopes; and (4) for semithin sections, the problems of using inclusion into resines (araldite or epon).

5. Conclusions

It is worth noting that the main troubleshooting of coupling *in situ* hybridization with other anatomical techniques is related to the compatibility of *in situ* hybridization with the immunohistochemical detection of additional markers. Several protocols have been proposed in the present chapter to couple *in situ* hybridization with retrograde tracing and immunohistochemistry. Despite the fact that some coupling procedures have been relatively generalized (e.g., in **Subheadings 3.4.** and **3.5.1.**), it is important to consider that there is no standard protocol, especially for the coupling with immunohistochemistry. Thus each particular coupling will have to take into account the specific experimental conditions to optimize the two types of detections within a single protocol. Interestingly, the development of fluorescent *in situ* hybridization will probably help to minimize the technical incompatibilities with immunohistochemistry. Today, the sensitivity of fluorescent *in situ* hybridization for mRNA detection in the mammalian central nervous system is still limited to highly expressed mRNAs. Nevertheless, it can be used for neuropeptide mRNA detection and coupling (e.g., **ref. 23**) and will have promising applications in the future. The progress in sensitivity and the multiplicity of detection systems will probably also help in the wide development of triple labeling procedures, as described by some authors *(8,16,26)*.

Acknowledgments

The author thanks Dr. Patricia Gaspar, Dr. Mohamed Jaber, and Dr. Véronique Bernard, who have participated in part to the work described in the present chapter, as well as Pr. Bertrand Bloch for his constant support. The author also thanks Claude Vidauporte and Loïc Grattier for their help with the photographic art. This work was supported by the Université Victor Segalen Bordeaux 2, the CNRS, and the Région d'Aquitaine.

References

1. Bloch, B., Popovici, T., Le Guellec, D., et al. (1986) In situ hybridization histochemistry for the analysis of gene expression in the endocrine and central nervous system tissues: a 3-year experience. *J. Neurosci. Res.* **16**, 183–200.
2. Darby, I. A. (ed.) (2000) In Situ *Hybridization Protocols.* Humana Press, Totowa, NJ.
3. Le Moine, C., Normand, E., and Bloch, B. (1995) Use of non-radioactive probes for mRNA detection by in situ hybridization: interests and applications in the central nervous system. *Cell Mol. Biol.* **41**, 917–923.
4. Le Moine, C. (2000) Quantitative in situ hybridization using radioactive probes to study gene expression in heterocellular systems. *Methods Mol. Biol.* **123**, 143–156.
5. Gerfen, C. R. and Young, W.S. D. (1988) Distribution of striatonigral and striatopallidal peptidergic neurons in both patch and matrix compartments: an in situ hybridization histochemistry and fluorescent retrograde tracing study. *Brain Res.* **460**, 161–167.
6. Hermanson, O., Ericson, H., Sanchez-Watts, G., Watts, A. G., and Blomqvist, A. (1994) Autoradiographic visualization of ^{35}S-labeled cRNA probes combined with immunoperoxidase detection of choleragenoid: a double-labeling light microscopic method for in situ hybridization and retrograde tract tracing. *J. Histochem. Cytochem.* **42**, 827–831.
7. Smith, M. D., Parker, A., Wikaningrum, R., and Coleman, M. (2000) Combined immunohistochemical labeling and in situ hybridization to colocalize mRNA and protein in tissue sections, in *Methods in Molecular Biology*, Vol. 123: In Situ *Hybridization Protocols* (Darby, I., ed.), Humana Press, Totowa, NJ, pp. 165–175.
8. Trembleau, A., Roche, D., and Calas, A. (1993) Combination of non-radioactive and radioactive in situ hybridization with immunohistochemistry : a new method allowing the simultaneous detection of two mRNAs and one antigen in the same brain tissue section. *J. Histochem. Cytochem.* **41**, 489–498.
9. Gerfen, C. R., Engber, T. M., Mahan, L. C., et al. (1990) D1 and D2 dopamine receptor-regulated gene expression of striatonigral and striatopallidal neurons. *Science* **250**, 1429–1432.
10. Le Moine, C. and Bloch, B. (1995) D1 and D2 dopamine receptor gene expression in the rat striatum: sensitive cRNA probes demonstrate prominent segregation of D1 and D2 mRNAs in distinct neuronal populations of the dorsal and ventral striatum. *J. Comp. Neurol.* **355**, 418–426.
11. Georges, F., Stinus, L., and Le Moine, C. (2000) Mapping of *c-fos* gene expression in the brain during morphine dependence and precipitated withdrawal, and phenotypic identification of the striatal neurons involved. *Eur. J. Neurosci.* **12**, 4475–4486.

12. Bernard, V., Dumartin, B., Lamy, E., and Bloch, B. (1993) Fos immunoreactivity after stimulation or inhibition of muscarinic receptors indicates anatomical specificity for cholinergic control of striatal efferent neurons and cortical neurons in the rat. *Eur. J. Neurosci.* **5,** 1218–1225.
13. Jaber, M., Cador, M., Dumartin, B., Normand, E., Stinus, L., and Bloch, B. (1995) Acute and chronic amphetamine treatments differently regulate neuropeptide messenger RNA levels and Fos immunoreactivity in rat striatal neurons, *Neuroscience* **65,** 1041–1050.
14. Le Moine, C., Normand, E., Guitteny, A. F., Fouque, B., Teoule, R., and Bloch, B. (1990) Dopamine receptor gene expression by enkephalin neurons in rat forebrain. *Proc. Natl. Acad. Sci. USA* **87,** 230–234.
15. Gaspar, P., Bloch, B., and Le Moine, C. (1995) D1 and D2 receptor gene expression in the rat frontal cortex: cellular localization in different classes of efferent neurons. *Eur. J. Neurosci.* **7,** 1050–1063.
16. Rocamora, N., Pascual, M., Acsady, L., de Lecea, L., Freund, T. F., and Soriano, E. (1996) Expression of NGF and NT3 mRNAs in hippocampal interneurons innervated by the GABAergic septohippocampal pathway. *J. Neurosci.* **16,** 3991–4004.
17. Le Moine, C. and Gaspar, P. (1998) Subpopulations of cortical GABAergic interneurons differ by their expression of D1 and D2 dopamine receptor subtypes. *Mol. Brain Res.* **58,** 231–236.
18. Hrabovszky, E., Vrontakis, M. E., and Petersen, S. L. (1995) Triple-labeling method combining immunocytochemistry and in situ hybridization histochemistry: demonstration of overlap between Fos-immunoreactive and galanin mRNA-expressing subpopulations of luteinizing hormone-releasing hormone neurons in female rats. *J. Histochem. Cytochem.* **43,** 363–370.
19. Landry, M., Trembleau, A., Arai, R., and Calas, A. (1991) Evidence for a colocalisation of oxytocin mRNA and galanin in magnocellular hypothalamic neurons: a study combining in situ hybridization and immunohistochemistry. *Mol. Brain Res.* **10,** 91–95.
20. Tison, F., Normand, E., Jaber, M., Aubert, I., and Bloch, B. (1991) Aromatic L-amino-acid decarboxylase (DOPA decarboxylase) gene expression in dopaminergic and serotoninergic cells of the rat brainstem. *Neurosci. Lett.* **127,** 203–206.
21. Aubert, I., Brana, C., Pellevoisin, C., et al. (1997) Molecular anatomy of the development of the human substantia nigra. *J. Comp. Neurol.* **379,** 72–87.
22. Griffond, B., Ciofi, P., Bayer, L., Jacquemard, C., and Fellmann, D. (1997) Immunocytochemical detection of neurokinin B receptor (NK3) on melanin-concentrating hormone (MCH) neurons in rat brain. *J. Chem. Neuroanat.* **12,** 183–189.
23. Senatorov, V. V., Trudeau, V. L., and Hu, B. (1995) Expression of cholecystokinin mRNA in corticothalamic projecting neurons: a combined fluorescence in situ hybridization and retrograde tracing study in the ventrolateral thalamus of the rat. *Mol. Brain Res.* **30,** 87–96.

24. Fujimori, K. E., Takeuchi, K., Yazaki, T., Uyemura, K., Nojyo, Y., and Tamamki, N. (2000) Expression of L1 and TAG-1 in the corticospinal, callosal, and hippocampal commissural neurons in the developing rat telencephalon as revealed by retrograde and in situ hybridization double labeling. *J. Comp. Neurol.* **417,** 275–288.
25. Jongen-Relo, A. L. and Amaral, D. G. (2000) A double labeling technique using WGA-apoHRP-gold as a retrograde tracer and non-isotopic in situ hybridization histochemistry for the detection of mRNA. *J. Neurosci. Methods* **101,** 9–17.
26. Acsady, L., Pascual, M., Rocamora, N., Soriano, E., and Freund, T. F. (2000) Nerve growth factor but not neurotrophin-3 is synthesized by hippocampal GABAergic neurons that project to the medial septum. *Neuroscience* **98,** 23–31.
27. Guitteny, A. F., Bohlen, P., and Bloch, B. (1988) Analysis of vasopressin gene expression by in situ hybridization and immunohistochemistry in semi-thin sections. *J. Histochem. Cytochem.* **36,** 1373–1378.
28. Le Moine, C., Tison, F., and Bloch, B. (1990) D2 dopamine receptor gene expression by cholinergic neurons in the rat striatum. *Neurosci. Lett.* **117,** 248–252.
29. Le Moine, C. and Bloch, B. (1991) Rat striatal and mesencephalic neurons contain the long isoform of the D2 dopamine receptor mRNA. *Mol. Brain Res.* **10,** 283–289.

10

Analysis of mRNA Expression Using Double In Situ Hybridization Labeling with Isotopic and Nonisotopic Probes

John Q. Wang

1. Introduction

A large number of in vivo studies in the last decade have confirmed that abused drugs are able to up-regulate gene expression in striatal neurons. For example, acute or chronic exposures to psychostimulants, cocaine and amphetamine, increased basal levels of mRNA and protein products of immediate early genes (c-*fos*, *zif/268*, Δ*FosB*, etc.) in the rodent striatum *(1–3)*. Similarly, acute administration of cocaine and amphetamines induced robust mRNA expression of preprodynorphin (PPD) and substance P (SP) in striatonigral projection neurons and preproenkaphalin (PPE) in striatopallidal projection neurons *(4–6)*. The increased immediate early gene and neuropeptide expression is mediated via selective activation of dopamine receptors (primary D_1 subtype) *(7,8)*. Glutamatergic transmission also significantly contributes to the genomic effect of stimulants *(9,10)*. The altered gene expression has been considered to be an essential molecular step for the development of psychoplasticity in the striatum related to addictive properties of drugs of abuse *(11,12)*.

Quantitative *in situ* hybridization with isotopic cDNA or cRNA probes has been utilized on mounting striatal sections on glass slides in most of the previous studies. The technology allows a quantifiable detection of a single mRNA of interest in particular cells. However, detection of coexpression of two different mRNAs in a cell is often required in order to validate their interactions at a single cell level. For instance, to clarify the role of D_1 receptors in mediating the drug-stimulated c-*fos* mRNA expression in striatal neurons, colocalization of D_1 receptors and c-*fos* in individual cells needs to be established. To assess the interaction between rapid, transient c-*fos* and delayed, prolonged neuropeptide gene expression in response to drug exposures, coexpression of the two mRNAs is a histological prerequisite. Double *in situ* hybridization labeling with isotopic and nonisotopic probes on the same brain section achieves this goal, as it has been successfully used to detect coexpression of glutamate receptor subtypes and neuropeptides in striatal neurons *(13,14)*. This chapter describes a detailed protocol of double *in situ* hybridization labeling. In this protocol, rodent brains are removed from anesthetized rats and cut into 12 μm thick sections in a cryostat. Brain sections are mounted on coated glass slides and fixed with 4% paraformaldehyde. *In situ* hybridization with an isotopic probe is then performed on the slides according to procedures outlined in Chapter 9 to detect the first mRNA of interest. Following the radioactive *in situ* hybridization, *in situ* hybridization with a nonisotopic probe is performed on the same slides to visualize a second mRNA of interest in the same cell.

2. Materials

1. Rat brain striatal sections on glass slides on which *in situ* hybridization with ^{35}S-labeled oligonucleotide probes has been performed (*see* Chapter 9 for detailed procedures).
2. 1X Phosphate-buffered saline (PBS): Diluted from 10X PBS (GIBCO) with dH_2O.
3. Oligodeoxynucleotide probes complementary to a given mRNA of interest, generally ~48 bases long.
4. 10 m*M* Tris-HCl–1 m*M* EDTA buffer (TE buffer): 1 mL of 1 *M* Tris-HCl, pH 7.5, and 200 μL of 0.5 *M* EDTA, pH 8.0, in 99 mL of dH_2O.
5. 4X Sodium chloride and sodium citrate (SSC): Diluted from 20X SSC (Sigma) with dH_2O.
6. Detergent solution: 1% Sarcosyl (Fisher), 1X Denhardt's solution (Sigma), and 4X SSC.
7. Detention buffer: 100 m*M* Tris-HCl, pH 7.5, 150 m*M* NaCl, and 0.6% (v/v) Triton X-100.
8. Hybridization buffer: 23.8 mL of formamide; 0.95 mL of 1 *M* Tris-HCl, pH 7.4, 0.19 mL of 250 m*M* EDTA, pH 8.0, 3.75 mL of 4 *M* NaCl, 9.52 mL of 50% (w/v) dextran sulfate, 0.95 mL of 50X Denhardt's solution. Add dH_2O to 40 mL and store indefinitely at –20°C.
9. Alkaline phosphatase buffer: 100 m*M* Tris-HCl, pH 9.5, 100 m*M* NaCl, 50 m*M* $MgCl_2$, and 0.1% Tween 20.
10. Nitroblue tetrazolium chloride (NBT) stock solution: 100 mg/mL of NBT (Roche) in 70% dimethylformamide.
11. 5-Bromo-4-chloro-3-indolyl phosphate (BCIP) stock solution: 50 mg/mL (Roche) in 100% dimethylformamide.

12. NBT/BCIP working solution: Dilute to 1:300 in alkaline phosphatase buffer. Prepare fresh just prior to use. Levamisole (Vector Laboratories) may be added to a concentration of 1 mM to block peripheral-type endogenous alkaline phosphatase.
13. Genius 6 digoxygenin oligo tailing kit (Roche).
14. Yeast tRNA.
15. Normal goat serum.
16. Alkaline phosphatase-conjugated sheep polyclonal antidigoxygenin antibodies (Roche).
17. Proteinase K solution: 10 µg/mL of Proteinase K (Sigma) in 1X PBS.
18. 70%, 80%, 95%, and 100% ethanol and xylenes.
19. DPX mounting medium (Electron Microscopy Sciences).
20. Kodak hypoclearing agent.

3. Methods

In situ hybridization detection of a first mRNA of interest with isotopic probes is performed according to procedures described in Chapter 9. After this, the same slides are used for the following *in situ* hybridization detection of a second mRNA of interest with nonisotopic probes, that is, digoxigenin-labeled probes. Digoxigenin-labeled probes are detected by using antibodies against the digoxigenin moiety. These antibodies are usually conjugated to alkaline phosphatase or horseradish peroxidase, either of which deposits a colored reaction product in the presence of the appropriate substrate at the site of the hybridization probe. The protocols here are for alkaline phosphatase-conjugated antibodies. All steps are preformed at room temperature unless otherwise indicated.

3.1. Pretreatment of Slides

1. Rinse slides in Kodak hypoclearing agent for 5 min (*see* **Note 1**). Rinse two times in sterile dH$_2$O (15 min each) and two times in 1X PBS (15 min each).
2. Prehybridization in the detergent solution (30 min). The prehybridization increases signal-to-noise ratio and access of probes to target mRNAs.
3. Rinse three times in 1X PBS (5 min each).
4. Incubate in the Proteinase K solution (10 min) to increase penetration of probes. Rinse three times in 1X PBS (5 min each).
5. Dehydrate slides through graded ethanol (2 min each) and air-dry for hybridization or store at −20°C until use.

3.2. Probe Labeling

Synthesized oligonucleotide probes for mRNAs that are frequently tested with *in situ* hybridization may be commercially available. If the probe for a particular mRNA is not available, quality oligonucleotide probes can be synthesized by a number of biochemical companies according to the sequence designed and provided by customers at a reasonable cost. We use a Genius 6 digoxygenin oligo tailing kit to label probes with digoxygenin.

1. Prepare the following reaction mix in a microcentrifuge tube (1.5 mL) on ice (final volume = 50 µL):

5X tailing buffer (vial 1)	10 µL
$CoCl_2$ (vial 2)	4 µL
Oligonucleotide probe (5 µM)	1 µL
Digoxygenin-11-dUTP (250 µM, vial 3)	1 µL
dATP (1 mM, vial 4)	1 µL (see **Note 2**)
Terminal deoxynucleotidyle transferase (TdT; vial 5)	2 µL
Sterile dH_2O	31 µL

2. Incubate for 2 h in a circulating water bath at 37°C.
3. Add 2 µL of diluted glycogen (1 µL of glycogen in vial 9 + 200 µL of TE buffer) into the tube to stop the reaction.
4. To precipitate the tailed probes, add 375 µL of TE buffer, 25 µL of 4 M NaCl, 50 µg of yeast tRNA, and 1 mL of prechilled (–20°C) 100% ethanol into the tube. Mix thoroughly, leave for 1–2 h at –20°C, and then microcentrifuge for 30 min at 17,000 rpm (you may see the precipitated oligos after centrifuge).
5. Remove and discard the supernatant. Rinse the pellet with 100 µL of cold 70% ethanol. Remove the ethanol and add 50 µL of TE buffer. Store indefinitely at 4°C.

3.3. Hybridization

1. Add 3–10 µL of digoxygenin-labeled probe to 100 µL of hybridization buffer (see **Note 3**). Pipet 25 µL of hybridization buffer containing digoxygenin-tailed probes onto one brain section. Cover it with a glass coverslip.
2. Incubate overnight at room temperature. The incubation conditions (time and temperature) can be adjusted empirically to optimize the signal-to-noise ratio.

3.4. Post-Hybridization Wash and Digoxygenin Detection

1. Rinse two times in 2X SSC (15 min each) and two times in 1X SCC (15 min each).
2. Rinse 5 min in detection buffer.
3. Incubate 30 min in 3% normal goat serum/detection buffer.
4. Incubate in alkaline phosphatase-conjugated antidigoxygenin antibody (1:500–1000 in 3% normal goat serum/detection buffer) for 5 h with gentle rocking (or overnight at 4°C) without shaking.
5. Rinse two times in detection buffer (15 min each).
6. Rinse 5 min in alkaline phosphatase buffer and 5 min in alkaline phosphatase with 5 mM levamisole (see **Note 4**).
7. Incubate in NBT/BCIP working solution in the dark overnight. Slides may be checked after 1–2 h to see if signals have appeared (blue reaction product)

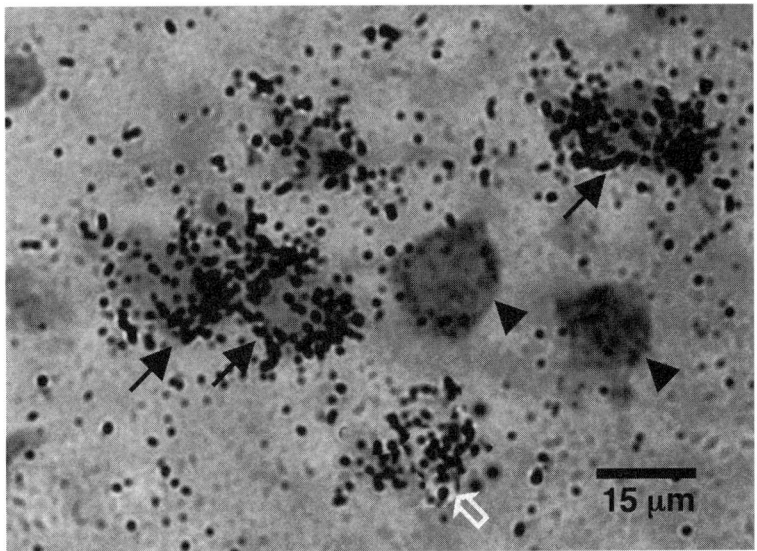

Fig. 1. Coexpression of preproenkephalin (silver grains detected by ^{35}S-labeled oligonucleotide probes) and metabotropic glutamate receptor 5 (alkaline phosphatase reaction products detected by digoxygenin-labeled oligonucleotide probes) mRNAs in striatal neurons. Some neurons contain both mRNAs *(solid arrows)*, whereas others contain only preproenkephalin *(open arrow)* or metabotropic glutamate receptor 5 *(arrowheads)* mRNA.

and silver grains still exist under a microscope with bright- and dark-field, respectively.
8. Rinse two times in 1X PBS (10 min each). Dip slides in dH$_2$O, air-dry, and mount. Loss of alkaline phosphatase may occur with exposure to ethanol. Visualize microscopic colocalization of hybridization signals (silver grains with isotopic probes and color products with digoxygenin-labeled probes) in the same cells (**Fig. 1**) (*see* **Note 5**).

4. Notes

1. The hypoclearing agent rinse removes any remaining fixative chemicals left on slides during development of emulsion-dipped slides. The rinse accordingly prevents severe alkaline phosphatase staining artifacts.
2. Molar ratio of digoxygenin-11-dUTP/unlabeled dATP can be modified empirically from 1:4 to 1:10.
3. The exact amount of probes must be determined empirically. Do not use the reducing agent dithiothreitol, which increases background and imparts a strong purplish color.
4. Color development artifacts may occur with alkaline phosphatase if endogenous peripheral-type alkaline phosphatase is present to an extent adequate to produce

detectable color product. The inhibitor selective for peripheral-type alkaline phosphatase, levamisole, can be used here to attenuate alkaline phosphatase activity.
5. Specificity of probes is an essential issue of *in situ* hybridization experiments. Unfortunately, there is no single, direct control for hybridization signals. Researchers will have to rely on as many different indirect controls as possible. These indirect controls (checks) have been outlined in an excellent protocol chapter *(15)*. They include mainly (1) comparing the intensity and distribution pattern of hybridization signals with antisense and sense probes; (2) comparing the intensity and distribution pattern of hybridization signals with probes against different portions of the same transcript; (3) comparing the intensity and distribution pattern of hybridization signals with probes against target and unrelated transcripts; (4) examining blockade of signals by prior hybridization with unlabeled probe; (5) correlating with immunocytochemical results, but knowing that an mRNA and its protein products are not always parallel; and (6) observing Northern analysis using the probe under the same degrees of stringency that shows band(s) of expected size(s).

Acknowledgments

This work was supported by the NIH Grants DA10355 and MH61469.

References

1. Graybiel, A. M., Moratalla, R., and Robertson, H. (1990) Amphetamine and cocaine induce drug-specific activation of the c-*fos* gene in striosome-matrix compartments and limbic subdivisions of the striatum. *Proc. Natl. Acad. Sci. USA* **87,** 6912–6916.
2. Kelz, M. B., Chen, J., Carlezon, W. A., Jr., (1999) Expression of the transcription factor ΔFosB in the brain controls sensitivity to cocaine. *Nature* **401,** 272–276.
3. Nguyen, T. V., Kosofsky, B. E., Birnbaum, R., Cohen, B. M., and Hyman, S. E. (1992) Differential expression of c-*fos* and *zif/268* in rat striatum after haloperidol, clozapine, and amphetamine. *Proc. Natl. Acad. Sci. USA* **89,** 4270–4274.
4. Hurd, Y. L. and Herkenham, M. (1992) Influence of a single injection of cocaine, amphetamine, or GBR 12909 on mRNA expression of striatal neuropeptides. *Mol. Brain Res.* **16,** 97–104.
5. Wang, J. Q., Smith, A. J. W., and McGinty, J. F. (1995) A single injection of amphetamine or methamphetamine induces dynamic alterations in c-*fos*, *zif/268* and preprodynorphin messenger RNA expression in rat forebrain. *Neuroscience* **68,** 83–95.
6. Wang, J. Q. and McGinty, J. F. (1995) Dose-dependent alteration in *zif/268* and preprodynorphin mRNA expression induced by amphetamine or methamphetamine in rat forebrain. *J. Pharmacol. Exp. Ther.* **273,** 909–917.
7. Cole, A. J., Bhat, R. V., Patt, C., Worley, P. F., and Baraban, J. M. (1992) D_1 dopamine receptor activation of multiple transcription factor genes in rat striatum. *J. Neurochem.* **58,** 1420–1426.

8. Wang, J. Q. and McGinty, J. F. (1996) D_1 and D_2 receptor regulation of preproenkephaline and preprodynorphin mRNA in rat striatum following acute injection of amphetamine or methamphetamine. *Synapse* **22,** 114–122.
9. Wang, J. Q., Daunais, J. B., and McGinty, J. F. (1994) NMDA receptors mediate amphetamine-induced upregulation of *zif/268* and preprodynorphin mRNA expression in rat striatum. *Synapse* **18,** 343–353.
10. Wang, J. Q. and McGinty, J. F. (1996) Intrastriatal injection of the metabotropic glutamate receptor antagonist MCPG attenuates acute amphetamine-stimulated neuropeptide mRNA expression in rat striatum. *Neurosci. Lett.* **218,** 13–16.
11. Nestler, E. J. (2000) Genes and addiction. *Nat. Genet.* **26,** 277–281.
12. Wang, J. Q., Mao, L., and Lau, Y. S. (2002) Glutamate cascade from metabotropic glutamate receptor to gene expression in striatal neurons: implications for psychostimulant dependence and medication, in *Glutamate and Addiction* (Herman, B. H., Frankenheim, J., Litten, R., Sheridan, P. H., and Weight, F. F., eds.), Humana Press, Totowa, NJ, pp. 157–170.
13. Kerner, J. A, Standaert, D. G., Penney, J. B., Young, A. B., Jr., and Landwehrmeyer, G. B. (1997) Expression of group I metabotropic glutamate receptor subunit mRNAs in neurochemically identified neurons in the rat neostriatum, neocortex, and hippocampus. *Mol. Brain Res.* **48,** 259–269.
14. Testa, C. M., Standaert, D. G., Landwehrmeyer, G. B., Penney, J. B., and Young, A. B. (1995) Differential expression of mGluR5 metabotropic glutamate receptor mRNA by rat striatal neurons. *J. Comp. Neurol.* **354,** 241–251.
15. Young, W. S., III and Mezey, E. (1997) Hybridization histochemistry of neural transcripts, in *Current Protocols in Neuroscience* (Crawley, J. N., Gerfen, C. R., McKay, R., Rogawski, M. A., Sibley, D. R., and Skolnick, P., eds.), John Wiley & Sons, New York, pp. 1.3.1–1.3.17.

11

Quantification of mRNA in Neuronal Tissue by Northern Analysis

Christine L. Konradi

1. Introduction

RNA is transcribed in the cell nucleus from a DNA template with ribonucleotides as the building blocks. As RNA is transported out of the nucleus, it is spliced; that is, predetermined sequences (introns) are cut out of the transcript. If certain introns serve as either exons or introns, "alternative splicing" can occur and one gene can give rise to several different size RNA transcripts. The spliced RNA, the mRNA, is released into the cytosol and translated into protein.

Levels of particular mRNA transcripts are assayed in Northern blots. For Northern blots, tissue is homogenized and RNA transcripts are separated from DNA, proteins, and other contaminants. The purified RNA is then size separated on a denaturing agarose gel by electrophoresis. RNA is transferred to a nylon membrane, which is reacted with a known, labeled, antisense RNA or DNA molecule ("probe"). This molecule will find its counterpart on the membrane and bind to it. The labeled probe is then visualized. Because the probe is presented in excess, the amount of labeling is a representation of the amount of RNA in the original sample. The most common way of labeling the probe is by incorporating radioactive nucleotides. Luminescence can be used as well, but it is less sensitive. Either way, X-ray film is used to visualize the probe location and probe quantity on the membrane. Alternatively, image analysis systems are used instead of X-ray films.

RNA is degraded by RNases. RNases are found within the same tissues as RNA. It is therefore important to separate RNases from RNA quickly and effectively. The methods used for RNA extraction should be chosen according to the RNase content of the tissue. A preparation that may work for one type of tissue may not work for another. Pancreatic tissue is particularly well known for high content of RNases (e.g., RNase A), while brain tissue seems to be less vulnerable to RNA degradation. Because the preparations that are the most conservative can also be the most time consuming and expensive, it is advised to try out different preparations.

RNases are not only in tissue, but also on the skin and in saliva. A common source of RNA degradation is contamination with RNases from the skin of the investigator. It is therefore imperative that gloves are worn when handling RNA or any type of chemicals and containers that will be used for RNA extractions. Because the first steps of RNA extraction are usually performed under RNase-denaturing conditions, reintroduction of RNases from outside sources is most detrimental in the latter stages of RNA preparation, when RNA is at its purest and least protected.

2. Materials

For common molecular biology solutions *see also* **ref. 1**.

2.1. RNA Extraction

2.1.1. RNA Extraction from Primary Neuronal Culture

Important: All buffers and stock solutions are treated with diethyl pyrocarbonate (DEPC) and autoclaved, or brought to final volume in DEPC-treated distilled water. Tris buffer is prepared with DEPC-treated distilled water.

1. Phosphate-buffered saline (PBS) (for optional washing of cells): 11.5 g of Na_2HPO_4, 3.0 g of $NaH_2PO_4 \cdot 2H_2O$, and 5.84 g of NaCl. Bring up to 1 L with distilled water, adjust pH to 7.4–7.6, and add 1 mL of DEPC. Stir for at least 1 h, and autoclave.
2. Tris-HCl buffer stock solution: 1 M Tris-HCl, pH 7.6. Bring Tris up to three fourths of final volume with DEPC-treated distilled water; adjust pH to 7.6 with 1 N HCl, and bring up to final volume with DEPC-treated distilled water. *Note*: Rinse pH meter with DEPC-treated distilled water before use.
3. Nonidet P-40 (NP-40) lysis buffer: 50 mM Tris-HCl, pH 7.6, 100 mM NaCl, 5 mM $MgCl_2$; and 0.5% (v/v) NP-40. Prepare with DEPC-treated distilled water.
4. 10% Sodium dodecyl sulfate (SDS): 1 g of SDS; bring to 10 mL with DEPC-treated distilled water.
5. 3 M Sodium acetate, pH 4.5: pH sodium acetate with acetic acid to pH 4.5. Prepare with DEPC-treated distilled water.
6. Phenol–chloroform–isoamyl alcohol (PCI): 25 mL of phenol, 24 mL of chloroform, and 1 mL of isoamyl alcohol. *Note*: Phenol needs to be saturated with distilled water before it is added to the mixture (*see* **item 7**). Store at 4°C.

7. Water-saturated phenol: Place crystalline phenol at 60°C until molten. Allow to cool down to room temperature and overlay with DEPC-treated distilled water. Mix carefully, and let phases separate. (**Caution:** If mixing in a closed bottle, pressure may be building up; frequently open the bottle cup to relieve pressure.) Replace the upper phase (aqueous phase) at least once with DEPC-treated distilled water. Store at 4°C. Use the lower phase (phenol) for the phenol–chloroform–isoamyl alcohol mixture. *Note*: Because phenol needs to be acidic, do not saturate with buffer.
8. 75% and 100% ethanol: Prepare 75% ethanol with DEPC-treated distilled water.
9. 10X Morpholinopropanesulfonic acid (MOPS) buffer, pH 7.0: 200 mM MOPS, 50 mM sodium acetate, 10 mM EDTA, adjust pH to 7.0, and add 0.1% DEPC. Stir for several hours, autoclave for 1 h, and store in light-tight bottles at room temperature.
10. RNA gel running buffer: 25% formamide, 1X MOPS buffer, 1 M formaldehyde (37% commercially available), 5% glycerol, and 0.4 mg/mL of bromophenol blue. Store at –20°C.

2.1.2. RNA Extraction from Brain Tissue

1. Sodium citrate stock solution, pH 7.0: 1 M sodium citrate; adjust pH to 7.0 with HCl.
2. Guanidine thiocyanate solution: 4 M guanidine thiocyanate, 0.5% sodium lauroyl sarcosinate, and 25 mM sodium citrate; adjust pH to 7.0. Store at 4°C. Immediately before use add 100 mM β-mercaptoethanol and 0.1% antifoam A (30% solution from Sigma Chemical).
3. Sodium citrate stock solution, pH 5.5: 1 M sodium citrate; adjust pH to 5.5 with HCl. Add 0.1% DEPC, stir for several hours, and autoclave for 1 h.
4. Cesium chloride solution, 5.7 M: 5.7 M cesium chloride, 25 mM sodium citrate; adjust pH to 5.5. Prepare with DEPC-treated distilled water.
5. Tris-HCl buffer stock solution: 1 M Tris base; adjust pH to 8.0 with HCl. Prepare with DEPC-treated distilled water. *Note*: Rinse pH meter with DEPC-treated distilled water before use.
6. EDTA stock solution: 0.5 M EDTA; adjust pH to 8.0 with sodium hydroxide. Add 0.1% DEPC, stir for several hours, and autoclave for 1 h. *Note*: EDTA does not go into solution below pH 8.0.
7. Tris-EDTA (TE) buffer, pH 8.0: 10 mM Tris-HCl, pH 8.0, and 1 mM EDTA, pH 8.0. Prepare from stocks with DEPC-treated distilled water.

2.2. Gel Electrophoresis of RNA and Transfer to a Nylon Membrane

1. Denaturing agarose gel: 1.2 g of agarose; 10 mL of 10X MOPS buffer, pH 7.0; and 82.5 mL of DEPC-treated distilled water. Microwave for 1–2 min, and swirl gently to make sure the agarose is completely dissolved. **Caution:** Use heat-protective gloves; solution tends to boil over when swirling. Microwave for another 30 s, and swirl carefully. Bring to chemical hood, let cool down for

5 min, and add 8.5 mL of 37% formaldehyde solution. Shake again and pour into horizontal gel tray in chemical hood. Let cool.
2. Agarose: Low electroendosmosis (EEO).
3. Formaldehyde solution (37%).
4. Nylon membrane for nucleic acid blotting: Zeta Probe membrane (Bio-Rad) or Genescreen membrane (Perkin-Elmer).
5. 50X TAE buffer: 242 g of Tris base, 57.1 mL of glacial acetic acid, and 18.6 g of EDTA sodium salt in 1 L of distilled water.

2.3. Labeling of Probes

Many companies sell optimized kits that contain all necessary reagents for labeling. When purchasing these kits, follow the instructions of the manufacturer. In the following subheadings I provide generic protocols that can be used when components are purchased individually. RNA polymerases are provided by the manufacturer with optimized buffers that are used in the reactions.

2.3.1. Labeling of Riboprobes

1. 100 mM Dithiothreitol (DTT): Prepared in DEPC-treated distilled water. Stored at –20°C.
2. Stock of rATP, rGTP, and rUTP: Purchase 10 mM stocks of all nucleotides in DEPC-treated water or buffer, and mix one volume each of rATP, rGTP, and rUTP and DEPC-treated distilled water (final volume of each nucleotide: 2.5 mM). Store at –20°C.
3. RNasin or any other type of commercially available RNase inhibitor (e.g., from Promega, Ambion, Invitrogen, Stratagene, etc.). Stored at –20°C.
4. RQ1 RNase-Free DNase (Promega). Stored at –20°C.
5. NucTrap push columns (Stratagene) or any other type of prepacked size-exclusion chromatography column that retains nucleotides and short oligonucleotides to separate them from larger pieces of DNA or RNA.
6. 1X STE buffer: 100 mM Sodium chloride; 20 mM Tris-HCl, pH 7.6; and 10 mM EDTA.

2.3.2. Labeling of cDNA Probes

1. Stock of dATP, dGTP, and dTTP: Commercially available at 100 mM concentrations of each nucleotide. Dilute one volume of dATP, one volume of dGTP, and one volume of dTTP with one volume of distilled water to make a 25 mM stock solution. Dilute 1:50 (500 µM). Store at –20°C.
2. Random oligonucleotide primers. Commercially available. Store at –20°C.

2.4. Hybridization of Probes to Nylon Membranes

1. 50X Denhardt's reagent: 1 g of Ficoll (type 400), 1 g of polyvinylpyrrolidone, and 1 g of bovine serum albumin (BSA) (fraction V). Bring to 100 mL with DEPC-treated distilled water. Store at –20°C.

2. 5X Buffer P: 5 M Sodium chloride, 250 mM Tris-HCl buffer, pH 7.4–7.6; and 0.5% sodium pyrophosphate. Prepare with DEPC-treated distilled water.
3. 50% Dextran sulfate: Bring 50 g of dextran sulfate to 100 mL with DEPC-treated distilled water, shake vigorously overnight at 4°C in a shaker, and make sure it is completely dissolved. Store at –20°C.
4. 20% SDS: 20 g of SDS, bring to 100 mL with DEPC-treated distilled water. **Caution:** Wear a mask.
5. Salmon sperm DNA: 10 mg of salmon sperm DNA/mL of DEPC-treated distilled water. Autoclave for 30 min, shear by pushing rapidly through a 20-gauge syringe 20 times, and store at –20°C.
6. Hybridization solution: 5 mL of 100% formamide, 2 mL of 50X Denhardt's reagent, 2 mL of 5X buffer P, 2 mL of 50% dextran sulfate, and 0.5 mL of 20% SDS. Add 100 µL of boiling salmon sperm DNA to hybridization bottle (final concentration: 100 µg/mL).
7. 20X Saline sodium citrate (SSC), pH 7.0: 175.3 g of sodium chloride and 88.2 g of sodium citrate. Bring up to 900 mL with distilled water, and measure the pH (it should be around 7.0). Bring up to 1 L; this solution does not need to be RNase free, as RNA bound specifically to complementary RNA (double-stranded RNA) is not subjected to degradation by RNases.

3. Methods
3.1. Setting Up the RNA Workplace

RNA is very susceptible to hydrolysis by RNases. RNases are present not only in tissues, but on the skin of investigators, and can thus be reintroduced into the preparation at any time. RNases are very difficult to inactivate. They are resistant to metal chelating agents, are active over a wide pH range, and can survive prolonged boiling or autoclaving. Indeed, if autoclaved reagents contain microorganisms, their RNases can be set free during autoclaving. Fortunately, RNases rely on histidine residues for catalytic activity and they can be inactivated by the alkylating agent DEPC, by 3% hydrogen peroxide solutions, 1% SDS solutions, or 1 N sodium hydroxide solutions. Glassware can also be baked at 180°C for several hours. The following steps should be taken before starting to work with RNA:

1. Always use gloves when working with RNA. If you need to communicate during the assay, use a face mask to avoid saliva contamination of your preparation.
2. Try to maintain a separate work area for RNA work, with its own set of RNase-free labware and pipets. Always use gloves in this area.
3. Use sterile, disposable plasticware and bake your glassware in an oven at 180°C for 3 h or longer. Many companies guarantee that their plasticware is RNase/DNase free. RNases can also be destroyed by soaking labware in 3% hydrogen peroxide solution, 1% SDS, or 1 N sodium hydroxide solution, and

rinsing with DEPC-treated distilled water. One of the latter three procedures should also be used to decontaminate electrophoresis tanks. Do not use DEPC in polycarbonate or polystyrene containers.
4. Label chemicals used for RNA work and keep separate from common chemical stocks. Always use gloves when handling these chemicals.
5. Decontaminate solutions by adding 1:1000 volume of DEPC, stir for a couple of hours, and autoclave for 1 h to hydrolyze/inactivate DEPC. Solutions containing Tris react with DEPC and cannot be decontaminated with DEPC. For these solutions, DEPC-treated distilled water (autoclaved after treatment) should be used. Solutions with thermolabile substances should be prepared with DEPC-treated distilled water and filtered through a 0.2-µm filter (e.g., syringe filter). DEPC destroys and should not be brought into contact with polycarbonate or polystyrene.

3.2. RNA Extraction

I present two different methods that we have used successfully and consistently in our laboratory, one for primary neuronal culture (*see* **Fig. 5A**), and one for brain tissue (*see* **Fig. 5B**). The difference in the preparations are (1) in the number of samples that can be worked up simultaneously, (2) in the speed of separation between RNA and RNases, (3) in the strength of RNase-denaturing conditions (*see* **Note 1**), and (4) the maximum amount of tissue that can be used in the preparation.

3.2.1. RNA Extraction from Primary Neuronal Culture

This RNA extraction is limited to small numbers of neurons or cells that do not have high levels of RNases. The RNase-denaturing conditions in this protocol are very gentle, and a quick separation of RNA from RNases is essential (*see* **Note 1**). All procedures need to be performed on ice at 4°C. Only cells in a dissociated state, such as they are in culture, should be used, as this protocol does not involve dissociation of tissue. We have used this protocol with good results for primary striatal culture (**Fig. 5A**), primary cortical culture, primary hippocampal culture, and cerebellar granule neurons *(2)*. In the first step of the protocol, nuclei with DNA must be separated from cytoplasm to keep contamination of the RNA preparation with DNA at a minimum. Acid extraction with phenol–chloroform–isoamyl alcohol (PCI) is used to extract RNA into the aqueous phase, while DNA and proteins partition into the organic phase or interphase *(3)* (*see* **Note 2**). RNA is then precipitated and cleaned from salt contaminants.

1. Primary neurons are grown on six-well plates for at least 1 wk, and each well has between 2 and 3 million neurons.

2. After exposure of neurons to the drugs of interest, plates are removed from the cell culture incubator, medium is quickly and completely aspirated, and plates are immediately put on wet ice.
3. Optional: One or two washes with 1X PBS (1 mL each wash). We generally do not wash.
4. For cell lysis, add 500 µL of NP-40 lysis buffer to each well. Leave the cells with lysis buffer on wet ice for 5 min for complete lysis, and then swirl plates to lift the cells off. Resuspend cells by repeated pipetting up and down. Transfer the cell solutions into Eppendorf tubes.
5. To separate cell nuclei, centrifuge the tubes at 4°C for 5 min at 8000g, then transfer supernatants to fresh Eppendorf tubes, and discard pellets (cell nuclei and membranes).
6. Add SDS to supernatants to a final volume of 0.2% SDS (add 10 µL of 10% SDS). Optional: Samples can be frozen at this point.
7. For acidic phenol–chloroform–isoamyl alcohol extraction, bring up supernatants to 0.3 M sodium acetate (add 50 µL of 3 M sodium acetate), add 550 µL of PCI (equal volume), and invert tubes a couple of times and vortex-mix for 30 s. Centrifuge the tubes for 10 min at 16,000g, 4°C. Transfer upper phase (aqueous) to a fresh tube. Do not touch the interphase; if you touch the interphase, repeat centrifugation.
8. For RNA precipitation, add two volumes of 100% ethanol to the aqueous phase and leave the samples at –20°C for 30 min (minimum time: 5 min). Optional: Samples in ethanol can be stored safely at –80°C for a prolonged time.
Centrifuge tubes for 10 min at 16,000g at 4°C and carefully aspirate supernatants and discard.
9. To desalt RNA, add 500 µL of 75% ethanol to the pellets, vortex-mix the pellets briefly, and leave on ice for 10 min to dissolve residual sodium acetate. Centrifuge tubes for 5 min at 16,000g at 4°C and carefully remove supernatants and discard.
10. Air-dry the pellets and take up in 12 µL of RNA gel running buffer.
11. Heat samples at 70°C for 10 min.
12. Place on wet ice.

We do not measure RNA content in these samples, and we use the entire sample to run Northern blots; because all wells have approximately the same number of cells, we have found very little loading differences in these gels. The amount of RNA is at least 1 µg/1 million neurons, 2–3 µg per well of a six-well plate.

3.2.2. RNA Extraction from Brain Tissue

This RNA preparation is modified from **ref. 4**, and can be used for up to 300 mg of tissue (**Fig. 5B**) (*see* **Note 3**). We use this protocol for preparing RNA from tissue *(5,6)*.

1. Homogenize tissue in 3 mL of guanidine thiocyanate solution for 1 min with a Polytron homogenizer at medium setting.
2. Layer homogenized tissue on top of 2 mL of cesium chloride solution in ultracentrifuge tube.
3. Pellet RNA in an SW55 rotor (Beckman Ultracentrifuges) for 12–16 h at 40,000g.
4. Carefully aspirate upper layer (upper 3 mL), and pour off the lower layer (lower 2 mL). Make sure that the solution does not run down to the pellet again. RNases flushed off the tube wall can contaminate the pelleted RNA.
5. Make sure the tubes are labeled at the bottom; cut 1 inch off the bottom of the tube to avoid contamination of pellet with RNases clinging to the walls at the top of the tubes. Discard tube tops and transfer bottoms of tubes to wet ice. RNA should form a translucent pellet at the bottom of the tube. If you can't see it do not be discouraged at this point but proceed.
6. To wash pellet, carefully add 500 µL of 95% ethanol, and make sure you do not dislodge the pellet. Let sit on ice for 5 min and drain carefully. RNA should still be at the bottom of the tube.
7. To collect RNA, add 200 µL of RNase-free TE buffer to the bottoms of tubes and resuspend RNA pellets by pipetting up and down. Transfer to labeled microcentrifuge tubes, and add another 200 µL of TE buffer to the bottom of tubes. Resuspend and add to the 200 µL in microcentrifuge tubes.
8. To precipitate and collect RNA, add one-tenth volume of 3 M sodium acetate, pH 4.5, and two volumes 100% ethanol to the microcentrifuge tubes, and mix well. Let RNA precipitate at –20°C for 30 min (minimum: 5 min). Centrifuge samples for 10 min at 16,000g at 4°C. Carefully aspirate and discard supernatants.
9. To desalt pellets, add 500 µL of 75% ethanol to the pellets, vortex-mix briefly, and leave on ice for 10 min to dissolve residual sodium acetate. Centrifuge for 5 min at 16,000g at 4°C. Carefully aspirate supernatant and discard. Air-dry pellets and take up in 50 µL of TE, making sure to resuspend RNA completely.
10. Determine RNA content by the standard spectrophotometric assay at 260 nm. An OD of 1.0 corresponds to approx 40 µg/mL of single-stranded RNA. The purity of the RNA sample is reflected in the A_{260}/A_{280} ratio, which should be between 1.8 and 2.0 (*see* **Note 4**).
11. To prepare the sample for gel electrophoresis, aliquot equal amounts of RNA (between 5 and 10 µg; same amounts within an experiment), and then dry in SpeedVac concentrator. Take each sample up in 10–15 µL of RNA gel running buffer and heat at 70°C for 10 min. Put samples on wet ice and load onto the gel.

3.3. Gel Electrophoresis of RNA and Transfer to a Nylon Membrane

1. Prepare a denaturing agarose gel (12 × 14 cm) in a horizontal gel electrophoresis apparatus and prepare the appropriate volume of 1X MOPS running buffer. Use

Northern Blots

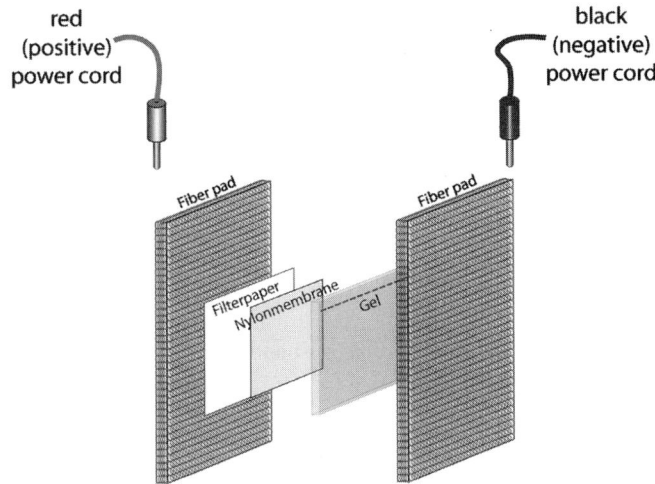

Fig. 1. Assembly of the electroblotter. Put the fiber pad provided with the transfer unit into a tray with gel transfer buffer (1X TAE), and remove all air bubbles by gently pressing on the fiber pad with a gloved hand. Add a piece of strong filter paper and let wet thoroughly. Add the precut nylon membrane and let wet. Put the gel on top followed by a second fiber pad. Make sure there are no air bubbles between any layers of the sandwich. Close the cassette and slide into the electroblotter. The red powercord (leading to the anode) needs to be on the side of the membrane, and the black powercord (leading to the cathode) on the side of the gel.

a gel box with a buffer circulation system or stir the gel on a magnetic stirrer during the run.
2. In each lane, load between 5 and 10 µg of total RNA, or the entire sample in the case of RNA from neuronal cultures. Load equal amounts of RNA within an experiment.
3. Run samples at 100–150 V for 90 min to 3 h (depending on the size of the RNA of interest and separation needed). If you do not have a buffer circulation system or cannot stir the gel during the run, reduce to 80–90 V and increase running time.
4. Turn off the power supply, disconnect the cables, remove the lid, and take out the gel.
5. Prepare the electroblotting tank (vertical buffer tank) with 1X TAE buffer, and cut the nylon membrane and filter paper to size of gel (*see* **Fig. 1**).
6. Assemble the electroblotting sandwich in a tray with transfer buffer as shown in **Fig. 1**.
7. Electroblot RNA onto nylon membrane (**Fig. 1**) at constant current of 1.2 amp for 2 h. RNA moves toward the anode (positive electrode).

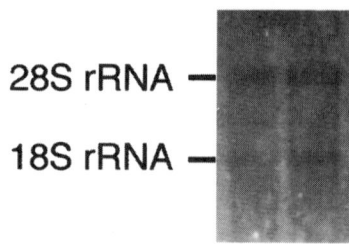

Fig. 2. UV shadowing. Examine the RNA on the nylon membrane under UV light (254 nm). After transfer, a handheld UV lamp is used to reflect UV light onto the membrane. The RNA can be seen as a darker band that stretches along the lanes. Two ribosomal bands are visible on top of the mRNA. Human cerebellar RNA is shown.

8. Hold a UV lamp (254 nm) over the membrane. You should see dark bands running down the lanes. 28S and 18S RNA bands should be visible above the background (**Fig. 2**). Mark 28S and 18S RNA bands with a soft pencil on the nylon membrane. **Caution:** Wear proper gloves and face shield to avoid skin exposure to UV light, and never look into the UV lamp (*see* **Notes 5–8**).
9. UV crosslink membrane at 254 nm, 120,000 µJ/cm^2.
10. Store the membrane in dry place between sheets of filter paper.

3.4. Labeling of Probes

We use ^{32}P- or ^{33}P- radiolabeled DNA or RNA probes for our Northern blots. In general, we prefer ^{32}P-labeled riboprobes over random-labeled DNA probes. The advantage of riboprobes is a higher sensitivity, higher specific activity (only the antisense strand is labeled, whereas both strands are labeled in cDNA probes), lower background, and higher stringency. However, riboprobes cannot be stripped off the blots. If blots need to be exposed to more than one probe of similar size, the blots can be reused only once the radioactivity is decayed. ^{33}P has a longer half-life than ^{32}P, yields crisper bands, but needs a longer exposure time to the film than ^{32}P. Kits for riboprobes are commercially available (e.g., Amersham-Pharmacia, Promega, etc.) (*see* **Note 9**).

3.4.1. Labeling of Riboprobes

Transcription of RNA is performed with the appropriate RNA polymerase (T3, T7, or SP6), depending on the RNA polymerase promoter sites present in the vector used. The plasmid is linearized at an appropriate restriction site at the opposite end of the antisense polymerase site, and "run-off" transcripts are obtained, which should not contain vector sequences (**Fig. 3**). Antisense RNAs are needed for the positive reactions, while sense RNAs are used as negative controls.

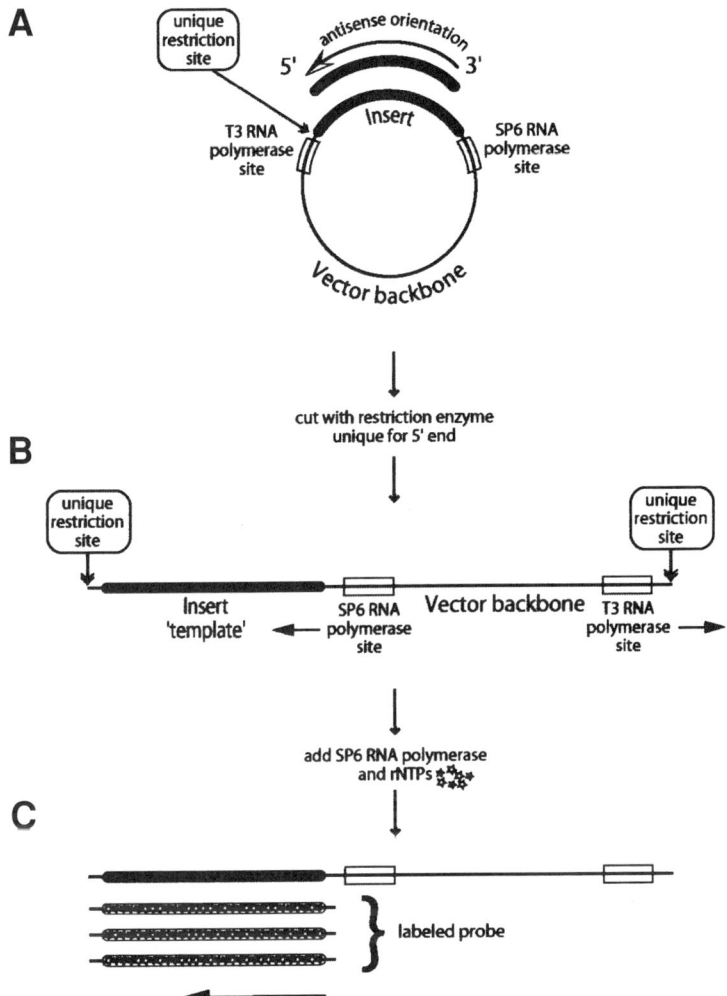

Fig. 3. Preparation of riboprobes. (**A**) The plasmid contains the vector backbone and the insert ("template") with the DNA of interest. Most plasmids have two RNA polymerase sites, one on each side of the vector. In the example provided, the plasmid has a T3 RNA polymerase site that yields sense RNA, and an SP6 RNA polymerase site that yields antisense RNA. Unique restriction sites should be available on both sides of the insert. For antisense RNA transcripts, a unique restriction site is needed on, or close to, the 5′ end of the insert. This site should be unique in the insert, but can cut in the vector backbone between the T3 and SP6 sites without consequences. (**B**) After the unique restriction site is cut with the appropriate endonucleases, a linearized plasmid is obtained. Amplification with SP6 RNA polymerase should yield labeled template sequence ("probe") (**C**), whereas T3 RNA polymerase should not yield any probe. Thus, this particular plasmid needs to be cut with the unique restriction enzyme, and labeled with SP6 RNA polymerase.

1. Add the following components at room temperature in the order listed:

5X Buffer specific for RNA polymerase provided with the polymerase by the manufacturer	4 µL
100 mM DTT	2 µL
Ribonuclease inhibitor (e.g., RNasin)	20–40 µg
Stock of rATP, rGTP, and rUTP (2.5 mM each)	4 µL
DEPC-treated dH$_2$O to a final volume of 20 µL (taking the entire reaction volume into account)	X µL
Linearized template DNA	0.2–1.0 µg
[α-^{32}P]rCTP (50 µCi at 10 µCi/µL, 800 Ci/mmol)	5 µL
SP6, T3, or T7 RNA polymerase	15–20 µg
Final volume	20 µL

   At low temperature, DNA can precipitate in the presence of the 5X buffer. It is therefore important to keep the order of addition as listed in the protocol, and to keep the mixture at room temperature during the addition. Specific activity of [α-^{32}P]rCTP can be increased to 3000Ci/mmol.

2. Incubate for 1 h at 37°C.
3. To digest the DNA template with DNase I after the transcription reaction, add RQ1 RNase-free DNase to a concentration of 1 µg/µg of template DNA and incubate for 15 min at 37°C.
4. Separate unincorporated oligonucleotides by size-exclusion chromatography (e.g., Nuctrap Push Columns, Stratagene). Equilibrate the column with 80 µL 1X TE buffer. Bring the sample volume to 80 µL with STE buffer; load onto the resin and elute sample with 80 µL of STE buffer. Measure radioactivity of 1 µL of eluate in a scintillation counter (see **Note 10**).

3.4.2. Labeling of cDNA Probes

The cDNA labeling is based on the method developed by Feinberg and Vogelstein (7), in which a mixture of random hexadeoxyribonucleotides is used for DNA synthesis from a linear double-stranded DNA template. Kits with all the necessary components are commercially available from a variety of companies (e.g., Stratagene, Promega, Invitrogen, Amersham-Pharmacia). A generic protocol is presented here (see **Note 11**). The insert has to be cut from the vector to ensure exclusive labeling of the DNA of interest ("template") (**Fig. 4**). The DNA template needs to be separated from the vector (e.g., by separating it in a low melting point agarose gel) and should have a concentration between 2 and

---

Fig. 4. *(opposite page)* Preparation of cDNA probes. (**A**) The plasmid contains the vector backbone and the insert ("template") with the DNA of interest. Unique restriction sites on both sides of the insert are needed to cut out the insert. The restriction sites can be identical, can cut in the vector backbone, but cannot cut in the insert. (**B**) After the insert is cut with the appropriate restriction enzyme(s), it is separated from the backbone (e.g., by electrophoresis). (**C**) The double-stranded cDNA

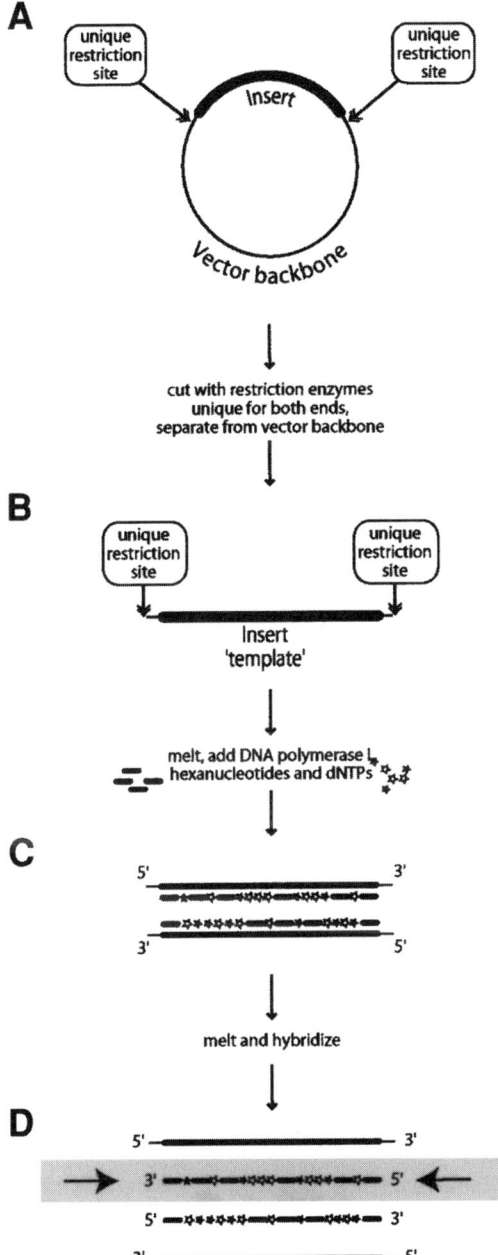

Fig. 4. *(continued)* is melted, separating both strands, and new complementary strands are synthesized with random hexamers and dNTPs. (**D**) Before hybridization to the membrane, the strands have to be melted again. Only one of the two labeled strands is complementary to the RNA on the membrane *(arrow)*.

25 µg/µL. The two strands of the template DNA have to be separated by heating to 100°C for 5 min in the presence of random oligonucleotide primers.

1. Add the following components at room temperature in the order listed:
   DEPC-dH$_2$O to a final volume of 50 µL
   (taking the entire reaction volume into account)   X µL
   Random oligonucleotide primers (hexamers)   0.3 A$_{260}$ units
   Template DNA   25 ng
2. Heat at 100°C for 5 min.
3. Centrifuge for 10 s to recollect sample at bottom of tube.
4. Add at room temperature (*see* **Note 12**):
   5X Buffer specific for DNA polymerase I,
   provided with the polymerase by the manufacturer   10 µL
   10 mg/mL Nuclease-free BSA   2 µL
   Stock of dATP, dGTP, and dTTP (500 µ*M* each)   2 µL
   [α-^{32}P]dCTP (50 µCi at 3000Ci/mmol)   5 µL
   DNA polymerase I, Klenow fragment   5 µg
   Final volume   50 µL
5. Mix and incubate the reaction tube at the temperature indicated by manufacturer of the DNA polymerase (room temperature) for 60 min.
6. To terminate the reaction, heat to 100°C for 2 min and chill on wet ice. Add EDTA to 20 m*M*.
7. Separate unincorporated oligonucleotides by size-exclusion chromatography (e.g., Nuctrap Push Columns, Stratagene).

3.5. Hybridization of Probes to Nylon Membranes

1. Wet the dry membrane in TE.
2. Prehybridize for 1–4 h at 42°C in 10 mL of hybridization solution in roller bottles or containers with tightly fitting lids. Alternatively, blots and hybridization solutions can be sealed in plastic bags and put in shaking water bath. Gentle agitate during prehybridization.
3. Heat radioactive probe for 5 min at 100°C and cool on wet ice. **Caution:** Use screw-capped lids to avoid the lid opening under the pressure.
4. Add radioactive probe to hybridization solution at 5×10^5 cpm per blot.
5. Hybridize over night at 65°C for riboprobes or at 42°C for cDNA probes with gentle agitation.
6. Wash twice (30 min each) with 2X SSC, 0.1% SDS at 65°C for riboprobes or at 42°C for cDNA probes with gentle agitation.
7. Wash twice (30 min each) with 0.2X SSC, 0.1% SDS at 65°C for riboprobes or at 42°C for cDNA probes with gentle agitation.
8. Scan with Geiger counter corners of blot. There should be little to no radioactivity detectable away from the specific bands, if blots are still radioactive. Wash twice (30 min each) with 0.2X SSC, 0.1% SDS at 65°C for riboprobes or at 42°C for cDNA probes.

A Primary striatal culture RNA

control   glutamate

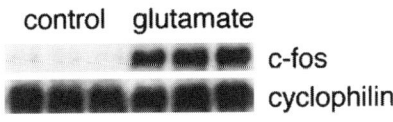
c-fos
cyclophilin

B Rat striatal RNA

control   amphetamine

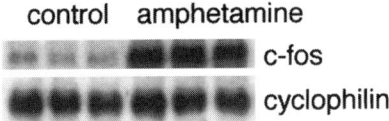
c-fos
cyclophilin

Fig. 5. Northern blots with the two RNA extraction methods described. (**A**) RNA from a primary striatal culture treated with glutamate. NA was prepared with the extraction protocol for primary neuronal culture. The membrane was developed with a c-*fos* riboprobe, stored dry to let the radioactivity decay, and subsequently probed with a cyclophilin random hexamer labeled cDNA probe. Cyclophilin is used to demonstrate equal loading *(11)*. (**B**) RNA from rat striatum prepared with the CsCl ultracentrifugation protocol. Rats were treated with either saline control or amphetamine. The membrane was developed with a c-*fos* riboprobe, stored dry to let the radioactivity decay, and subsequently probed with a cyclophilin random hexamer labeled cDNA probe.

9. Wrap wet membrane in saran wrap and expose to X-ray film with intensifying screens at –80°C (*see* **Notes 13** and **14**). Alternatively, if you have access to a phosphorimager system, use this system
10. Representative Northern blots for primary striatal culture and rat striatum are shown in **Fig. 5**.

4. Notes

Following is a list of problems that may be encountered, with suggestions on how to uncover and correct these problems and how to avoid them in subsequent preparations.

1. RNA is degraded by RNases from brain tissue and external sources. Tissue needs to be frozen immediately after harvest at –80°C and should not be subjected to freeze–thaw cycles. Primary neurons from culture cannot be frozen before nuclei have been separated. These cells need to be harvested rapidly in lysis buffer on wet ice, centrifuged in the cold room, and supernatants can then be frozen or worked up further. RNases need to be inactivated rapidly by denaturing solutions (e.g., guanidine thiocyanate) and separated from RNA at low temperature (4°C).

Be aware that guanidine thiocyanate does not irreversibly denature RNases. RNases may renature when denaturing agents are removed and degrade the sample.
2. Brinkmann/Eppendorf sells a product called "phase lock gel," which forms a stable, gelatinous barrier between the organic and aqueous phases. This allows an easy aspiration and transfer of the aqueous phase.
3. The protocol provided is for up to 300 mg of tissue. For higher amounts of tissue, the protocol can be upgraded to larger ultracentrifuge tubes and rotors. *See* **ref. 4** for increased amounts of tissue. Many companies provide RNA purification systems. These systems are most often based on guanidine thiocyanate methods *(9,10)* that do not require ultracentrifugation steps. In many instances and for many applications, these systems work well for purifying RNA from brain tissue. However, for larger amounts of tissue or for purest types of RNA, we prefer to use the cesium chloride gradient method.
4. A low A_{260}/A_{280} ratio may be caused by protein contamination. Perform an additional phenol–chloroform–isoamyl alcohol (25:24:1) extraction on the purified RNA, followed by ethanol precipitation and ethanol wash. Loss of RNA should be expected. A low A_{260}/A_{280} ratio may also caused by contamination of the aqueous phase with the phenol phase. Thus, during the extraction do not completely remove the aqueous layer. Reextract with chloroform–isoamyl alcohol (49:1) to remove phenol, followed by ethanol precipitation and ethanol wash. Loss of RNA should be expected. Finally, acidic water used for the spectrophotometric measurement (we use DEPC-treated distilled water for the spectrophotometric measurement, and high acidity has turned out to be most often responsible for low A_{260}/A_{280} ratios) may cause low A_{260}/A_{280} ratio. In this case, use Tris-HCl or phosphate buffer, pH 7.6–8.6, instead of distilled water for spectrophotometric measurement. *See* **ref. 8** for further information on low pH and A_{260}/A_{280} ratios.
5. The integrity of the purified RNA can be determined by denaturing agarose gel electrophoresis. Prepare a miniformaldehyde gel, load 1 µg of RNA per lane, run at 70 V for 1 h, stain in ethidium bromide (10 µg/mL)–TE bath for 15 min, destain in TE for 5 min, and visualize under UV light (254 nm) (**Caution:** Ethidium bromide is a carcinogen). You should see two bands, the larger 28S ribosomal RNA band and the smaller 18S ribosomal RNA band (**Fig. 6**). The ratio of 28S to 18S ribosomal RNAs should be approx 2:1. If RNA samples have been partially degraded, the bands will be less sharp and the intensity of the 28S RNA will be diminished. A total degradation of RNA will result in the loss of the 28S and 18S RNA bands, and the occurrence of a "smear" of smaller, degraded size RNA in the lanes.
6. Salt contamination can retard the movement of RNA in the gel. To solve this problem, use ethanol washes to remove any residual salt from the pellet. Add 70–75% ethanol to the RNA pellet, vortex-mix vigorously, and centrifuge for 10 min to recollect the pellet. Carefully aspirate off the ethanol.

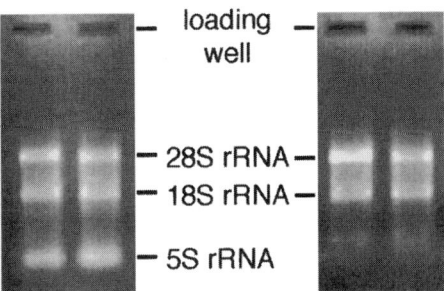

Fig. 6. Staining of an RNA gel shows 28S and 18S ribosomal RNA bands. Human hippocampus RNA was loaded onto a denaturing gel at 1 µg of RNA per lane. Gel was stained with ethidium bromide and visualized under UV light (254 nm). 28S and 18S ribosomal RNAs (rRNA) can be seen in both blots, with a size of 4.7 and 1.9 kb, respectively. The molar ratio of 28S to 18S rRNA is 1. *Left blot*: In some preparations, a smaller, 5S rRNA band can be seen, with a size of 120 nucleotides. 5S rRNA does not come from the same precursor as 28S and 18S rRNAs, and is therefore independent from them in its molar concentration.

7. Genomic contamination of RNA causes an overestimation of RNA loaded per well and may contribute to uneven loading. We have never observed this problem with the cesium chloride protocol, but have with the cell culture protocol. Nuclei need to be separated from the cytosol before the samples can be frozen. If nuclei are not intact and DNA leaks out it may contaminate the preparation. A solution is to perform an additional acidic phenol–chloroform–isoamyl alcohol (25:24:1) extraction on the purified RNA to separate DNA into the interphase, followed by ethanol precipitation and ethanol wash. Loss of RNA should be expected.
8. No RNA visible by UV shadowing of the membrane is caused by either too little RNA loaded per well, or the electrophoretic transfer performed in the wrong direction. In the first case, radiolabeled riboprobes may still yield a visible band. In the second case, the RNA is lost. Make sure to have the membrane at the anode side, the gel at the cathode side (**Fig. 1**).
9. Many companies sell systems that include enzymes, buffers, nucleotides, and control reactions for the labeling of RNA riboprobes (e.g., Promega, Amersham-Pharmacia, etc.) or random-primed DNA probes (e.g., Amersham-Pharmacia, Invitrogen, Promega, Stratagene, etc.). We prefer to use these systems to label probes. Systems that use alternatives to radioactive nucleotides are available as well. However, they are generally less sensitive.
10. There could be various reasons for low yield of radioactive labeled riboprobes: (1) degradation of newly synthesized RNA due to contamination with RNases, (2) high salt concentration of template could inhibit RNA polymerase, (3) template may be precipitated by the polymerase buffer, (4) the template may be

too short or the wrong polymerase was used. To solve the problem, use RNase-free working area and tools, and precipitate template with 75% ethanol at –20°C for 20 min, centrifuge for 10 min at 16,000g at 4°C, aspirate supernatant, wash pellet with 1 mL of 75% ethanol, centrifuge for 5 min at 16,000g at 4°C, aspirate the supernatant, and let the pellet air-dry. This procedure should remove residual salts from the template. You may have to redetermine the amount of DNA. Use all solutions at room temperature, do not use ice, and make sure the restriction enzyme used for cutting the vector leaves at least a 300-basepair sequence from the polymerase site.

11. For the purpose of reprobing the membrane, cDNA probes can be stripped off by adding boiling 0.1% SDS in distilled DEPC water, then letting it come down to room temperature while shaking for 30 min. Riboprobes cannot be stripped. These blots can be reused only after the radioactivity is decayed.
12. If the DNA template is in low melting point agarose, the agarose may inhibit the labeling reaction. It is also important to stay at room temperature or higher during the labeling to avoid gelling of agarose. Extract DNA template from the agarose to minimize this problem.
13. The sticky probe may make X-ray film come out too dark. Rewashing the blot may solve this problem.
14. Visible ribosomal band is a problem if the specific band is too close to the ribosomal bands. Decreasing the amount of radiolabeled probe added to the hybridization solution may solve the problem.

References

1. Sambrook, J., Fritsch, E. F., and Maniatis, T. (1989) *Molecular Cloning: A Laboratory Manual*, 2nd edit. Cold Spring Harbor Laboratory Press, Cold Spring Harbor, NY.
2. Rajadhyaksha, A., Barczak, A., Macías, W., Leveque, J. C., Lewis, S., and Konradi, C. (1999) L-type Ca^{2+} channels are essential for glutamate-mediated CREB phosphorylation and c-fos gene expression in striatal neurons. *J. Neurosci.* **19,** 6348–6359.
3. Perry, R. P., La Torre, J., Kelley, D. E., and Greenberg, J. R. (1972) On the lability of poly(A) sequences during extraction of messenger RNA from polyribosomes. *Biochim. Biophys. Acta* **262,** 220–226.
4. Berger, S. L. and Chirgwin, J. M. (1989) Isolation of RNA. *Methods Enzymol.* **180,** 3–13.
5. Leveque, J. C., Macías, W., Rajadhyaksha, A., et al. (2000) Intracellular modulation of NMDA receptor function by antipsychotic drugs. *J. Neurosci.* **20,** 4011–4020.
6. Konradi, C., Leveque, J. C., and Hyman, S. E. (1996) Amphetamine and dopamine-induced immediate early gene expression in striatal neurons depends on postsynaptic NMDA receptors and calcium. *J. Neurosci.* **16,** 4231–4239.
7. Feinberg, A. P. and Vogelstein, B. (1983) A technique for radiolabeling DNA restriction endonuclease fragments to high specific activity. *Analyt. Biochem.* **132,** 6–13.

8. Wilfinger, W. W., Mackey, K., and Chomczynski, P. (1997) Effect of pH and ionic strength on the spectrophotometric assessment of nucleic acid purity. *Biotechniques* **22,** 478–481.
9. Chomczynski, P. and Sacchi, N. (1987) Single-step method of RNA isolation by acid guanidinium thiocyanate–phenol–chloroform extraction. *Analyt. Biochem.* **162,** 156–159.
10. Chirgwin, J. M., Przybyla, A. E., MacDonald, R. J., and Rutter, W. J. (1979) Isolation of biologically active ribonucleic acid from sources enriched in ribonuclease. *Biochemistry* **18,** 5294–5299.
11. Danielson, P. E., Forss-Petter, S., Brow, M. A., et al. (1988) p1B15: a cDNA clone of the rat mRNA encoding cyclophilin. *DNA* **7,** 261–267.

12

Analysis of Gene Expression in Striatal Tissue by Multiprobe RNase Protection Assay

Neil M. Richtand

1. Introduction

Studies of RNA expression within the central nervous system (CNS) have contributed significantly to increased understanding of both the acute and long-term neurological effects of drugs of abuse. Before beginning such an investigation, however, one must first determine the method of RNA analysis most applicable to the problem at hand. The four most commonly used techniques include *in situ* hybridization, Northern blot, reverse transcriptase-polymerase chain reaction (RT-PCR), and ribonuclease (RNase) protection assay. Each method has its own strengths and limitations, and these factors must be evaluated carefully to determine the method most appropriate for the given problem.

A major but frequently overlooked limitation of all four methods is the fact that these methodologies measure mRNA expression. In contrast, the scientific question of greatest interest in determination of the neurological effect of a given drug is often determination of the effect on protein, rather than mRNA expression. There are several known examples of single proteins in the mammalian CNS in which changes in mRNA level reliably predict later change in protein expression *(1)*. Evaluation of large numbers of yeast genes for which both mRNA and protein expression data are available, however, demonstrates little correlation between mRNA and protein expression levels, with protein expression changing 20-fold, in some cases, in the absence of any appreciable change in mRNA level *(2)*. Whether the same is true in mammalian systems has not been clearly determined. The importance of information regarding both mRNA and protein levels in evaluating regulation of protein function therefore cannot be overestimated.

The major advantage of Northern blot analysis lies in the ease of the technique, which is generally less problematic than any of the other methods. The major disadvantages of Northern blot lie in the lack of sensitivity and the lack of anatomic specificity. The major advantage of *in situ* hybridization lies in the anatomic specificity provided. This is of particular importance for studies in the CNS, in which discrete populations of cells within a given brain region may play an important regulatory role. The disadvantage of *in situ* hybridization lies in the lower sensitivity compared to RT-PCR, and the lack of absolute specificity when studying closely homologous mRNAs. In contrast, the advantage of RT-PCR lies in the exquisite sensitivity resulting from the multiple amplification steps of this method.

One of the major advantages of the RNase protection assay is the absolute specificity of this technique in determining expression of the mRNA of interest. This advantage is particularly powerful in studies examining expression of closely homologous transcripts, or of alternatively spliced transcripts arising from a single gene. Second, RNase protection assay is highly sensitive. In practice, it is possible to detect reliably approx 1–5 pg of target RNA with the RNase protection assay employing high specific activity probes. The third major advantage of this technique is the ease with which multiple samples may be assayed, once a given assay is established. This makes it feasible to measure mRNA expression in individual animals from several reasonably sized groups, an approach frequently employed in studies simultaneously measuring the behavioral and cellular effects of drugs of abuse *(3,4)*.

2. Materials
2.1. cDNA Insert Plasmid Construction

Plasmids used for synthesis of both synthetic RNA standards used for standard curves and cRNA probes may be obtained by synthesizing your own (*see* **Subheading 3.3.**), or "clone-by-phone." It's generally worth the effort it takes to send an e-mail to try and obtain an existing clone from another researcher or commercial source prior to embarking on your own synthesis. If synthesized in-house, the following materials are needed.

1. M-MLV reverse transcriptase (Invitrogen).
2. Restriction enzymes (Invitrogen).
3. Yeast tRNA (RNase-free, Invitrogen cat. no. 15401-011).
4. Random hexamers (Invitrogen).
5. Oligo(dT)-cellulose spin columns (Clontech).
6. pGEM-4 Vector and RNasin (Promega).

7. 5X Plasmid buffer: 250 mM Tris-HCl, pH 8.3, 375 mM KCl, and 15 mM MgCl$_2$.
8. Dithiothreitol (DTT).
9. 10X Proteinase K buffer: 10 mM Tris-HCl, pH 7.8, 5 mM EDTA, 0.5 % sodium dodecyl sulfate (SDS).
10. Proteinase K (Invitrogen) dissolved into 10 mg/mL solution, divided into 10-µL aliquots, and stored frozen at –20°C. It should not be freeze–thawed more than five times.

2.2. mRNA Standard Production

1. Deoxyribonucleoside triphosphates (dNTPs): store at –70°C (they may be purchased individually, or premixed as ribonucleoside triphosphate sets from Roche).
2. Diethyl pyrocarbonate (DEPC)-treated water *(5)*.

2.3. Radiolabeled cRNA Probe Preparation

Probes may be synthesized as described in **Subheading 3.3.** using the following individual reagents.

1. DNA-dependent RNA polymerases (SP6, T7; Promega).
2. RNasin (Promega).
3. RQ1 DNase (Promega).
4. Nuctrap columns (Stratagene)
5. [α-^{32}P]UTP (800 Ci/mmol; New England Nuclear).
   Alternatively, commercial kits are available containing nonradioactive reagents required for probe preparation, such as MAXI script kit (Ambion, Austin, TX).

2.4. mRNA Sample Preparation

Commercial kits containing needed reagents, and directions for their use, are available (TRI REAGENT, Molecular Research Center, Cincinnati, OH). Alternatively, the following recipe may be prepared.

1. Stock denaturing solution (100 mL): 50 g of guanidine isothiocyanate (GIBCO, cat. no. 5535UA), 59.48 mL of H$_2$O, 2.64 mL of 1 M sodium citrate, pH 7, and 5.28 mL of 10% sarcosyl, heat at 62.5°C to dissolve. Store at room temperature up to 3 mo.
2. Working denaturing solution: Add 350 µL of β-mercaptoethanol and one drop of antifoam A (Sigma) to 50 mL of stock denaturing solution. Store refrigerated up to 1 mo.
3. Coronal brain matrix for brain tissue dissection (Research Instruments and Manufacturing, Corvalis, OR).

2.5. Ribonuclease Protection Assay

The assay can be performed using individual reagents including RNase ONE (Promega). Alternatively, commercial RNase protection assay kits such as RPA II Kit (Ambion) include all reagents required for assay.

1. Nucleotide triphosphates (NTPs) may be purchased individually, or premixed as ribonucleotide triphosphate sets from Roche. They are labile at room temperature and should be stored in small aliquots at –70°C.
2. Vacuum sampling manifold Model 1225 (Millipore).
3. Hybridization buffer: 200 mM piperazine-1,4-*bis*(2-ethanesulfonic acid) (PIPES), pH 6.4, 2 M sodium acetate, and 5 mM EDTA.
4. Digestion buffer: 100 mM Tris-HCl, pH 7.5, 50 mM EDTA, and 2 M sodium acetate.

3. Methods
3.1. Plasmid Construction

If cDNA plasmids for synthesis of cRNA riboprobes are not already in hand, these can readily be constructed by RT-PCR amplification. Construction of dopamine receptor and cyclophilin plasmids is described in **step 9**. Total RNA is extracted from rat brain as described in **Subheading 3.4.** mRNA is next poly(A) selected using oligo(dT)-cellulose. This is best performed following the specific manufacturer's protocol (Clontech). cDNA is synthesized from 15 µg of poly(A)-selected mRNA in a 750-µL reaction mixture as follows, as adapted from **ref. 5**:

1. Mix 150 µL of 5X plasmid buffer, 75 µL of 0.1 M DTT, 75 µL of 10 mM dNTP mix, 37.5 µL of random hexamer, 75 µL of RNasin, 15 µL of RNA (1 µg/µL in DEPC-treated water), and 285 µL of DEPC-treated water to a final total volume of 750 µL.
2. Add 37.5 µL of M-MLV reverse transcriptase.
3. Incubate for 2 h at 37°C. Then precipitate by addition of 0.1 volume of 3 M NaOAc, pH 5.2. Add two volumes of ice-cold 100% ethanol.
4. Store on ice 30 min, and then centrifuge for 15 min in the cold. Rinse pellet by addition of 70% ethanol.
5. Digest pellet with Proteinase K as follows:
   a. Suspend DNA pellet in 90 µL of TE, and then add 10 µL of 10X Proteinase K buffer.
   b. Heat at 68°C for 3 min, and then add 1 µL of Proteinase K (10 mg/mL).
   c. Incubate at 65°C for 1 h, and then add 100 µL of TE, pH 8.0.
   d. Extract once with phenol–CHCl$_3$–isoamyl alcohol (25:24:1) and once with CHCl$_3$ isoamyl alcohol (24:1).
6. Add 10 µL (0.04 volumes) of 5 M NaCl, vortex-mix, and centrifuge. Add two volumes of ice-cold 100% ethanol. Store on ice for 30 min, centrifuge in the cold at 12,000g for 15 min, and rinse pellet with cold 70% ethanol.
7. The pellet is redissolved in a small volume of 10 mM Tris-HCl, pH 8.0. The purified rat brain cDNA may then be used as a template for amplification of a sequence of interest.

Table 1
PCR Primer Sequences

Gene	Primer sequence	Annealing temperature
Cyc	5′-**GGGA**ATTCCAGGATTCATGTGCCAG-3′	61°C
	5′-**CCGGA**TCCTGAGCTACAGAAGGAATGG-3′	
D1	5′-**GGGAATT**CAGCTAAGCTGGCACAAGGCAA-3′	61°C
	5′-**GGC**TGCAGAATGGCTGGGTCTCC-3′	
D2	5′-**GGGAATT**CGCAGCAGTCGAGCTTTCAGA-3α	61°C
	5′-**GGCTGCA**GCTCATCGTCTTAAGGGAGGT-3′	
D3	5′-**GGGAATT**CTCACTCGACAGAACAGCCA-3′	63°C
	5′-**GGGGATCC**GGACGTGGATAACCTGCCGT-3′	

"GC clamp" and added restriction endonuclease sites are noted in boldface.

8. PCR primer sequences may be designed with "GC clamps" at the 5′ end to facilitate annealing. It is convenient to include also restriction endonuclease sites. **Table 1** provides sets of PCR primers for amplification of D1, D2, and D3 dopamine receptors and cyclophilin. Each sequence is amplified in a 20-µL PCR reaction solution containing 20 mM Tris-HCl, pH 8.3, 25 mM KCl, 1.5 mM MgCl$_2$, 0.2 µM each primer, 0.1 mM each dNTP, and 2 U of *Taq* polymerase. PCR reactions for D1 and D2 receptors and cyclophilin employed an initial denaturation at 94°C for 1.5 min, followed by 30 cycles at 94°C for 30 s, 30 s at 61°C, and 72°C for 30 s. The D3 receptor PCR reaction employed an annealing temperature of 63°C. Each reaction was terminated following a 3-min elongation at 72°C. Amplified PCR products were then separated by agarose gel electrophoresis, and bands corresponding to the products of interest may be purified onto DEAE-cellulose membranes (*5*). PCR amplification across alternative splice site junctions provides products of different size, corresponding to each splice isoform. To provide sufficient product for subsequent steps, it is generally necessary to reamplify the gel-purified products by PCR under identical conditions. The final product may then be extracted once with phenol–CHCl$_3$–isoamyl alcohol (25:24:1) and once with CHCl$_3$–isoamyl alcohol (24:1), precipitated with sodium acetate, and then washed with 70% ethanol.
9. Plasmids are constructed by resuspending the purified PCR product in restriction digest buffer. D1 and D2 receptor inserts were digested with *Eco*R I and *Pst* I, and D3 and cyclophilin inserts digested with *Eco*R I and *Bam*H I. Following purification by agarose gel electrophoresis, restriction digest products are ligated into the corresponding restriction sites of PGEM-4 with T4 DNA ligase according to standard protocols (*[5]*, Ligation Reactions, Ligation of Cohesive Termini 1.68). Plasmids are grown in bacterial culture, purified, and plasmid DNA sequenced to confirm the identity of the insert in each plasmid construct.

3.2. Standard Production

Use of a standard curve for each ribonuclease protection assay provides confirmation that the assay is measuring accurately the unknown input mRNA. Synthetic RNAs for use as standards are easily synthesized using a "cold runoff" of the opposite strand to that used to synthesize radioactive riboprobes. In practice, it is easiest to "spike" the cold runoff with a trace amount of radioactive dNTP, so as to calculate the yield of mRNA.

Synthetic RNA for use as standards in ribonuclease protection assays is made by DNA-dependent RNA polymerase reaction using plasmids constructed as described in **Subheading 3.1.** as template. Plasmids are cut so that the antisense strand is used as template for the synthesis reaction for standard (while the sense strand is used as template for synthesis of riboprobes; *see* **Subheading 3.3.**).

1. Digest plasmid DNAs with *Bam*HI for D3 receptor and cyclophilin, and with *Hin*dIII for D1 and D2 receptor.
2. Treat plasmids with Proteinase K (*see* **Subheading 3.1.**), extract with phenol–$CHCl_3$–isoamyl alcohol (25:24:1) and $CHCl_3$–isoamyl alcohol (24:1), precipitate with sodium chloride, and wash with 70% ethanol.
3. Purify plasmid DNAs by agarose gel electrophoresis and electrophoresis onto DEAE-cellulose membranes *(5)*.
4. RNA for standard curves in nuclease protection assays (**Subheading 3.5.**) is synthesized with SP6 DNA-dependent RNA polymerase from the appropriate plasmid construct digests. Reactions contain transcription buffer supplied by the manufacturer, 5 m*M* DTT; 20 U of RNasin, 7.5 U of SP6 DNA-dependent RNA polymerase, 2.5 m*M* ribonucleoside-triphosphates, and 0.4 µCi of [α-^{32}P]UTP added as tracer (final volume = 10 µL).
5. Incubate for 1 h at 37°C. Add 10 µL of tRNA (10 mg/mL) and 5 U of DNase.
6. Incubate for 30 min at 37°C. The RNA product is then purified by phenol–chloroform–isoamyl alcohol (25:24:1) extraction and elution through a Nuctrap column.
7. The yield of RNA is calculated from the amount of radioactivity in an aliquot of the product using the known specific activity of [α-^{32}P]UTP. An aliquot of RNA standards should also be electrophoresed by denaturing polyacrylamide gel electrophoresis to verify that the product is a single band of the expected size.

3.3. Probe Preparation

If a commercial kit (such as MAXI script kit) is used for probe production, it is recommended that the manufacturer's instructions be followed. Detailed instructions are given in the following subheadings for probe synthesis using individually purchased reagents (*see* **Note 1**).

3.3.1. Plasmid Linearization

1. For riboprobe synthesis, we linearize approx 20 µg each of the plasmid constructs listed in **Table 1**. by *Eco*RI digestion. After digestion, plasmids are treated with Proteinase K (*see* **Subheading 3.1.**), extracted with phenol–CHCl$_3$–isoamyl alcohol (25:24:1) and CHCl$_3$–isoamyl alcohol (24:1), precipitated with sodium chloride, and washed with 70% ethanol.
2. Electrophorese linearized plasmid DNA through agarose gel and purify by electrophoresis onto a DEAE-cellulose membrane *(5)*.
3. Resuspend the final plasmid pellet at a concentration of 1 mg/mL in TE buffer, pH 8.0, and store frozen at –20°C. The sample may be thawed and aliquots removed as needed.

3.3.2. Probe Synthesis

1. Add the following components at room temperature: 2 µL of 5X transcription buffer (as supplied with DNA-dependent RNA polymerase); 0.4 µL of linearized plasmid (containing ~0.4 µg of DNA); 0.5 µL of 0.1 *M* DTT; 0.5 µL of 10 m*M* ATP, CTP, and GTP; 0.5 µL of RNasin (20 U); 5.6 µL of [α-^{32}P]UTP (800 Ci/mmol, 50 µ*M*; final reaction concentration 28 µ*M*); and 0.5 µL of T7 DNA-dependent RNA polymerase (7.5 U).
2. Incubate for 1 h at either 30°C or 37°C. Incubation temperature should be individualized for each sequence. Probe synthesis at 30°C gives lower yield, but tends to produce more full-length product, and in our hands was often preferable. Probe synthesis at 37°C gives higher yield, but depending on specific sequence may have a larger percentage of incomplete length product.
3. Add 10 µL of tRNA (10 mg/mL), and then incubate 30 s at 68°C.
4. Add 5 µL of RQ1 DNase, and then incubate 30 min at 37°C.
5. Add 55 µL of DEPC-treated water. Extract with 80 µL of phenol–CHCl$_3$–isoamyl EtOH.
6. Unincorporated NTP nucleotides are separated from the RNA riboprobe using NucTrap columns or a comparable method *(5)*. For a NucTrap column, 70 µL of extracted reaction product is applied to a NucTrap column previously equilibrated with 70 µL of 1β STE buffer (0.1 *M* NaCl; 10 m*M* Tris-Cl, pH 8.0; and 1 m*M* EDTA). The column is rinsed with 70 µL of 1X STE buffer and the eluate collected.
7. If the probe is mainly full-length transcript (determined by running an aliquot on gel electrophoresis), it may be used for RNase protection assay without further purification. Gel purification is needed in cases in which full-length probe is not the major reaction product.

3.4. mRNA Sample Preparation

One of the important factors in planning studies of mRNA expression is determination of whether the method employed is sufficiently sensitive to

Table 2
Tissue Weight and RNA Yield from Rat by Brain Region

Tissue	Approximate weight (mg)	Approximate yield (µg) total RNA/animal
Caudate/ putamen	35	20
Nucleus accumbens	35	30
Substantia nigra/ventral tegmentum	30	20
Locus coeruleus	20	12
Prefrontal cortex	110	95
Hypothalamus	75	50
Olfactory tubercle	55	55
Hippocampus	75	63

Dissection was made by rodent coronal brain matrix *(7)*.

detect the transcript being measured. This will be determined by two variables: transcript prevalence and mass of total RNA in the tissue studied. The weight of tissue studied will determine the yield of total RNA extracted, which will in turn guide details of the study being planned. We dissect brain tissue using a coronal brain matrix, as has been described previously *(7)*. The mass of individual brain section and total RNA yield obtained using this dissection method are listed in **Table 2**. We extract RNA according to the method of Chomczynski and Sacchi *(6)*, modified only slightly from *(8)* (*see* **Note 2**).

1. Dissect tissue quickly on ice. Immediately freeze tissue in liquid nitrogen. Homogenize individual tissue sample (weight 30–75 mg) in 810 µL of denaturing solution. Transfer to a 2-mL microfuge tube containing 81 µL of 2 M sodium acetate, pH 4.0.
2. Mix, add 810 µL of water-saturated phenol, and mix.
3. Add 162 µL of $CHCl_3$–isoamyl EtOH (49:1), mix, and allow to stand on ice for 15 min.
4. Centrifuge for 20 min at 10,000g at 4°C.
5. Transfer the upper phase to a 2-mL microfuge tube containing 810 µL of isopropanol. Incubate for 30 min at –20°C. Centrifuge for 10 min at 10,000g at 4°C, and resuspend pellet in 300 µL of denaturing solution.
6. Transfer to a 1.5-µL microfuge tube containing 300 µL of isopropanol. Incubate for 30 min at –20°C. Centrifuge for 10 min at 10,000g at 4°C.
7. Wash pellet two times with 700 µL of 75% EtOH. Vortex-mix, allow to stand for 15 min at room temperature, and then precipitate by centrifugation for 5 min at 10,000g.
8. Air-dry the pellet, being careful not to overdry. Resuspend pellet in 100 µL of TE.

3.5. Ribonuclease Protection Assay

Ribonuclease protection assays may be performed by a variety of methods. Several commercial kits are available that include reagents and protocol for the assay. In all cases radiolabeled riboprobe, in at least a 10-fold excess, is allowed to equilibrate with the mRNA of interest in a solution hybridization reaction. Single-stranded, unhybridized riboprobe is digested by ribonuclease, while riboprobe hybridized with the target mRNA is protected from degradation. Undigested riboprobe, proportional to the mass of target mRNA in the assay sample, may be ascertained by polyacrylamide gel electrophoresis, followed by determination of radioactivity in the band of interest (by Phosphor-Imager, or by cutting out the band and scintillation counting). Alternatively, the hybridized, undigested riboprobe may be trichloroacetic acid (TCA)-precipitated unto GF/C filters. The latter method, as we have modified from previously published protocols *(9,10)*, is quick and efficient for processing large numbers of samples, as is often the case in studies of neurological effects of drugs of abuse, and is described in detail in **step 9** (*see* **Note 3**).

1. Set up assay tubes containing 50 µg of RNA in a total volume of 200 µL. tRNA (10 mg/mL) is added to samples to achieve the needed mass. A standard curve with hybridizations containing 2–20 pg of RNA standard should also be included, as well as samples omitting nuclease and RNA as control hybridizations.
2. Samples are precipitated with ammonium acetate by addition of 52 µL of 10 M ammonium acetate. Vortex-mix. Add 520 µL of cold 100% EtOH. Vortex-mix. Place at –20°C for 30 min. Precipitate by microfuging in the cold (maximal speed) for 20 min.
3. Rinse pellet with 1 mL of 70% EtOH. Allow to stand at –20°C for 30 min. Microfuge in the cold at maximal speed for 20 min. Remove supernatant. Allow pellet to air-dry at room temperature, or if absolutely necessary vacuum dry for 50 s on rotovap (do not allow pellet to dry completely).
4. Hybridize with 0.5 µCi of probe in the presence of 40 mM PIPES, pH 6.4; 0.4 M sodium acetate; 1 mM EDTA; and 80% formamide as follows: Add 6 µL of 5X hybridization buffer (200 mM PIPES, pH 6.4; 2 M sodium acetate; 5 mM EDTA) containing 0.5 µCi of probe. Vortex-mix if needed to dissolve sample. Add 24 µL of 100% formamide. **Make certain sample is fully dissolved.** Incubate in a water bath at 85°C for 10 min. Hybridization time and temperatures must be determined for individual mRNAs. For D1, D2, D3 receptor, and cyclophilin mRNA, reduce temperature and hybridize at 55°C overnight (at least for 14 h).
5. After hybridization, place tubes on ice. Add 300 µL per assay tube of a mix containing 1 µL RNaseONE (10 U), 33 µL of 10X digestion buffer, and 267 µL of H$_2$O. Vortex-mix gently, and incubate at 37°C with shaking for exactly 1 h.
6. Place tubes on ice, and add SDS to 0.1% and 30 µg of carrier tRNA by adding 10 µL per tube of mix containing: 3.7 µL of DEPC-treated water/tube, 3.3 µL of 10% SDS/ tube, and 3 µL of tRNA (10 mg/ml per tube). Vortex-mix.

7. Ethanol precipitate by adding 700 μL of cold 100% EtOH per tube. Vortex-mix. Place at –20°C for 30 min. Microfuge at maximal speed in the cold for 20 min. Wash the pellet with 700 μL of 70% EtOH.
8. Resuspend the final pellet for polyacrylamide/urea gel electrophoresis or quantitation by liquid scintillation counting. For electrophoresis, resuspend sample in 5 μL of DEPC-treated water and add 5 μL of load dye. Dilute hotter assay samples without nuclease by resuspending in 300 μL of DEPC-treated water, and adding 5 μL of the dilution to 5 μL of load dye for electrophoresis.
9. For quantitation of mRNA samples by scintillation counting, resuspend pellet in 300 μL of 1X RNase digestion buffer and vortex-mix. Add 700 μL of ice-cold 10% TCA containing 20 mM sodium pyrophosphate, and place on ice for 15 min. Precipitate onto GF/C glass fiber filters on a vacuum sampling manifold, and rinse tube two times with 10% TCA containing 20 mM sodium pyrophosphate. Rinse filter with 2 mL of 10% TCA containing 20 mM sodium pyrophosphate and 10 mL of 5% TCA containing 20 mM sodium pyrophosphate. Count the filter paper in 10 mL of scintillation fluid.

3.6. Data Analysis

A major advantage of the RNase protection assay method is that it allows calculation of the actual mass of sample mRNA assayed (*see* **Note 4**). This value may be calculated using the specific activity of the probe employed, as has been previously described *(9)*. The relationship between actual mass of the protected fragment and sample radioactivity is described by the formula:

mass protected fragment = (dpm per sample/specific activity [α-^{32}P]UTP)
× (molecular weight protected fragment/moles of ^{32}P per mole of probe)

4. Notes

1. Probe preparation: The preparation of clean, high-quality probe is one of the more critical aspects of the RNase protection assay. Use of fresh reagents of the highest quality cannot be overemphasized. In particular, NTPs are a labile reagent. The K_m for RNA polymerase enzymes is generally close to the highest achievable concentration of radiolabeled NTP in the runoff reaction *(11)*. Therefore, decreases in NTP concentration due to degraded NTPs will have an adverse effect on probe quality and quantity. For this reason, it is prudent to check probe quality with a polyacrylamide gel to ensure production of a full-length runoff product before each RNase protection assay. In many cases, perhaps related to specific plasmid sequence, production of a homogeneous full-length probe is impossible. In these cases the probe should be run on a polyacrylamide gel, and the probe gel purified prior to use for the RNase protection assay.
2. mRNA sample preparation: High-quality, nondegraded RNA is critical to the success of this method. The quality of total RNA may be assessed by electrophoresis

on an 0.8% agarose gel. Apply approx 5 μg of total RNA to see 18S and 28S bands clearly. For undegraded total RNA, the intensity of 28S (higher molecular weight) RNA is approximately twice that of 18S RNA (lower molecular weight).
3. Ribonuclease protection assay: All laboratory methods involving use of RNA are prone to RNA degradation by RNases, which are ubiquitous, highly resistant to inactivation, and may be used in milligram quantities as part of the RNase protection assay. For this reason all reagents in contact with RNA should be RNase-free. It is generally good practice to set aside a section of the lab specifically devoted to RNA work. Reagents should be reserved specifically for RNA work, and working solutions aliquoted into disposable RNase-free plasticware and kept separate from stock containers. If RNase contamination occurs, it is often simpler to discard all working solutions than to track down the specific source of contamination. Procedures for making the lab "RNase-free" *(5)* should be followed prior to initiating development of an RNase protection assay. Particular care should be taken when handling powdered or concentrated RNase solutions. This work should be done in a ventilation hood to avoid contamination of other parts of the lab.
4. Data analysis: In practice, a standard curve is determined for each probe using the synthesized RNA standards with each RNase protection assay. It is critical to verify in each case that the measured mRNA level is linearly proportional to the input mRNA over the entire range assayed. This confirms the requirement for probe concentration to be in great excess of the measured transcript (≥ 10-fold) throughout the entire assay range *(9)*. As an additional check, one should calculate the mass of mRNA detected in the assay samples from the specific activity of the protected radiolabeled probe. This value should be in close agreement to the magnitude of input mass of the mRNA standard.

Acknowledgments

This work was supported by the Department of Veterans Affairs Medical Research Service and by a NARSAD Essel Investigator Award.

References

1. Wang, H. Y., Runyan, S., Yadin, E., and Friedman, E. (1995) Prenatal exposure to cocaine selectively reduces D1 dopamine receptor-mediated activation of striatal Gs proteins. *J. Pharmacol. Exp. Ther.* **273,** 492–498.
2. Gygi, S. P., Rochon, Y., Franza, B. R., and Aebersold, R. (1999) Correlation between protein and mRNA abundance in yeast. *Mol. Cell Biol.* **19,** 1720–1730.
3. Richtand, N. M., Kelsoe, J. R., Kuczenski, R., and Segal, D. S. (1997) Quantification of dopamine D1 and D2 receptor mRNA levels associated with the development of behavioral sensitization in amphetamine treated rats. *Neurochem. Int.* **31,** 131–137.
4. Hondo, H., Spitzer, R. H., Grinius, B., and Richtand, N. M. (1999) Quantification of dopamine D3 receptor mRNA level associated with the development of

amphetamine-induced behavioral sensitization in the rat brain. *Neurosci. Lett.* **264,** 69–72.
5. Sambrook, J., Fritsch, E. F., and Maniatis, T. (1989) *Molecular Cloning: A Laboratory Manual,* 2nd edit. Cold Spring Harbor Laboratory Press, Cold Spring Harbor, NY.
6. Chomczynski, P. and Sacchi, N. (1987) Single-step method of RNA isolation by acid guanidinium thiocyanate-phenol-chloroform extraction. *Analyt. Biochem.* **162,** 156–159.
7. Segal, D. S. and Kuczenski, R. (1974) Tyrosine hydroxylase activity: regional and subcellular distribution in the brain. *Brain Res.* **68,** 261–266.
8. Chomczynski, P. and Sacchi, N. (1994) Single-step RNA isolation from cultured cells or tissues, in *Current Protocols in Molecular Biology* (Ausubel, F. M., Brent, R., Kingston, R. E., et al., eds.), John Wiley & Sons, New York, pp. 4.2.4–4.2.8.
9. Lee, J. J. and Costlow, N. A. (1987) A molecular titration assay to measure transcript prevalence levels. *Methods Enzymol.* **152,** 633–648.
10. Brewer, G., Murray, E., and Staeben, M. (1992) RNase ONE: advantages for nuclease protection assays. *Promega Notes* **38,** 1–7.
11. Melton, D. A., Krieg, P. A., Rebagliati, M. R., Maniatis, T., Zinn, K., and Green, M. R. (1984) Efficient in vitro synthesis of biologically active RNA and RNA hybridization probes from plasmids containing a bacteriophage SP6 promoter. *Nucleic Acids Res.* **12,** 7035–7056.

13

Analysis of mRNA Expression in Striatal Tissue by Differential Display Polymerase Chain Reaction

Joshua D. Berke

1. Introduction
1.1. What Is DDPCR?

Differential display polymerase chain reaction (DDPCR) is a technique that allows comparisons between the expressed mRNA population in two or more tissues, or in the same tissue under two or more different conditions. For example, it has been used to discover striatal genes whose expression is altered following a drug injection *(1,2)*. The essential idea behind DDPCR is to generate an RNA "fingerprint," a pattern of bands corresponding to the expressed mRNA population. By performing parallel sets of reactions for each experimental condition and displaying the resulting fingerprints side by side, bands that differ between conditions can be noted and the corresponding mRNA identified.

1.2. Why Use DDPCR, Rather than Something Else?

DDPCR has a number of advantages and disadvantages when compared to other techniques. In contrast to a standard Northern blot, *in situ* hybridization, RNase protection assay, or reverse transcriptase-PCR (RT-PCR), it allows simultaneous analysis of a large number of mRNA species. Compared to other mass-screening techniques such as subtractive hybridization, when done properly DDPCR can be extremely sensitive, detecting mRNA differences as low as twofold or even less. Also, unlike most forms of microarray analysis, DDPCR makes few assumptions about which mRNAs might be different; DDPCR can detect wholly novel mRNA species, or novel splice variants of previously characterized genes. This is especially important now because it has been recognized that alternative splicing is a major mechanism for dynamically generating mRNA diversity in the central nervous system (*see* **ref. 3** and references therein). On the downside, the DDPCR procedure can be quite laborious, especially if the intention is to assess the majority of the expressed mRNA population. It has an (only partially deserved) reputation for producing numerous false-positive results that can clog up the process of confirming differential expression. Because the isolated mRNA species may be novel, they can require a great deal of work for proper characterization, and may not be the type of genes that were being sought out. Many of these disadvantages can be minimized with proper experimental design. In the discussion that follows, it will be assumed that the reader is considering using DDPCR on brain tissue samples, although most of the advice holds whether or not this is true.

1.3. How DDPCR Works

Since the original report by Liang and Pardee *(4)*, a great number of DDPCR variants have been described in the scientific literature. The version described here (**Fig. 1**) is modified from Liang et al. *(5)* and has been used to identify dopamine- and cocaine-induced striatal genes *(2,3)*.

First, RNA is extracted from each condition of interest. Complementary DNA (cDNA) copies of these RNA samples are created using reverse transcriptase (RT). These reactions are initiated at the 3′ end of the RNA using one of three "anchored" downstream primers, each of which has a string of thymidine bases designed to bind to the poly(A)$^+$ tail of mRNA transcripts—specifically, at the 5′ end of the poly(A)$^+$ tail. The resulting cDNAs are then used as the template for low-stringency PCR, using the same downstream primers and also one of a set of "random" primers that will bind unpredictably but consistently to upstream sites in some fraction of the cDNA population. In both sets of primers, an extra sequence is included that serves two functions: to allow the stringency of the PCR to be increased after the first few cycles and in subsequent reamplification, and to include useful sites that aid later subcloning, RNA probe generation, and/or sequencing of the DDPCR fragments. Each primer combination used for PCR will generate a distinct mix of products of various lengths, which can be separated out with a sequencing gel; because the PCR reaction contains a small amount of a radioactive DNA base, the gel can be dried and opposed to film to visualize the resulting RNA fingerprint.

Any band in the fingerprint that consistently varies between conditions can be excised from the gel and reamplified with PCR. At this stage differential expression between conditions can be confirmed in a variety of ways (of which the best is usually *in situ* hybridization) and the DDPCR fragment sequenced and/or subcloned into a standard plasmid vector.

Differential Display PCR

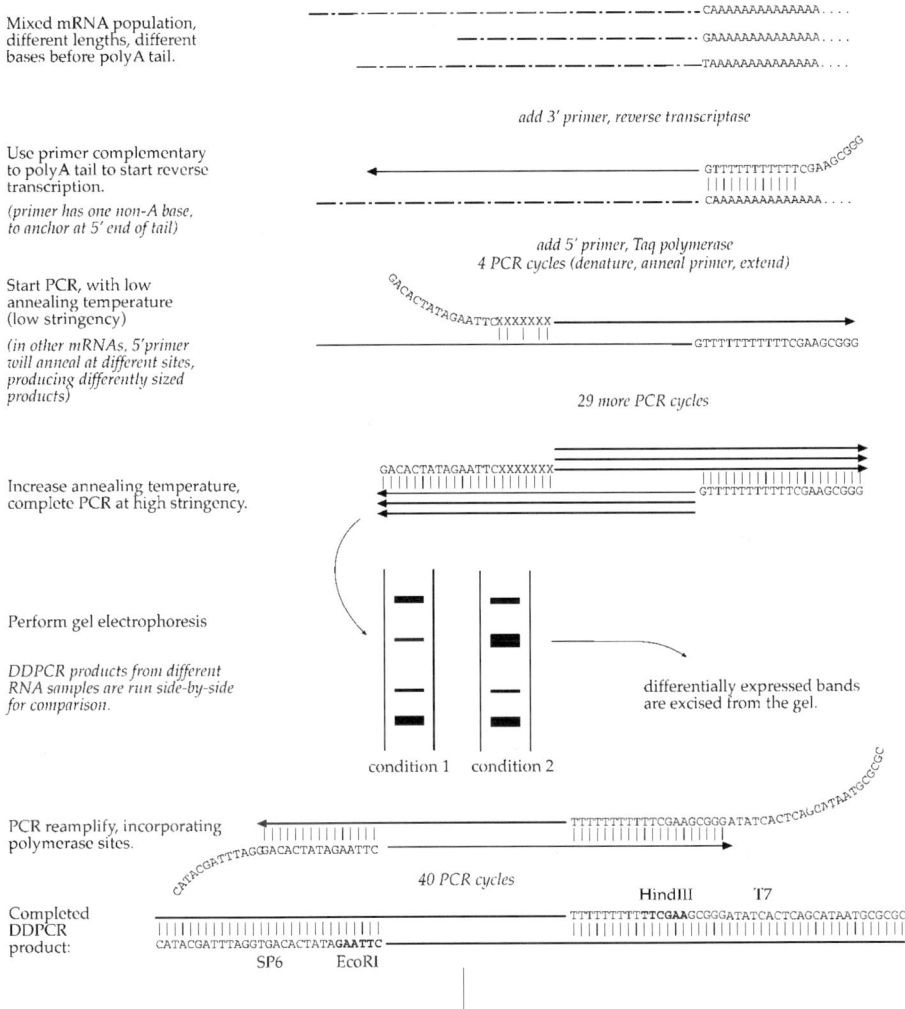

Fig. 1. Outline of DDPCR procedure.

1.4. Designing a DDPCR Experiment

Careful design of a DDPCR experiment involves multiple tradeoffs—mostly between the amount of work involved and the probability of obtaining a useful result.

The first issue to consider is which, and how many, conditions to compare. Reasonably enough, most people start with a specific scientific question, which naturally leads to a specific comparison. But bear in mind alternative hypotheses, especially if they can be readily incorporated into the DDPCR experiment. One great advantage of DDPCR is that multiple comparisons are easy to perform, as the RNA fingerprints for each condition are laid out side by side (by contrast, in subtractive hybridization the amount of work required grows geometrically with the number of conditions to be compared to each other). A major concern in starting a DDPCR experiment is that a substantial amount of effort and resources may prove fruitless; by intelligent incorporation of extra conditions, the chance of finding something interesting can be greatly increased. Another reason for increasing the number of comparison conditions is greater ability to detect subtle differences in band patterns—for this purpose, the more lanes that use the same primer combination, the better. However, more comparisons means more work. A further way to improve your yield is to compare states that differ as much as possible, within the framework of your experiment. For example, pick a larger dose of a gene-inducing drug rather than a smaller dose. It is easier to have many differentially expressed genes and examine the subtleties of their expression in subsequent screens, rather than have too few.

The next issue is how many RT-PCR reactions to run per condition (for a given primer combination). This should generally be a minimum of four in order to avoid a high rate of false-positives. It is a good idea to perform separate RT reactions from distinct animals' RNA in each condition—that way the chance of any particular individual differences in mRNA expression leading to a "false positive" result is minimized. This is particularly important when working with a genetically out-bred animal strain (e.g., several lines of lab rats). Each RT reaction should be used for duplicate PCR reactions (band pattern differences between these duplicates is a sure sign of contamination or some other procedural problem).

Another tradeoff involves the stringency at which the PCR is performed. The key parameter is the annealing temperature in the first few PCR cycles, which effectively determines how many bases in the upstream primer are required to match the target cDNAs before extension can take place (*see* **Note 1**). The lower the temperature, the more cDNAs will bind the primer and the more bands will be generated in each lane. As this number increases, it becomes harder to detect differences in a particular band, and also increases the chance that nondifferentially expressed cDNAs will be also be excised and reamplified, potentially yielding "false-positives" and requiring subcloning to separate them from the intended target. On the other hand, a small number of bands per lane indicates that few mRNAs are being sampled, and therefore presumably increases the number of primer combinations required.

This primer combination number is one of the major determinates of the time required for a DDPCR experiment. If there are lots of differentially expressed genes appearing in the screen, the DDPCR experiment can be stopped after just a few primers' worth of gels (perhaps 1–2 mo work, including the initial setup time) and attention focused on the often more difficult task of trying to understand the significance of the found genes. In other experiments there will be few or no differentially expressed genes found in the initial primer combinations. There is a gradually diminishing rate of return with increasing numbers of gels, as each primer combination does not sample a distinct mRNA population. DDPCR cannot be used to prove that there is no mRNA difference between two conditions, as there is no way of ensuring 100% coverage of the mRNA populations. In my experience, the use of about 50 upstream primers with each of the three anchored primers yielded about half of the genes that were previously known to be altered, and therefore presumably sampled about half of the expressed mRNAs. It is worth emphasizing that DDPCR is not a good technique for assessing changes in certain specific transcripts—if something is known about the gene some other strategy (perhaps degenerate PCR) is a better choice.

DDPCR can be performed with little more equipment than a thermal cycler and a standard sequencing gel box. However, a higher quality gel apparatus such as the Genomyx-LR, although expensive, can be highly cost-effective because of improved data quality. The main advantage is that it allows DDPCR to obtain larger cDNA fragments (up to 2 kb, as compared with up to 0.6 kb on a standard rig). This can greatly reduce the amount of work required to obtain the full-length sequence of the differentially expressed mRNA. Although this has become somewhat less critical in an era of fully sequenced genomes, such equipment is still generally worth obtaining if sufficient funds are available.

Finally, a key consideration is the means that will be used to confirm differential expression. Many who perform DDPCR find that this step is more time consuming than the fingerprinting procedures; it is therefore essential that it should be as streamlined as possible. In principle, any standard method can be used, including Northern blots, reverse Northerns (dot-blots), RT-PCR, nuclease protection assays, or *in situ* hybridization; the choice of method is guided by the amounts of RNA available, location of brain structure, and which techniques are already routinely performed in the laboratory. In practice, Northern blots are generally a poor choice, because they require substantial amounts of RNA, are time consuming, and have poor sensitivity. They are primarily useful for establishing the size of the full-length mRNA *after* differential expression has been confirmed. For studies examining the striatum, *in situ* hybridization is usually the preferred method for confirmation screening, because it is sensitive and because so many 12-μm sections can be taken through a single striatum. More than one animal should be used for confirmation screening, and if at all possible these particular animals should themselves be "confirmed" (e.g., by hybridizing a probe for a known differentially expressed gene). The conditions used for the confirmation screen should naturally include those used for DDPCR, but it may be useful to add conditions. For example, if the DDPCR screen assays changes 1 h following a drug injection, the screen might include baseline, 1-h, and 2-h timepoints. DDPCR is very sensitive, and changes that are barely visible by *in situ* may be clearer under slightly altered circumstances.

2. Materials

1. An RNase-free working environment (*see* **Note 2**).
2. –80°C, –20°C Freezers.
3. Ice buckets, ice, dry ice.
4. Thermal cycler with hot top (preferably one that can take a 96-well plate of 0.2-mL tubes).
5. 96-Well PCR plates of 0.2-mL thin-wall tubes, with mat-style sealing lids.
6. Apparatus for running small-medium sized (e.g., 10 cm) agarose gels (Owl Scientific).
7. Apparatus for running large polyacrylamide sequencing gels (preferably Genomyx-LR or equivalent).
8. Gel dryer and large sheets of filter paper (if not using Genomyx-LR).
9. (Optional) Small benchtop centrifuge for touch-spinning 0.2-mL tubes/tube strips.
10. (Optional) Large benchtop centrifuge with adaptor for spinning 96-well PCR plates.
11. Small vortex mixer.
12. Pipetters: 1-µL, 20-µL, 200-µL, and 1-mL.
13. RNase-free, DNA-free, aerosol-block pipet tips for above.
14. Multichannel pipetter: 0.5–10 µL × 8 channels (may be different depending on experimental design).
15. Repeater pipetter (e.g., Eppendorf) for dispensing 2–200-µL aliquots with sterile tips (Eppendorf Combitip Plus).
16. (Optional) Multichannel Hamilton syringe for loading sequencing gels.
17. DNase I (must be RNase-free; RQ1 DNase I from Promega).
18. RNasin RNase-inhibitor (Promega; RNase inhibitors from other companies also work well, e.g., Epicentre).
19. MMLV reverse transcriptase (Promega or others).
20. *Taq* DNA polymerase (any major supplier).
21. Deoxyribonucleoside triphosphates (dNTPs), Mg^{2+} (PCR grade).
22. Expand High Fidelity DNA polymerase blend (Roche).
23. [α-^{33}P]dATP (e.g., New England Nuclear cat. no. NEG612H, stable at 4°C).
24. Radioactive waste bucket.
25. 5X DNA loading dye (standard bromophenol blue/xylene cyanol).
26. Kodak BioMax MR film (same size as DDPCR gel).
27. Visible light box for examining DDPCR films.
28. Ethidium bromide (EtBr).

29. UV light box, camera for visualizing EtBr-stained agarose gels.
30. Scalpel with disposable blades.
31. Primers:
    Primer sets can be ordered commercially (Genomyx or GenHunter) or custom-ordered. A set of custom primers might include:
    One-base anchored primers:
    GGGCG AAGCT TTTTT TTTTT A
    GGGCG AAGCT TTTTT TTTTT C
    GGGCG AAGCT TTTTT TTTTT G
    Upstream "random" primers of the form:
    GACAC TATAG AATTC **NNNNN NN**, where **N** is any base.
    (As a starting point, try GTTGTGC, AACGAGG, TCTCTGG, and TGGTCAG)
    Reamplification primers:
    Downstream: CGCGC GTAAT ACGAC TCACT ATAGG GCGAA GCTTT TTTTT TTT
    Upstream: CATAC GATTT AGGTG ACACT ATAGA ATTC
    (These reamplification primers complete the T7, SP6 RNA polymerase sites on either side of the reamplified DDPCR fragment).

3. Methods
3.1. RNA Extraction

Total RNA can be extracted using any of a variety of standard methods. It is not necessary to extract further only the poly(A)$^+$ (mRNA) component, but it is important to obtain high-quality (i.e., undegraded) RNA. The RNA should also be treated with RNase-free DNase I to remove any DNA contamination (*see* **Notes 3** and **4**). Run 0.5 or 1 µg of RNA per sample in a simple (nondenaturing) agarose/ethidium bromide gel to confirm that the RNA is undegraded (**Fig. 2**). This gel also confirms that the relative amounts of RNA in each sample are even (accurate quantitation), and should be performed immediately before the reverse transcription reactions.

3.2. RT Reactions

(Total of 20 µL; can be scaled up or down by a factor of 2.)

1. Assemble all components on ice (*see* **Notes 5** and **6**).

RT Mix:	(per reaction)	(example: 27 reactions-worth)
H$_2$O	6.9 µL	186.3 µL
5X RT buffer	4.0 µL	108 µL
dNTP (250 µM)	1.6 µL	43.2 µL
DTT (0.1 M)	2.0 µL	54 µL
RNasin	0.5 µL	13.5 µL
Total	15 µL	

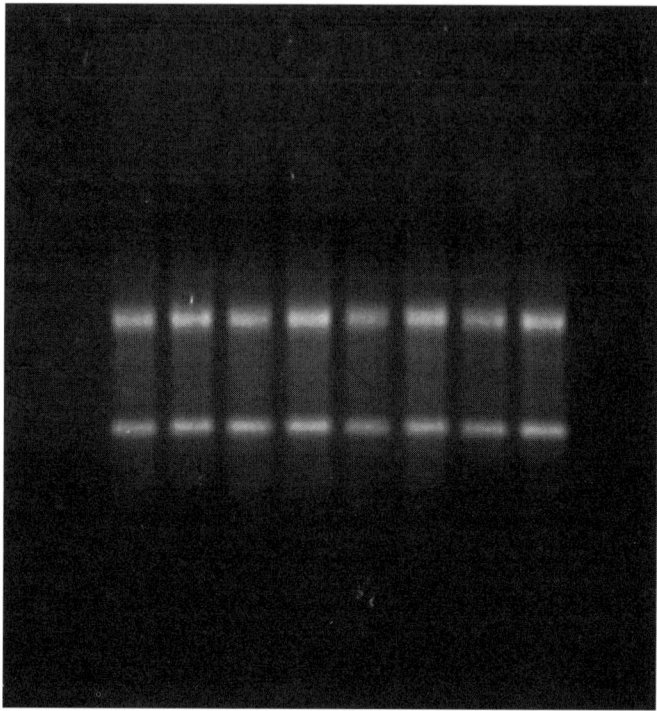

Fig. 2. RNA quality check before starting RT reactions. These are striatal RNA samples from eight different rats, extracted using TriReagent and treated with DNase I. The 28S and 18S ribosomal bands should be clearly visible, the background mRNA smear should be of similar size to the rRNA bands, and there should be little or no lane-to-lane variation.

2. For each RNA sample, there will be three RT reactions—one for each of the three anchored 3′ primers ("A," "C," "G"). In each 0.2-mL tube, add the appropriate RT primer (2 µL of 2 µM), 0.2 µg of RNA, and H$_2$O to 4 µL total. Then add 15 µL of RT mix, vortex-mix, touch-spin, and immediately place into a preheated (65°C) thermal cycler. Start following program: 65°C for 5 min, 37°C for 60 min, 95°C for 5 min, → 4°C. After 10 min of the 37°C period, carefully add 1 µL of the RT enzyme directly into each reaction, then rapidly pipet up and down a few times to mix (*see* **Note 7**). RT products are best used immediately, but can also be frozen (at −20°C) and reused for several weeks. Depending on your experimental design, there will generally be enough cDNA for four or more sets of PCR reactions.

3.3. PCR Reactions

(Total of 11.1 µL; this works as well as 20 µL but is less expensive.)

1. Assemble all components on ice.
2. It is essential that all the PCR reactions using a given primer combination be as identical as possible in their preparation and composition (except for the distinct cDNA templates). For each primer combination make up a mix of the primers and all common components; once these mixes have been distributed to individual wells, the RT products are added to initiate the PCR reaction (*see* **Note 8**). For example, the following array of 96 reactions might compare four conditions, each with three downstream primers ("A," "C," "G") and two upstream primers ("1," "2") (*see* **Notes 9 and 10**). Each row will receive the products of a distinct RT reaction. The products of the 96 PCR reactions can be run on two 48-lane gels, one for each upstream primer.

```
A1  A1  C1  C1  G1  G1  A2  A2  C2  C2  G2  G2
A1  A1  C1  C1  G1  G1  A2  A2  C2  C2  G2  G2
A1  A1  C1  C1  G1  G1  A2  A2  C2  C2  G2  G2
A1  A1  C1  C1  G1  G1  A2  A2  C2  C2  G2  G2
A1  A1  C1  C1  G1  G1  A2  A2  C2  C2  G2  G2
A1  A1  C1  C1  G1  G1  A2  A2  C2  C2  G2  G2
A1  A1  C1  C1  G1  G1  A2  A2  C2  C2  G2  G2
A1  A1  C1  C1  G1  G1  A2  A2  C2  C2  G2  G2
```

In this example, make up six mixes ("A1, A2, C1, C2, G1, G2"), each as follows (*see* **Note 11**):

PCR mix	18 reactions-worth
H$_2$O	99.5 µL
10X PCR buffer	20 µL
dNTP (250 µ*M*)	16 µL
5′ Primer ("1" or "2"), 2 µ*M*	20 µL
3′ Primer ("A," "C," or "G"), 2 µ*M*	20 µL
[α-^{33}P]dATP	2.5 µL
Taq polymerase	2 µL
Total	180 µL

3. Vortex each mix well and touch-spin. For each mix, pipet out 10 µL into the appropriate wells of the PCR plate (as in the preceding chart). Once all the mixes have been aliquoted out, use the multichannel pipetter to transfer 1.1 µL from the eight-well strips of RT reactions to the appropriate columns of the PCR plate (i.e., the RT rxns that used the "A" downstream primer are used in the "A" columns in the figure) (*see* **Note 12**). Pipet up and down a few times to mix.

4. Place the PCR plate in a preheated (94°C) thermal cycler and immediately commence the following sequence:
94°C, 2 min; 4 cycles of {94°C, 15 s, 40°C, 4 min, 72°C, 3 min}; 29 cycles of {92°C, 15 s, 55°C, 1 min, 72°C, 1.5 min + 3 s/cycle}; 68°C, 5 min, 4°C hold (*see* **Note 13**).
5. Once the PCR is complete, add 2.5 µL of 5X DNA loading dye directly to each sample in the PCR plate. Samples can be stored at 4°C for up to a week, although ideally they are run in the gel immediately.

3.4. Gel Electrophoresis

1. Set up a large denaturing polyacrylamide sequencing-style gel, following the instructions of the manufacturer. Typically 6% polyacrylamide gels are used (4.5% for Genomyx-LR). If using a multichannel syringe or pipet for gel loading, use a gel comb with appropriate spacing (i.e., half of microtiter spacing, so that PCR duplicates from adjacent columns in the PCR plate are run side by side). Once gel is *fully* set and ready to load, denature samples by heating at 94°C for 2 min, then load 1.5–2.5 µL per lane (depending on the width of each lane) and apply the voltage (*see* **Notes 14** and **15**).
2. After the gel has run, carefully separate the glass plates. If using a conventional sequencing rig, transfer the gel to filter paper and dry using a flat-bed vacuum gel dryer. If using Genomyx-LR, dry the gel directly onto the glass plate and wash thoroughly and repeatedly with water, and allow to dry.
3. Once the gel is dried, you'll need to make spots on the gel that will be picked up on the film and used to accurately align the two together after identification of differentially expressed bands. "Radioactive ink" can be easily made by mixing standard DNA loading dye with any convenient radioactive source (an older vial of [^{33}P]dATP that has lost too much activity to be useful for the DDPCR reaction itself is perfect). Mark each corner of the gel with one to three small spots of this ink (*see* **Note 16**).
4. Allow the ink to dry, then in a dark room oppose the gel to the appropriate side of the film and sandwich both tightly between glass plates. Typical exposure time is overnight to 2 d (*see* **Note 17**).
5. Inspect the film carefully on a light table, band by band (*see* **Fig. 3** and **Note 18**). If a particular band is of interest, use a needle or thumbtack to punch a small hole through the film, on either side of the band. Mark out just one lane for a given band. This will provide plenty of DNA for reamplification, and leave backups in case of a problem.
6. Realign the film with the dried gel, and use a sharp pencil to poke through the holes and make small marks on the gel.
7. Remove the film, and with a scalpel blade carefully scrape a narrow strip of dried gel off the glass plate and into a PCR tube (if your gel is dried onto filter paper, carefully slice through it, removing as little extraneous paper as possible). If

Differential Display PCR

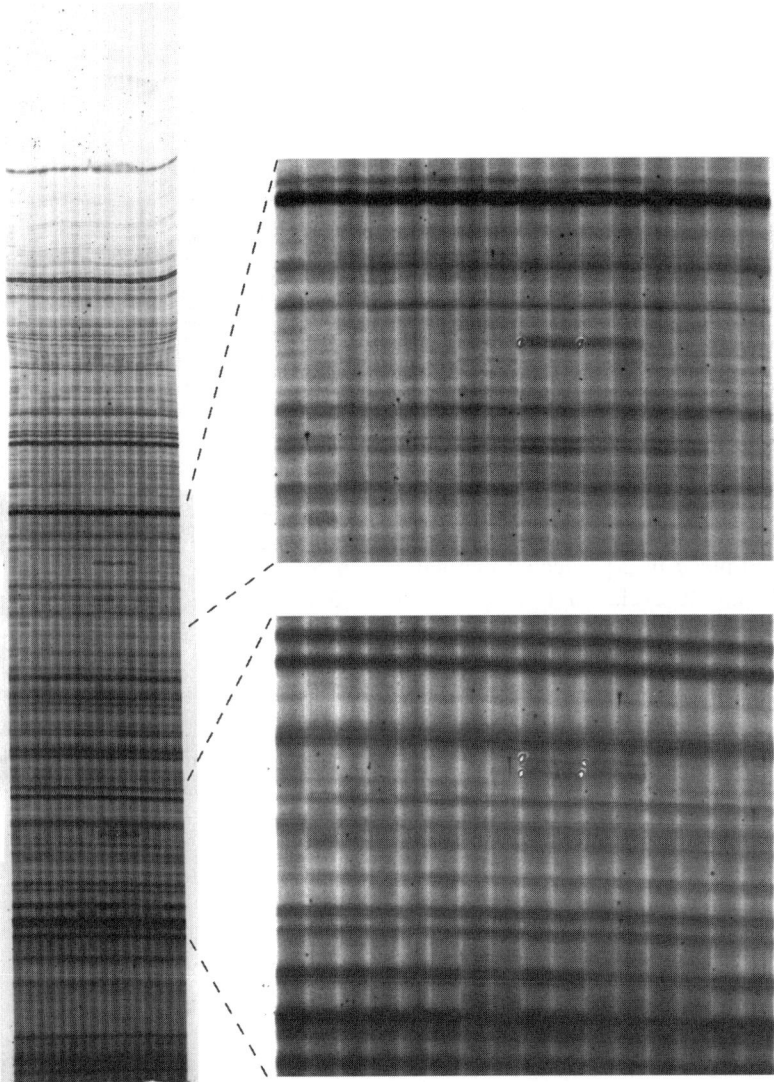

Fig. 3. A portion of an example DDPCR film. On *left*, RNA fingerprints arising from a single primer combination, with four comparison conditions. Each condition is represented by four adjacent lanes (2 animals × 2 PCR duplicates). On *right*, magnification of two regions of the film indicating drug-induced genes in the striatum (condition 3; for details, *see* **ref. 2**). Note that pairs of lanes often seem to show changes in band density—hence the importance of using multiple RNA sources for each condition to avoid excessive false positives.

desired, the accuracy of the band cut can be checked by reexposing the remainder of dried gel to film.

3.5. Reamplification

1. Reamplification reaction components are as follows (see **Note 19**):

H$_2$O	21.2 µL
10X Expand HiFi standard buffer	4.0 µL
dNTP (250 µM)	6.4 µL
Upstream reamp primer (2 µM)	4.0 µL
Downstream reamp primer (2 µM)	4.0 µL
Expand High Fidelity polymerase	0.4 µL
	(40 µL total per reaction)

2. The reamplification primers are designed to work with bands generated using any combination of DDPCR primers. Therefore a single mix suffices for all reamplification reactions. Add 10% extra to compensate for pipetting losses. Add 40 µL of this mix directly into each PCR tube containing a dried gel fragment and place in a preheated thermal cycler. Start the following program: 95°C, 2 min; 40 cycles of {92°C, 15 s; 60°C, 5 s, 72°C, 2 min + 2 s/cycle}; 68°C, 7 min; 4°C, hold.

3. When completed, add 10 µL of 5X DNA loading dye to each of the PCR tubes, and mix by pipetting up and down. Run 20 µL from each tube on a standard 1% agarose-EtBr gel, including a (nonradioactive) DNA ladder in a spare lane. Take a picture of the resulting bands on a UV light box, and compare their sizes to those on the DDPCR gel. Cut out the succesfully reamplified bands and purify the DNA (see **Notes 20** and **21**).

3.6. Confirming Differential Expression

At this stage, the DNA can be directly used as a template for generating an antisense RNA probe (by in vitro transcription using T7 polymerase) and differential expression confirmed by *in situ* hybridization, using standard protocols. This is the preferred approach if your DDPCR films show significant separation between bands. If the DDPCR bands are packed tightly together, it will be hard to cut out a single band cleanly (and a single position on the gel may well contain a mixture of cDNA species). In this situation, subclone the reamplified DNA by ligation into a standard vector with antibiotic resistance (see **Note 22**), transform competent cells, grow on an agar/antibiotic plate overnight, pick colonies (eight colonies is usually enough), grow up a few milliliters of culture overnight, and perform minipreps. If you have access to a mass-sequencing facility, sequence all the minipreps; otherwise try using the first one to confirm differential expression and proceed from there (see **Note 23**).

3.7. After DDPCR

Once differential expression has been confirmed, sequence the cDNA if you have not already done so (well-isolated cDNAs can be directly cycle sequenced using the T7, SP6 sites or custom primers; alternatively, subclone and obtain minipreps as described earlier). Isolate the portion of the sequence that is not vector or primer and compare it to a public resource such as GenBank (www.ncbi.nlm.nih.gov). The cDNA may match a gene that is already known to have differential expression (from other experiments or because it already showed up earlier in the DDPCR experiment). In such cases, be sure to check carefully that the sequences match along their full length—DDPCR can turn up novel or unexpected splice variants that may have important functional roles. For this reason it is also a good idea to perform a Northern blot, with an RNA ladder (e.g., Ambion), to establish the full-length size of mRNAs whose differential expression has been confirmed (*see* **Note 24**). If the gene is novel or poorly characterized, the next step is to obtain the sequence of the full-length mRNA sequence. This can often be obtained directly from an established GenBank entry; if not, often the sequence can be inferred by assembly from many GenBank EST (expressed sequence tag) sequences. Barring that, if the genomic sequence is known, a number of PCR primers can be designed and used for RT-PCR in combination with PCR primers antisense to the 5′ end of the DDPCR sequence (this is a quick but hit-and-miss approach). Finally, the remainder of the mRNA sequence can be obtained "the hard way"—by performing rapid amplification of cDNA ends (RACE; e.g., *6*) or by screening cDNA libraries for clones that contain the DDPCR fragment (*see* **Note 25**).

4. Notes

1. It is only the bases at the 3′ end of the primer that participate in the annealing to the cDNA population. Some investigators have found that it is desirable to have C or G bases, which form more stable bonds than A or T bases, in the most 3′ one or two positions.
2. It is essential to avoid RNase contamination during the initial stages of the procedure, and widespread DNA contamination thereafter. If you are already routinely performing successful Northern blots, your materials-handling techniques are probably fine. Otherwise, follow the following steps. Keep all compounds, fluids, and so forth used in RNA handling separate from other lab materials and handle them only while wearing gloves. Any solutions, buffers, and so forth should either be prepurchased in sealed, RNase-free form, or made up with diethyl pyrocarbonate (DEPC)-treated H_2O. The source of water used in RT-PCR is particularly important—unfortunately, nuclease-free water provided by different manufacturers can give variable results. Find a water source that works and use it consistently. Use RNase-free pipet tips with an aerosol blocking filter.

Before starting an RT or PCR procedure, clean off your benchtop and pipetters throughly (solutions such as RNase-ZAP from Ambion can be helpful). During all procedures, use a different pipet tip each time you draw off any enzyme or RNA, to avoid cross-contamination. Because PCR is so sensitive, if practical it is best to keep the products of PCR reactions in a different refridgerator or freezer from PCR primers, buffers, and so forth.

3. When working with rodent brains, following decapitation I rapidly dissect out striatum on an ice-cold surface (e.g., roughened, up-ended culture plate in an ice bucket), then immediately place the tissue into a prechilled microcentrifuge tube sitting in dry ice. When all samples were collected they are placed in a –80°C freezer. RNA extraction (which can be weeks or even months later) involves rapid homogenization in a guanidine thiocyanate based solution, such as TriReagent (Molecular Research Center, Inc.). It can be helpful to include a coprecipitant (e.g., Microcarrier TR, Molecular Research), especially if working with a smaller amount of RNA. Follow the manufacturer's protocol for RNA isolation, redissolve RNA in DEPC-treated H_2O (e.g., 50 μL for one rat striatum's worth of RNA). Do *not* allow the RNA pellet to dry completely before adding DEPC-treated water or it will be very difficult to dissolve; incubation in a warm (58°C) shaking incubator (thermomixer) can help. Place in a 37°C bath or incubator and remove any DNA by adding to each 50-μL sample the following: 0.5 μL of RNasin, 5.7 μL of 10X DNAse buffer, and 1 μL of DNase I. Incubate for 1 h, then remove the enzyme followed by phenol–chloroform (3:1), chloroform extractions (the Phase-Lock Gel I tubes provided by FivePrime-ThreePrime Inc. are useful for avoiding waste and contamination during these extractions). Precipitate out RNA by adding 5 μL of 3 *M* sodium acetate and 250 μL of ethanol and placing on dry ice for 30–60 min followed by cold centrifugation at high speed (12,000*g* or more, 15 min). Carefully remove supernatant, wash with 70% ethanol, allow pellet to dry *briefly* (until all visible liquid is gone; this time, redissolve RNA in a smaller volume (10–20 μL for a rat striatum), then carefully measure RNA concentration in duplicate with a spectrophotometer. The A_{260}/A_{280} ratio of the RNA should be approx 1.8 or higher. Avoid repeated freeze–thawing of your RNA samples, so if the same samples will be used many times, adjust each of the samples to the same concentration (preferably 1 μg/μL or higher) with DEPC-treated water and split into aliquots before freezing at –80°C.

4. In some experimental designs a test can be performed at this stage to ensure that the animal(s) exhibited the desired genetic response. For example, if it is already known that gene X is induced by a drug at a particular timepoint, and the object of the DDPCR experiment is to find other genes that respond in a similar fashion, the extracted RNA could be used for a Northern blot to ensure that gene X is in fact being induced in these particular samples. Although this creates extra work, RNA from a given rat striatum can be used for months of DDPCR gels, so it is often worth confirming that you are not wasting your time.

5. It is usually most convenient to use a full ice bucket as a work surface. Many kinds of pipet tips come prearrayed in microtiter racks; the tops of these racks can be detached and used to keep strips of 0.2-mL tubes or 96-well PCR plates chilled on the ice surface. If you have a thermal cycler that accepts 0.2-mL tubes, it can be convenient to assemble the RT reactions in strips of 8×0.2 mL attached tubes with individual lids.
6. Remember to make enough mix to compensate for pipetting losses; for example, if using three RT primers and eight comparison animals (24 combinations), make up a 27-reaction mix.
7. It is also possible to try slightly higher temperatures than 37°C toward the end of the RT reaction. At higher temperatures (e.g., 42 or 45°C) the enzyme will be degraded faster but there will be less secondary structure in the mRNA; such secondary structure may impede the enzyme and cause some cDNA transcripts to be truncated. At the end of the RT reaction some protocols call for a brief incubation with RNase H, to remove the RNA from the newly formed cDNA. However, omitting this step does not seem to make a substantial difference.
8. It is usually convenient to make up one "supermix," which lacks primers, and split it into submix tubes to which specific primers have been added.
9. If you're only comparing two conditions with two animals/condition, you could use four upstream primers/96-well plate (each upstream primer would be used in a 4×6 well quadrant).
10. Remember to mark, snip, or otherwise indicate which is the top left corner of the PCR plate; otherwise you may turn it through 180° by mistake.
11. ^{33}P is vastly superior to ^{35}S for DDPCR. ^{35}S requires longer gel exposure times, but more importantly forms volatile products during DDPCR that contaminate the thermal cycler and potentially the rest of the laboratory.
12. With a large number of components to be added to specific combinations of PCR tubes, it is easy to make a time-consuming and expensive mistake—for example, adding RT reactions to the eighth column when the seventh was intended. To help avoid this, it helps to use a marker pen to divide the PCR plate visually into halves or thirds. Starting with a fresh array of 96 pipet tips and using the missing tips as a visual guide to which corresponding tubes have been already worked on is also useful.
13. This protocol is designed to generate DDPCR products of up to 2 kb, which can be resolved on a Genomyx sequencer; if using a conventional metal-backed sequencing gel apparatus, the polymerase extension times (the 72°C periods) can be substantially shortened.
14. It is a very good idea to label radioactively a DNA ladder (e.g., 100-basepair (bp) ladder from Gibco) by standard methods (e.g., T4 polynucleotide kinase with [γ-^{33}P]ATP) and run it alongside the DDPCR products. This will allow a reasonable estimate of band size, so that problems with reamplification can be more readily noticed.

15. For Genomyx-LR, the following program works well: 16 h at 1000 V, 50°C (with a 4.5% HR-1000 gel, 340 µm thick). This allows visualization of cDNAs ranging in size from approx 400 bp to 2 kb. If desired, the same DDPCR products can also be run on a shorter duration gel, to allow the shorter cDNAs to also be seen (see manufacturer's instructions).
16. It is useful to make different shapes or numbers of spots on each corner, to prevent confusion over orientation. I also use the radioactive ink to indicate the name and date of each gel, further reducing the chance of errors.
17. Passing a Geiger counter over the surface of the dried gel can (with a little experience) serve as a good guide to an appropriate exposure time.
18. Various problems may contribute to a DDPCR band pattern that is unreadable. If the pattern is highly variable from one lane to the next, that may indicate that uneven pipetting caused variable reaction conditions. When using the multichannel pipet to add RT products to the PCR plate, visually inspect the fluid levels each time; loose fitting pipet tips can cause variability. Contamination can also cause problems. Keep your work area and equipment very clean, and unwrap new boxes of presterilized RNase- and DNA-free pipet tips each time. You may also find it useful to expose your pipetters to UV light (e.g., in a Stratalinker or specialized PCR work area) for 10–15 min immediately before commencing the RT or PCR procedure.
19. *Taq* polymerase is not a very accurate enzyme. Therefore, when reamplifying bands it is better to limit the number of sequence errors introduced by using a more accurate polymerase blend such as Expand High-Fidelity (Roche), which is also more reliable than *Taq* at reamplifying bands larger than 1 kb. Such more accurate enzymes can also be used for the main DDPCR reactions, but the extra expense may not be worthwhile.
20. Keep the exposure of the DNA to the UV light as short as possible. A hand-held, low-power UV bulb is best when actually cutting bands.
21. Any standard method can be used to purify the DNA. I prefer the rapid Supelco columns (Supelco EtBr-, available from Sigma) which remove the EtBr and leave the DNA ready for subcloning or in vitro transcription of RNA probes.
22. For subcloning I have used either pCRII-TOPO (Invitrogen), which readily accepts PCR products in random orientation, or pGEM-7z for directional cloning (after restriction digest of PCR products with *Eco*RI and *Hin*dIII).
23. An unusual phenomenon that occasionally arises is that following *in situ* hybridization the sections are intensely and uniformly stained black. One way this can occur is if there is a repetitive element in the cDNA sequence, that binds to a large number of genes. If sequencing of the clone indicates such an element, design oligonucleotide probes that match a nonrepetitive part of the sequence and use that to confirm differential expression.
24. Although the DDPCR technique is intended to produce bands corresponding to 3′ ends of expressed mRNAs, the anchored primers used for the reverse transcription reaction can also readily bind to any significant string of adenosine bases—sometimes even if there is a mismatch or two. For that reason, it cannot

be assumed that a gene fragment found using DDPCR is at the 3' end of the expressed mRNA, or that two distinct DDPCR fragments with nonoverlapping sequence come from distinct genes.
25. If screening a cDNA library, it is a good idea to pick or make one from a tissue that is likely to have reasonable representation of your mRNA (especially if it has low baseline expression in the tissue of interest). It may be helpful to perform a multiple-tissue Northern blot first; for example, interesting brain genes are often highly expressed in testis.

Acknowledgments

Optimization and use of the DDPCR procedure was performed in the laboratories of Steven Hyman and Charles Gerfen at NIH. The author is currently supported by NIDA.

References

1. Douglass, J., McKinzie, A. A., and Couceyro, P. (1995) PCR differential display identifies a rat brain mRNA that is transcriptionally regulated by cocaine and amphetamine. *J. Neurosci.* **15,** 2471–2481.
2. Berke, J. D., Paletzki, R. F., Aronson, G. J., Hyman, S. E., and Gerfen, C. R. (1998) A complex program of striatal gene expression induced by dopaminergic stimulation. *J. Neurosci.* **18,** 5301–5310.
3. Berke, J. D., Sgambato, V., Zhu, P-P., et al. (2001) Dopamine and glutamate induce distinct striatal splice forms of Ania-6, an RNA polymerase II-associated cyclin. *Neuron* **32,** 277–287.
4. Liang, P. and Pardee A. B. (1992) Differential display of eukaryotic messenger RNA by polymerase chain reaction. *Science* **257,** 967–971.
5. Liang, P., Zhu, W., Zhang, X., et al. (1994) Differential display using one-base anchored oligo-dT primers. *Nucleic Acids Res.* **22,** 5763–5764.
6. Schaefer, B. C. (1995) Revolutions in rapid amplification of cDNA ends: new strategies for polymerase chain reaction cloning of full-length cDNA ends. *Analyt. Biochem.* **227,** 255–273.

14

Semiquantitative Real-Time PCR for Analysis of mRNA Levels

Stephen J. Walker, Travis J. Worst, and Kent E. Vrana

1. Introduction

The reverse transcription-polymerase chain reaction (RT-PCR) has become a standard tool in gene expression analysis studies *(1,2)*. Starting with a very small amount of material (usually total RNA), the investigator is able to copy the RNA by reverse transcription (RT) to produce single-stranded, complementary DNA (also known as first-strand cDNA). The cDNA, which is much less prone to degradation than RNA, can then be amplified by PCR and quantified to determine the relative abundance of expressed genes within and between sample groups. Two of the most common uses of this technology in the field of drug abuse are: (1) to compare relative or absolute levels of expression of genes between control and treatment groups or, more recently, (2) for confirmation of gene expression (mRNA) changes identified in a primary screen (e.g., hybridization array analysis). The great utility of the technique is that if one has access to a real-time PCR thermocycler and is interested in comparing *relative* mRNA levels, the experiments are neither difficult nor expensive to perform. Indeed, attempts to be absolutely quantitative provide a very significant set of technical concerns *(1)* that are not the subject of the present discussion.

Quantitative PCR is one way to document changes in RNA or protein levels. In a study that compared several techniques—semiquantitative noncompetitive RT-PCR, Northern blot analysis, and enzyme-linked immunosorbent assay (ELISA)—results showed that all the methods detected, qualitatively, the same differences in gene expression and in exactly the same order *(3)*. Moreover, the authors of that study concluded that real-time kinetic RT-PCR offers several

From: *Methods in Molecular Medicine, vol. 79: Drugs of Abuse: Neurological Reviews and Protocols*
Edited by: J. Q. Wang © Humana Press Inc., Totowa, NJ

distinct advantages, namely: (1) increased sensitivity to measure baseline mRNA levels in unstimulated samples; (2) the lowest interassay variability of all the techniques investigated; and (3) this method eliminates the need to perform post-PCR manipulations because PCR product identification (i.e., a single product with a single T_m) can be accomplished at the same time by melting curve analysis.

Semiquantitative real-time RT-PCR involves multiple individual steps. These include: (1) preparation of RNA; (2) first strand DNA synthesis (cDNA production); (3) optimization of reaction conditions (primer concentration, T_ms, etc.); (4) the kinetic real-time PCR; and (5) data analysis. By using examples from an experimental series performed to validate results from an array experiment, each of these steps is discussed in the context of real-time PCR using an ABI 7000 instrument and SYBR Green dye for detection.

2. Materials
2.1. Reagents (see Note 1)
1. Source of RNA (tissue, cell culture, etc.).
2. RNA extraction reagent: TRI Reagent (Molecular Research Center, Inc., Cincinnati, Ohio).
3. Nuclease-free water (Promega Corp., Madison, WI).
4. Agarose.
5. 50X Tris–acetate buffer (TAE): 242 g of Tris base, 57.1 mL of glacial acetic acid, and 100 mL of 0.5 M EDTA, pH 8.0. Bring to a final volume of 1 L with dH_2O.
6. 5X Tris-borate buffer (TBE): 54 g of Tris base, 27.5 g boric acid, and 20 mL of 0.5 M EDTA, pH 8.0. Bring to a final volume of 1 L with dH_2O *(4)*. A precipitate forms when concentrated solutions of TBE are stored for long periods of time. Discard any stock solution that contains precipitate.
7. Ethidium bromide.
8. Formamide.
9. Formaldehyde.
10. Bromophenol blue.
11. 10X 3-(*N*-morpholino)Propanesulfonic acid (MOPS) buffer: purchased from Invitrogen (Life Technologies, Carlsbad, CA) as a 10X stock solution *(4)*.
12. Reverse transcriptase: Omniscript RT kit (Qiagen, Valencia, CA).
13. DNase I (optional; GeneHunter Corp., Nashville, TN).
14. Oligonucleotide primers (Integrated DNA Technologies, Carlsbad, CA).
15. Optically clear PCR tubes and/or 96-well plates (Applied Biosystems, Foster City, CA).
16. QuantiTect SYBR Green PCR kit (Qiagen, Valencia, CA) containing: *Taq* polymerase, SYBR Green PCR buffer, deoxyribonucleoside triphosphate (dNTP) mix, SYBR Green, ROX (passive reference dye), and 5 mM $MgCl_2$.

2.2. Equipment and Supplies

1. Agarose gel apparatus and power supply.
2. Light box (with UV lamp).
3. Gel-documentation workstation (Alpha Innotech, San Leandro. CA).
4. Spectrophotometer (with UV lamp).
5. Thermocycler (for reverse transcription reactions).
6. Water bath(s).
7. Real-time PCR thermocycler. Some examples: ABI Prism® 7000 Sequence Detection System (Applied Biosystems, Foster City, CA), iCycler (Bio-Rad, Hercules, CA), Smart Cycler System (Cephid, Sunnyvale, CA), or LightCycler System (Roche Molecular Biochemicals Indianapolis, IN).

2.3. Software

1. Adobe Photoshop Image capture (San Jose CA).
2. RNA quantitation (TINA, Raytest, Wilmington, NC; helpful but not absolutely necessary).
3. Primer design: ABI Prism® Primer Express™ from Applied Biosystems; or Primer3 from the Whitehead Institute at MIT (http://www.genome.wi.mit.edu/genome_software/other/primer3.html).
4. Kinetic real-time PCR sequence detection and analysis (ABI Prism® 7000 supplied by Applied Biosystems).

3. Methods

The methods outlined in this subheading describe a typical experimental series using RNA isolated from brain regions of experimental animals, and then assayed using the ABI 7000 thermocycler (other real-time instruments, e.g., ABI's 7700, the Bio-Rad iCycler, or Roche LightCycler can also be used for kinetic real-time PCR). This particular platform is chosen for discussion because the authors have this instrument on-site, and have experience of using it (*see* **Note 2**).

3.1. RNA Preparation

The single most important factor in the ability to generate quality results from a quantitative real-time PCR experiment, assuming a valid experimental design for the generation of "control" and "treated" samples, is the preparation and use of high-quality RNA. Total RNA (or mRNA) can be reliably and repeatedly generated from any number of biological samples including whole blood, tissues (e.g., brain, liver, etc.), and cells in culture. In this subheading, we discuss the isolation and assay of total RNA from brain tissue.

3.1.1. Tissue Dissection

The most reliable preparations of RNA come from tissue samples that have been rapidly dissected and "flash-frozen" in liquid nitrogen. Because brain regions are not absolutely discrete, it is best to have the same individual perform the dissections on all samples for a given experiment (and any subsequent repeat experiments) to ensure uniformity in sample collection. Once the whole brain has been removed from an animal's skull, and working as rapidly as possible, individual brain regions are dissected and immediately put into prelabeled/tared tubes. The samples in the pretared tubes are then reweighed, without thawing, and dropped into liquid nitrogen. Tissue collected and frozen in this fashion can then be stored for extended periods of time at –80°C until use.

3.1.2. Creating a Tissue Powder

Tissue that has been previously dissected and stored frozen at –80°C can be readied for use by the creation of a tissue powder under liquid nitrogen. Such a powder represents a homogeneous mixture consisting of all of the cell types originally present in that tissue sample. At the same time, aliquots of this preparation can be removed and diverted to quite distinct assay protocols (DNA arrays, RT-PCR, Western blots, etc.) without concern for differentially analyzing distinct subregions.

1. A stainless steel mortar and pestle is chilled on dry ice.
2. The frozen tissue "block" is placed in the mortar and covered with liquid nitrogen. The sample is then ground to a fine powder (keeping it covered with liquid nitrogen at all times). The frozen powder at the bottom of the bowl can be retrieved with a (prechilled) stainless steel spatula and quickly deposited into prechilled, preweighed cryovials.
3. In our experiments, half of the tissue is routinely placed into each of two tubes: one to be used for RNA extraction and the other for protein preparation.
4. The tubes containing tissue are quickly reweighed (without thawing) and immediately stored at –80°C.

3.1.3. Extracting RNA

RNA is extracted from frozen tissue powder using a modification of the single-step method originally developed by Chomczynski and Sacchi *(5,6)*. Of the variety of commercially available reagents, this laboratory routinely uses TRI Reagent (*see* **Note 3**). This reagent combines phenol and guanidine isothiocyanate in a monophase solution for the effective inhibition of RNase activity during cell lysis. Following addition of TRI Reagent to the tissue powder and the resulting cell lysis (via sonication), the homogenate is separated into aqueous and organic phases by the addition of the nontoxic organic

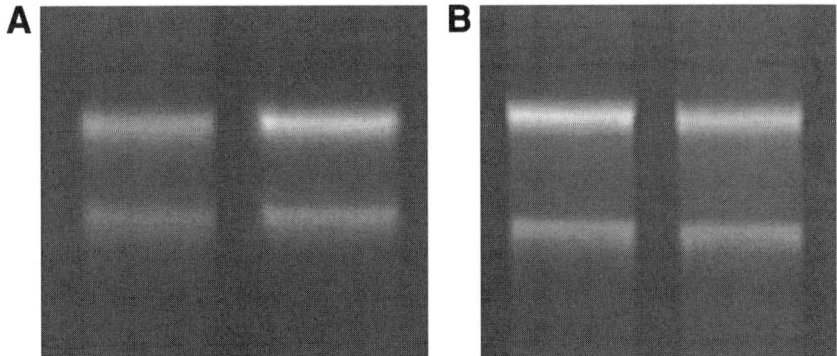

Fig. 1. RNA qualification. (**A**) Poor quality RNA on the left, showing little difference in 28S *(top)* and 18S *(bottom)* band intensities, and broad bands, and smearing between bands. Compare this with the lane on the *right*; note that the 28S intensity is approximately twice that of the 18S. It should be noted that many laboratories might consider the RNA on the *left* to be acceptable, whereas this laboratory has had the greatest success using RNA of the quality on the *right*. (**B**) Good quality total RNA showing the 18S band at half the intensity of the 28S band and sharp bands. In addition, these two samples were loaded at approximately equal concentrations which has been confirmed by the gel electrophoresis.

compound bromochloropropane (alternatively, chloroform may be used). RNA is precipitated from the aqueous (upper) phase with isopropanol. Following a washing step in 75% ethanol, the final RNA pellet is dried (by inverting the tube for 5–15 min at room temperature) and resuspended in nuclease-free water. The step-by-step protocol is provided with the commercial product and can be followed essentially as written.

3.1.4. RNA Quality and Quantity Determination

Quantity and relative quality of RNA can be determined spectrophotometrically by diluting an aliquot in TE buffer and measuring the absorbance at 260 and 280 nm. High-quality RNA should have an A_{260}/A_{280} ratio of 1.9–2.0 and one can expect to obtain a yield of 1–1.5 µg RNA/mg of brain tissue. Although the A_{260}/A_{280} ratio is a good relative measure of purity, it is essential to resolve an aliquot (0.5–1.0 µg) of the RNA on a denaturing 1% agarose gel to verify the presence of two distinct bands (representing the predominant 28S and 18S RNA species; they should be present in a quantitative ratio of approx 2:1) and the lack of degraded (occurs as a smear) RNA. **Figure 1** depicts a high-quality RNA sample and a sample that contains a modest amount of degradation. The next generation analysis platform for validating RNA quality is typified by the Agilent 2100 Bioanalyzer Automated Analysis System (Agilent Technologies). This instrument provides for automated analysis of RNA with high sensitivity and small sample volume (as little as 5 ng/μL).

3.1.5. Protocol for Formaldehyde Denaturing Agarose Gel Electrophoresis

1. Prepare a denaturing mix: 200 μL of formamide, 30 μL of 10X MOPS buffer, 80 μL of formaldehyde, and 15 μL of a 1 mg/mL solution of ethidium bromide (exercise caution and wear gloves when handling ethidium bromide, a potent mutagen).
2. Combine RNA (0.5–1.0 μg) and nuclease-free water to a final volume of 7 μL. Add 7 μL denaturing mix, vortex-mix the sample, and place in a 65°C water bath for 5 min.
3. Remove the sample from the water bath. Add 2 μL of 0.2% bromophenol blue dye solution.
4. Combine 36.5 mL of nuclease-free water, 0.5 g of agarose, and 5.0 mL of a 10X MOPS buffer solution. Heat in a microwave or on a hot plate until the agarose is completely in solution.
5. In a fume hood, add 8.5 mL of formaldehyde, pour molten solution into a gel manifold, insert comb, and allow gel to solidify (approx 30 min). Assemble in a submarine.
6. Format with the electrophoresis apparatus.
7. Load the entire sample (16 μL) onto 1% denaturing agarose gel and resolve in a 1X MOPS buffer solution at 50 V for 30–60 min.
8. Visualize bands under UV light in a UV light box. Prepare a photographic image (Polaroid or gel-documentation workstation).

3.2. cDNA Production

To eliminate potential DNA contamination in the RNA preparation, total RNA can be treated with DNase I (0.4 U/μg of RNA) according to the reagent supplier's instructions. Alternatively, in lieu of DNase I treatment, and/or as an additional control, it is advisable to add a "no reverse transcriptase" reaction to the experiment. To accomplish this, simply prepare one reaction tube or well exactly as the others, but leave out the RT enzyme. Subsequently, use the resulting "reaction product" for PCR (no PCR product = no genomic DNA contamination). When performing quantitative or semiquantitative PCR, it is important to rule out contamination of the RNA preparations with genomic DNA.

3.2.1. Priming for First-Strand Synthesis

There are three types of oligonucleotide priming protocols used with the RT reaction. The method of choice, in our experience, is gene-specific primers (*see* **Note 4**). If the primer set is well designed and working optimally, the PCR product resulting from the use of this primer pair will be a single product of the predicted sequence. A major drawback, however, with the use of gene specific primer(s) for first-strand synthesis is that the cDNA can then be used only for amplification of those specific sequences. This could be a severe limitation if the RNA is in limiting quantity. The second priming method most often used for RT is with oligo-dT. This primer choice relies on a unique feature of most eukaryotic mRNA molecules, namely the presence of a poly-adenylated "tail" at the 3′ end—the oligo-dT is complementary to (and therefore binds to) the "tail." A drawback with this priming method is that target sequences far removed from the 3′ end of a gene may be under-represented in the resulting cDNA owing to failure of the enzyme to transcribe the entire gene sequence. A third and common method for first-strand priming involves the use of a mixture of random hexamers. In this case, full coverage of the RNAs is assured (many primers for each RNA); however, instances where there is only partial coverage of the target sequence of interest will present a problem for quantitation. In other words, depending on where on a specific mRNA a hexamer lands and the RT reaction begins, it may or may not include copying through both primer sequences necessary to "see" the target in the subsequent PCR reaction. This would result in an under-representation of that gene.

No matter what priming method is used, it is highly recommended that before any extensive real-time experiments are set up, a test PCR reaction is run with the cDNA. The resulting product should be sequenced to verify the correct cDNAs have been made.

3.2.2. Gene-Specific Primer Design and Concentration Considerations

Primer design is a very important consideration for performing successful and reproducible quantitative RT-PCR. Points to consider when designing primers to be used in these real-time assays with SYBR Green dye detection are that the primer pair (1) are gene-specific, (2) have similar melting temperatures, and (3) produce a product that is in the range of 100–150, but <500, nucleotides. The real-time PCR systems are supplied with primer design software for this purpose; however, numerous other primer design programs are also available both commercially (e.g., GENSET OLIGOS, OligoVersion 4; GENSET Corp., La Jolla CA or PRIMERSELECT of LASERGENE Software; DNASTAR Inc., Madison, WI) and in the public domain (e.g. Primer3 from the Whitehead Institute at MIT (http://www.genome.wi.mit.edu/genome_software/other/primer3.html). Some general rules for designing gene-specific primers are as follows:

1. Primers should be kept to between 18 and 30 nucleotides in length.
2. Keep the GC content between 40% and 60%.
3. Keep melting temperatures (T_m) of the primers matched.

4. Avoid complementarity of two or more bases at the 3' ends of the primer pairs (to decrease primer–dimer formation).
5. Avoid runs of more than three Gs or Cs at the 3' end.
6. If feasible, and particularly if you plan to assay the same target over and over, create several primer pairs and test all the combinations to find the best pair.

In special cases, where you are performing quantitative RT-PCR for the validation of results from nylon-based macroarray experiments from Clontech, it is very convenient to use the primer sets provided by Clontech (they have already done your "bioinformatic homework" for you). PCR using these primers yields amplicons that are 200–400 basepairs in length and correspond exactly to the gene fragments spotted on their nylon arrays (*see* **Note 5**). Although the resulting amplicons are somewhat longer than is optimal, they have been shown to work well for doing kinetic real-time PCR validation of array results *(7,8)*.

Finally, a word about primer concentration is in order. Primer concentration can influence product formation (as measured by C_t), but also reaction specificity. Higher primer concentrations can result in lower C_t values, but can also increase nonspecific product formation (e.g., primer–dimers). This can be especially important when using SYBR Green for signal determination because the dye binds to all double-stranded DNA, including primer–dimers. It is advisable, therefore, to use the recommended primer concentrations initially and to perform a dissociation curve analysis (**Subheading 3.5.4.** and **Fig. 4**). If a primer–dimer peak is present, it will be necessary to titrate primer concentration.

3.2.3. Reverse Transcription Reaction

One microgram of total RNA is used for cDNA synthesis with the OmniScript Reverse Transcriptase kit.

1. For each reaction, the following are combined in a 200-µL thin-walled PCR tube (or into one well of a 96-well plate): 1 µg of total RNA (volume dependent on RNA concentration), 1 µL of dNTPs (5 m*M* each one, supplied in the kit), 2 µL of 10× RT buffer (supplied in the kit), 1 µL of (reverse) primer (0.1 µg/µL, you supply; see **Subheading 3.3.1.**), nuclease-free water (to bring the reaction mixture volume to 19.5 µL), and 0.5 µL of RT enzyme (4 U/µL). If multiple samples are being analyzed simultaneously, then to reduce the potential for error due to pipetting, it is advisable to prepare a "master mix" of all common reagents (e.g., everything except RNA and primer). The master mix can then be aliquoted to each tube or well in a single delivery, followed by the addition of template and/or primer.

2. The reaction tubes are placed into a standard thermocycler and the RT reaction is set to proceed at between 42 and 50°C for 45 min *(2)*.
3. Following the RT reaction, it is absolutely essential to terminate the RT activity (to prevent RT inhibition of the real-time PCR reaction) by placing the reactions at 95°C for 10 min.

3.3. Real-Time PCR

Real-time kinetic PCR (two-step; *see* **Note 6**) is typically performed in 25- or 50-µL reaction volumes prepared by first combining the following three components: (1) cDNA template (5 µL of some dilution, typically 1:5, of the cDNA prepared above), (2) gene-specific primers (50–200 n*M*), and (3) nuclease-free water (to bring the total volume to one half the reaction volume). Next, add 2X SYBR Green Master Mix (one half the total reaction volume; so, e.g., in a 25-µL reaction, add 12.5 µL of 2X Master Mix) (*see* **Note 7**). Reactions can be carried out in individual 200-µL PCR tubes, strips of tubes, or in a 96-well plate manufactured with optically clear plastic. The standard PCR reaction conditions (default ABI settings to match their kit components) are: one cycle of 50°C for 2 min followed by one cycle of 95°C for 10 min and then 40–45 cycles of: 95°C, 15 s; 60°C, 1 min. These parameters can be changed (edited on-screen) prior to the start of the experiment to conform to the particular target(s) being amplified or to the reagents being used. A typical experiment takes 2.0–2.5 h to complete.

In semiquantitative PCR, a comparison is being made between samples, based on where the threshold cycle (C_t) occurs. The concept of threshold cycle is at the heart of accurate and reproducible quantitation in a real-time assay using fluorescence-based RT-PCR *(2)*. Fluorescence values are recorded at every step of the amplification and represent product formation; therefore, the more target present in the starting material, the fewer cycles it takes to reach the cycle where fluorescence is statistically above background. The cycle at which a given template reaches this point is the threshold cycle, C_t and this always occurs during the exponential phase of the PCR.

3.4. Assay Optimization

The amount of assay optimization you perform and the degree to which it is necessary is dependent upon the specific application. If you are interested in comparing only a few genes across numerous experimental samples, it would be useful to optimize primer concentration, cDNA dilution, and PCR conditions for each gene. If, however, your primary use will be to validate array results, it may quickly become overly cumbersome to do extensive optimization and therefore using the "default" parameters initially may be sufficient. Some hints on assay validation include:

1. Test multiple primer pairs in all combinations with a known template.
2. Use standard assay conditions: 300–400 nM primers; 100 nM probe, and 3 mM MgCl$_2$.
3. Choose the primer pair that gives the lowest C_t.
4. Make a dilution of a template (e.g., from a "control" sample) for a standard curve.
5. An ideal standard curve in this assay will have a slope of –3.3 (*see* **Fig. 3**). If the slope of the standard curve of the best primer is around –3.5, increase MgCl$_2$ to 5 mM; if the slope is >–3.6, find another primer pair.
6. Demonstrate consistent amplification efficiencies for different reactions with the same primer pairs.

3.5. Data Analysis

Generally, two types of quantification in real-time RT-PCR are possible: (1) relative quantification based on the relative expression of a target gene compared to a reference gene (or control = calibrator), and (2) an absolute quantification based on an internally or externally derived standard curve *(9,10)*. Given that one of the most common uses for real-time RT-PCR experiments is for validation of array results, a task that does not require absolute measures of initial mRNA concentration, we will limit our discussion to relative quantification methods (*see* **Note 8**).

3.5.1. Relative Quantification

Using this method for quantification, the ratio between the amount of target molecule and reference molecule within the same sample is determined. The most common type of reference molecule used is a housekeeping gene, such as glyceraldehyde-3-phosphate dehydrogenase (GAPDH). The use of a housekeeping gene as a reference standard assumes that the RNA level for that gene remain relatively invariant across experimental samples; however, this is not always the case *(11)*. Another possibility, in the situation in which the comparison is between gene expression in a "control" set of samples vs a "treated" set, is to use the control sample as the reference quantity, and determine the change in "treated" based on the ratio between the two. The success of this method depends on the efficiency with which the targets in the two samples are amplified. If they are amplified with comparable efficiencies, then the control sample can serve as the standard or calibrator.

3.5.2. Determination of Amplification Efficiency

To assess the amplification efficiency of two target (cDNA) sequences, "control" and "treated," it is first necessary to prepare a dilution series for each sample. A typical dilution series may consist of the following amounts of cDNA: 1, 1:2, 1:4, 1:8 in triplicate wells. Each dilution series is then amplified in a real-time PCR instrument (**Fig. 2**). The next step is to subtract the average C_t values for the controls from the average C_t values for the treated at each dilution and plot the differences in C_t (ΔC_t) values against the logarithm of the (relative) template amount. The plot would then be the difference in C_t values against the log of 1, 0.5, 0.25, and 0.125. If the slope of the resulting line is <0.1, the amplification efficiencies are comparable. If amplification efficiencies between the samples are comparable, one can either create a single standard curve (with either sample; **Fig. 3**) or alternatively, one can use the comparative method ($\Delta\Delta C_t$ method; *12*) which relies on comparing the differences in C_t values to estimate -fold differences between "control" and "treated." In this case, standard curves are not necessary.

Converting ΔC_t into a ratio:

$$\text{Target A/ Target B} = 2^{-\Delta C_t}$$

For example, if the ΔC_t between RNA species A and B (at a given dilution = five cycles, (i.e., it took five more cycles to see amplification of A), then there is $2^{-5} = 1/2^5 = 1/32$ as much A as B. This assumes the amplification efficiency is equal to 2. Insert the *actual* amplification efficiency value, determined experimentally, to obtain a more valid ratio.

3.5.3. Relative Standard Curve

It is not feasible to prepare a standard curve for each individual target on an array, so one solution is to use a "calibrator" to compare the relative quantitation of the other samples. In this case the calibrator becomes 1X sample, and all of the quantities are expressed as an *n*-fold difference relative to the calibrator. When an experiment involves comparing control and treated condition, for example, looking at gene expression as a function of a drug treatment, the untreated control sample would be an appropriate calibrator. The target C_t is compared directly with the calibrator C_t and is recorded as containing either more or less mRNA *(10)*.

To accomplish this, a relative standard curve can be constructed by preparing twofold dilutions of the cDNA such that the units could be the dilution values 1, 0.5, 0.25, 0.125, etc. A sample standard curve derived in this way, using triplicate reactions, is depicted in **Fig. 3**. It is possible now to use this DNA standard for relative quantification of RNA if we determine that reverse transcription efficiency is the same for all samples.

3.5.4. Dissociation Curve

Following a real-time experiment, the machine will perform a melting curve analysis if this option is chosen at the start of the run. The data represented are

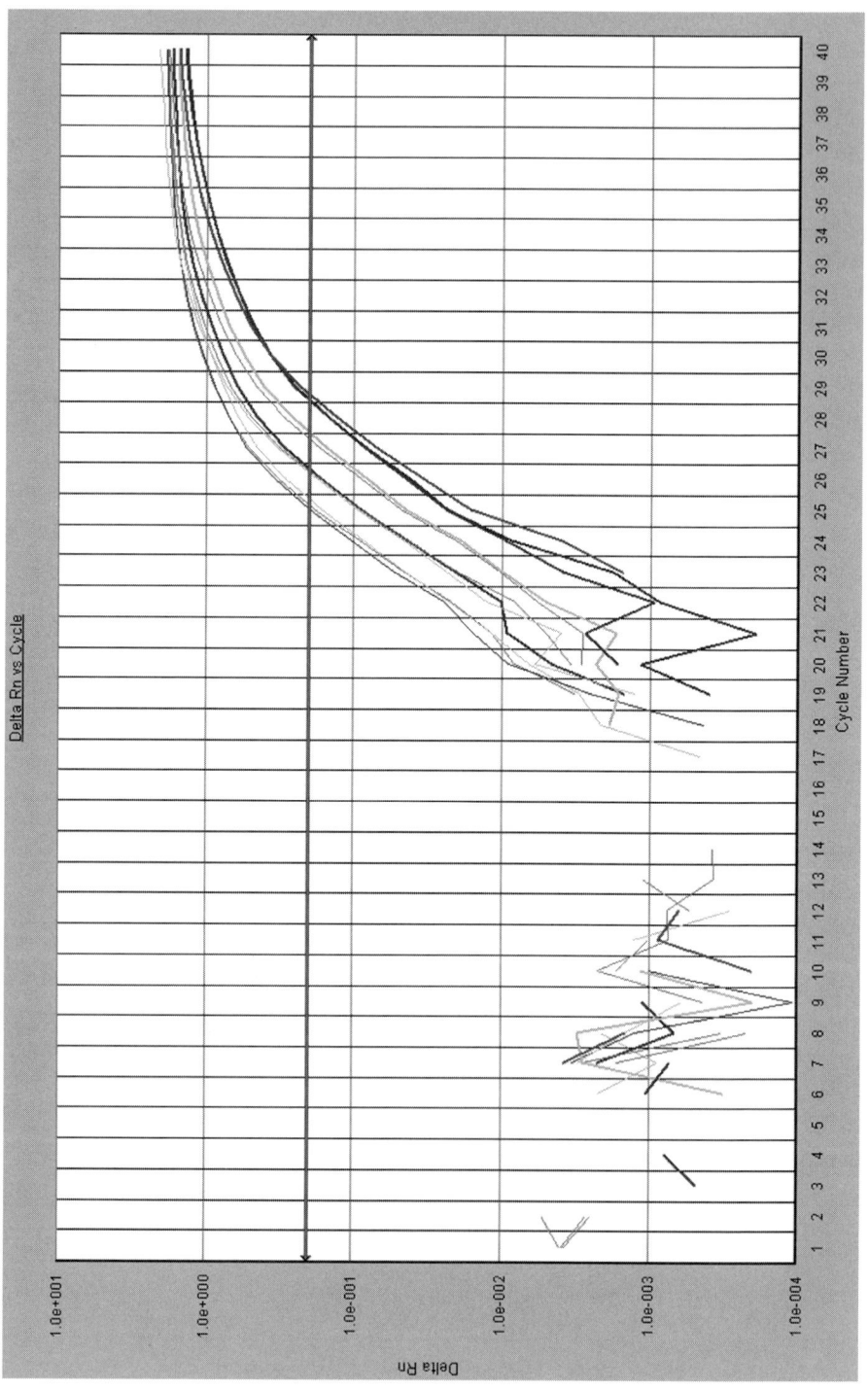

the change in fluorescence as a function of temperature from a dye or probe interacting with the double-stranded DNA. This is especially relevant if SYBR Green is the dye being used for product detection, as the real-time profiles do not distinguish between specific amplicon and nonspecific products (e.g., primer–dimers). The dissociation profile can also be useful in primer concentration optimization.

4. Notes

1. Reagents listed in **Subheading 2** are by no means the only reagents that can be used. There are multiple vendors for RNA purification reagents, RT and PCR enzymes and reagents, and SYBRGreen master mix solutions. The authors of this chapter make no specific endorsement of any one reagent. We present these only as a point of reference for how the experiments herein were performed.
2. There are multiple protocols available to do quantitative PCR (*13,14*; a simple Medline search for *real-time PCR* and *quantitation* gave >800 citations), and which one you chose will depend on your specific application.
3. RNA is notoriously unstable and prone to degradation, so the utmost care must be taken when working with it. Any RNA isolation reagent will likely be provided with detailed instructions for working with RNA.
4. The efficiency of the reverse transcription reaction is by no means uniform across all samples and using all methods. In addition, depending on the report, people use oligo-dT, random hexamers, or gene-specific primers to produce the cDNA. Each choice comes with advantages and disadvantages. We find that if RNA is not limiting (increasing signal-to-noise ratios and decreasing nonspecific signals), then gene-specific reverse priming is the most desirable.
5. ABI recommends primers that produce amplicons of ~100 basepairs (bp) for maximum efficiency of the PCR reaction; however, we routinely use primers designed by Clontech (amplicons are 200–400 bp) without any apparent problems.
6. RT-PCR can be performed as a one-step or two-step reaction. We prefer two-step RT-PCR, largely because it makes maximal use of precious RNA samples (a single cDNA can be used in multiple subsequent PCR reactions). In addition, a relative standard curve can be easily derived from a cDNA preparation by performing a dilution series. Finally, two-step reactions are superior in detecting low abundance transcripts and therefore best to use when doing new assays with unknown abundance transcripts.

Fig. 2. *(previous page)* Real-time PCR analysis of rat brain RNA levels (of the nerve growth factor [NGF] gene). Plot of a dilution series (1, 1:2, 1:4, 1:8) of cDNA derived in the first-strand synthesis reaction. Each dilution was assayed in triplicate. The line at δ Rn = 1 (center) represents the C_t for each reaction. Average C_t values: 1:1 C_t = 24.1 ± 0.169; 1:2 C_t = 25.1 ± 0.024; 1:4 C_t = 26.2 ± 0.109; and 1:8 C_t = 27.3 ± 0.143.

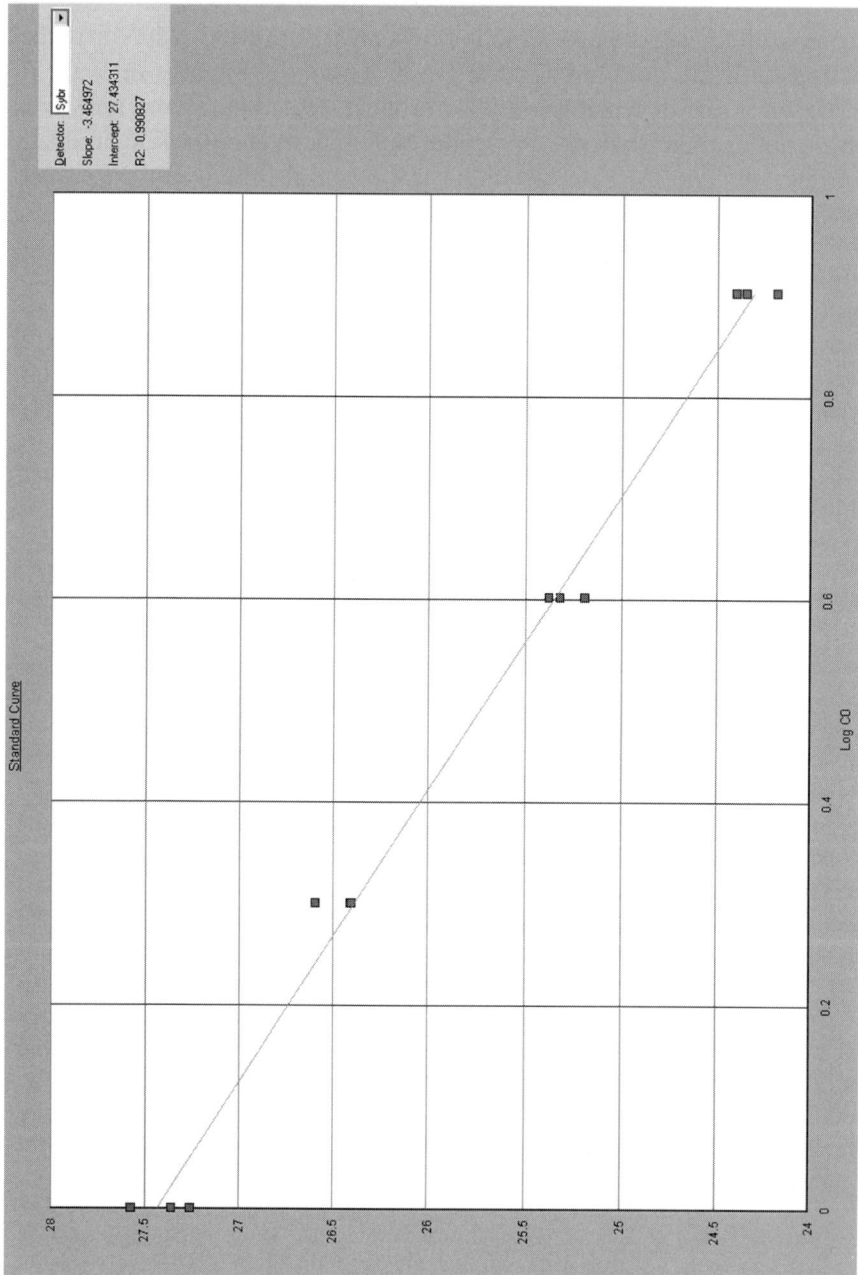

Fig. 3. Relative standard curve. Aliquots of a control cDNA sample (1:1, 1:2, 1:4, and 1:8) were assayed in triplicate and plotted as logarithm of the dilution versus C_t.

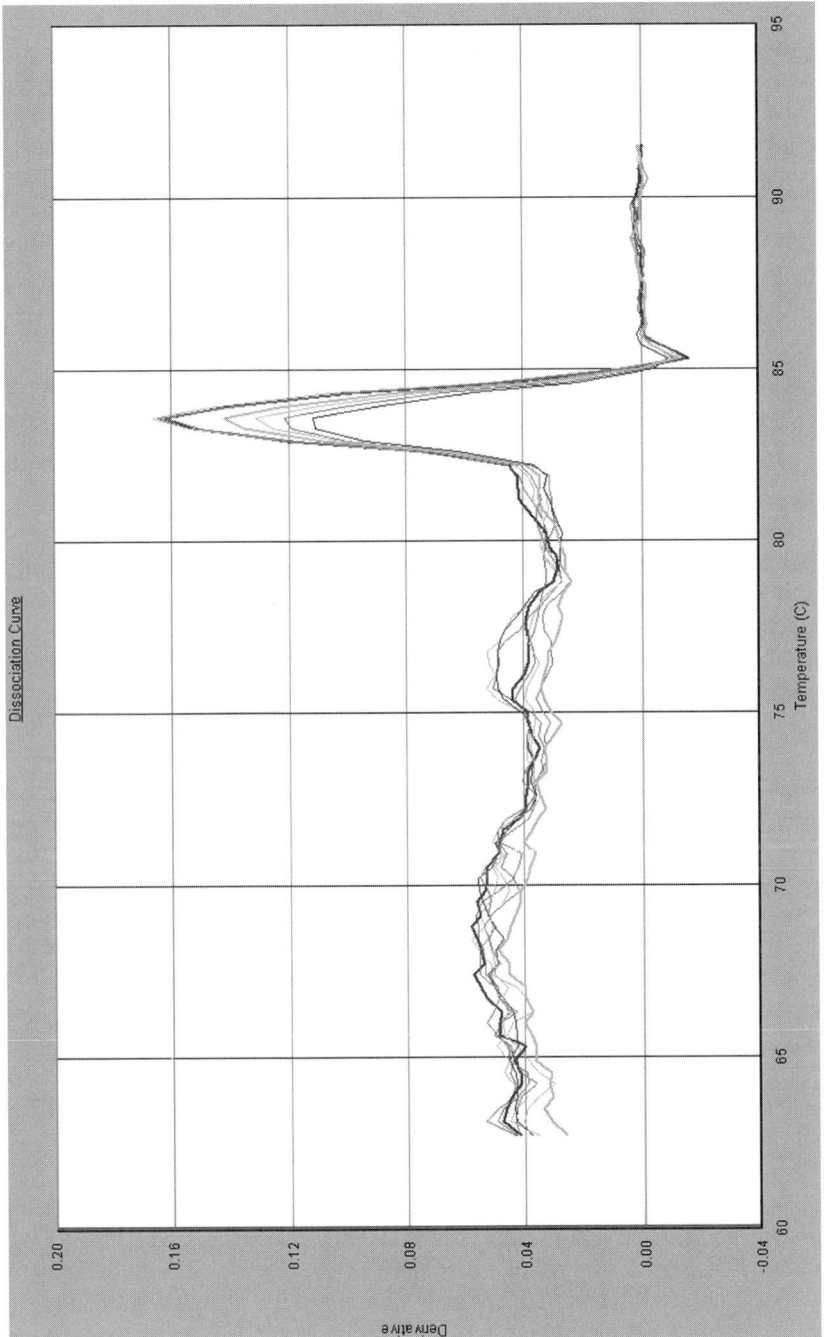

Fig. 4. Heat dissociation protocol. Following the final PCR protocol (40 cycles), the samples from a primer titration experiment were subjected to a heat dissociation protocol over the indicated temperature range. Note the single major peak, representing the melting temperature (83°C) of the amplicon.

7. Real-time PCR can be accomplished using SYBRGreen (described here) or molecular beacons that contain two dyes (one is a quencher) on the primer pair. The advantage of these beacon primers is that one does not have to be concerned with fluorescent signal due to nonspecific amplification, or with primer dimers. The major disadvantage, however, is the cost (much more expensive, per reaction, than SYBR Green). If an investigator has only a handful of genes to be quantified on a routine basis, the molecular beacon approach will prove superior.
8. Methods for quantification of RT-PCR results range from the relatively simple to the very complex *(13,14)*. Presented here is an example of how to do measurements of relative quantitation in a straightforward fashion.

References

1. Freeman, W. M., Walker, S. J., and Vrana, K. E. (1999) Quantitative RT-PCR: pitfalls and potential. *BioTechniques* **26,** 112–125.
2. Freeman, W. M., Vrana, S. L., and Vrana, K. E. (1996) Use of elevated reverse transcription reaction temperatures in RT-PCR. *BioTechniques* **20,** 782–783.
3. Hein, J., Schellenberg, U., Bein, G., and Hackstein, H. (2001) Quantification of murine IFN-(mRNA and protein expression: impact of real-time kinetic RT-PCR using SYBR Green I dye. *Scand. J. Immunol.* **54,** 285–291.
4. Sambrook, J., Fritsch, E. F., and Maniatis, T. (1989) *Molecular Cloning: A Laboratory Manual.* Cold Spring Harbor Laboratory Press, Plainview, NY.
5. Chomczynski, P. and Sacchi, N. (1987) Single step method of RNA isolation by acid quanidinium thiocyanate-phenol-chloroform extraction. *Analyt. Biochem.* **162,** 156–159.
6. Chomczynski, P. (1993) A reagent for the single-step simultaneous isolation of RNA, DNA and proteins from cell and tissue samples. *BioTechniques* **15,** 532–537.
7. Rajeevan, M. S., Vernon, S. D., Taysavang, N., and Unger, E. R. (2001) Validation of array-based gene expression profiles by real-time (kinetic) RT-PCR. *J. Mol. Diagn.* **3,** 26–31.
8. Rajeevan, M. S., Ranamukhaarachchi, D. G., Vernon, S. D., and Unger, E. R. (2001) Use of real-time quantitative PCR to validate the results of cDNA array and differential display PCR technologies. *Methods* **25,** 443–451.
9. Higuchi, R., Fockler, C., Dollinger, G., and Watson, R. (1993) Kinetic PCR analysis: real time monitoring of DNA amplification reactions. *Biotechnology* **11,** 1026–1030.
10. Bustin, S. A. (2000) Absolute quantification of mRNA using real-time reverse transcription polymerase chain reaction assay. *J. Mol. Endocrinol.* **25,** 169–193.
11. Schmittgen, T. D. and Zakrajsek, B. A. (2000) Effect of experimental treatment on housekeeping gene expression: validation by real-time, quantitative RT-PCR. *J. Biochem. Biophys. Methods* **46,** 69–81.
12. Livak, K. J. and Schmittgen, T. D. (2001) Analysis of relative gene expression data using real-time quantitative PCR and the $2^{-\Delta\Delta}$ method. *Methods* **25,** 402–408.

13. Gentle, A., Anastasopoulos, F., and McBrien, N. A. (2001) High-resolution semiquantitative real-time PCR without the use of a standard curve. *BioTechniques* **31,** 502–508.
14. Vuillaume, I., Schraen-Maschke, S., Formstecher P., and Sablonniere, B. (2001) Real time RT-PCR shows correlation between retinoid-induced apoptosis and NGF-R mRNA levels. *BBRC* **289,** 647–652.

15

Application of TaqMan RT-PCR for Real-Time Semiquantitative Analysis of Gene Expression in the Striatum

Andrew D. Medhurst and Menelas N. Pangalos

1. Introduction

TaqMan reverse transcription-polymerase chain reaction (TaqMan RT-PCR) is a recently developed technique *(1)* that has been used to study gene expression in tissues of the central nervous system (CNS) including the striatum. For example, TaqMan has been used to profile mRNA distribution patterns across the brain for γ-aminobutyric acid-$_B$ (GABA$_B$) receptor subunits *(2)*, 5-hydroxytryptamine$_4$ (5-HT$_4$) receptor splice variants *(3)*, novel G-protein-coupled receptors *(4)*, and ion channels including vanilloid receptors *(5)* and two pore potassium channels *(6)*. More specifically, TaqMan RT-PCR studies have demonstrated a huge enrichment of dopamine D$_2$ *(7)* and D$_3$ *(8)* receptors in human striatal tissues compared to other brain regions. In addition, the technique has been utilized for the analysis of gene expression changes in animal models of CNS diseases including Parkinsons's disease *(9)*, stroke *(10)*, and migraine *(7)*.

The TaqMan RT-PCR technique enables the measurement of an accumulating PCR product in real time by utilizing a dual-labeled TaqMan fluorogenic probe *(1,11,12;* **Fig. 1A**). It allows a more accurate comparison of mRNA expression levels between large numbers of samples, compared to other techniques such as Northern blotting, standard RT-PCR, and *in situ* hybridization. It is also useful for validating samples prior to microarray analysis.

The chemistry of TaqMan RT-PCR is based around the endogenous 5′-3′ exonuclease activity of *Thermus aquaticus* (Taq) polymerase *(1,11,12)*. During polymerization, the gene-specific TaqMan probe anneals to the specific DNA

From: *Methods in Molecular Medicine, vol. 79: Drugs of Abuse: Neurological Reviews and Protocols*
Edited by: J. Q. Wang © Humana Press Inc., Totowa, NJ

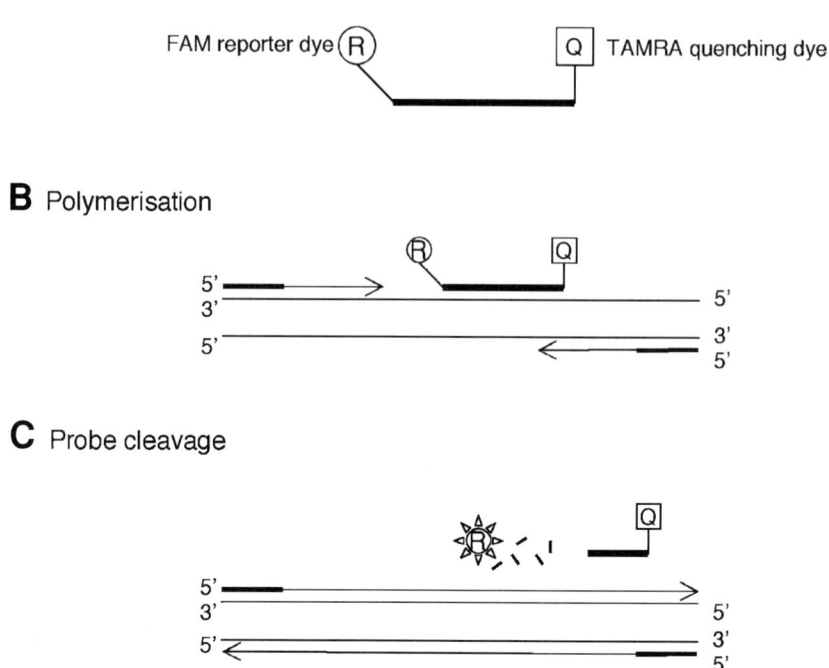

Fig. 1. Schematic representation of TaqMan PCR assay. (**A**) TaqMan probe with fluorescent reporter dye FAM (6-carboxyfluorescein) attached to the 5′ end and a quencher dye TAMRA (6-carboxytetramethylrhodamine) attached to the 3′ end. (**B**) During polymerization the TaqMan probe containing a reporter dye (R) and a quencher dye (Q) specifically anneals to one strand of the DNA template between forward and reverse PCR primers. During amplification, the fluorogenic probe is displaced by Taq polymerase and then cleaved (**C**), releasing the reporter dye. After cleavage, the shortened probe dissociates from the template, allowing polymerization of the strand to complete. Fluorescence is proportional to the amount of product accumulated.

template between forward and reverse PCR primers (**Fig. 1B**). When the probe is intact the reporter fluorophore emission is suppressed by a quencher fluorophore. The 5′-3′-nuclease activity of Taq polymerase releases the reporter dye from the vicinity of the quencher dye, resulting in the generation of reporter fluorescence (**Fig. 1C**). The remaining probe fragment dissociates from the target sequence, allowing polymerization to continue. The intensity of fluorescence is proportional to the amount of DNA amplified, and this reaction occurs during every cycle of PCR. The fluorescent signal is captured using an ABI Prism 7700 sequence detection system (PE Applied Biosystems) which allows the rapid analysis of 96 PCR samples simultaneously, without any time-consuming downstream gel electrophoresis steps. Data are easily captured using Sequence Detector Software (Applied Biosystems) linked to the ABI Prism 7700, and can then be exported to Microsoft Excel workshcets for further analysis.

The standard techniques used to generate mRNA expression data in human and rat CNS tissues using TaqMan RT-PCR are described, including protocols we have used for RNA extraction, cDNA synthesis, primer and probe design and quality control, TaqMan PCR assays, and data analysis *(7)*.

2. Materials

1. Trizol reagent (Life Technologies).
2. Chloroform, isopropanol, and 75% ethanol.
3. RNase-free/DNase-free sterile water.
4. Poly(A)$^+$ mRNA samples (Clontech).
5. Superscript II RNase H$^-$ reverse transcriptase (200 U/µL) supplied with 5X first-strand buffer and 100 mM dithiothreitol (DTT) (Life Technologies).
6. Oligo(dT)$_{12-18}$ primer (0.5 µg/µL; Life Technologies).
7. 10 mM dATP, dTTP, dCTP, and dGTP (Life Technologies).
8. RNAsOUT (40 U/µL; Life Technologies).
9. 2X TaqMan Universal Mastermix (Applied Biosystems).
10. Custom synthesized gene specific oligonucleotides (10 µM) and fluorogenic TaqMan probes (5 µM).
11. Genomic or plasmid DNA for standard curve generation.
12. ABI Prism 7700 or similar thermocycler with fluorescence detection system.

3. Methods

The methods described below outline (1) RNA extraction, (2) cDNA synthesis, (3) primer and fluorogenic probe design and quality control, (4) TaqMan PCR assay, and (5) data analysis *(7)*.

3.1. RNA Extraction

There are several methods for extracting total RNA from tissues but we have successfully used the Trizol method for numerous TaqMan RT-PCR studies.

1. Quickly dissect striatal tissues from rats after they are killed, wrap samples in labeled aluminum foil, snap freeze in liquid nitrogen, and store at –80°C until required for further processing.
2. Directly homogenize frozen rat striatal tissues in 1 mL of Trizol reagent (Life Technologies) per 50–100 mg of tissue. Do not leave Trizol in tubes on ice, as this greatly reduces RNA quality, and ensure the Trizol reagent has reached room temperature after removing it from 4°C storage. Trizol aliquots can be stored

for several months at –80°C, but again allow frozen aliquots to thaw to room temperature before proceeding with the RNA extraction protocol.
3. Extract total RNA following the manufacturer's suggested protocol but with some additional minor modifications to obtain purer RNA.
4. Add 0.2 mL of chloroform per 1 mL of Trizol, shake, and leave for 2–3 min at room temperature, and then centrifuge for 15–20 min (4°C, 12,000g). Transfer the upper aqueous layer containing RNA fraction to a fresh Eppendorf tube and perform an additional chloroform extraction step by adding another 0.5 mL of Trizol, then another 0.2 mL of chloroform, mix, centrifuge as before, and again transfer the aqueous layer to a fresh tube.
5. Precipitate RNA by adding 0.5 mL of isopropanol (per 1 mL of Trizol) and wash the pellet twice in 1 mL of 70% ethanol.
6. Remove the ethanol and after air-drying, resuspend RNA in PCR-grade water (Sigma) and calculate the concentration of each sample from A_{260} measurements.
7. Confirm RNA integrity using standard agarose gel (1%) electrophoresis techniques. Keep RNA samples on ice prior to cDNA synthesis, or aliquot for storage at –80°C.
8. For human distribution studies, poly(A)$^+$ mRNA samples extracted from striatal and other CNS tissues can be obtained commercially from Clontech. Take care to aliquot the stock solutions of these samples to avoid mRNA degradation due to freeze thawing.
9. If preparing RNA directly from human brain samples, total RNA tends to be more highly contaminated with genomic DNA than fresh animal tissue, and will require DNase treatment or preparation of poly(A)$^+$ mRNA directly from frozen tissue.

3.2. cDNA Synthesis

There are also a wide variety of reverse transcriptase systems that can be used for first-strand cDNA synthesis. We found the Superscript II system to be particularly effective for TaqMan RT-PCR.

1. Thaw RNA samples on ice and perform cDNA synthesis in duplicate or triplicate from each RNA sample where possible.
2. For each 20 µL of reverse transcription reaction, mix 200 ng of human poly(A)$^+$ mRNA or 1 µg of rat total RNA (in 9 µL of water) with 1 µL of oligo(dT)$_{12-18}$ primer (0.5 µg) and incubate for 5 min at 65°C.
3. After cooling on ice for a couple of minutes, mix the solution with 4 µL of 5X first-strand buffer, 2 µL of 0.1 M DTT, 0.5 µL each of dATP, dTTP, dCTP, and dGTP (each 10 mM), 1 µL RNAseOUT (40 U), and 1 µL of SuperScript II reverse transcriptase (200 U). If sample numbers are large, then preparation of a core mastermix with all required reagents in sufficient volume for the number of reactions being run (+10%) is recommended, to reduce sample variation and preparation time.

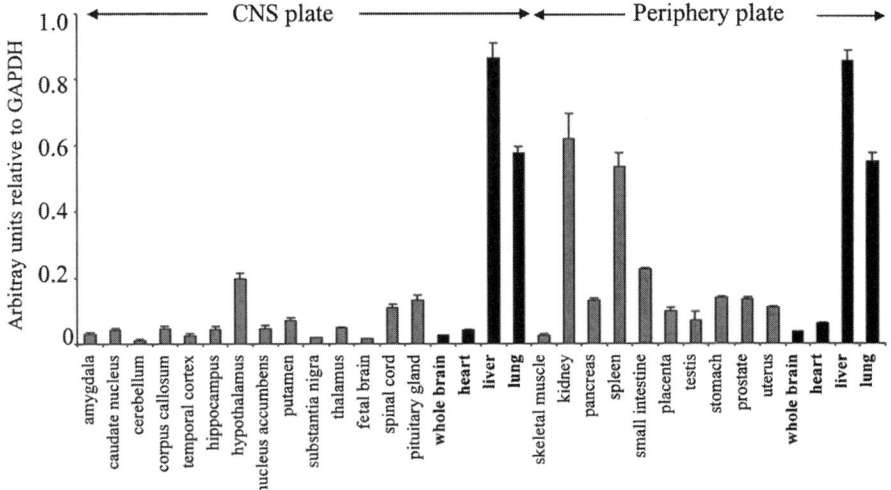

Fig. 2. Comparison of TaqMan data across two plates. Four identical samples (*black bars*) were included on both plates to allow a correction factor to be calculated from the means of these four tissues.

4. Perform reactions for 60 min at 42°C on a PCR machine and terminate by incubating for 15 min at 70°C.
5. Run parallel reactions for each RNA sample in the absence of SuperScript II (no amplification control) to assess the degree of any contaminating genomic DNA, particularly if you are not treating your RNA samples with DNase (*see* **Note 1**).
6. Run a preliminary TaqMan assay (as described in **Subheading 3.4.**) on a housekeeping gene (e.g., cyclophilin, β-actin, GAPDH) to assess sample integrity and reproducibility of replicates.
7. cDNA can then be aliquoted into replicate 96-well MicroAmp optical plates using Hydra96 (Robbins Scientific) or a similar dispensing system.
8. If your experiment involves more samples than can be run on a single 96-well plate, ensure that you have at least two identical cDNA samples on each plate, so that expression data for a particular gene can be appropriately compared across the plates (*see* **Note 2** and **Fig. 2**).
9. Apply optical caps or sheets tightly to each plate and store at –20°C until required for TaqMan PCR assays.

3.3. Primer and Fluorogenic Probe Design and Quality Control

3.3.1. Primer and Probe Design

There are various ways to design primer/probe sequences for TaqMan studies depending on sequence information available. Given that information on genomic structure is often unknown for novel genes of interest, we do not routinely target primers and probes to intron/exon boundries. The approach we routinely use for primer and probe design is described.

1. Design primer and probe sequences using Primer Express software (Applied Biosystems) as close as possible to the 3' coding region of target gene sequences obtained from Genbank or other sources.
2. Whenever possible, follow the specific primer and probe design criteria suggested by the manufacturers of TaqMan reagents (Applied Biosystems).
3. Pick sequences with 30–80% GC content and avoid runs of more than three consecutive Gs in either primers or probes. Pick a T_m of 59–61°C for primers and 66–68°C for probes.
4. Choose probe from the strand with more Cs than Gs, but avoid any probes with G at the 5' end, as this can exert a quenching effect on the reporter dye (*13*).
5. Position forward and reverse primers as close as possible to each other without overlapping the probe, and aim to have fewer than three Cs and Gs in the five most 3' nucleotides of these primers.
6. Primers and probes were purchased from Applied Biosystems and each probe was synthesized with the fluorescent reporter dye FAM (6-carboxyfluorescein) attached to the 5' end and a quencher dye TAMRA (6-carboxytetramethylrhodamine) to the 3' end.

3.3.2. Primer/Probe Quality Control

1. Check gene-specific primer pairs for efficient amplification using genomic DNA in standard TaqMan PCR assays (described in **Subheading 3.4.**) but replacing the probe with water. If testing primers and probes for splice variants, plasmid or cDNA templates may be required.
2. Run products on 2% agarose gels to confirm correct size of product amplified.
3. Once TaqMan probes are obtained, check these in standard TaqMan reactions using genomic DNA as template. We found this standard genomic DNA test a valuable way of assessing the relative sensitivity and efficiency of each primer/probe set from the threshold cycle (C_t) values and standard curve slopes, respectively (*7*; *see also* data analysis in **Subheading 3.5.**). We consistently found that the C_t values for hundreds of primer/probe sets against a standard concentration of genomic DNA were very similar, reflecting the tight criteria used for primer/probe design (**Table 1**).

3.4. TaqMan PCR Assay

In recent years, a number of manufacturers have started to produce specific all inclusive buffers for TaqMan assays. We have primarily used the 2X TaqMan Universal PCR Mastermix from Applied Biosystems for our studies which are now described.

1. Whenever possible, prepare duplicate or triplicate TaqMan PCR samples in 96-well optical plates.

Table 1
Primer/Probe Sensitivity (Threshold Cycle C_t) and Efficiency (Standard Curve Slope) Parameters Assessed from Amplification of Genomic DNA

Gene	C_t	Slope
Rat		
BDNF	22.8	−3.4
IL-1ra	26.0	−3.7
D2 receptor	21.6	−3.3
GAPDH	16.0	−3.46
Cyclophilin	21.6	−3.39
β-Actin	23.1	−3.31
Human		
D2 receptor	21.2	−3.57
GAPDH	22.3	−3.43
Cyclophilin	24.2	−3.38
β-Actin	25.2	−3.2

2. For each 25 µL of TaqMan reaction, mix 1 µL of cDNA (or genomic DNA standard) with 11.25 µL of PCR-grade water, 11.25 µL of 2X TaqMan Universal PCR Master Mix (PE Applied Biosystems), 0.5 µL of sense primer (10 µM), 0.5 µL of antisense primer (10 µM), and 0.5 µL of TaqMan probe (5 µM).
3. If cDNA samples are already aliquoted into 96-well plates, a mastermix of water, TaqMan Universal mastermix, primers, and probe can be added directly to the samples. Volumes required will depend on which automated dispensing system is used for dispensing (*see* **Note 3**).
4. Run reactions on an ABI Prism 7700 Sequence Detection system (PE Applied Biosystems) using the PCR parameters of 50°C for 2 min, 95°C for 10 min, 40 cycles of 95°C for 15 s, and 60°C for 1 min.

3.5. Data Analysis

Data are automatically captured by Sequence Detector software (Applied Biosystems) following completion of the TaqMan PCR cycling protocol. For every cDNA sample, an amplification plot is generated, showing an increase in fluorescence (ΔRn) with each cycle of PCR. Examples of amplification plots showing reproducibility of triplicate samples are shown in **Fig. 3**. A threshold cycle (C_t) value is then calculated from each amplification plot which represents the PCR cycle number at which fluorescence is detectable above a set threshold, usually 0.1 fluorescence units, based on the variability of baseline data in the first 15 cycles (*see* **Note 4**).

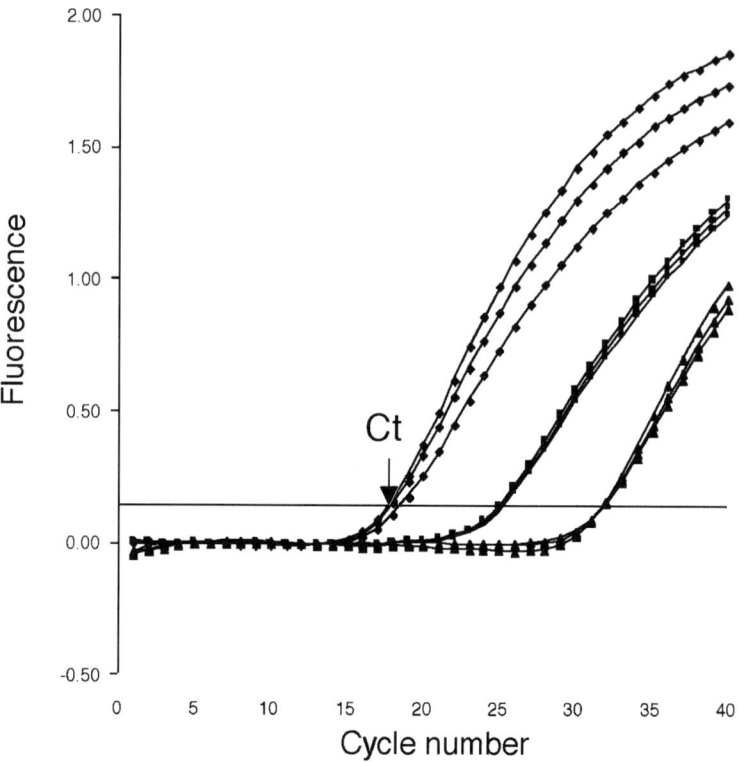

Fig. 3. TaqMan amplification plots showing triplicate dilutions of genomic DNA. Threshold cycle (C_t) values are determined as described in **Subheading 3**.

Using known dilutions of genomic DNA, standard curves are constructed by plotting C_t values against input copy number. The standard curve is then used to calculate copy numbers for cDNA samples from C_t values (*see* **Note 5**). Where "no reverse transcription" controls have been run for each sample, these low copy number values are subtracted from the cDNA triplicate values to correct for any small amounts of contaminating genomic DNA. The mean and standard error are then calculated for each triplicate in Microsoft Excel and plotted graphically as copies per nanogram. Care should be taken when expressing data as copies per nanogram, as this can be misleading if primers and probes are not properly characterized (*see* **Note 6** and **Fig. 4**). In addition, copies per nanogram values do not reflect potential differences in reverse transcription efficiencies, and it may be more appropriate to give values as arbitrary units.

Another difficulty with expressing data as copies per nanogram is that it does not take into consideration relative quantity and integrity of RNA samples. To

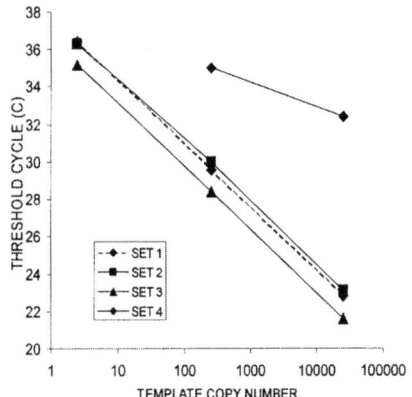

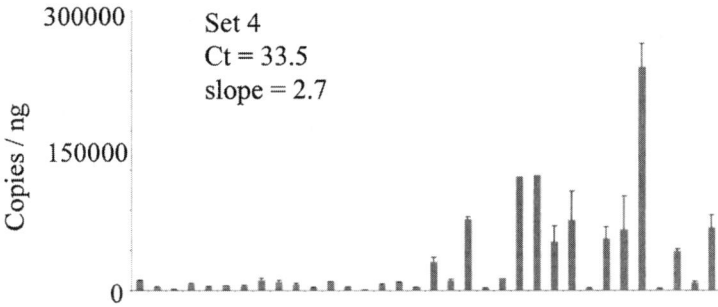

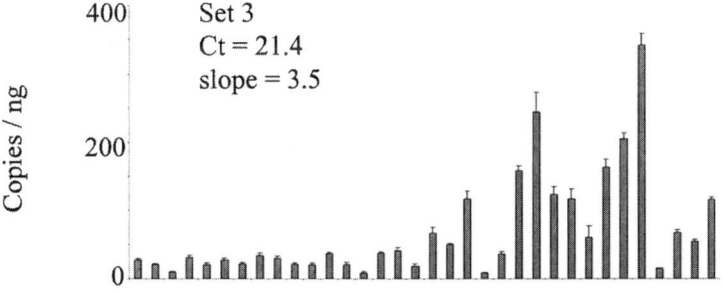

Fig. 4. Optimization of primer/probe sets. (**A**) Probe set 4 showed much lower PCR efficiency and sensitivity compared to the other three sets when tested against genomic DNA as reflected by shallow slope and higher C_t values obtained from comparable copy numbers of template. The distribution pattern obtained with primer/probe set 4 (**B**) was very similar to that obtained with the good primer/probe set 3 (**C**), but the copy per nanogram values were dramatically different.

correct for this, each value can be divided by the corresponding copy number obtained for an appropriate housekeeping gene determined in a replicate plate, and data plotted graphically as a proportion of that reference gene. The appropriate housekeeping gene for normalization is dependent on the model system being investigated, and is best determined by running at least three different genes across samples to find the one most consistently expressed. In some cases, there is little difference between data when expressed as copies per nanogram or relative to a housekeeping gene as seen with the D_2 receptor enrichment in caudate and putamen (**Fig. 5**).

4. Notes

1. We do not routinely treat our RNA samples with DNase, as experiments performed in our laboratories showed little effect of DNase treatment on TaqMan PCR signals, reflecting the minimal presence of contaminating genomic DNA following the modified Trizol RNA extraction protocol. Instead we used the subtraction method to correct for minimal genomic DNA contamination as described in **Subheading 3.4.** It is also important to be careful which DNase system is used. For example, DNase reactions run at 37°C for 60 min with high concentrations of $MgCl_2$ in the buffer result in Mg^{2+}-dependent RNA hydrolysis as demonstrated by increased C_t values obtained using TaqMan RT-PCR analysis. When required, we have used DNase I amplification grade (Life Technologies) successfully, which involves only 15 min of incubation at room temperature, followed by termination of the reaction with EDTA, which did not cause any RNA hydrolysis or affect TaqMan signals downstream.
2. When large numbers of cDNA samples were compared across different plates plates, four samples in triplicate were included on both plates. Comparison of the mean values from these four parallel samples allowed the calculation of a correction factor between plates, which took into account any variation between different TaqMan runs. If two identical plates are run for the same gene in parallel, we observed similar expression profiles across the plates, but some difference in actual fluorescence levels between plates, hence the use of a correction factor.
3. When aliquoting our cDNA into 96-well plates using the Hydra96 (Robbins Scientific), we first dilute the 20 μL of cDNA by adding 60 μL of sterile water to each well. This allows an increase in dispensing volume to 4 μL per well. This routinely gives approx 18–19 plates from 20 μL of cDNA. We have also

---

Fig. 5. *(opposite page)* Classical enrichment of D_2 receptor expression in human caudate and putamen. Data are expressed as (**A**) arbitrary units, (**B**) normalized to GAPDH, and (**C**) normalized to cyclophilin. With a discrete distribution like this one, the data is not really affected by normalization to the housekeeping genes.

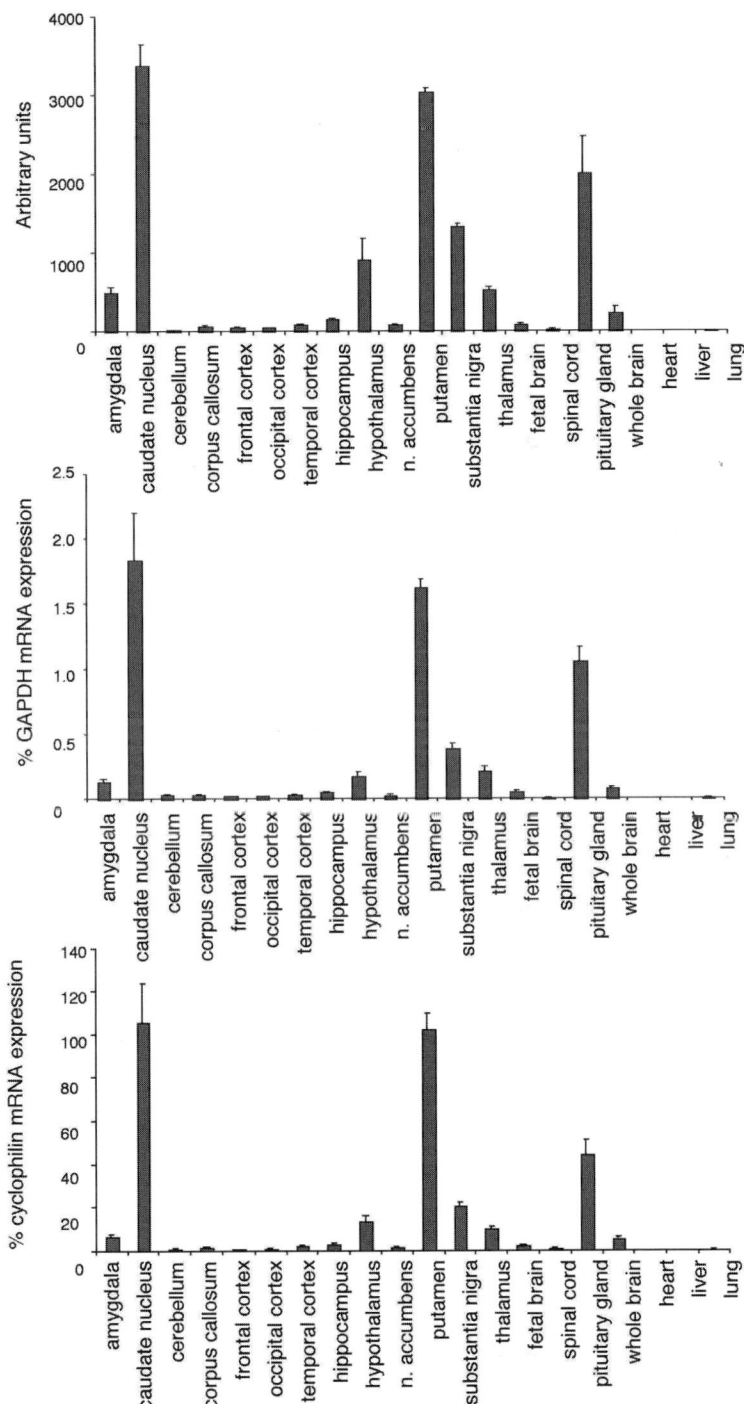

diluted cDNA further to allow 40 plates to be produced, with little change in data quality.

4. We routinely set the threshold level at 0.1, which is fine for the majority of assays as this is in the exponential phase of PCR where there are no rate-limiting components. Occasionally, the flourescent signal obtained with certain primer/probe sets is too low, making the 0.1 threshold level unsuitable as it would fall outside the linear portion of the amplification plot. In this case, the threshold level is set by eye at an appropriate lower level within the linear portion of the curve, but above background noise at early cycle numbers. Probes generating low fluorescence can give valuable results, but the variation between replicates was greater than when a probe generating higher fluorescence was used.

5. Standard curves can also be generated using known dilutions of gene specific plasmid constructs (when available), reverse transcribed RNA or cDNA from a tissue known to express the gene of interest. Given the small variation in standard curves we observed for hundreds of primer/probe sets it is possible to calculate copy number from the theoretical equation (copy number = $10^{[(C_t-37.4)/-3.3]}$), as long as absolute quantification is not required and only relative comparisons across samples are being made.

6. We observed an interesting problem when using a primer/probe set from an external lab. In one of our assays we obtained unexpectedly high copies per nanogram values of approx 300,000. When tested against genomic DNA, the characteristics of this primer/probe set were quite different from most others we had designed (Fig. 4). When we designed a new primer/probe set showing more typical sensitivity and efficiency, the copies per nanogram values became more biological relevant (hundreds of copies) but interestingly the relative expression between tissues was barely different from that obtained with the poor primer set. This highlighted how copies per nanogram potentially can be misleading if primer/probe sets are not carefully characterized each time. If primer/probe sets are old, it is also advisable to run a quick TaqMan assay against genomic DNA or other standard to check they are still working efficiently.

References

1. Heid, C. A., Stevens, J., Livak, K. J., and Williams, P. M. (1996) Real time quantitative PCR. *Genome Methods* **6,** 986–994.
2. Calver, A. R., Medhurst, A. D., Robbins, M. J., et al. (2000) Expression of $GABA_{B1}$ and $GABA_{B2}$ receptor subunits in the central nervous system differs from that in peripheral tissues. *Neuroscience* **100,** 155–170.
3. Medhurst, A. D., Lezoualch, F., Fischmeister, R., Middlemiss, D. N., and Sanger, G. J. (2001) Quantitative mRNA analysis of five c-terminal splice variants of the human 5-HT4 receptor in the central nervous system by TaqMan real time RT-PCR. *Mol. Brain Res.* **90,** 125–134.
4. Robbins, M. J., Michalovich, D., Hill, J., et al. (2000) Molecular cloning and characterisation of two novel retinoic acid inducible orphan G protein coupled receptors. *Genomics* **67,** 8–18.

5. Hayes, P., Meadows, H., Harries, M., et al. (2000) Cloning and expression of a human homologue to rat vanilloid receptor-1. *Pain* **88,** 205–215.
6. Medhurst, A. D., Rennie, G., Chapman, C. G., et al. (2001) Distribution analysis of human two pore domain potassium channels in tissues of the central nervous system and periphery. *Mol. Brain Res.* **86,** 101–114.
7. Medhurst, A. D., Harrison, D. C., Read, S. J, Campbell, C. A., Robbins, M. J., and Pangalos, M. N. (2000) The use of TaqMan RT-PCR assays for semiquantitative analysis of gene expression in CNS tissues and disease models. *J. Neurosci. Methods* **98,** 9–20.
8. Reavill, C., Taylor, S. G., Wood, M. D., et al. (2000) Pharmacological actions of a novel, high-affinity, and selective human dopamine D(3) receptor antagonist, SB-277011-A. *J. Pharmacol. Exp. Ther.* **294,** 1154–1165.
9. Medhurst, A. D., Zeng, B. Y., Charles, K. J., et al. (2001) Up-regulation of secretoneurin immunoreactivity and secretogranin II mRNA in rat striatum following 6-hydroxydopamine lesioning and chronic L-DOPA treatment. *Neuroscience* **105,** 353–364.
10. Harrison, D. C., Medhurst, A. D., Bond, B. C., Campbell, C. A., Davis, R. P., and Philpott, K. L. (2000) The use of quantitative RT-PCR to measure mRNA expression in a rat model of focal ischemia—caspase-3 as a case study. *Mol. Brain Res.* **75,** 143–149.
11. Lang, R., Pfeffer, K., Wagner, H., and Heeg, K. (1997) A rapid method for semiquantitative analysis of the human Vβ-repertoire using taqMan PCR. *J. Immunol. Methods* **203,** 181–192.
12. Lie, Y. S. and Petropoulos, C. J. (1998) Advances in quantitative PCR technology: 5′ nuclease assays. *Curr. Opin. Biotech.* **9,** 43–48.
13. Livak, K. J. (1999) Allelic discrimination using fluorogenic probes and the 5′-nuclease assay. *Gen. Analyt. Biomol. Engin.* **14,** 143–149.

16

Systematic Screening of Gene Expression Using a cDNA Macroarray

Travis J. Worst, Willard M. Freeman, Stephen J. Walker, and Kent E. Vrana

1. Introduction

In the field of drug abuse research, a critical need is to identify the events that underlie the transition from recreational drug use to drug abuse and addiction. It is clear that the mechanisms will involve psychosocial, behavioral, neuroanatomical, and molecular biological components. This last facet, the molecular component, is the subject of the present discussion and the motivation for the use of DNA arrays. Although the contributions of social and behavioral components to the etiology of drug abuse are substantial, growing evidence shows that long-term drug administration produces persistent changes in brain gene expression. The current thinking is that these changes in gene expression may contribute to an allostatic state that is manifested in physiological and behavioral phenomena such as physical dependence, tolerance, withdrawal, craving, sensitization, and psychological addiction *(1–4)*.

This concept of drug-induced allostasis can be described in the following manner. Every cell in an organism contains the same DNA. Tissue identity and the function of particular cells is therefore largely determined by the unique pattern and identities of the genes expressed within the cells of a tissue. For example, neuronal cells are created using one set of genes while liver cells use another set. The sets of genes used for creation of these different cell types may overlap, but will not be identical. If a pattern of gene expression is taken to represent a normal or homeostatic pattern of expression for that cell, a disease state may manifest (be caused by) an altered, state-specific pattern of gene expression. The identity of the genes may remain the same, but their levels of expression have been changed. Determining which genes' expression levels change in response to drug administration in various animal models will provide insights into the epigenetic imprinting by drugs of abuse. Moreover, gene expression changes identified in these studies may provide important molecular targets for development of pharmacotherapeutic agents.

1.1. General Approach to DNA Array Technologies

With the completion of the human and other genome projects (and the identification of more than 30,000 human genes), traditional models of examining one gene at a time are being supplemented by large-scale screening technologies. DNA hybridization arrays are a common form of screening technology and allow the analysis of hundreds to thousands of genes in parallel *(5,6)*. In the past, Northern blotting, dot blots, *in situ* hybridization, and quantitative reverse transcription-polymerase chain reaction (QRT-PCR) were the common methods for investigating changes in gene expression. Although these approaches remain in common use, low throughput is a pervading problem.

Hybridization array technology, on the other hand, offers to bypass many of the limitations of these techniques by simultaneously creating labeled copies of multiple RNAs and then hybridizing them to many different, gene-specific, fixed DNA molecules (**Fig. 1**). The nomenclature has developed whereby the labeled sample RNA is termed the target and the individual gene sequences placed on the array are termed probes.

Although arrays are increasingly used for gene expression analysis, one limitation to this technology is that arrays only measure relative levels of mRNA expression. That is, the relative levels of RNAs can be described (e.g., sample A has 50% more of the specific RNA than sample B), but absolute amounts (e.g., sample A has 1000 copies of the RNA and sample B has 500 copies of the transcript) cannot be reported. As well, most hybridization arrays are not designed to differentiate between alternatively spliced transcripts of the same gene and, in some cases, between highly homologous members of a gene family. Finally, a change in messenger RNA does not necessarily correlate with a change in protein expression, and the translated protein often requires further modification to realize its full activity. These latter two points are a common and legitimate criticism of array technology because it measures an intermediate step (mRNA levels) and not functional product (active protein). However, until proteomic technologies become universally accessible to the research community, hybridization arrays are the best opportunity for studying gene expression on a genome-wide scale.

DNA Macroarray Screening

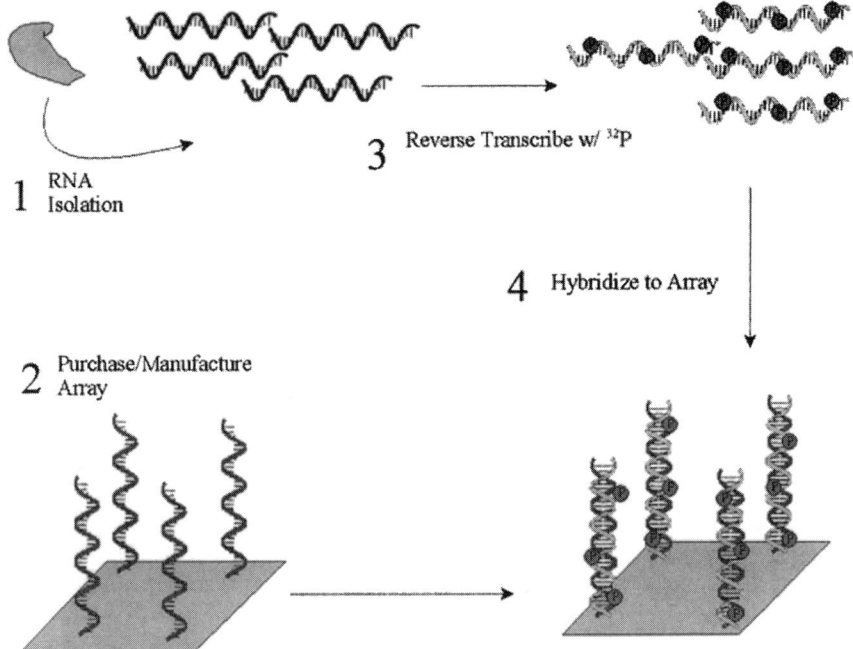

Fig. 1. Basic experimental design from tissue isolation to hybridization.

1.2. Platforms

The hybridization array, an ingenious inversion of the Northern blot, has spawned a number of different formats. Current array formats can be categorized into four groups: macroarrays, microarrays, high-density oligonucleotide arrays (Gene Chips), and spotted oligonucleotide arrays. Although terms such as microarray and gene chip are sometimes used interchangeably, we use these terms to describe distinct DNA array formats. Although "microarray" is often used to describe the technology in general, we use the terms hybridization array or DNA array instead and leave the term microarray to describe a specific subset of hybridization arrays. These varying platforms differ according to the material used to construct the platform (matrix), type of probe on the array, probe number/density, array size, and type of label. An understanding of the strengths and weakness of each platform is necessary to decide which is appropriate for an individual investigator's research aims. Although each of the approaches is considered in the next section, this laboratory uses the macroarray format and the experimental detail is presented for this type of hybridization array.

1.2.1. Macroarray

Macroarrays are generally defined by the deposition of probes deposited onto membranes or plastic and by the use of radioactivity for detection. The term macroarray, as opposed to microarray (discussed next), also refers to the lower probe density on these arrays. Although density varies among arrays, the term macroarray is useful because of other inherent differences of membrane-based arrays. Currently, DNA clones, PCR amplicons, or oligonucleotides are spotted onto membranes using spotting robots or ink-jet-like printers. Macroarrays are unique among hybridization arrays in that they generally use radioactive target labeling. After the target is radioactively labeled, control and experimental samples are hybridized to individual and separate arrays. Phosphorimagers (or less frequently X-ray film) are then used to detect the bound target. These arrays, typically containing between 200 and 8000 genes, are commercially available for a wide variety of organisms and genes and can be obtained from a number of companies. "Custom" macroarrays can also be constructed in-house and may contain as few as a dozen or as many as thousands of genes. The experimental protocols described in this chapter are conducted exclusively with membrane-based macroarrays obtained from Clontech (www.clontech.com).

1.2.2. Microarray

Microarrays can be differentiated from macroarrays in three ways. First, microarrays generally use silicon/glass as a matrix and, second, they use fluorescent dye-labeling detection. Microarrays also tend to have a larger number and higher density of probes than macroarrays. As with macroarrays, probes are made from clones, PCR amplicons, or oligonucleotides spotted robotically onto the matrix surface. This approach has advantages in that hybridization takes place in a flow cell or small hybridization chamber, which uses a much smaller hybridization solution volume as compared to macroarrays, thereby increasing the relative target concentration and decreasing costs. Moreover, a competitive fluorescent scheme allows both sample groups (control and treated) to be hybridized to the same array using distinct fluorescent dyes (typically Cy3 and Cy5). Like macroarrays, an ever-expanding number of microarrays are commercially available. Many research institutions are currently investing heavily in the equipment to produce custom microarrays in-house.

1.2.3. High-Density Oligonucleotide Arrays (Affymetrix)

High-density oligonucleotide arrays differ from other formats in that the probe is generated *in situ* on the surface of the matrix. The leader in this type of array is Affymetrix (Santa Clara, CA), a company that uses a unique combinatorial synthesis method. This method makes use of a process called photolithography to construct probes on the array surface by making oligonucleotides one base at a time. Because the combinatorial synthesis scheme has a finite efficiency at each step, synthesis of oligonucleotides longer than 25 bases is problematic. As a result of using these 25-mer oligonucleotides for gene expression analysis, mismatches and spurious target-probe binding can take place because of the limited specificity and binding affinity for a 25-residue oligonucleotide. To overcome this problem, a series of oligonucleotides that differ by a one-base mismatch from the gene-specific probe are also included on the array and can be used to determine the amount of mismatch hybridization, which can then be subtracted from the signal. These arrays, which are available only from Affymetrix, contain between 40,000 and 500,000 probes (including multiple mismatch controls for each gene) and provide the highest density of probes of any array.

This laboratory focuses most of its efforts on the use of the macroarray platform for the identification of gene expression changes in drug abuse. The macroarray experiment requires that the two populations of RNA be radioactively labeled and then used to query identical, but separate, membranes. This has an obvious disadvantage because the use of two different membranes provides an opportunity for hybridization error and therefore experimental variability. Macroarrays do, however, offer increased sensitivity over fluorescence-based methods, a feature that can be very advantageous—especially when looking for rare genes. In addition, we find that this platform has an increased ability to detect small changes in gene expression (50% increases and 33% decreases). The experimental description that follows details how to perform a hybridization array using the macroarray platform.

2. Materials

2.1. Reagents

1. Source of RNA (tissue, cell culture, etc.).
2. RNA extraction reagent, for example, TRI Reagent (Molecular Research Center, Inc., Cincinnati, OH); Trizol (Gibco, Carlsbad, CA).
3. Nuclease-free water or diethyl pyrocarbonate (DEPC)-treated water.
4. Agarose (for denaturing nucleic acid gels).
5. 3-(N-morpholino) propanesulfonic acid (MOPS) buffer.
6. TE buffer: 10 mM Tris-HCl, pH 8.0, 1 mM EDTA, pH 8.0.
7. Ethidium bromide.
8. Formamide.
9. Formaldehyde.
10. cDNA labeling kit (Clontech Laboratories, Palo Alto, CA).
11. α-[^{32}P]dATP (ICN).

12. Prehybridization and hybridization solutions (Express Hyb, Clontech).
13. Salmon sperm DNA.
14. 20X standard saline citrate (SSC) stock solution.

2.2. Equipment and Supplies

1. Agarose gel apparatus and power supply.
2. Light box (with UV lamp).
3. Spectrophotometer (with UV lamp).
4. Gel documentation/image capture camera (UVP ImageStore 7500, UVP, Upland, CA).
5. Arrays (nylon-based cDNA macroarrays, Atlas Arrays, Clontech Laboratories, Palo Alto, CA).
6. Thermocycler.
7. Size-exclusion columns (Chroma Spin-200 DEPC-water column, Clontech).
8. Hybridization bottles.
9. Hybridization oven.
10. Water bath(s).
11. Saran wrap.
12. Laminate plastic.
13. Bag sealer.
14. Phosphorimaging cassette.
15. Phosphorimager.

2.3. Software

1. RNA quantification, for example, TINA (Raytest; Wilmington, NC) for providing analysis and quantification of RNA—the Agilent Bioanalyzer 2100 performs a similar function with less material.
2. Array image capture (Photoshop; Adobe; San Jose, CA).
3. Array analysis (AtlasImage 2.01).

3. Methods

The methods outlined in this section describe a typical experiment using the macroarray format and include a brief discussion of *post hoc* analysis following the generation of results from the array experiment. This format is chosen for discussion because the authors have extensive experience with these methods, and because the equipment, supplies, and expertise needed to perform this type of array experiment are readily available to most laboratories that do molecular biology work (especially, e.g., Northern blots). With this platform, an array consists of hundreds to thousands of unique cDNAs (200–500 basepairs in length) spotted onto nylon, or, alternatively, thousands of oligonucleotides, representing a nonredundant unique gene set, spotted onto plastic. Because radioactivity is used for the labeling step, each array "experiment" consists of two separate, but identical, membranes for each sample or group of samples. This unitary ($n = 1$) array experiment is then replicated twice to assess variability and then followed by an appropriate *post hoc* experiment such a quantitative RT-PCR and/or quantitative Western blot analysis to confirm changes.

3.1. Tissue Selection

Sample collection is a basic element of experimental design for many molecular biological experiments, but it is worth reiterating. In the field of drug abuse, it is most often imperative to use dissected tissue from treated animals. Therefore, it is often unavoidable that the samples will contain multiple cell types. In complex samples such as brain tissue there is routinely a heterogeneous cell population. Therefore, observed changes may represent a change in one cell type or multiple cell types. Similarly, smaller changes occurring in only one type of cell may be hidden if the expression of this gene in other cell types is unchanged. Thus, researchers must be mindful of this heterogeneous population when drawing conclusions. A promising technological solution is laser capture microdissection, which allows very small and identified cellular populations to be dissected. The timing of tissue collection will also prove to be important and should be driven by the biological question. For instance, the point at which samples are collected should be chosen to represent some particular period in the time course of drug abuse. Samples collected while the drug is still in the bloodstream will differ from samples collected after a period of withdrawal.

3.2. RNA Isolation

The single most important factor in the ability to generate quality results from an array experiment, assuming first a solid experimental design for the generation of "control" and "treated" samples, is the preparation and use of high-quality RNA. Total RNA (or mRNA) can be reliably and repeatedly generated from any number of biological samples including whole blood, other tissues (e.g., brain, liver), and from cells in culture. In this section we discuss the isolation and assay of total RNA from brain tissue.

3.2.1. Tissue Dissection and Storage

The most reliable preparations of RNA come from brain tissue samples that have been rapidly dissected and "flash-frozen" in liquid nitrogen. For retrieving tissue, because brain regions are not absolutely discrete, it is best to have the same individual perform the dissections on all samples for a given experiment (and any subsequent repeat experiments) to ensure relative uniformity in sample collection. Once the whole brain has been removed from an animal's skull, and working as rapidly as possible, individual brain regions are dissected, immediately put into prelabeled and weighed tubes, tissue weights are determined, and the samples dropped into liquid nitrogen or buried in crushed dry ice. Tissue collected and frozen in this fashion can then be stored at –80°C for long periods of time.

3.2.2. Creating a Tissue Powder

Tissue that has been previously dissected and stored frozen at –80°C can be readied for array and *post hoc* analysis by the creation of a frozen tissue powder. This procedure is especially useful in that a homogeneous powder is formed that can be aliquoted for subsequent RNA or protein purification.

1. A stainless steel mortar and pestle is chilled on dry ice.
2. Frozen tissue is transferred, without thawing, into the mortar and covered with liquid nitrogen.
3. The sample is grounded to a fine powder (keeping it covered with liquid nitrogen at all times). The frozen powder at the bottom of the bowl is then retrieved with a (prechilled) stainless steel spatula and quickly deposited into prechilled, preweighed, cryovials. We routinely place half of the tissue into each of two tubes: one to be used for RNA extraction, the other for protein preparation.
4. The tubes, containing tissue, are quickly reweighed (without thawing), returned to liquid nitrogen or dry ice, and stored at –80°C until further processing can be performed.

3.2.3. RNA Isolation

RNA is extracted from frozen tissue powder using a modification of the single-step method originally developed by Chomczynski and Sacchi *(7,8)*. Although there are a number of commercially available reagents, this laboratory routinely uses TriReagent from Molecular Research Center, Inc. This reagent combines phenol and guanidine isothiocyanate in a monophase solution for the effective inhibition of RNase activity during cell lysis.

1. Following addition of TriReagent to the tissue powder (10 µL of TriReagent/mg tissue weight), the tissue is immediately homogenized at 40–80% sonicator power (3 × 5 s with 20-s rest intervals).
2. Leave at room temperature for 5 min, then centrifuge at 12,000*g* for 15 min at 4°C. Remove and use the aqueous layer (top portion).
3. Add one fifth of the original TriReagent volume of chloroform and vortex-mix.
4. Repeat **step 2**.
5. Add one half of the original TriReagent volume of isopropanol and vortex-mix.
6. Sit overnight or for at least 4 h at –20°C.
7. Centrifuge at 12,000*g* for 15 min at 4°C. Remove supernatant wash pellet with one volume (1 µL/µL of TriReagent) of 75% ethanol. Centrifuge at 12,000*g*

DNA Macroarray Screening

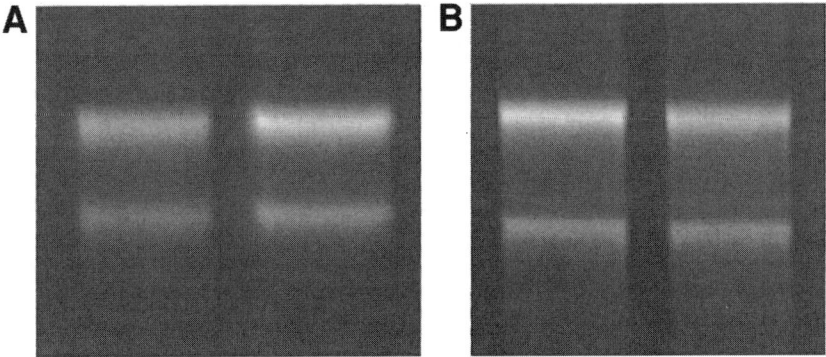

Fig. 2. RNA qualification. (**A**) Poor quality RNA on the *left*, showing little difference in 18S and 28S band intensities, broad bands, and smearing between bands. Compare this with the lane on the *right*, noticing that the 28S intensity is approximately twice that of the 18S. It should be noted that many laboratories might consider the RNA on the *left* to be acceptable, whereas this laboratory has had the greatest success using RNA of the quality on the *right*. (**B**) Good quality total RNA showing the 18S band at half the intensity of the 28S band and sharp bands. In addition, these two samples were loaded at approximately equal amounts, which has been confirmed by gel electrophoresis.

for 15 min at 4°C. Remove supernatant and dry pellet for 2–3 min in SpeedVac (can air-dry as well).
8. Resuspend pellet in DEPC-treated water. Note, however, that if overdried, RNA will frequently not dissolve).
9. Place in a 55–60°C water bath for 10 min. Store at –80°C.

3.2.4. RNA Quality Assessment and Quantification

Quantity and relative quality of RNA can be determined spectrophotometrically by diluting an aliquot into TE-buffered water (unbuffered water can produce spurious results) and measuring the UV absorbance at 260 nm and 280 nm (*see* **Note 1**). High-quality RNA, essentially free of DNA and proteins, should have an A_{260}/A_{280} ratio of 1.9–2.0 and one can expect to obtain a yield of 1–1.5 µg of total RNA/mg of brain tissue. Although the A_{260}/A_{280} ratio is a good relative measure of purity, if the RNA is being prepared for use in a hybridization array experiment, it is imperative to run an aliquot (0.5–1.0 µg) of the RNA on a denaturing 1% agarose gel to verify the presence of two distinct bands (representing the predominant 28S and 18S RNA species; they should be in a concentration ratio of approx 2:1) and minimal degradation (**Fig. 2**). Although this laboratory routinely employs denaturing agarose gel electrophoresis for determining quality, new technology from Agilent

Technologies (Palo Alto, CA) provides a workstation that will quantify RNA and produce information on degradation and quality from as little as 5 ng of total RNA.

3.2.5. RNA Equilibration

Once the quantity and quality have been determined, RNA pools of individual treatment groups can be made. The intent, at this stage, is to create a population mean of individual samples. In this manner, genes illuminated by an array experiment will more likely represent the biological phenomenon rather than animal-to-animal variability.

1. An equal amount of RNA is added from each sample to create pooled RNA.
2. The pools will generally be composed of different volumes and so they should be adjusted to 1 µg of RNA/µL with DEPC-treated water. This may require vacuum concentration of the sample pools.
3. The Clontech macroarrays require 4 µg of total RNA to be hybridized to the membranes. Therefore, pools of 6 µg of total RNA are generally prepared. This permits an excess of 2 µg of RNA to run on a denaturing gel (1 µg at a time, as described in **Subheading 3.2.4.**) to ensure an equal quantity of RNA is being loaded on each membrane. The 28S bands should be with within 10% of each other as determined by a densitometry program, such as TINA (Raytest).
4. The pooled RNA is now and ready for use in the experiment.

3.3. Reverse Transcription

1. Prepare a "master mix" with all of the reagents except the [^{32}P]dATP and Moloney Murine Leukemia Virus (MMLV).
2. Create RNA pool and primer mixture using 4 µg pooled RNA in 4 µL and 2 µL of a CDS primer provided by Clontech. This mixture contains gene-specific reverse transcription primers for each gene represented on the array. This mixture is placed in a thermocycler for 2 min at 70°C and then for 2 min at 55°C. During this 4-min time period, the [^{32}P]dATP and MMLV can be added to the master mix. At the end of the 4 min, 14 µL of the master mix is added to each reaction. The reaction then proceeds at 50°C for an additional 25 min.
3. Add 2 µL of a termination mixture to stop the reaction.

3.4. Probe Purification and Equilibration

The RNA has now been reverse transcribed to create radioactive cDNA targets. The next step is to remove the unincorporated [^{32}P]dATP using size-exclusion columns. Clontech includes spin columns in the hybridization array package, but this laboratory prefers the gravity-fed Chroma Spin-200 columns from the same company.

1. Drain size-exclusion chromatography columns by snapping off bottom and allowing to drain into waste tube.
2. The radioactive target mixture is applied to the column and eluted in 100-µL fractions using sterile water.
3. The fractions are rapidly assessed directly in the scintillation counter without scintillation fluid ("Cerenkov" counting), and provide immediate feedback on the resolution of target and unincorporated radionucleotide.
4. The three fractions comprising the first radioactive peak serve as the basis for the hybridization experiment. Successful chromatography will completely separate this peak from the larger peak of unincorporated radioactivity. Spot 1 µL from each of the three fractions on Whatman filter paper in duplicate and determine the radioactivity by liquid scintillation spectrometery (with scintillation fluid). This will provide a much more accurate and high-efficiency estimate of radioactive incorporation. Following scintillation counting, the amount of radioactivity per microliter of fraction can be determined from the average of the replicates. Optimal hybridization results are achieved using a total of 3.5–8.0 million cpm per sample. This will generally require pooling two fractions.
5. The two experimental target samples must now be equilibrated so that each sample has the same number of cpms. Equivalent amounts of radioactivity should be added to two separate tubes, which are then brought up to an equal volume and 1 µL counted again in duplicate as before. The average sample counts should be within 1–2% of each other for use in hybridization.
6. Once the radioactivity has been equilibrated, the samples can be stored in a protective container at −20°C for 1–2 d until the remainder of the experiment is prepared (i.e., prehybridization of the filters).

3.5. Prehybridization

Prior to the actual hybridization experiment, the membranes are prehybridized to block nonspecific interaction with radiolabeled target. The importance of effective prehybridization is illustrated in **Fig. 3**.

1. Remove array membranes from freezer. Wet membranes in DEPC-treated water and insert into bottles. The bottles should be well washed and the interior surfaces coated with a siliconization agent (Sigmacote) to prevent the membranes from sticking to the glass surface. The membranes should be touched as little as possible with gloves, using tweezers to hold the edges and manipulate them. Arrange the membranes in separate bottles with the face of the membrane containing the probe facing the inside of the tube.
2. Pour off water and briefly allow to drain.
3. Freshly make prehybridization buffer: 25 mL of the hybridization buffer provided by Clontech and 250 µL of a 10 mg/mL denatured (boiled) salmon sperm DNA (ssDNA) solution. The buffer must be mixed thoroughly while heating in a 68°C water bath.

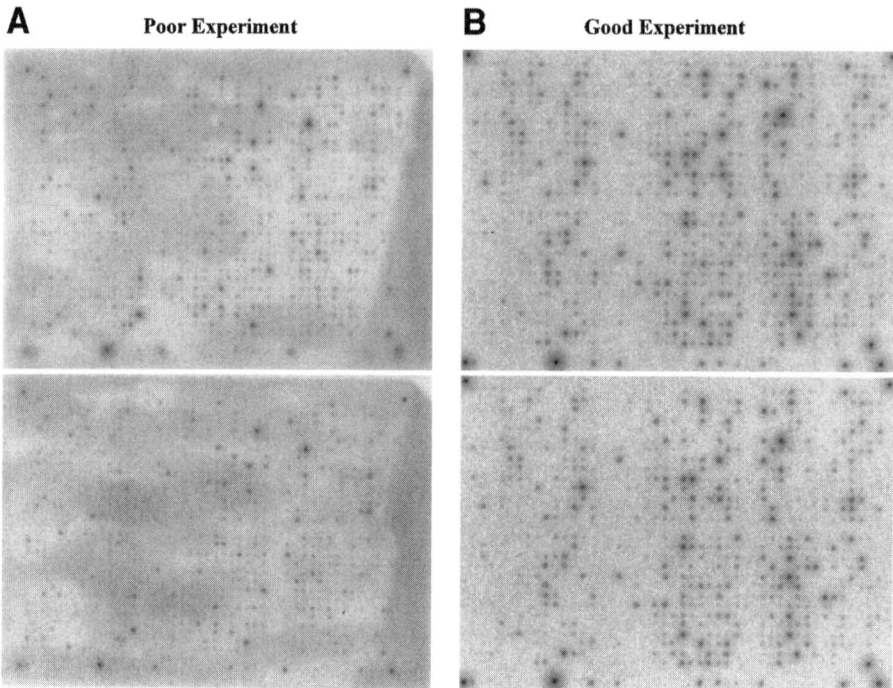

Fig. 3. Hybridization arrays showing normal and clouded backgrounds. (**A**) An example of poor array experiment in which the two arrays had inconsistent and prominent background problems. This clearly results from nonspecific binding and poor membrane washing. Frequent "shaking" of the prehybridization and hybridization steps minimizes this problem. Note that high background decreases the specific signal and produces diffuse images. (**B**) An example of two arrays showing low background and high intensity binding.

4. Add 12.5 mL of the prehybridization solution to each bottle. Replace the lid to the bottles and then shake in a circular motion so that the membrane glides freely within the tube.
5. Prehybridize at 68°C in an oven at least 4 h or overnight for best results. Also for improving distribution of buffer and therefore background values, the bottles should be shaken often (six to eight times on average) over the course of the prehybridization (*see* **Note 2**).

3.6. Hybridization

After prehybridizing the arrays, the radiolabeled targets must be prepared before hybridization.

DNA Macroarray Screening

1. Thaw the samples.
2. Add one tenth of the total volume of denaturing buffer into the samples and incubate in a 68°C water bath for 20 min.
3. Add each volume of neutralizing buffer and 5 μL of Cot-DNA (provided by the manufacturer) and incubate for 10 min.
4. Add probes to appropriate labeled bottles and hybridize overnight in an incubation oven.

3.7. Washing and Phosphorimaging

Wash solutions 1 and 2 should be kept at 68°C in a water bath for the duration of wash.

1. Remove the tubes from the oven and empty the contents into a radioactive container.
2. Add 50 mL of wash solution 1, shake quickly, and empty into the same container. Add 50 mL again and incubate at 68°C in the oven for 30 min. Repeat two more times with wash solution 1.
3. Wash two more times with the more stringent wash solution 2. A final wash in 2X SSC is at room temperature and occurs in a container on a desktop shaker. This is a 15-min wash followed by mounting onto laminate plastic.
4. Remove the membranes from the shaker and allow one membrane at a time to drip dry quickly (about 30 s). Place the membrane on the mounting surface with the probe surface facing upward. Cover the mounting surface with plastic wrap, making sure to remove any air bubbles and seal with a thermal bag sealer.
5. Mount the sealed membranes into a phosphorimaging cassette and exposed to a phosphorimaging plate for 3–5 d, depending on the amount of radioactivity present.
6. Read and analyze the plate using an imaging program such as Clotech's Atlas Image 2.01.

3.8. Data Analysis

Data analysis is a critical component of hybridization array experiments and poses a number of challenges owing to the large amount of data generated even in a single experiment. The most basic goal of array data analysis is to identify genes that are differentially regulated. Hybridization arrays are prone to yielding false-positive results and therefore analysis strategies attempt to decrease this error rate without increasing false-negatives. The analysis method described here is a simple, empirical approach. There are now more advanced statistical approaches to data analysis *(9)*.

3.8.1. Background Subtraction

The first steps in data analysis are background subtraction and normalization. The principals of both are similar to the techniques used with conventional nucleic acid or protein blotting. Background subtraction corrects for nonspecific background noise and permits comparison of specific signals. For illustration, if the signal intensities for the control and experimental spots are 4 and 6, respectively, it would appear that the experimental is 50% higher. However, if a background of 2 is subtracted from both signal intensities, the experimental value is actually 100% higher than control (2 vs 4). Background is often taken from the blank areas on the array. A complication to background subtraction is that differences in background across the array can affect some spots more than others. An alternative is to use either a local background for the area around each spot or designate spots with the lowest signal intensities for background determination. The latter may be a more accurate determination of nonspecific background because it represents the nonspecific binding of targets to probe. Background intensities from blank areas (no nucleic acids) do not contain probe, and therefore are arguably a different form of background.

3.8.2. Normalization

Normalization is the process by which differences between separate arrays are accounted for. All macroarray experiments require the use of normalization for accurate comparisons. For example, when a pair of macroarrays representing control and treated samples show a difference in overall or total signal intensity, such differences can arise from unequal starting amounts of RNA or cDNAs, from labeling reactions of different efficiencies, or from differences in hybridization. Any of these factors can skew the results. One common method of normalization to use a housekeeping gene(s)—a gene thought to be invariant under experimental conditions—for comparison. If the signal for this gene is higher on one array than the other it can be used to normalize the data. Housekeeping genes may be problematic because they themselves vary under some experimental conditions. To overcome the variability of these genes, some researchers have turned to a "basket" or "sum" approach for normalization. This strategy is based on the precept that the global radiolabeled target synthesis should be constant. That is, even though the compared samples will have selected increases and decreases, on balance the total hybridization signal should be constant. Therefore, arrays can be normalized by equilibrating the sum of the intensities for all control and experimental spots. In a similar vein, the median value of signal intensities can be used. This value is less susceptible to distortions caused by outliers.

3.8.3. Gene Calling and Differential Expression

The next step is to determine which genes were detected by the array analysis. When using large-scale arrays there will be probes for genes that are not expressed in the tissue sample or are expressed below the level of detection. In fact, arrays are inherently much less sensitive than RT-PCR for example. This means that there must be some method for "calling" a gene as being detected by the array. The simplest method for use with macroarrays is to set a threshold, expressed as a percentage above background, that a signal must exceed for the gene to be called as present. In our experience, a 50% above background threshold has worked well. This threshold can be increased or decreased. Genes with low signal intensities are more variable, so decreasing the gene-calling threshold will increase the number of genes analyzed (as well as increasing the attendant variability). The converse is true when increasing the gene-calling threshold.

Once genes have been called as being detected by the array, the next step is to determine which genes where changed in their expression by the experimental condition. Differential expression calls are more problematic and there are no commonly accepted standards. We will describe the most basic method. The normalized signal for each gene is converted into a ratio of the treated signal intensity over the control signal intensities. Previous work by our laboratory has shown that macroarrays can reliably detect a 50% increase and greater or a 33% decrease and greater *(10–12)*. These amounts are equivalent on a natural log scale (+0.4 and –0.4, respectively). Genes whose expression ratios exceed these cutoffs are potential differentially regulated genes. Relying on only one array experiment is likely to result in a number of false-positives. Therefore, it is highly recommended that array experiments be repeated one or two times. Genes that consistently exceed the differential expression ratio across multiple experiments are less likely to be the result of random error and more likely to represent real changes in mRNA abundance (**Fig. 4**).

4. Notes

1. RNA quality is the number one predictor of a successful hybridization array experiment. Make sure that the tissue is kept on dry ice or in the presence of liquid nitrogen while extracting the RNA to ensure minimal tissue thawing. Thawing will compromise the RNase protection provided by the extreme temperatures and cause RNA degradation. When removing extracted RNA from the –80°C freezers for use in an experiment, always thaw the RNA on wet ice and not at room temperature for the same reason. In addition, repeated freeze–thawing will ultimately lead to degradation of the RNA, so this must be kept to a minimum.

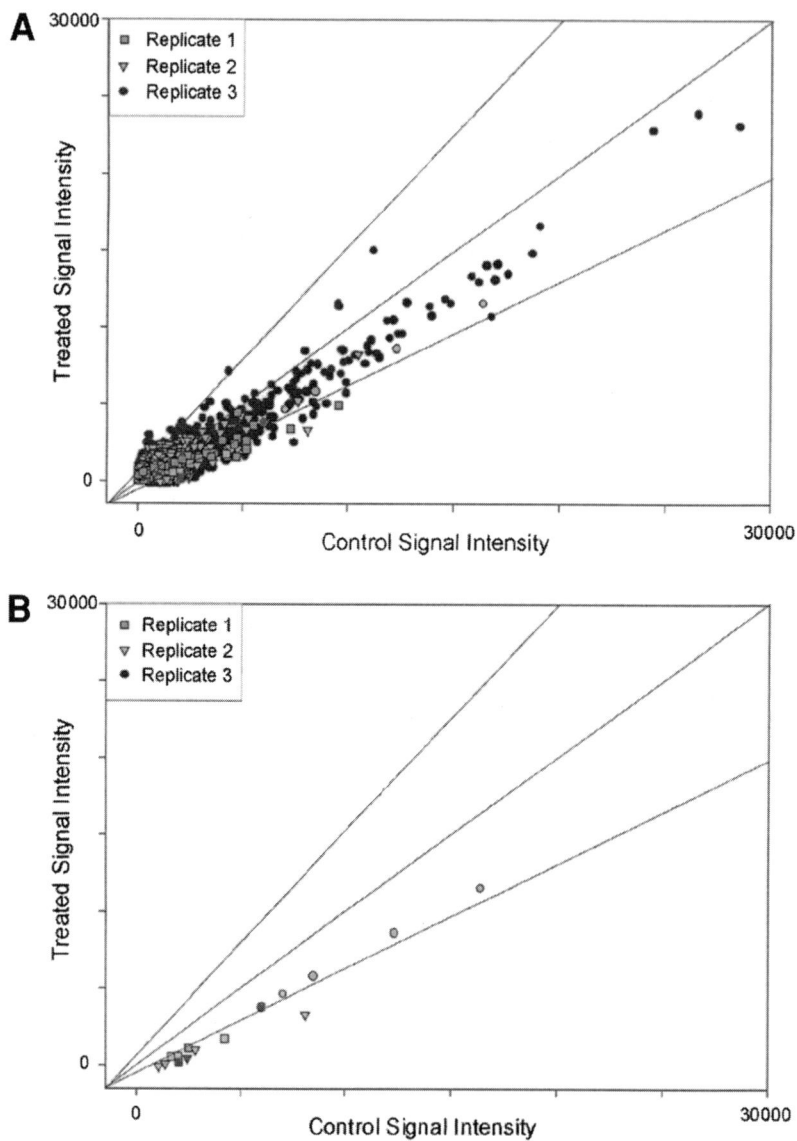

Fig. 4. Macroarray analysis of treated cerebellar granule neuron cultures. (**A**) Scatterplot representing the ratio determined by comparing treated signal intensity to control signal intensity from three independent hybridization experiments. (**B**) Restricted scatterplot generated by removing the nonreplicating data points.

Finally, never perform an experiment with RNA that has not been carefully and fully characterized.
2. Shaking is such an important step in the process that it bears mentioning again. During both prehybridization and hybridization, shaking ensures that "clouds" do not appear at corners of edges of the membranes, which can complicate analysis. Examples of this problem are illustrated in **Fig. 3**.

Acknowledgments

This work was supported by NIH Grants R01-DA13770, P50-DA06634, P50-AA15773 (to K. E. V.), T32-DA07246 (to W. M. F.), and T32-AA07565 (to T. J. W.).

References

1. Freeman, W. M., Vacca, S. E., Dougherty, K., and Vrana, K. E. (2002) An interactive database of cocaine-responsive gene expression. *Sci. World. J.* **2,** 701–706.
2. Hyman, S. E. and Malenka, R. C. (2001) Addiction and the brain: the neurobiology of compulsion and its persistence. *Nat. Rev. Neurosci.* **2,** 695–703.
3. Koob, G. F. and LeMoal, M. (2001) Drug addiction, dysregulation of reward, and allostasis. *Neuropsychopharmacology* **24,** 97–129.
4. Nestler, E. J. (2001) Molecular basis of long-term plasticity underlying addiction. *Nat. Rev. Neurosci.* **2,** 119–128.
5. Freeman, W. M., Robertson, D. J., and Vrana, K. E. (2000) Fundamentals of DNA hybridization arrays for gene expression analysis. *BioTechniques* **29,** 1042–1055.
6. Lockhart, D. J. and Barlow, C. (2001) Expressing what's on your mind: DNA arrays and the brain. *Nat. Rev. Neurosci.* **2,** 63–68.
7. Chomczynski, P. and Sacchi, N. (1987) Single step method of RNA isolation by acid quanidinium thiocyanate–phenol–chloroform extraction. *Analyt. Biochem.* **162,** 156–159.
8. Chomczynski, P. (1993) A reagent for the single-step simultaneous isolation of RNA, DNA and proteins from cell and tissue samples. *BioTechniques* **15,** 532–537.
9. Tusher, V. G., Tibshirani, R., and Chu, G. (2001) Significance analysis of microarrays applied to the ionizing radiation response. *Proc. Natl. Acad. Sci. USA* **98,** 5116–5121.
10. Freeman, W. M., Nader, M. A., Nader, S. H., et al. (2001) Chronic cocaine-mediated changes in non-human primate nucleus accumbens gene expression. *J. Neurochem.* **77,** 542–549.
11. Freeman, W. M., Brebner, K., Lynch, W. J., Robertson, D. J., Roberts, D. C. S., and Vrana, K. E. (2001) Cocaine-responsive gene expression in rat hippocampus. *Neuroscience* **108,** 371–380.
12. Freeman, W. M., Brebner, K., Lynch, W. J., Robertson, D. J., Roberts, D. C. S., and Vrana, K. E. (2002) Changes in rat frontal cortex gene expression following chronic cocaine. *Mol. Brain Res.* **104,** 11–20.

III

DETECTION OF PROTEIN EXPRESSION IN THE STRIATUM

17

Quantification of Protein in Brain Tissue by Western Immunoblot Analysis

Christine L. Konradi

1. Introduction

Western blots are designed to determine protein levels and their patterns of modification in homogenized tissue samples. Western blots are quantifiable, but unlike immunohistochemistry, the cellular integrity is lost. Both Western blots and immunohistochemistry depend on the availability of antibodies against the protein of interest. Antibodies may be directed not only against a protein, but can also be directed against a chemical modification of the protein, such as phosphorylation of specific amino acid residues or glycosylation. For Western blots, the proteins in the sample are denatured, size-separated on a denaturing acrylamide gel, and transferred to a nylon membrane. An antibody is reacted with the proteins on the membrane and binds to its specific antigen. The resulting antibody–antigen complex is visualized with the help of a chemiluminescent assay system that darkens X-ray films.

2. Materials

2.1. Preparation of Samples

1. 2X Laemmli buffer: 5.0 mL of 0.5 M Tris-HCl buffer, pH 7.0, 5.0 mL of distilled water, 1.2 g of sodium dodecyl sulfate (SDS), 4.0 mL of glycerol, 2.0 mL of bromophenol blue solution, and 620 mg of dithiothreitol (DTT). Bring to 20 mL with distilled water and freeze in aliquots at –20°C. The buffer needs to be warmed up and vortex-mixed before use, as SDS may fall out in the freezer; ensure there is no white precipitate before use.
2. Bromophenol blue solution: 1 mg of bromophenol blue in 1 mL of distilled water. Store at room temperature.

3. 4X Upper buffer: 0.5 M Tris-HCl, 0.4% SDS, pH 6.8. Bring 6.05 g of Tris base up in 40 mL of distilled water, and adjust pH to 6.8 with 1 N HCl. Add distilled water to 100 mL, filter, and add 0.4 g of SDS.
4. Sonication buffer: 10 mM Tris-HCl, pH 7.5, 50 mM NaF, 2 mM Na$_3$VO$_4$, 1 mM EDTA (from a 0.5 M sodium-EDTA solution, pH 8.0), and 1 mM EGTA.

2.2. Preparation of Gel and Gel Electrophoresis

1. 10% Ammonium persulfate (APS) in distilled water, good for approx 1 mo at 4°C.
2. 30% Acrylamide–bis-acrylamide solution (37.5:1): 29.2 g of acrylamide and 0.8 g of bis-acrylamide are dissolved into 100 mL of distilled water.
3. 4X Lower buffer: 1.5 M Tris-HCl and 0.4% SDS, pH 8.8. Bring 91 g of Tris base up in 300 mL of distilled water, adjust pH to 8.8 with 1 N HCl, add distilled water to 500 mL, filter, and add 2 g of SDS.
4. Water-saturated butanol: Carefully mix 20 mL of distilled water with 100 mL of 100% n-butanol. (**Caution:** If mixing in a closed bottle, pressure may be building up; open bottle cup frequently to relieve pressure.) Let phases separate, with butanol making up the upper phase. Solution is stored at room temperature.
5. 5X SDS running buffer: 125 mM Tris-HCl, 960 mM glycine, and 0.5% SDS, pH 8.3. 15.1 g of Tris base, 72 g of glycine, 5 g of SDS; bring to 800 mL with distilled water, adjust pH to 8.3, and bring to 1 L with distilled water.

2.3. Transfer of Protein to Polyvinylidene Difluoride (PVDF) Membrane

1. Transfer buffer: 25 mM Tris, 190 mM glycine, and 20% methanol. This buffer can also be made up as 5X stock solution without methanol, and diluted with one volume of 5X stock, three volumes of distilled water, and one volume of methanol.

2.4. Antibody Reaction

1. PBST: 11.5 g of Na$_2$HPO$_4$, 3.0 g of NaH$_2$PO$_4$•2H$_2$O, and 5.84 g of NaCl. Dissolve in 1 L of distilled water, and add 1 ml/L of 0.1% Tween 20.
2. Blocking solution: 5% Nonfat dry milk in PBST.

3. Methods

3.1. Preparation of Samples

3.1.1. Cell Culture Samples

We do not find it necessary to determine the protein concentration in our cell culture samples. Within an experiment, all cultures are plated from the same cell suspension, and all wells have approximately the same number of cells. However, we do not exchange samples from different platings. Because often we are interested in phosphorylation events, the rapid aspiration of medium and addition of Laemmli buffer ensures the conservation of the state of phosphorylation of proteins.

1. Aspirate medium.
2. Add 100 µL of 1X boiling Laemmli buffer to each well of a 12-well plate (approx $1–1.2 \times 10^6$ neurons/well). Scrape cells into microcentrifuge tubes.
3. Sonicate for 10 s with an ultrasonic dismembrator at medium setting (*see* **Note 1**).
4. Heat samples to 80°C for 5 min.
5. Proceed to loading the gel; start with 8–15 µL of sample per lane.

3.1.2. Brain Samples

1. Dissect brain area of interest.
2. Weigh and take up in sonication buffer at 10 mg of tissue/100 µL of sonication buffer.
3. Sonicate for 20 s with an ultrasonic dismembrator at medium setting. Make sure no tissue clumps are left.
4. Determine the protein concentration (*see* **Note 2**).
5. Dilute to 4 µg of protein/µL of sonication buffer.
6. Add equal volume of 2X Laemmli buffer.
7. Heat samples to 80°C for 5 min.
8. Centrifuge for 10 min at 16,000*g* at room temperature.
9. Transfer supernatant to new tube, and discard pellet.
10. Load samples onto the gel, and start with 2–20 µg of protein per lane.

3.2. Preparation of Gel and Gel Electrophoresis

The upper gel (stacking gel) is usually 4% acrylamide–*bis*-acrylamide. The concentration of the lower gel (running gel) depends on the molecular mass of the protein of interest. As a general rule, we use 8% for proteins above 100 kDa, 10% for proteins between 60 and 100 kDa, and 12% for proteins between 20 and 60 kDa. Proteins with lower molecular masses can be run with higher concentrations of acrylamide–*bis*-acrylamide, but improved results may be obtained with gradient gels that are commercially available.

1. Thoroughly wash glass plates, spacers, and combs. Both plates and spacers need to be washed with water and rinsed with ethanol. The plates need to be spotless, but soaps should be avoided because they may interact with the gel matrix during the run and degrade the gel.
2. Assemble glass plates and spacers. We use 0.75-mm thick spacers and combs.
3. Prepare lower gel with:
   a. Acrylamide–*bis*-acrylamide solution to the concentration desired (8%, 10%, 12%)

b. Lower buffer to a final concentration of 1×
   c. Distilled water to the desired final volume
      (e.g., for 15 mL 10% gel: 3.75 mL of lower buffer, 5 mL of acrylamide–*bis*-acrylamide solution, and 6.25 mL of distilled water). To start the polymerization reaction, add 7 µL of APS/mL of gel solution and 1.5 µL of N,N,N',N'-tetramethyl-1,2-diaminomethane (TEMED)/mL of gel solution (*see* **Note 3**).
4. Mix and pour between glass plates to a level approx 1 inch below the lower level of the comb.
5. Overlay with 0.25–0.5 inches of water-saturated *n*-butanol (from the upper phase).
6. Let polymerize for 30 min.
7. Decant butanol and quickly rinse with tap water to remove excess butanol.
8. Decant water thoroughly.
9. Overlay lower gel with upper gel:
   a. Acrylamide–*bis*-acrylamide solution to 4%
   b. Upper buffer to a final concentration of 1X
   c. Distilled water to desired final volume
      To the start polymerization reaction, add 8 µL of APS/mL of gel solution and 1.5 µL of TEMED/mL of gel solution.
10. Insert comb and let gel polymerize for 1 h. Make sure no air bubbles are trapped between the comb and the gel.
11. Place gel in electrophoresis unit.
12. Add 1X running buffer to the upper and lower chambers.
13. Load samples and start electrophoresis (*see* **Note 4**).
14. Run through the upper gel at 15 mA per gel (10 cm long gel) at constant amperage.
15. Increase to 20 mamp/gel after dye reaches the interphase between upper and lower gel.
16. Run until the lower dye reaches the bottom of the gel.
17. Turn off the power supply, disconnect the cables, remove the lid, and take out the gel.

3.3. Transfer of Protein to PVDF Membrane in an Electroblotter

1. Disassemble the gel by removing the spacers and sliding a spatula between both plates, remove one plate, and cut the stacking gel off with a spacer (*see* **Fig. 1**) by using the second plate as support for the gel. Do not use sharp tools as they will scratch the plate.
2. Cut a piece of blotting paper and PVDF membrane to the size of the running gel.
3. Immerse PVDF membrane in 100% methanol for 1 min to wet. PVDF is hydrophobic and will not wet in aqueous solutions.
4. Place the lower part of the gel transfer cassette with the fiber pad into a pan with transfer buffer to wet. Add the blotting paper and the methanol-soaked PVDF membrane. Make sure no air bubbles are trapped inside the "sandwich." Mark the

Western Blots

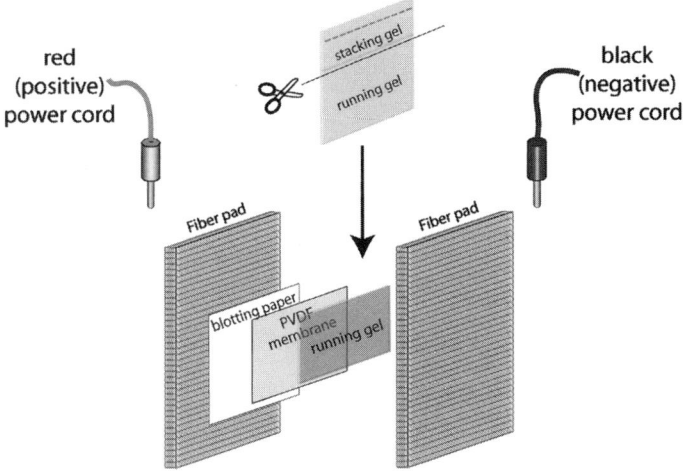

Fig. 1. Assembling the electroblotter: In a tray with transfer buffer, wet the fiber pad and the blotting paper. Prewet the PVDF membrane in methanol and put on top of the blotting paper. Cut the stacking gel from the running gel with a gel spacer and discard. Put the running gel on top of the PVDF membrane, add the second fiber pad, assemble the cassette, and slide into the transfer unit. When attaching the power cords, make sure the red cord (positive cord) is on the side of the membrane, whereas the black cord (negative cord) is on the side of the gel.

membrane with a soft pencil to keep track of the orientation (*see* **Fig. 1**).
5. Place the gel on top of the PVDF membrane.
6. Add a second fiber pad, close the cassette, and transfer to the electrophoresis unit (*see* **Notes 5** and **6**). The gel should be on the side of the cathode and membrane on the side of anode (*see* **Fig. 1**).
7. Transfer for 2–3 h at 300 mA.
8. Disassemble transfer, discard the gel, and keep the membrane. Quickly rinse the membrane under tap water to wash off residual gel pieces. Proceed either to the antibody reaction, or store the membrane in blocking solution overnight at 4°C, or let the membrane dry. If the membrane dries, it has to be rewetted in 100% methanol.

3.4. Antibody Reaction (Fig. 2)

1. Place the wet membrane into blocking solution for 1 h at room temperature with gentle shaking (*see* **Note 7**).
2. Wash for 7–10 min in PBST at room temperature with gentle shaking.
3. Incubate in primary antibody, made up in PBST, for 2 h at room temperature with gentle shaking.
4. Wash four times for 7–10 min in PBST at room temperature with gentle shaking.
5. Incubate in secondary antibody, made up in PBST, for 1 h at room temperature

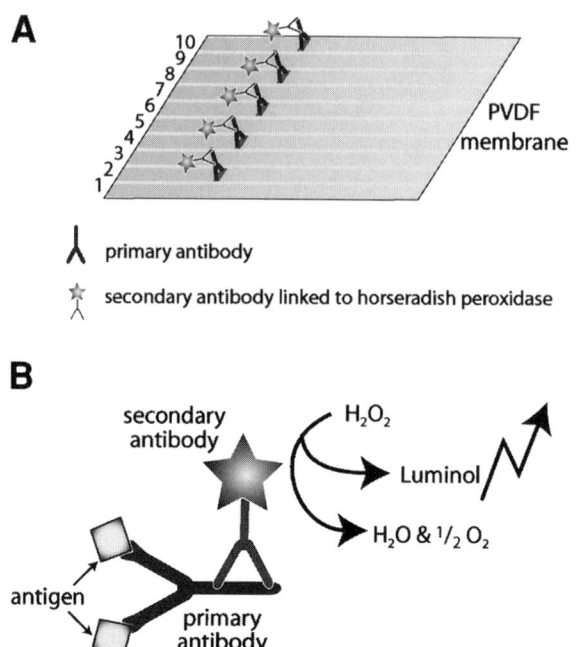

Fig. 2. Development of the PVDF membrane with the antibody sandwich and luminol. (**A**) Ten samples were loaded onto a gel, and proteins were transferred to a PVDF membrane. Antigen for the primary antibody is present in all even lanes, but not odd lanes. The primary antibody binds to the antigen, which is immobilized on the membrane. The secondary antibody has the primary antibody as antigen (*1*). The secondary antibody is linked to horseradish peroxidase. (**B**) Sandwich of the antigen, primary antibody, and secondary antibody linked to horseradish peroxidase. After addition of hydrogen peroxide and luminol (in the substrate bottles provided by the manufacturer), horseradish peroxidase proceeds to metabolize hydrogen peroxide, a reaction that causes the oxidation/ excitation of luminol. When luminol converts to the ground state, light is emitted.

with gentle shaking. (*Note*: Dilution of secondary antibody is usually between 1:10,000 and 1:30,000.) Secondary antibody is directed against IgG from the host species of the primary antibody, and conjugated to horseradish peroxidase. For more information on antibody reactions, *see* **ref.** *1*. *Note:* If the background is too high, primary and secondary antibody can be diluted in blocking solution.
6. Wash four times for 7–10 min in PBST at room temperature with gentle shaking.

3.5. Development of Antibodies

Different reagents are available from various companies which are provided with the appropriate protocols. The following protocol is for Luminol Reagent from New England Nuclear, which is used with horseradish peroxidase coupled antibodies (*see* **Note 8**). The reagent, which is based on a commonly known chemical reaction (**Fig. 2**), comes with two different solutions that are mixed immediately before use.

1. Mix an equal volume of solution 1 and solution 2. The final volume required is 0.125 mL/cm^2 of membrane.
2a. For an image analysis system: Distribute premixed solution onto the glass surface of the image analyzer, and place the membrane face down into the solution. Be careful to avoid air bubbles. Start the scanning procedure.
2b. If you use X-ray film: Drain off excess reagent and wrap blot in plastic wrap or put inside clear plastic sheet protector. Put in light tight film cassette in the dark room, face of blot up, and put X-ray film on top of it. Start with a 30-s exposure and adjust thereafter. **Caution:** The development system wears out quickly and can be exhausted within 20–30 min. For short exposures, stay in the darkroom and do not close the cassette. Make sure the film and membrane are in close apposition. Try to work without safety light. Take multiple exposures.

An example of the protocol used in primary striatal neurons is shown in **Fig. 3**. For further reading *see* **ref.** *1*. (*see* **Notes 9–11**).

4. Notes

1. If samples have a high viscosity, the sonication was too short and samples need to be resonicated.
2. Protein concentration cannot be determined in Laemmli buffer owing to the high concentration of SDS and the bromophenol blue.
3. Deaerate gel solutions before adding APS and TEMED. The gel solutions can be deaerated under vacuum for 20 min to reduce interference during the run.
4. We mark the bottom of the wells with marker on the glass plates before we remove the comb to make the loading easier. Alternatively, a template of the wells can be drawn on a transparency, which is cut out and affixed to the glass plate with a small amount of water. After loading, the template is removed and reused.
5. Transfer to the membrane can also be performed in a semidry electroblotter. These systems use less transfer buffer than the regular electroblotters.
6. Many protocols use several sheets of blotting paper in the electroblot transfer. We use only one sheet (*see* **Fig. 1**) to reduce the possibility of introducing air bubbles in the "sandwich."
7. If working with phosphorylation-specific antibodies, use 3% bovine serum albumin (radioimmunoassay grade) instead of nonfat dry milk in the blocking solution.
8. Alternative detection systems: Although horseradish peroxidase is most commonly used in antibody detection systems, other systems can be applied, such as alkaline phosphatase systems.

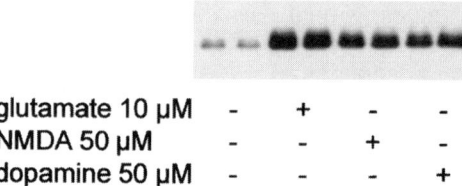

glutamate 10 μM	-	+	-	-
NMDA 50 μM	-	-	+	-
dopamine 50 μM	-	-	-	+

Fig. 3. Western blot of primary striatal neurons, developed with a phosphorylation-specific cAMP response element binding protein (CREB) antibody. Treatment of the cultures for 15 min with either glutamate, N-methyl-D-aspartate, or dopamine leads to the phosphorylation of the transcription factor CREB at ^{133}Ser. The blot was developed with an antibody that detects CREB only when it is phosphorylated at the ^{133}Ser phosphorylation site.

9. Too little or too much staining: Weak staining can be caused by (a) insufficient concentration of primary or secondary antibody; (b) the absence of antigen in the sample, or the concentration of the antigen is below the detection level; or (c) problems with the transfer. There are several solutions to resolve the problem. First, various concentrations of antibody both higher and lower dilutions can be tried. Second, the experimental protocol can be validated with either commercially available antigen or in an experimental system that has been reported to give good results; for example, retaining phosphorylation in in vivo experiments can be difficult, and phosphorylation-specific antibodies may have not antigens to react with. When using phosphorylation-specific antibodies, in vitro phosphorylation of the antigen as a positive control can be considered. Third, if the antigen in the sample is too low, immunoprecipitation before running the Western blot can be performed. Make sure the antibody used for immunoprecipitation is from a different host than the antibody used for the Western blot. Finally, make sure (a) there are no air bubbles between the gel and the membrane in the transfer, (b) the electrotransfer has been set up correctly and in the correct orientation (**Fig. 1**), (c) to use enough current and time for the transfer; you can use prestained protein size markers, which are available from commercial sources, to monitor the transfer

10. High background may be caused by: (a) antibodies used are too concentrated, (b) blots are not washed enough, (c) blocking solutions and blocking times are insufficient, (d) protein samples are overloaded, and (e) exposure time with substrate is too long. If high background happens, one can (a) try lower concentrations of antibody (higher dilutions), lower temperatures during incubation (e.g., 4°C) and lower exposure times to primary antibody; (b) increase washing time and number of washes; try washing after the primary antibody overnight at 4°C with slightly shaking; (c) try different blocking solutions and different detergents in the blocking solution; add blocking solution to primary

and secondary antibodies; (d) dilute samples and load less protein per lane; and (e) try a less sensitive chemiluminescent detection system
11. Stripping and reprobing blots: If membrane is dry, soak in methanol for 2 min. Rinse blots with PBST. Incubate for 20 min at 50°C in a solution containing: 2% SDS; 100 mM β-mercaptoethanol; and 50 mM Tris-HCl, pH 6.8. Wash 3 × 5 min in PBST before reprobe by starting with the blocking step.

References

1. Harlow, E. and Lane, D. (1998) *Using Antibodies: A Laboratory Manual.* Cold Spring Harbor Laboratory Press, Cold Spring Harbor, NY.

18

Immunohistochemical and Immunocytochemical Detection of Phosphoproteins in Striatal Neurons

Limin Mao and John Q. Wang

1. Introduction

Phosphorylation of protein kinases and transcription factors represents a major biological mechanism to activate those imperative intracellular effectors. Examination of such phosphorylation in response to extracellular stimulation can assess the functional participation of those proteins in a particular signaling cascade. For example, the Ca^{2+} signal facilitates the phosphorylation of Ca^{2+}/calmodulin-dependent protein kinase II (CaMKII), a multifunctional protein kinase regulating diverse cellular functions *(1)*. Phosphorylated CaMKII (pCaMKII) can in turn phosphorylate a nuclear transcription factor, cAMP response element-binding protein (CREB), to regulate target DNA transcription *(2)*. The first subclass of the mitogen-activated protein kinase (MAPK), the extracellular signal-regulated kinase (ERK), is another protein kinase that is catalyzed by MAPK kinase (MEK) to undergo the phosphorylation *(3)*. Once phosphorylated, pERK further induces the phosphorylation of a nuclear transcription factor, Elk-1 (a member of the ternary complex factor family of Ets domain proteins that bind serum response elements), to alter DNA transcription *(4)*. Apparently, detection of the phosphorylation state of those proteins can provide valuable information on their functional profiles in a given cellular activity of interest.

Detection of the phosphorylated enzymes and factors in striatal neurons can be accomplished by immunohistochemical and immunocytochemical approaches. With recently developed anti-active antibodies specific for the phosphorylated form of the proteins, the phosphorylation can be visualized in cells

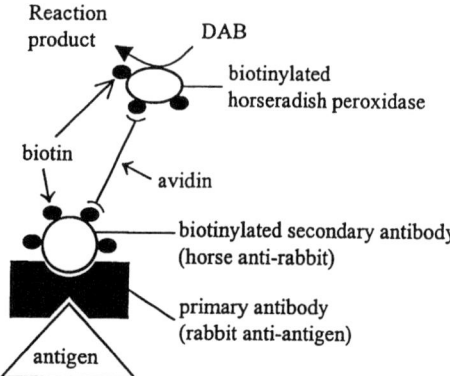

Fig. 1. Schematic diagram illustrating typical immunoperoxidase steps of immunohistochemistry. After the binding of a primary antibody to its target antigen, the biotinylated secondary antibody is added to bind to the primary antibody. Through avidin, the biotinylated horseradish peroxidase binds to the biotinylated secondary antibody. Finally, horseradish peroxidase catalyzes DAB to give rise to visible brown reaction products.

of brain sections from living animals or culture slides. This chapter describes a detailed protocol of immunohistochemistry employed on floating brain sections and immunocytochemistry on culture slides in this laboratory in detecting pCaMKII, pCREB, pERK, and pElk-1 immunoreactivity in rat or mouse striatal neurons (5–8). According to the protocol, brain sections (30–40 µm) cut through the striatum on a freezing sliding microtome or a vibratome or culture striatal neurons are fixed with a fixative, usually 4% paraformaldehyde, to preserve tissue morphology and the protein in its native site. The sections or cultured neurons are then permeablized, and nonspecific sites are preblocked with normal animal serum from the species in which the secondary antibody was raised. Striatal neurons are incubated with the primary antibody to allow the antibody to bind to the enzyme or factor, followed by incubation with the secondary antibody conjugated to biotin. After the formation of antigen–primary antibody–secondary antibody complex, avidin–biotinylated horseradish peroxidase is added. The peroxidase substrate diaminobenzidine (DAB) is finally added as a chromagen to produce a visible reaction product, an indirect indication of the phosphorylation of the protein tested (**Fig. 1**).

2. Materials

2.1. Animals and Equipment

1. Adult male or female rats (weighing 200–250 g) or mice (20–25 g) from commercial vendors. For immunostaining on cultured cells, primary striatal

neuronal cultures are prepared on eight-chamber glass slides according to procedures outlined in Chapter 25. Neurons are usually cultured for 10–14 d before use *(5,6)*.
2. Surgical tools for the transcardial perfusion and removing of brains (from Roboz Surgical or Fine Science Tools), including blunt 15-gauge (rat) and 22-gauge (mouse) hypodermic needles, scalpel, scissors, clamps, forceps, hemostat, and bone rongeur.
3. Costar netwells (12-well polystyrene inserts and plates) from Cole-Parmer.
4. SAMCO transfer pipets (Fisher).
5. Fine-tipped paint brushes.
6. 24 × 60 mm glass coverslips and gelatin-subbed 25 × 75 mm glass slides.
7. Netwell mounting and reagent trays (Cole-Parmer) and slide racks.
8. Glass transfer pipets with the ends heat-sealed and bent to form a hook.
9. Platform orbital shakers.
10. A Leica SM2000R sliding microtome with freezing stage or 1000 plus vibratome (The Vibratome Company).
11. A peristaltic perfusion pump (7520-00 Masterflex L/S pump with variable-speed standard drives and quick load pump head from Cole-Parmer).

2.2. Solutions and Reagents

1. 1X (~0.15 M, pH 7.2) phosphate-buffered saline (PBS).
2. 4% (w/v) Paraformaldehyde: In a chemical fume hood, add paraformaldehyde granules (EM grade, Electron Microscopy Sciences) into 1X PBS when PBS is stirred and heated up to 80–85°C and filter. Store at 4°C and use within 2–3 d.
3. 10% Sucrose: Store at 4°C.
4. Cryoprotectant solution (25% ethylene glycol–25% sucrose in 1X PBS). Store at 4°C.
5. Triton X-100.
6. Hydrogen peroxide (H_2O_2).
7. Blocking solution: 3% (v/v) normal animal (goat) serum from a VECTASTAIN Elite ABC kit (Vector), 1% (v/v) bovine serum albumin (Vector), and 0.4% (v/v) Triton X-100 in 1X PBS. Stir until dissolved. Store at 4°C.
8. Primary antibody dilution buffer: 1% (v/v) normal goat serum (Vector) and 0.4% (v/v) Triton X-100 in 1X PBS. Store at 4°C.
9. Rabbit polyclonal primary antibodies against antigens of interest: anti-CaMKII, anti-phospho-CaMKII (pThr^{286}), anti-CREB, anti-phospho-CREB (pSer^{133}), anti-ERK, anti-phospho-ERK (pThr^{202} and pTyr^{204} for both p44 [ERK1] and p42 [ERK2] MAPKs), anti-Elk-1, and anti-phospho-Elk-1 (pSer^{383}). These quality antibodies, which were validated to produce satisfactory immunostaining in experiments conducted in this laboratory and others, are available from the following four companies: Upstate Biotechnology (Lake Placid, NY), Santa Cruz Biotechnology (Santa Cruz, CA), Cell Signaling Technology (Beverly, MA), and Promega (Madison, WI) (*see* **Notes 1** and **2**).
10. Secondary antibody dilution buffer: 0.02% Triton X-100 in 1X PBS.

11. Secondary antibody (biotinylated goat anti-rabbit IgG), avidin (reagent A), and biotinylated horseradish peroxidase (reagent B) are provided by the kit (Vector, PK-6101).
12. 1 M Tris-HCl stock solution: 121 g of Tris base (TRIZMA base) in 800 mL of distilled H_2O. Adjust pH to 7.4 with concentrated HCl (~70 mL of 1 N HCl is needed to achieve pH 7.4). Adjust volume to 1 L with distilled H_2O. Store up to 6 mo at room temperature.
13. Staining solution: 0.25 mg/mL of 3,3′-diaminobenzidine tetrahydrochloride (DAB, as chromagen to produce reaction products), 0.01% H_2O_2 and 0.04% nickel chloride ($NiCl_2$). The heavy metal (nickel) is used to amplify the DAB reaction product *(9)*. Dissolve 2.5 mg of DAB; 3.33 µL of 30% H_2O_2, and 4 mg of $NiCl_2$ in 10 mL of 50 mM Tris-HCl, pH 7.4. Shake well and use immediately. Because DAB is carcinogenic, wear gloves and work in a fume hood.
14. 0.1 M Sodium nitrate.
15. 70%, 80%, 95%, and 100% ethanol and xylenes.
16. DPX mounting medium (Electron Microscopy Sciences).

3. Methods
3.1. Fixation
3.1.1. Transcardial Perfusion for Fixation of Brain Tissue

1. After a lethal dose of anesthesia (chloral hydrate, 500 mg/kg, i.p.), the animal is moved to a fume hood and an incision on the upper abdomen is made to expose the diaphragm. Carefully incise the diaphragm to expose the beating heart. To open the thoracic cavity, two parallel cuts through the rib on either side of the heart are made, and fold the cut rib flap headward with a hemostat. Insert a blunt 15-gauge (rat) or 22-gauge (mouse) hypodermic needle upward from the bottom apex of the left ventricle through the ventricle and the aortic valve so that the needle is placed approx 3–5 mm inside the ascending aorta. Clamp the needle in place with a curved clamp (*see* **Note 3**).
2. Perfuse 300 (rat) or 70 (mouse) mL of 4% paraformaldehyde over a period of 5–7 min at 4°C. (Cold paraformaldehyde helps reduce possible blood coagulation during the initial perfusion.) The peristaltic pump is set at 35–40 (rat) or 10 (mouse) mL/min. Immediately after perfusion, cut the right atrium to create an escape route for the blood and perfusion fluid. Muscle twitching can be seen if the fixative is properly perfused into the animal. After several minutes, stiffening of the forelimbs and head occurs, a further indication that the fixative is being properly perfused (*see* **Note 3**).
3. When the perfusion is finished, stop the pump and remove the brain from the skull using a bone rongeur. The brain is postfixed for 2 h at 4°C in a vial containing 10% sucrose and 4% paraformaldehyde. The brain is then infiltrated overnight at 4°C in the cryoprotectant solution until the floating brain sinks, indicative of sucrose infiltration of the brain. The infiltration is to prevent freezing artifacts in the tissue during sectioning.

4. Section the brain on a Leica sliding microtome with freezing stage or a vibratome into 30- or 40-μm sections. Collect sections in 1X PBS for immediate use or in the cryoprotectant and store at −20°C until future use (*see* **Note 4**).

3.1.2. Fixation of Cultured Cells

The medium is aspirated. Ice-cold 4% paraformaldehyde is added to each chamber (0.5 mL/chamber). Following 10 min of fixation with paraformaldehyde, cells are rinsed in 1X PBS twice (5 min each). Save cells in 1X PBS at 4°C for immunostaining.

3.2. Blockade of Nonspecific Immunostaining

All of the following steps are performed on a platform shaker at room temperature unless otherwise indicated. The floating sections are processed in 12-well Costar netwells. When exchanging solutions, the carriers containing brain sections are transferred to a separate netwell plate filled with the next-step solution. Cultured cells are processed on the chamber slides. To preserve maximally the morphology of the tissue, the tissue should always be treated in buffered solutions with appropriate molarity and pH.

1. Rinse twice (5 min each) in 1X PBS to remove excessive fixative. Incubate in 1% Triton X-100 for 10 min to increase the tissue permeability. Rinse twice in 1X PBS (10 min each).
2. To block endogenous peroxidase, the section is incubated in 0.6% H_2O_2 in the dark for 30 min. Rinse for 5 min in 0.1% Triton X-100 and 5 min in 1X PBS.
3. Incubate for 30 min in the blocking solution (3% normal goat serum, 1% bovine serum albumin and 0.4% Triton X-100) to block nonspecific staining, and rinse for 10 min in 1X PBS. It is crucial to select the normal (nonimmune) animal serum from the same species in which the secondary antibody was raised (*see* **Note 5**). Because the anti-rabbit IgG antibody raised from goats (goat anti-rabbit) is used here as the secondary antibody, the normal goat serum should be selected to block nonspecific binding sites. Likewise, if the horse anti-rabbit is used as the secondary antibody, normal horse serum should be chosen as the blocking serum.

3.3. Incubation with Primary and Secondary Antibodies

1. Dilute primary antibodies to desired concentrations in the primary antibody dilution buffer. The following dilutions of primary antibodies have produced the satisfactory signal-to-noise ratios in our studies *(5–8)*: anti-phospho-CaMKII (1:500); anti-CREB (1:1000); anti-phospho-CREB (1:1000); anti-phospho-ERK (1:500); and anti-phospho-Elk-1 (1:2000) (*see* **Notes 3** and **6**).
2. Incubate overnight in the diluted primary antibody solution in a cold room (at 4°C). The incubation conditions (time, temperature, and antibody concentration) can be adjusted empirically for each antibody to optimize the signal-to-noise ratio.

3. On the next day, continue the incubation with primary antibody at room temperature for 0.5–1 h. Rinse four times in 0.02% Triton X-100–1X PBS (10 min each) to wash off free primary antibodies.
4. Dilute the secondary antibody (biotinylated goat anti-rabbit from the Vector ABC kit) 1:200 in the secondary antibody dilution buffer (0.02% Triton X-100–1X PBS).
5. Incubate in the diluted second antibody solution for 1 h. Rinse three times in 1X PBS (10 min each) to wash off free secondary antibodies.

3.4. ABC Complex, Staining and Mounting

1. Mix one drop each of reagent A (avidin) and B (biotinylated horseradish peroxidase) (from the Vector ABC kit) in 10 mL of 0.02% Triton X-100–1X PBS for 15–30 min. This allows the formation of the avidin–biotinylated horseradish peroxidase complex (*see* **Note 7**).
2. Incubate in the mixture for 1 h and rinse three times in 1X PBS (10 min each).
3. Incubate in the freshly prepared staining solution (0.25 mg/mL of DAB, 0.01% H_2O_2, and 0.04% $NiCl_2$) for 5–6 min. Closely monitor the intensity of the reaction to ensure that background immunostaining remains at a minimal level (*see* **Note 8**).
4. Once the desired level of staining has been visualized, stop the reaction in 1X PBS. Rinse the section three times in 1X PBS (10 min each).
5. Remove floating sections to a Netwell mounting tray half filled with 0.1 M sodium nitrate using a fine-tipped paint brush. Mount sections onto the gelatin-coated glass slide. Several drops of 10% Triton X-100 can be added to facilitate the moving of tissue onto the slide. For culture cells, remove mounted chambers from glass slides carefully.
6. Slides are air-dried at room temperature to allow the adhesion of sections to the surface of the glass slide. The DAB reaction product is essentially insoluble and very stable. Immerse slides in distilled H_2O for 1 min to remove the salts remaining from the buffers previously used. If necessary, a counterstaining can be performed here (*see* **Note 9**). Dehydrate sections in graded ethanol (70%, 80%, 95%, and 100%; 1 min each). Ethanol is generally used to remove the water in the tissue to allow the mounting media that is not miscible with water to coverslip slides.
7. Place the slide in xylene to prevent the tissue from drying out and coverslip with the DPX mounting media. Let slides dry overnight before examination under a light microscope.

3.5. Anticipated Results in the Striatum

Using the procedures outlined in the preceding, the following characteristics can be anticipated (for details, *see* **refs. 5–8**). It should be mentioned that experimental manipulations that can enhance striatal cellular activity need

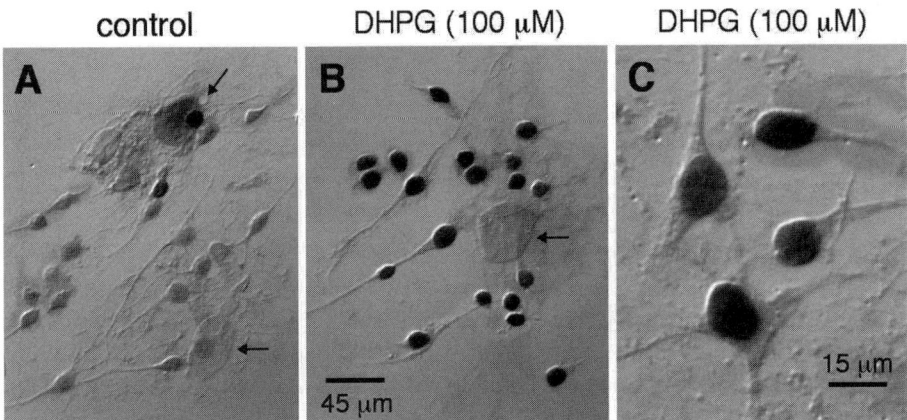

Fig. 2. CREB phosphorylation in rat primary cultures of striatal neurons following incubation of the group I mGluR agonist DHPG for 10 min (**A** vs **B**). The increased pCREB immunostaining occurred in the medium-sized neuronal cells rather than the astroglial cells with large, flat, and phase-dark nuclei (*arrows*). Moreover, pCREB staining was exclusively expressed within the nuclear compartment (**C**).

to be applied prior to the immunostaining procedures to visualize a high throughput of cells containing phosphorylated proteins. Such manipulations can be achieved by application of psychostimulants (cocaine or amphetamine) or the metabotropic glutamate receptor agonist *(5–8)*.

1. The pCREB immunoreactivity is seen throughout the striatal region (both the dorsal and ventral striatum) following a systemic injection of an indirect dopamine agonist, amphetamine *(8)*, or in most cultured striatal neurons *(5)*. The pCREB immunostaining seems to be confined to in neuronal cells rather than non-neuronal glial cells. Moreover, the dense pCREB staining is exclusively and evenly displayed over the entire nuclear compartment of each perikaryon (**Fig. 2**). The detectable specific labeling is hardly seen on the other parts of neurons, for example, cytoplasm and fibers.
2. The distribution pattern and density of pElk-1 immunoreactivity in the striatal area of brain sections and cultured striatal neurons share the same characteristics as pCREB *(5,8)*. At the subcellular level, the strong pERK immunoreactivity is present in the nucleus of perikarya in this brain area. In addition to the nuclear staining, diffuse pERK staining exists in the cytoplasm and expands to the proximal segment of axonal or dendritic processes. Moreover, the number of pERK-positive cells is significantly less than the population of pCREB- or pElk-1-immunoreactive cells.
3. The dense pCaMKII immunoreactivity is evident in the cytoplasm and, to a much lesser extent, the nucleus in striatal neurons. Moderate fiber staining is also

revealed, which appears as widespread varicosities and small fibers. Like pERK, pCaMKII-positive cells are scattered in the entire striatum.

4. Notes

1. Specificity of antibodies is critical in interpreting data. Numerous studies performed recently with Western blotting, immunoprecipitation, immunohistochemistry in brain sections from living rodents, or immunocytochemistry in cultured cells have demonstrated the specificity of the antibodies that are sold commercially from Upstate Biotechnology, Santa Cruz Biotechnology, and Cell Signaling Technology. However, despite their demonstrated selectivity, it is recommended that the investigator gain full knowledge of the selectivity and cross-reaction with closely related proteins of the antibody by a careful review of the literature and adequate pilot tests *(10,11)*.
2. Prior to detection of a given protein, the control procedures of immunostaining need to be conducted to ensure the selectivity of primary and secondary antibodies. A purified protein can be used to preabsorb the primary antibody, resulting in a negative staining opposed to a positive staining when the primary antibody alone is used. In another negative control, the primary antibody is omitted in the procedures to evaluate the specificity of the secondary antibody.
3. Numerous factors may affect the quality and quantity of immunostaining. Readily controlled factors include tissue fixation, the application of the primary antibody (dilution and incubation duration), and the molarity and pH of buffers. Tissue fixation needs to be done quickly with cold paraformaldehyde after anesthesia and no saline solution should be performed prior to paraformaldehyde solution because delayed fixation and prior saline perfusion seem to cause dephosphorylation of kinases and factors, which may result in a failure to detect phosphorylated targets because of sizable loss of the signal. In a direct and quick paraformaldehyde perfusion, we found no effects of the two phosphatase inhibitors, pervanadate (10 μM; Sigma) inhibiting the tyrosine phosphatase and okadaic acid (1 μM; Tocris) inhibiting the protein phosphatase 1 (PP-1), applied in the first 50 mL of the paraformaldehyde perfusion on immunohistochemical detection of pCaMK II, pCREB, and pElk-1 in the rat striatum in response to stimulation of metabotropic glutamate receptors. This indicates that a minimal dephosphorylation occurred in our perfusion procedures. We also tested immunohistochemical detection of the phosphoproteins under anesthesia with different anesthetics. We found comparable levels of striatal pCaMK II and pCREB among different groups of rats treated with three different anesthetics (chloral hydrate, sodium pentobarbital, and Equithesin). Thus, the use of anesthetics seems to cause no evident dephosphorylation that may prevent us from detecting the phosphoproteins. The dilution and incubation duration of the primary antibody need to be determined in pilot studies even though valuable information can be obtained from the literature. It is noted in some of the literature that the listed dilution concentration of a primary antibody in the article is the reflection of the

dilution from the original solution of commercial antibody regardless of its real concentration. In this case, 1:1000 dilution from a 0.1 mg/mL antibody package actually means 1:10,000 dilution (0.1 μg/mL). Checking with the original author could help verify this issue.
4. Variation in immunostaining density exists among the procedures performed at different times even though the identical procedures are followed. It is therefore highly recommended to conduct the detection of a given protein simultaneously on the sections or culture slides from different experimental groups. By performing the same procedures at the same time, the density variation is expected to be minimal. Changes in the immunostaining density can therefore be compared quantitatively among groups of animals treated with different drugs.
5. The secondary species serum (normal goat serum) incubation allows for the binding of serum protein to charged sites on the brain section. The concentration of the normal goat serum and the duration of incubation may be adjusted to reduce nonspecific background staining.
6. Primary antibodies should not be thawed and refrozen frequently. They should be stored in aliquots with or without dilution of the original package and thawed once and used. However, most antibodies used in this study can be saved in solution at –20°C, which avoids freeze–thaw cycles during their frequent uses.
7. Some frozen tissues, most commonly kidney and liver, show intrinsic biotin-binding activity, which may increase the nonspecific background staining. However, such a problem is not significant when striatal sections are proceeded for immunohistochemistry. By using the Blocking Kit from Vector to block the intrinsic biotin-binding activity (adding avidin from the kit to the brain section to bind to biotin, followed by biotin to bind avidin and rinse), we observed a high signal-to-noise ratio comparable to that obtained without the blocking treatment.
8. We add nickel chloride to the DAB solution to increase the overall sensitivity of the DAB reaction. The DAB reaction product after nickel chloride addition is dark blue to black, instead of brown produced by the utilization of DAB alone. The dark reaction product seems preferable for black and white photography of stained slides.
9. If the counterstaining is desired, this stain is better remained weak. Strong counterstaining can markedly mask the principal reaction product. Usually, the counterstaining is controlled to intensity sufficient just to visualize the structure.

Acknowledgments

This work was supported by NIH Grants DA10355 and MH61469.

References

1. Churn, S. B. (1995) Multifunctional calcium and calmodulin-dependent kinase II in neuronal function and disease. *Adv. Neuroimmunol.* **5**, 241–259.

2. Shaywitz, A. J. and Greenberg, M. E. (1999) CREB: a stimulus-induced transcription factor activated by a diverse array of extracellular signals. *Annu. Rev. Biochem.* **68,** 821–861.
3. Sweatt, J. D. (2001) The neuronal MAP kinase cascade: a biochemical signal integration system subserving synaptic plasticity and memory. *J. Neurochem.* **76,** 1–10.
4. Davis, R. J. (1995) Transcriptional regulation by MAP kinases. *Mol. Reprod Dev.* **42,** 459–467.
5. Mao, L. and Wang, J. Q. (2002) Glutamate cascade to cAMP response element-binding protein phosphorylation in cultured striatal neurons through calcium-coupled group I mGluRs. *Mol. Pharmacol.* **62,** 473–484.
6. Mao, L. and Wang, J. Q. (2002) Metabotropic glutamate receptor-regulated phosphorylation of Elk-1 and immediate early gene expression in striatal neurons. *Mol. Pharmacol.,* in press.
7. Choe, E. S. and Wang, J. Q. (2001) Group I metabotropic glutamate receptor activation increases phosphorylation of cAMP response element-binding protein, Elk-1 and extracellular signal-regulated kinases in rat dorsal striatum. *Mol. Brain Res.* **94,** 75–84.
8. Choe, E. S. and Wang, J. Q. (2002) Amphetamine increases phosphorylation of extracellular signal-regulated kinase and transcription factors in the rat striatum via group I metabotropic glutamate receptors. *Neuropsychopharmacology* **27,** 565–575.
9. Adams, J. C. (1981) Heavy metal intensification of DAB-based HRP reaction product. *J. Histochem. Cytochem.* **29,** 775.
10. Ince, E. and Levey, A. (1997) Immunohistochemical localization of neurochemicals, in *Current Protocols in Neuroscience* (Crawley, J. N., Gerfen, C. R., McKay, R., Rogawski, M. A., Sibley, D. R., and Skolnick, P., eds.), John Wiley & Sons, New York, pp. 1.2.1–1.2.12.
11. Javois, L. C., ed. (1999) *Immunocytochemical Methods and Protocols.* Humana Press, Totowa, NJ.

19

Analysis of Protein Expression in Brain Tissue by ELISA

Steffany A. L. Bennett and David C. S. Roberts

1. Introduction

The enzyme-linked immunosorbent assay (ELISA) technique offers a sensitive, simple, and versatile method for quantifying as little as 100 pg of target protein in mixed cell or tissue lysates. A further advantage to the protocol is the ability to process rapidly and reproducibly large numbers of samples with minimal equipment requirements. The underlying principle depends on formation of an antigen–antibody complex immobilized on plastic microtitre plates. **Figure 1** illustrates the basic methodology. In the antibody-sandwich ELISA, a primary antibody directed against a protein of interest is bound (adsorbed) to the bottom of a polystyrene well (**Fig. 1A**). A mixed protein lysate is added to the well and the target protein is "captured" onto the solid phase by interaction with the capture antibody (**Fig. 1B**). A second antibody recognizing a different antigenic determinant on the target protein is added (**Fig. 1C**). The resulting antibody–antigen–antibody sandwich is detected with an enzyme-linked secondary antibody (**Fig. 1D**) followed by incubation with a suitable enzyme substrate. Substrate hydrolysis results in a detectable color change (**Fig. 1E**) and is proportional to the amount of captured protein. At each step, the antibody sandwich is separated simply and effectively from unbound conjugate by repeated washes. Sensitivity can be increased further by secondary and tertiary immunogenic enhancement.

In this chapter, we describe standard protocols for protein detection in lysates prepared from microdissected brain tissue using antibody-sandwich ELISAs. To illustrate methods, cerebral changes in expression of connexin32 (Cx32) following chronic cocaine self-administration are presented. Connexins

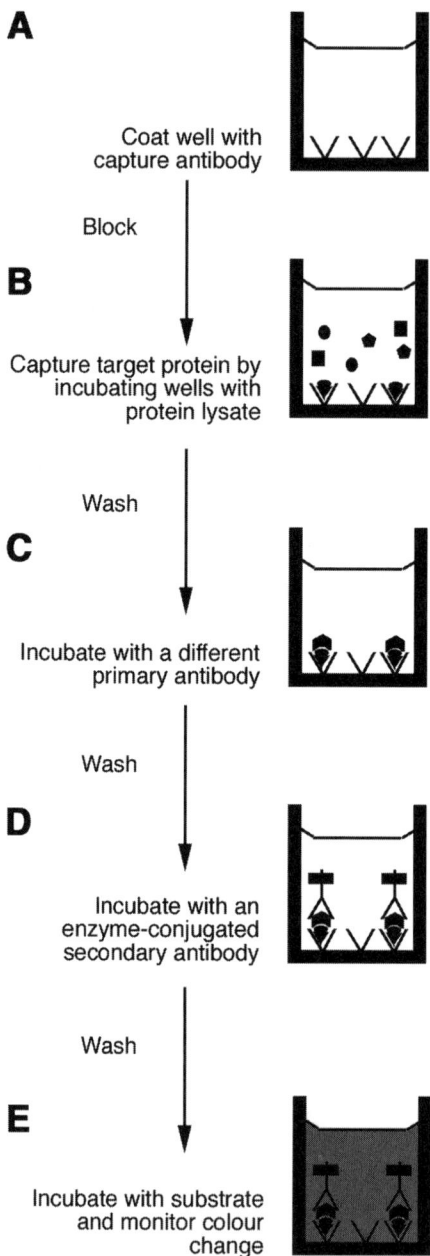

Fig. 1. Basic principles of the antibody-sandwich ELISA. The antibody-sandwich protocol is the most sensitive of the ELISA methods.

are a multigene family of more than 20 proteins that form the structural units of gap junction channels. These intercellular channels directly connect the cytoplasm of adjacent cells. Gap junctional intercellular communication (GJIC) allows coupled cells to synchronize their responses to extracellular cues by passive diffusion of ions, metabolites, and second messengers from one cell to another *(1)*. Administration of drugs of abuse has been shown to influence functional measures of GJIC between neurons in the striatum *(2,3)* although the individual connexin proteins responsible for these changes have only begun to be elucidated *(4)*. ELISA methodology represents one means of identifying these connexins.

2. Materials

1. Sodium pentobarbital (Somnotol, MTC Pharmaceuticals, Cambridge).
2. Dry ice.
3. Isopentane (2-methylbutane).
4. Razor blades.
5. Disposable borosilicate glass culture tubes (12 × 75 mm, VWR).
6. Tissue-Tearer hand-held small sample laboratory homogenizer (Fisher).
7. 10 mM Phosphate-buffered saline (PBS): 10 mM sodium phosphate; 150 mM NaCl, pH 7.4.
8. RIPA buffer: 10 mM PBS, 1% Nonidet P-40 (NP-40), 0.5% sodium deoxycholate, 0.1% sodium doedecyl sulfate (SDS). Buffer can be stored at 4°C. Add protease inhibitors and vortex-mix vigorously immediately before use. Protease inhibitors: Aprotinin (30 μL/mL from Sigma stock, stored at 4°C), 10 mM sodium orthovanadate (add 10 μL/mL of 100 mM stock to RIPA buffer, stored at –20°C), and phenylmethanesulfonyl fluoride (PMSF) (add 100 μL/mL of 10 mg/mL stock dissolved in isopropanol to RIPA buffer, stored at –20°C). **Warning**: PMSF is a neurotoxin. Take appropriate safety precautions when dissolving stock powder.
9. Bio-Rad DC protein assay kit (Bio-Rad, Mississauga, ON).
10. 96-Well microtitre plates.
11. Adhesive covers (available from any company supplying microtiter plates) or plastic wrap.
12. Capture primary antibody (rat monoclonal R5, kindly provided by Dr. David Paul, Harvard Medical School), detection primary antibody (mouse monoclonal M12.13, kindly provided by Dr. David Paul, Harvard Medical School), peroxidase-linked anti-mouse IgG secondary antibody (Jackson Immunolabs, Mississauga, ON).
13. 3,3′,5,5′-Tetramethylbenzidine (TMB). TMB can be purchased as a ready to use solution (BM Blue POD Substrate, Roche).
14. Coating buffer: 0.1 M NaHCO$_3$–Na$_2$CO$_3$, pH 9.5.
15. Blocking solution: 1% bovine serum albumin (BSA) (w/v) in 10 mM PBS, pH 7.4, and 0.02% thimerosol (w/v).

16. Multichannel pipet.
17. Plate reader with filters for detection at 450 nm and 650 nm.

3. Methods

Methodologies are described for extraction of protein and RNA from rodent striatum (**Subheading 3.1.**), protein capture onto microtiter plates (**Subheading 3.2.**), detection of bound antibody–antigen complex (**Subheading 3.3.**), and data analysis (**Subheading 3.4.**).

3.1. Protein Extraction from Rat Brain

This protocol includes a description of the euthanasia procedure (**Subheading 3.1.1.**), dissection parameters (**Subheading 3.1.2.**), and protein extraction methodology (**Subheading 3.1.3.**).

3.1.1. Animal Euthanasia

Male Wistar rats (275–300 g, Charles Rivers, St Foie, PQ) were lethally injected intraperitoneally with 80 mg/kg of sodium pentobarbital (Somnotol) and euthanized by decapitation. Brains were rapidly removed for dissection. In the present example, rats had been previously allowed to self-administer 38 injections of 1.5 mg/kg of cocaine over a 14-d period *(4)*. Animals were killed at 2, 7, or 21 d after the last cocaine self-administration session and Cx32 protein expression was compared to drug-naive, age-matched controls *(4)*.

3.1.2. Tissue Dissection

The medulla, parietal/temporal cortex, cerebellum, thalamus, dorsal striatum, nucleus accumbens, and globus pallidus/ventral lateral striatum (not including the nucleus accumbens) were removed under a dissecting microscope, weighed, flash-frozen in dry-ice chilled isopentane, and stored at –85°C until processing.

3.1.3. Protein Extraction

Lysates of soluble protein were prepared from pooled tissue samples ($n = 2$ animals/protein lysate). The steps in this process involve homogenization, separation of unhomogenized tissue and insoluble proteins from soluble protein lysate, and determination of protein concentration. The following protocol gives the highest protein yield from small samples with minimal extraction steps (*see* **Note 1**).

1. Place tissue on a clean glass plate chilled on ice.
2. Dice tissue into small pieces with a razor blade. Wipe plate with 100% ethanol (EtOH) between samples.
3. Immediately transfer diced tissue to disposable borosilicate glass culture tubes (12 × 75 mm) (VWR) and add RIPA buffer with protease inhibitors (3 mL/g of tissue). As it is difficult to homogenize tissue in <500 µL of RIPA (~170 mg of tissue), it may be necessary to pool multiple dissections from different animals to

achieve ~170 mg of tissue. Samples should be maintained on ice for the duration of the extraction procedure.
4. Homogenize tissue using a hand-held or stand-mounted homogenizer. The settings and number of strokes must be empirically determined depending on the unit employed. Using a Tissue-Tearer hand-held small sample laboratory homogenizer (Fisher) with a probe length of 8.3 × 0.7 cm, 8–10 up-and-down motions (strokes) through the tissue solution at setting 3 is sufficient to homogenize small samples of myelinated tissue (~170 mg tissue/500 µL of RIPA buffer). Large samples such as cortex or cerebellum may require additional homogenization. Care should be taken to ensure that the probe is not removed from the liquid or excess frothing will occur. Samples are subjected to three to five strokes and solid tissue and froth bubbles are allowed to settle on ice prior to renewed homogenization. Carefully clean pestle with 70% EtOH and double-distilled H_2O (2X H_2O) between samples to eliminate cross-contamination.
5. Transfer homogenized lysates to 1.5-mL microcentrifuge tubes and centrifuge at 12,000 rpm for 20 min at 4°C. Remove supernatant and microfuge again. The second supernatant is the total cell lysate used in the ELISA protocol.
6. Combine lysates from the same sample. Determine protein concentration using standard laboratory protocol such as the Bio-Rad DC protein assay kit (Mississauga, ON). If homogenization has been performed properly, the final protein concentration should be between 0.5 and 1 µg/µL. A yield of ≥50 µg of total protein/200 mg of starting tissue is commonly achieved.
7. Aliquot samples for storage. Aliquots can be frozen at –20°C for up to 3 mo without significant degradation of connexin proteins. Repeated freeze–thaw should be avoided.

3.2. Protein Capture onto Microtiter Plates

In the protocol described, Cx32 protein was immobilized on 96-well polystyrene plates coated with rat monoclonal R5 (provided by Dr. David Paul, Harvard Medical School). This assay was designed not to quantify absolute protein levels but to establish whether relative Cx32 expression changes over time in various tissues following cocaine self-administration *(4)*. As a result, a standard curve was not included. If a sufficient amount of purified protein is available, this protocol can be modified to quantify absolute protein concentration by inclusion of a standard curve with known amounts of target protein (*see* **Subheading 3.2.2., step 1**).

3.2.1. Adsorb the Capture Antibody onto the Solid Phase of a Microtiter Plate

1. Dilute capture primary antibody to 2.5 µg/mL in coating solution (*see* **Note 2**).
2. Add 50 µL per well using a multichannel pipet (12-channel 20–200-µL Costar Pipettor). Agitate plate to cover the surface of the well evenly with coating solution.

3. Cover the plate with plastic adhesive covers or plastic wrap and allow antibody to bind to the plastic by incubating at 4°C overnight.
4. Remove the coating solution by turning the plate upside down over a sink and shaking. Strike the plate against a paper towel three times to remove as much of the remaining coating solution as possible. A sharp, definitive striking motion is optimal.
5. Wash plates with 10 mM PBS using a multichannel pipet. Add 100 µL per well. Remove wash solution as described in **step 4**. Repeat for a total of three washes.
6. Block the residual binding capability of the wells with 100 µL/well in 1% BSA in 10 mM PBS containing 0.02% thimerosol as a preservative. Cover the plate with plastic adhesive covers or plastic wrap and incubate at 37°C for 1 h. Remove the plate to room temperature if more time is required to prepare subsequent reagents. Plates can also be prepared in advance and left in blocking solution overnight at 4°C.

3.2.2. Immobilize Target Protein

1. Prepare 20 µg of protein lysate per well. Adjust all samples to the same volume of RIPA buffer before addition of coating buffer. Bring each sample to a final volume of 50 µL with coating buffer. If a standard curve is required, it should be prepared at this step (*see* **Note 3**).
2. Remove the blocking solution as described in **step 4**.
3. Add 50 µL of test samples to antibody-coated wells in duplicate or triplicate to control pipetting errors. A sample plate layout is provided in **Table 1**. Include a standard curve on each plate if required (*see* **Note 3**). Cover the plate with plastic adhesive covers or plastic wrap and incubate ≥ 2 h at room temperature or overnight at 4°C. Include multiple wells incubated with coating buffer only as blanks **(Table 1)**.

3.3. Detection of Bound Antigen–Antibody Complex

Detection of bound antibody–antigen complex can be direct or indirect. In both cases, the captured antibody–protein complex is incubated with a primary detection antibody capable of recognizing different antigenic determinants on the target protein than the capture antibody. The epitopes must remain available after the target protein is immobilized on the microtiter plate and, as a result, not all antibody pairs are compatible (*see* **Note 2**).

Direct detection is accomplished using a primary detection antibody that is already conjugated to peroxidase or alkaline phosphatase. Advantages are a faster and easier assay with fewer steps as well as the freedom to use capture and detection antibodies raised in the same species. Disadvantages include reduced sensitivity relative to the indirect method and practical concerns in that an appropriate conjugated antibody may not be available.

Table 1
Microplate Layout for Experimental Detection of Cx32 in Medulla, Parietal/Temporal Cortex, Cerebellum, Thalamus, Dorsal Striatum, Nucleus Accumbens, and Globus Pallidus/Ventral Lateral Striatum in Drug-Naive (Control) Animals and in Chronic Cocaine Self-Administrators After 2, 7, and 21 d of Cocaine Withdrawal

	1	2	3	4	5	6	7	8	9	10	11	12
A	Blank	0 μg protein	M Control 20 μg	M Control 20 μg	CB Control 20 μg	CB Control 20 μg	DS Control 20 μg	DS Control 20 μg	VS Control 20 μg	VS Control 20 μg		
B	Blank	0 μg protein	M 2 d 20 μg	M 2 d 20 μg	CB 2 d 20 μg	CB 2 d 20 μg	DS 2 d 20 μg	DS 2 d 20 μg	VS 2 d 20 μg	VS 2 d 20 μg		
C	Blank	0 μg protein	M 7 d 20 μg	M 7 d 20 μg	CB 7 d 20 μg	CB 7 d 20 μg	DS 7 d 20 μg	DS 7 d 20 μg	VS 7 d 20 μg	VS 7 d 20 μg		
D	Blank	0 μg protein	M 21 d 20 μg	M 21 d 20 μg	CB 21 d 20 μg	CB 21 d 20 μg	DS 21 d 20 μg	DS 21 d 20 μg	VS 21 d 20 μg	VS 21 d 20 μg		
E	Blank	0 μg protein	Cx Control 20 μg	Cx Control 20 μg	T Control 20 μg	T Control 20 μg	NA Control 20 μg	NA Control 20 μg				
F	Blank	0 μg protein	Cx 2 d 20 μg	Cx 2 d 20 μg	T 2 d 20 μg	T 2 d 20 μg	NA 2 d 20 μg	NA 2 d 20 μg				
G	Blank	0 μg protein	Cx 7 d 20 μg	Cx 7 d 20 μg	T 7 d 20 μg	T 7 d 20 μg	NA 7 d 20 μg	NA 7 d 20 μg				
H	Blank	0 μg protein	Cx 21 d 20 μg	Cx 21 d 20 μg	T 21 d 20 μg	T 21 d 20 μg	NA 21 d 20 μg	NA 21 d 20 μg				

Layout of a 96-well plate and loading of experimental samples as described in **Subheading 3.2.2.** is illustrated. Blank: Wells were coated with R5 capture antibody and blocked with blocking solution. In all subsequent steps, wells were incubated with the appropriate buffers in the absence of protein lysate or antibodies. These values are set to 0 by the microplate reader during the run. 0 μg protein: Wells were treated exactly as described for experimental samples except protein was not included in the 40 μL of RIPA and 10 μL of coating buffer (**Subheading 3.2.2., step 1**). This control determines the lower limit of assay sensitivity and establishes nonspecific binding of all reagents in the absence of target protein. Experimental values that fall below this value cannot be interpreted. If this absorbance is high, increase the number of washes. M, Medulla; Cx, parietal–temporal cortex; CB, cerebellum; T, thalamus; DS, dorsal striatum; NA, nucleus accumbens; VS, ventral striatum (not including the nucleus accumbens). Control: Samples obtained for drug-naive animals. 2 d: Protein extraction performed 2 d after the last cocaine injection. 7 d: Protein extraction performed 7 d after the last cocaine injection. 21 d: Protein extraction performed 21 d after the last cocaine injection. Experimental samples represent protein lysates extracted from tissue pooled from two animals.

Indirect detection is accomplished using an unconjugated detection antibody raised in a different species to the capture antibody followed by incubation with an enzyme-conjugated secondary antibody that recognizes the detection antibody. Sensitivity can be enhanced further by tertiary reaction (*see* **Note 4**). The advantage to this method lies in its sensitivity. Disadvantages include increased assay processing time and labor as well as the requirement to identify capture and detection antibodies raised in different species.

The protocol provided below is an example of secondary enhancement. Cx32 protein, immobilized on 96-well polystyrene plates coated with monoclonal R5 raised in rat (*see* **Subheading 3.2.**), was reacted with mouse monoclonal M12.13. Antibody–antigen–antibody complexes were detected with a peroxidase-linked anti-mouse IgG (1:2000) and TMB.

1. Remove unbound lysate by turning the plate upside down over a sink and shaking. Strike the plate against a paper towel several times to remove as much of the remaining lysate as possible.
2. Wash plates with 10 mM PBS using a multichannel pipet. Add 100 µL per well. Remove wash solution as described in **step 4**. Repeat for a total of three washes.
3. Dilute detection primary antibody to 2.5 µg/mL in blocking solution (*see* **Note 2**).
4. Add 50 µL per well using a multichannel pipet (12-channel 20–200-µL Costar Pipettor). Agitate plate to cover the surface of the well evenly with solution. Cover the plate with plastic adhesive or plastic wrap and incubate ≥2 h at room temperature.
5. Remove unbound antibody and wash plates as described in **steps 1** and **2**.
6. Detect the antibody–antigen–antibody complexes with peroxidase-linked anti-mouse IgG (1:2000) or appropriate secondary antibody raised against the Ig (i.e., Ig, IgG, IgM, etc.) and species (i.e., mouse, rabbit, sheep, etc.) of the detection primary antibody. Add 50 µL per well. Cover with plastic adhesive or wrap and incubate for 2 h at 37°C.
7. Remove unbound antibody and wash plates as described in **steps 1** and **2**.
8. Add 75 µL per well of TMB substrate and incubate for 10–30 min at room temperature.
9. Stop reaction by addition of 25 µL of 1 M H_2SO_4. The solution will turn a yellow color.
10. Read the plate on a microplate reader at 450 nm against a reference wavelength of 650 nm.

3.4. Data Analysis

In the example provided, -fold changes in Cx32 expression were assessed in medulla, parietal/temporal cortex, cerebellum, thalamus, dorsal striatum, nucleus accumbens, and globus pallidus/ventral lateral striatum (not including

Table 2
Raw Data from a Representative Antibody-Sandwich ELISA Experiment

	1	2	3	4	5	6	7	8	9	10	11	12
A	Blank	0 µg protein	M Control	M Control	CB Control	CB Control	DS Control	DS Control	VS Control	VS Control		
	0.000	0.027	0.850	0.851	0.641	0.834	0.651	0.525	0.602	0.891		
B	Blank	0 µg protein	M 2 d	M 2 d	CB 2 d	CB 2 d	DS 2 d	DS 2 d	VS 2 d	VS 2 d		
	0.000	0.022	0.947	0.913	1.009	0.983	0.577	0.576	0.972	1.016		
C	Blank	0 µg protein	M 7 d	M 7 d	CB 7 d	CB 7 d	DS 7 d	DS 7 d	VS 7 d	VS 7 d		
	0.001	0.019	0.938	0.913	0.855	0.675	0.644	0.615	0.972	0.986		
D	Blank	0 µg protein	M 21 d	M 21 d	CB 21 d	CB 21 d	DS 21 d	DS 21 d	VS 21 d	VS 21 d		
	−0.003	0.022	0.971	0.774	0.785	0.582	0.560	0.704	0.923	0.834		
E	Blank	0 µg protein	Cx Control	Cx Control	T Control	T Control	NA Control	NA Control				
	−0.001	0.019	0.629	0.938	0.572	0.813	0.976	0.974				
F	Blank	0 µg protein	Cx 2 d	Cx 2 d	T 2 d	T 2 d	NA 2 d	NA 2 d				
	0.002	0.078	0.829	0.821	0.897	0.952	0.938	0.959				
G	Blank	0 µg protein	Cx 7 d	Cx 7 d	T 7 d	T 7 d	NA 7 d	NA 7 d				
	0.000	0.022	0.815	0.715	0.960	1.011	0.916	0.913				
H	Blank	0 µg protein	Cx 21 d	Cx 21 d	T 21 d	T 21 d	NA 21 d	NA 21 d				
	0.001	0.021	0.954	0.677	0.998	0.843	0.843	0.824				

Layout and abbreviations are as described in **Table 1**. Data represent well absorbance read at 450 nm minus absorbance read at a reference wavelength of 650 nm. For subsequent data analysis, the mean of replicate wells defines a single data point.

nucleus accumbens) at various time points after cocaine withdrawal. **Table 2** depicts raw data from a single experiment. Examples of data analysis and interpretation are presented in **Figs. 2** and **3**. Graphs depict averaged data from two independent experiments (two antibody-sandwich ELISAs) using different protein lysates. In **Fig. 2A–C**, mean absorbances and data standardized to Cx32 levels in drug-naive control animals are presented. To perform standardization calculations, each data point/experiment/tissue was divided by the mean of the control group (drug-naive animals). This transformation sets control expression levels to 1.0 and permits calculation of both average -fold change relative to control and standard error of the mean for each condition. Standardization simplifies comparisons of changes in protein expression between tissues that express different amounts of protein under basal conditions (compare **Fig. 2A** with **2B**). If a standard curve is included in the experimental design, -fold

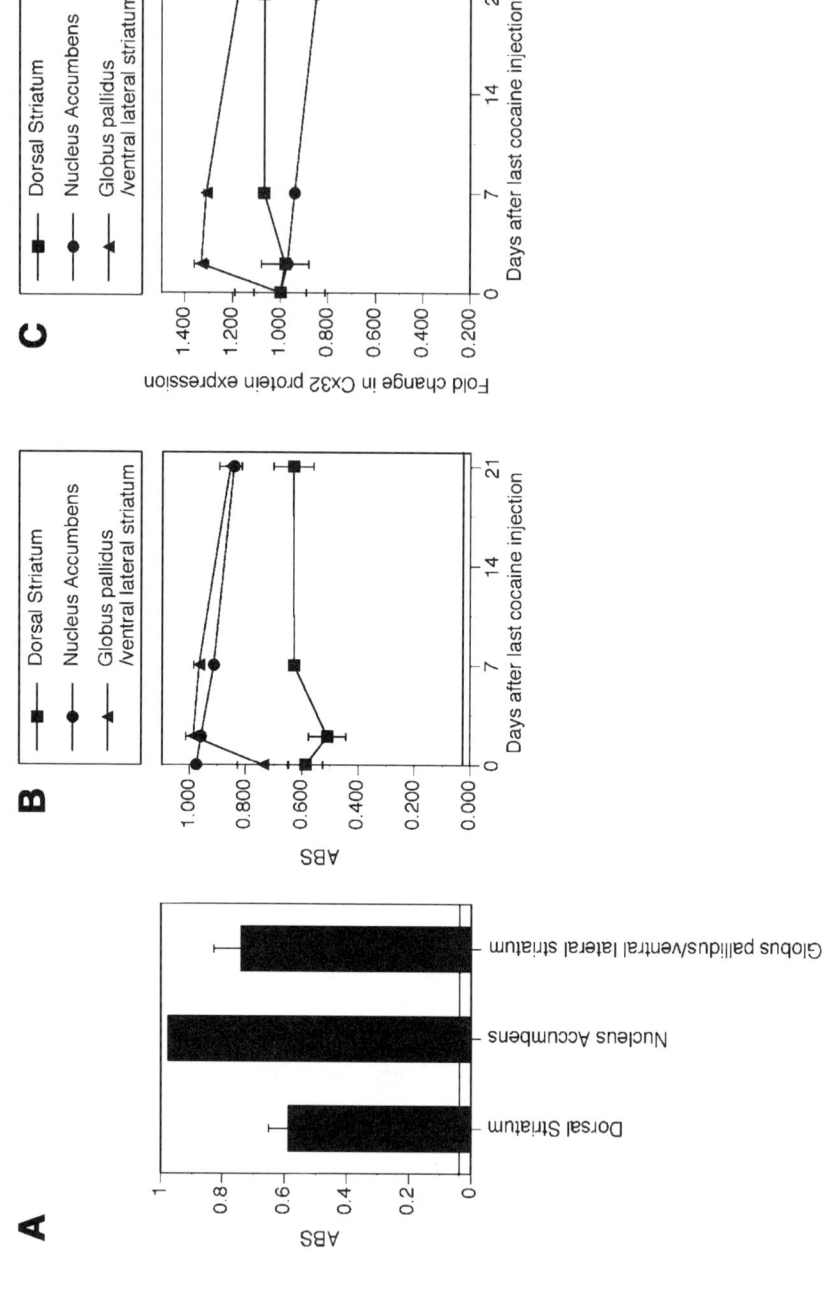

change can be replaced with precise quantitation of the protein concentration in each sample (*see* **Note 2**).

4. Notes

1. Alternative protocols for protein extraction include the use of Trizol Reagent (Invitrogen, Burlington, ON) permitting simultaneous extraction of RNA and protein from the same tissue sample and minimizing the number of animals required to perform both RNA and protein studies *(4)*. However, because Trizol-extracted protein is solubilized in 1% SDS, care should be taken to ensure that the final concentration of SDS does not exceed 0.5% when samples are diluted in coating buffer (**Subheading 3.3.2., step 1**). Stripping of capture antibody from the microtiter plates and loss of antibody–antigen complex can occur at elevated SDS concentrations.
2. The capture antibody can be either monoclonal or polyclonal and is diluted to a concentration between 0.5 and 10 µg/mL. As a rule, we have observed that 2.5 µg/mL is a convenient starting concentration and that sensitivity does not significantly increase when concentrations exceed 10 µg/mL. If the capture and detection antibodies have not previously been used for ELISA analysis, it may be necessary to establish the limits of sensitivity using serial dilutions of both reagents detecting a known positive control. Note that the sensitivity of the

---

Fig. 2. *(previous page)* Analysis and interpretation of antibody-sandwich ELISA results. **(A)** Distribution of Cx32 protein in the dorsal striatum, nucleus accumbens, and globus pallidus/ventral lateral striatum (minus nucleus accumbens) in drug-naive animals. Data represent the mean absorbance ± SEM of two independent experiments. Note that the expression levels of Cx32 vary under control conditions. The limit of sensitivity (i.e., the average of wells labeled 0 µg of protein in **Table 2**) is illustrated by the line at the bottom of the graph. Absorbances below this limit can not be distinguished from nonspecific binding (*see* Table 1). **(B)** Alterations in Cx32 expression following withdrawal from chronic cocaine self-administration. Data are expressed as the mean absorbance ± SEM of two independent experiments. The limit of sensitivity is indicated by the horizontal line at bottom of the graph. As presented, it is difficult to compared changes in Cx32 expression following cocaine withdrawal. **(C)** -Fold changes in Cx32 expression in the dorsal striatum, nucleus accumbens, and globus pallidus/ventral lateral striatum during cocaine withdrawal. Standardization permits a clearer comparison of trends in Cx32 expression at various time points after cocaine self-administration. Note that relative protein expression does not change in the dorsal striatum during cocaine withdrawal but that expression decreases over time to below that observed in drug-naive animals in the nucleus accumbens. These data replicate previously published results *(4)*. In the globus pallidus/ventral lateral striatum, a sharp increase in Cx32 expression is observed immediately after cocaine withdrawal followed by a gradual decrease to basal protein levels.

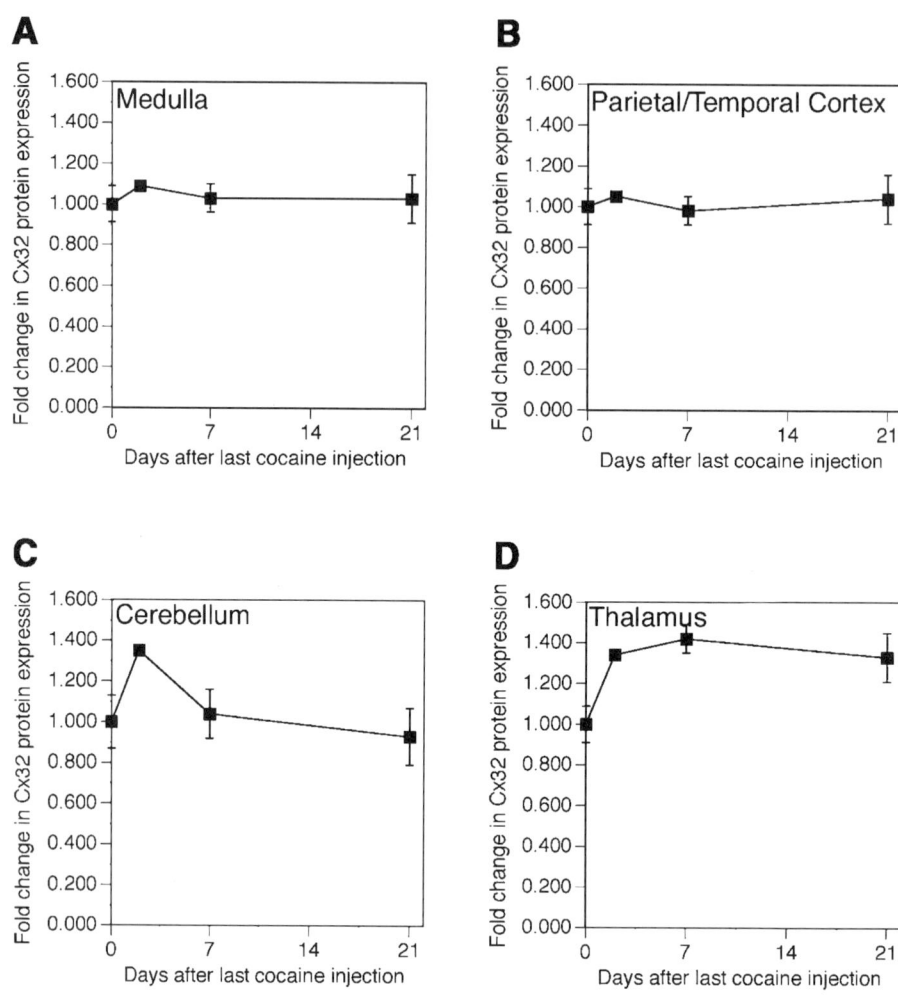

Fig. 3. Fold changes in Cx32 expression in the medulla, parietal/temporal cortex, cerebellum, and thalamus. (**A**) Medulla. (**B**) Parietal/temporal cortex. (**C**) Cerebellum. (**D**) Thalamus. Note that there is no change in Cx32 expression in the medulla or parietal/temporal cortex relative to drug-naive animals during cocaine withdrawal. A transient increase in expression is observed 2 d after cocaine withdrawal in the cerebellum while sustained increases in Cx32 expression are noted in thalamic nuclei 2–21 d after cocaine withdrawal.

ELISA protocol depends on use of high-affinity antibodies with minimal cross-reactivity to other proteins and reagents and it is worth the time to optimize these reagents.
3. If the absolute amounts of protein are to be quantified, then a standard curve with serial dilutions of known amounts of target protein is required. However, binding

efficiency does vary from plate to plate and we have found that (a) a standard curve is required on each microtiter plate for accuracy and (b) it is essential that all of the experimental conditions are kept constant (i.e., incubation times). The standard curve establishes the dynamic range of detection (i.e., the amount of target protein required to produce linear differences in substrate hydrolysis) and permits quantitative assessment of target protein concentration in experimental lysates. The concentration of target protein in mixed protein lysates must fall within this dynamic range for accurate quantitation.

4. In the most sensitive of assays, detection is achieved by incubation with an unconjugated detection antibody, a biotinylated secondary antibody, and tertiary reaction with enzyme-conjugated streptavidin. This method enhances detection of a single antigen–antibody interaction up to eightfold. If tertiary enhancement is required, the modifications to the protocol outlined in **Subheading 3.3.** should be made:

1–5. These steps remain unchanged.

6. Amplify the antibody–antigen–antibody complex signal with biotinylated anti-mouse IgG (1:200,000, Sigma) or appropriate secondary antibody. Add 50 µL per well. Cover with plastic adhesive or wrap and incubate for 1 h at 37°C.
7. Remove unbound antibody and wash plates as described in **steps 1** and **2**.
8. Detect this complex with extravidin peroxidase (2 µg/mL, Sigma). Add 50 µL per well. Cover with plastic adhesive or wrap and incubate for 1 h at 37°C.
9. Remove unbound antibody and wash plates as described in **steps 1** and **2**.
10. Add 75 µL per well of TMB substrate and incubate for 10–30 min at room temperature.
11. Stop reaction by addition of 25 µL of 1 M H_2SO_4. The solution will turn a yellow color.
12. Read the plate on a microplate reader at 450 nm against a reference wavelength of 650 nm.

References

1. Bruzzone, R., White, T. W., and Paul, D. L. (1996) Connections with connexins: the molecular basis of direct intercellular signaling. *Eur. J. Biochem.* **238,** 1–27.
2. Onn, S. P. and Grace, A. A. (1999) Alterations in electrophysiological activity and dye coupling of striatal spiny and aspiny neurons in dopamine-denervated rat striatum recorded in vivo. *Synapse* **33,** 1–15.
3. Onn, S. P. and Grace, A. A. (2000) Amphetamine withdrawal alters bistable states and cellular coupling in rat prefrontal cortex and nucleus accumbens neurons recorded in vivo. *J. Neurosci.* **20,** 2332–2345.
4. Bennett, S. A. L., Arnold, J. M., Chen, J., Stenger, J., Paul, D. L., and Roberts, D. C. S. (1999) Long-term changes in connexin32 gap junction protein and mRNA expression following cocaine self-administration in rats. *Eur. J. Neurosci.* **11,** 3329–3338.

20

Dopamine Receptor Binding and Quantitative Autoradiographic Study

Beth Levant

1. Introduction

The central nervous system (CNS) dopamine system plays an important role in mediating the reinforcing effects of drugs of abuse *(1)*. In addition, dopamine receptors have been the principal target of drugs employed in the treatment of neuropsychiatric disorders such as schizophrenia and Parkinson's disease. Until 1990, the dopamine receptor population in the brain and periphery was believed to consist of two subtypes, D_1 and D_2, which were distinguished by their pharmacology and coupling to signal transduction systems (for review *see* **ref. 2**). D_1 and D_2 receptors exhibit similar distributions in brain with the highest densities in the striatum (for review *see* **ref. 3**). A number of selective radioligands have been synthesized and extensively used to characterize the classical dopamine receptor subtypes. These radioligands include the D_1-selective ligands [^{3}H]SCH 23390 and [^{125}I]SCH 23982. D_2 receptor-selective ligands include the antagonists [^{3}H]spiperone, [^{3}H]YM 09151-2, and [^{125}I]iodosulpiride and the agonists [^{3}H]propylnorapomorphine and [^{3}H]quinpirole.

Molecular cloning efforts have revealed additional dopamine receptor subtypes: the D_3 and D_4 subtypes, which have homology with D_2; and the D_5 which has homology with D_1 *(4–6)*. These novel subtypes are expressed in roughly 10- to 100-fold lower density than the classical dopamine receptor subtypes. D_3 sites are preferentially localized in limbic brain regions such as the nucleus accumbens and islands of Calleja. D_4 sites are of greatest abundance in regions such as the frontal cortex. The distribution of D_5 receptors, as assessed by D_5 mRNA, appears to be similar to that of the D_1 receptors. Because of the

From: *Methods in Molecular Medicine, vol. 79: Drugs of Abuse: Neurological Reviews and Protocols*
Edited by: J. Q. Wang © Humana Press Inc., Totowa, NJ

pharmacological similarity of these sites with the D_1 and D_2 receptors, it is often not possible to distinguish between these sites in a heterogeneous tissue with currently available radioligands. As such the site labeled by any given radioligand in a such a tissue (i.e., brain) will, of course, be dependent on the relative affinities of that ligand for the related subtypes and the population of sites present in the tissue employed. Thus, although interactions with specific D_1- or D_2-like subtypes may be inferred, it is likely that sites labeled by these radioligands in brain tissue represent a mixture of the related subtypes. Several radioligands that appear to exhibit selectivity for the D_3 receptor have been synthesized including [^{3}H]7-hydroxy-2-(di-*n*-propylamino)tetralin ([^{3}H]7-OH-DPAT) and [^{3}H]PD 128907. These compounds label similar populations of sites in rat brain which are generally consistent with that reported for the D_3 receptor *(7–9)*.

This chapter outlines protocols for radioligand binding and autoradiographic assays for different subtypes of dopamine receptors in brain. The scope of the chapter is limited to the widely used, well established protocols for binding and autoradiographic assays that employ commercially available, tritiated and iodinated ligands. Radioligand binding assays, or "grind and bind" studies, can be used to measure receptor density and affinity in homogenized tissue samples. This method is rapid and highly quantitative; however, localization of binding sites is limited by the accuracy of dissection and any variation in binding within the brain region of interest will be obscured by homogenization. Receptor autoradiography, on the other hand, has a high degree of anatomical resolution. However, even when performed in a quantitative manner, the method is only semiquantitative compared to radioligand binding. Accordingly, receptor autoradiography is most suited to determining receptor localization and relative densities in various brain areas or between different treatment groups. In addition, receptor autoradiography can require extended periods of time for the exposure of autoradiograms.

Protocols are outlined for saturation analysis of radioligand binding for D_1-like receptors using the antagonist ligand [^{3}H]SCH 23390 and D_2-like receptors using the antagonist ligand [^{3}H]spiperone (*see* **Note 1**). A protocol for the putatively selective labeling of D_3 sites using [^{3}H]PD 128907 is also presented. For autoradiographic studies, protocols are outlined for D_1-like receptors using the antagonist ligand [^{3}H]SCH 23390, D_2-like receptors using the antagonist ligand [^{125}I]iodosulpiride, and D_3 sites using agonist [^{3}H]PD 128907. Procedures are generally similar for the respective assays; however, specific details, such as buffer composition, radioligand concentration, or incubation time may vary for each ligand. Ligand-specific details for each assay are presented at the relevant step of the protocol.

2. Materials
2.1. Radioligand Binding Assays
2.1.1. Membrane Homogenate Preparation
1. Brain tissue: Fresh or fresh-frozen and stored at −70°C. Caudate–putamen is most commonly used for D_1-like and D_2-like receptors, ventral striatum (nucleus accumbens and olfactory tubercles) for D_3.
2. Homogenizer (e.g., Brinkman Polytron, Tekmar Tissuemizer, or PRO).
3. Centrifuge tubes.
4. High-speed, refrigerated centrifuge.
5. Assay buffers. Store at 4°C for up to 2 wk.
    a. [^{3}H]SCH 23390: 50 mM Tris-HCl, 120 mM NaCl, 5 mM KCl, 2 mM CaCl$_2$, 1 mM MgCl$_2$, pH 7.4 at 23°C.
    b. [^{3}H]Spiperone: 50 mM Tris-HCl, 5 mM KCl, 2 mM CaCl$_2$, 1 mM MgCl$_2$, pH 7.4, at 23°C.
    c. [^{3}H]PD 128907: 50 mM Tris-HCl, 1 mM EDTA, pH 7.4, at 23° C.

2.1.2. Binding Assay
1. Radioligands:
    a. D_1-like receptors: [^{3}H]SCH 23390 (Amersham) (*see* **Note 2**).
    b. D_2-like receptors: [^{3}H]Spiperone (Amersham) (*see* **Note 3**).
    c. D_3 receptors: [^{3}H]PD 128907 (Amersham) (*see* **Note 4**).
2. Assay buffers. Store at 4°C for up to 2 wk.
    a. [^{3}H]SCH 23390: 50 mM Tris-HCl, 120 mM NaCl, 5 mM KCl, 2 mM CaCl$_2$, 1 mM MgCl$_2$, pH 7.4, at 23°C.
    b. [^{3}H]Spiperone: 50 mM Tris-HCl, 5 mM KCl, 2 mM CaCl$_2$, 1 mM MgCl$_2$, pH 7.4, at 23°C.
    c. [^{3}H]PD 128907: 50 mM Tris-HCl, 1 mM EDTA, pH 7.4, at 23°C.
3. Competing ligands (Sigma). A 1 mM stock may be prepared in ethanol and stored at −20°C.
    a. [^{3}H]SCH 23390: (+)-Butaclamol.
    b. [^{3}H]Spiperone: (+)-Butaclamol.
    c. [^{3}H]PD 128907: Spiperone.
4. Membrane homogenate.
5. 12 × 75 Polystyrene culture tubes.
6. Cell harvester (*see* **Note 5**).
7. Whatman GF/B filters.
8. Wash buffer: 50 mM Tris-HCl, pH 7.4, at 23°C. Store at 4°C.
9. Scintillation cocktail suitable for use with filters.
10. Scintillation vials.
11. Scintillation counter.
12. Protein assay kit.
13. Radioligand binding analysis software.

2.2. Receptor Autoradiography

2.2.1. Preparation of Slide Mounted Sections

1. Cryostat.
2. Chrome-alum-coated, precleaned, glass microscope slides (see **Note 6**).
3. Fresh-frozen rat brains. Stored at –70° C (see **Note 7**).

2.2.2. Labeling of Sections

1. Radioligands:
   a. D_1-like receptors: [^{3}H]SCH 23390 (Amersham).
   b. D_2-like receptors: [^{125}I]Iodosulpiride (Amersham) (see **Note 8**).
   c. D_3 receptors: [^{3}H]PD 128907 (Amersham).
2. Assay buffer. Store at 4°C for up to 2 wk.
   a. [^{3}H]SCH 23390: 50 mM Tris-HCl, 120 mM NaCl, 5 mM KCl, 2 mM CaCl$_2$, 1 mM MgCl$_2$, pH 7.4, at 23°C.
   b. [^{125}I]Iodosulpiride: 50 mM Tris-HCl, 120 mM NaCl, 5 mM KCl, 2 mM CaCl$_2$, 1 mM MgCl2, pH 7.4, at 23°C.
   c. [^{3}H]PD 128907: 50 mM Tris-HCl, 1 mM EDTA, pH 7.4, at 23°C.
3. Competing ligands (Sigma). A 1 mM stock may be prepared in ethanol and stored at –20°C.
   a. [^{3}H]SCH 23390: (+)-Butaclamol.
   b. [^{125}I]Iodosulpiride: (+)-Butaclamol.
   c. [^{3}H]PD 128907: Spiperone.
4. Coplin jars or plastic slide mailers (Thomas Scientific)—depending on the number of slides to be run.
5. ^{3}H-Hyperfilm (Amersham) (see **Note 9**).
6. X-ray cassettes.
7. [^{3}H] or [^{125}I]methylmethacrylate autoradiography standards (Amersham).
8. Dark room.
9. Developing trays.
10. Developer and fixer suitable for X-ray film such as Kodak D-19 developer and rapid fixer.
11. Plastic sheet protectors.
12. Image analysis system.

3. Methods

The methods described in the following subheadings outline procedures for assay of dopamine receptors in brain tissue by (1) radioligand binding and (2) quantitative receptor autoradiography. Procedures for preparation of membrane homogenates and slide-mounted brain sections for use in the respective assays are also presented.

3.1. Radioligand Binding Assays

A protocol is outlined for the determination of receptor density and affinity using a 10-point saturation analysis design. If analysis of a larger or smaller number of data points is desired (i.e., triplicates, multiple samples, etc.), the quantity of each reagent required should be scaled up or down accordingly.

3.1.1. Preparation of Membrane Homogenates

These assays typically use brain regions in which the receptor of interest is expressed in greatest abundance. Accordingly, caudate–putamen is most commonly used for assay of D_1-like and D_2-like receptors, whereas ventral striatum (nucleus accumbens and olfactory tubercle) is used for D_3 receptors.

1. Dissect a sufficient quantity of brain tissue for assay to be performed (*see* **Note 10**). A 10-point saturation analysis with [^{3}H]SCH 23390 or [^{3}H]spiperone will require roughly 35 mg of caudate tissue. A 10-point saturation analysis with [^{3}H]PD 128907 will require roughly 220 mg of ventral striatal tissue.
2. Homogenize tissue in 20 volumes (w/v) of the appropriate assay buffer for 10 s using a homogenizer (*see* **Note 11**).
3. Centrifuge the homogenate for 15 min at 48,000g at 4° C, then discard supernatant and resuspend the pellet in 20 volumes (w/v) of assay buffer using the homogenizer.
4. Centrifuge the homogenate for 15 min at 48,000g at 4° C, then discard the supernatant.
5. Final pellets may be used immediately or stored at –70°C for up to 1 mo for future use.

3.1.2. Assay Protocol

1. Using a homogenizer, resuspend the final pellet from membrane preparation in assay buffer to yield the desired concentration. Keep the membrane homgenates on ice while preparing other reagents.
   [^{3}H]SCH 23390: 3 mg original wet weight/ mL.
   [^{3}H]Spiperone: 3 mg original wet weight/ mL.
   [^{3}H]PD 128907: 20 mg original wet weight/ mL.
2. Prepare working solutions of radioligand. The assay is designed to test a range of concentrations spanning from roughly 10-fold above to 10-fold below the K_D for the respective radioligand. Working solutions are prepared at five times the desired final concentration to yield the desired final concentration after the addition of all reagents. For each assay, prepare 2 mL of radioligand in the respective assay buffer at the concentration shown below.
   [^{3}H]SCH 23390: 15 nM.
   [^{3}H]spiperone: 5 nM.
   [^{3}H]PD 128907: 15 nM.

Table 1
Dilution Scheme to Produce 10 Concentrations of Radioligand Over a 100-fold Range

Dilution number	Relative concentration	Preparation
1	100	Prepare 2 mL of radioligand in assay buffer at the highest concentration desired.
2	60	Mix 1.2 mL of dilution 1 with 0.8 mL of assay buffer.
3	36	Mix 1.2 mL of dilution 2 with 0.8 mL of assay buffer.
4	22	Mix 1.2 mL of dilution 3 with 0.8 mL of assay buffer.
5	13	Mix 1.2 mL of dilution 4 with 0.8 mL of assay buffer.
6	7.8	Mix 1.2 mL of dilution 5 with 0.8 mL of assay buffer.
7	4.7	Mix 1.2 mL of dilution 6 with 0.8 mL of assay buffer.
8	2.8	Mix 1.2 mL of dilution 7 with 0.8 mL of assay buffer.
9	1.7	Mix 1.2 mL of dilution 8 with 0.8 mL of assay buffer.
10	1	Mix 1.2 mL of dilution 9 with 0.8 mL of assay buffer.

This dilution scheme enables the preparation of 10 concentrations of radioligand in sufficient volume to allow for duplicate determination of total binding, nonspecific binding, and total counts added.

Serially dilute the radioligand (1.2 mL of radioligand + 800 µL of buffer) to produce 10 concentrations **(Table 1)**. Count duplicate aliquots of each dilution to allow determination of actual concentration.

3. Prepare 3 mL of the appropriate competing ligand (5 µM) in assay buffer.
    [^{3}H]SCH 23390: (+)-Butaclamol.
    [^{3}H]Spiperone: (+)-Butaclamol.
    [^{3}H]PD 128907: Spiperone.
4. To perform the binding assay, assemble the following reactions in duplicate 12 × 75 mm polystyrene culture tubes **(Fig. 1)**:
    Total binding tubes:
        50 µL of assay buffer
        100 µL of radioligand (concentration 1 – 10)
        100 µL of assay buffer
    Nonspecific binding tubes:
        50 µL of assay buffer
        100 µL of radioligand (concentration 1 – 10)
        100 µL of competing ligand
    Then, add 250 µL of membrane homogenate to all tubes. Gently vortex-mix all assay tubes. Incubate at 23°C to attain steady-state binding.
        [^{3}H]SCH 23390: 90 min.
        [^{3}H]Spiperone: 90 min.
        [^{3}H]PD 128907: 3 h.

Dopamine Receptors

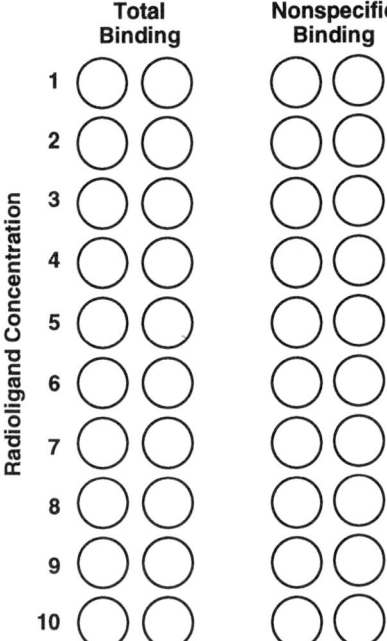

Fig. 1. Schematic of the layout for a 10-point saturation radioligand binding experiment.

5. Terminate reaction by rapid filtration over Whatman GF/B filters using a cell harvester. Wash filters three times with 3 mL of ice-cold wash buffer.
6. Place filters in scintillation vials, add cocktail, and count in a scintillation counter (*see* **Note 12**).
7. Determine protein concentration of membrane homogenate.

3.1.3. Analysis of Radioligand Binding Data (see **Note 13**)

The first step in the analysis of binding data is to determine the mean of the duplicate values obtained from the scintillation counter. If the scintillation counter expresses these data as counts per minute (cpm), convert the data to disintegrations per minute (dpm) using the following equation:

$$\text{cpm/counter efficiency for } [^3H] = \text{dpm}$$

The next step is to determine the exact concentration of each concentration of radioligand used using the following equation:

$$\frac{\text{dpm}}{2.22\text{e}12 \text{ dpm/Ci} \times \text{specific activity (Ci/mmol)} \times \text{volume counted (mL)}} = \text{Conc. } (M)$$

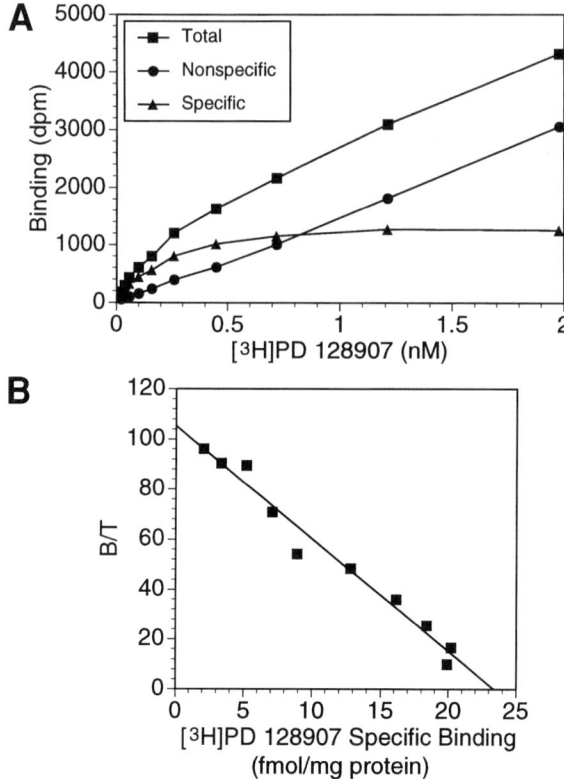

Fig. 2. Representative data from a 10-point saturation radioligand binding experiment (see **Table 1**). Data shown are for [^{3}H]PD 128907 binding in rat ventral striatal membranes. Saturation data, expressed in dpm, is shown in **(A)**. A Rosenthal (Scatchard) plot of the same data is shown in **(B)**. Results of this experiment, as analyzed by LIGAND, were K_D = 0.21 nM, B_{max} = 23 fmol/mg of protein.

Data can then be analyzed using any of a variety of computer programs for analysis of radioligand binding data such as LIGAND, EBDA, etc. Alternatively, data may be transformed an analyzed manually.

For manual analysis, determine the amount of specific binding at each concentration (see **Note 14**):

Specific binding = total binding − nonspecific binding

Binding data at each concentration of radioligand can then be plotted to generate a saturation curve (**Fig. 2A**).

Data are then transformed and plotted in a Rosenthal (Scatchard) plot (**Fig. 2B**). First, convert specific binding counts into moles bound using the following equation:

$$\frac{\text{dpm}}{2.22\text{e}12 \text{ dpm/Ci} \times \text{specific activity (Ci/mmol)} \times 1000 \text{ mmol/mol}} = \text{moles}$$

This value is divided by the quantity of protein per tube to yield the concentration of specific binding expressed as moles per milligram of protein. Because this concentration will be quite low, data may be converted into more convenient units such as picomoles or femtomoles. This value is then divided by the concentration of radioligand to yield specific binding/total (B/T) (*see* **Note 15**). Finally, specific binding (moles permilligram of protein) is plotted vs B/T with specific binding on the x-axis and B/T on the y-axis. The density of binding sites (B_{max}) is indicated at the x-intercept. The affinity, or K_D value, is equal to -1/slope.

Representative analysis of radioligand binding is shown in **Table 2** and **Fig. 2**.

3.2. Receptor Autoradiography

3.2.1. Preparation of Slide-Mounted Brain Sections

Using a cryostat, cut 10–20-μm sections containing the brain areas of interest (*see* **Note 16**). Regardless of the section thickness chosen, it is essential that the sections be of uniform thickness. Thaw-mount sections onto gelatin-chrome alum-coated glass microscope slides. Mount one or more sections near the bottom edge of the slide, away from the frosted end (**Fig. 3**). Because you will need a total binding and a nonspecific binding slide for each data point, two slides with adjacent sections should be collected for each level of the brain to be analyzed. Allow the sections to dry at room temperature. Slide-mounted sections can then be stored at −70°C for months.

3.2.2. Receptor Autoradiography

1. Remove slides to be assayed from the freezer. Remember that you will need pairs of slides with adjacent sections to allow you to determine total binding and a nonspecific binding. Allow slides to dry thoroughly at room temperature (at least 15 min) (*see* **Note 17**). While slides are drying, designate and label slides to indicate which slide of each pair will be used for total and nonspecific binding using a pencil or solvent-proof marking pen. It is often helpful to include several extra pairs of slides in the assay to use to generate "test films" for evaluating the adequacy of exposure of the autoradiograms.
2. Prepare radioligand solutions. First determine how much radioligand solution you will need. This depends on the size and number of incubation vessels you plan to use (i.e., Coplin jars or slide mailers), which, in turn, depends on the number slides to be incubated. It is generally best to determine empirically the volume required by measuring the amount of solution required to completely cover the your slide-mounted sections in an incubation vessel filled with slides

Table 2
Representative Analysis of Data from a 10-Point Saturation Radioligand Binding Experiment

Radioligand concentrations

	A Mean dpm	B Working conc. (M)	C Final conc. (nM)
1	254585	9.80E-09	1.96
2	156335	6.02E-09	1.20
3	92810	3.57E-09	0.71
4	57685	2.22E-09	0.44
5	34080	1.31E-09	0.26
6	21235	8.18E-10	0.16
7	12935	4.98E-10	0.10
8	7520	2.90E-10	0.06
9	4830	1.86E-10	0.04
10	2800	1.08E-10	0.02

Binding

Protein conc.: 245 µg/tube

Radioligand conc. (nM)	A Total (dpm)	A Nonspecific (dpm)	D Specific (dpm)	E Specific (mol)	F Specific (fmol/mg of protein)	G B/T
1.96	4317	3060	1257	4.84E-15	19.75	10.08
1.20	3100	1824	1276	4.91E-15	20.05	16.66
0.71	2173	1012	1161	4.47E-15	18.24	25.53
0.44	1635	615	1020	3.93E-15	16.03	36.09
0.26	1207	397	810	3.12E-15	12.73	48.51
0.16	805	241	564	2.17E-15	8.86	54.20
0.10	607	158	449	1.73E-15	7.06	70.84
0.06	434	104	330	1.27E-15	5.19	89.56
0.04	296	82	214	8.24E-16	3.36	90.42
0.02	190	58	132	5.08E-16	2.07	96.21

Data are those resulting from saturation analysis using [^{3}H]PD 128907 (specific activity = 117 Ci/mmol) in rat ventral striatal membranes (protein concentration 245 µg/tube). Counts from the working solutions of radioligand and total and nonspecific binding are obtained from the scintillation counter and represent the mean of duplicate determinations. All other values are calculated from these data. Graphical representation of this data is shown in **Fig. 2**.

A: Average of duplicates from scintillation counter expressed in dpm.

B: Working conc. (M) = $\dfrac{\text{dpm}}{2.22\text{e}12 \text{ dpm/Ci} * 117 \text{ Ci/mmol} * 0.1 \text{ mL}}$

C: Final conc. (nM) = $\dfrac{\text{Working conc. (M)} * \text{e}9}{5}$

D: Specific binding (dpm) = Total binding (dpm) − nonspecific binding (dpm).

E: Specific binding (mol) = $\dfrac{\text{Specific binding (dpm)}}{2.22\text{e}12 \text{ dpm/Ci} * 117\text{Ci/mmol} * 1\text{e-3 mmol/mol}}$

F: Specific binding (fmol/mg of protein) = $\dfrac{\text{Specific binding (mol)} * 1\text{e}15 \text{ fmol/mol}}{0.245 \text{ mg of protein/tube}}$

G: B/T = $\dfrac{\text{Specific binding (fmol/mg of protein)}}{\text{Radioligand concentration (nM)}}$

Dopamine Receptors

Fig. 3. Incubation of slide-mounted brain sections in radioligand solution. Note the placement of sections at the lower edge of the slide and that the radioligand solution is of sufficient depth to completely cover the sections.

(**Fig. 3**). Remember that you will need enough radioligand solution for incubating both total binding and nonspecific binding slides. For example, if you plan to assay 14 pairs of slides and you need 30 mL of radioligand solution to cover your sections completely, you will need to make 60 mL of radioligand solution. This will allow enough solution for two Coplin jars. One, containing only radioligand, will generate total binding sections. Competing ligand will be added to the second Coplin jar to generate nonspecific binding sections.

Prepare the radioligand in the respective assay buffer. The concentrations of radioligand used in these are designed to minimize exposure time and are therefore near-saturating for the tritiated ligands.

[^{3}H]SCH 23390: 0.7 nM.

[^{125}I]Iodosulpiride: 0.3 nM.

[^{3}H]PD 128907: 0.7 nM.

Transfer half of the radioligand solution into the Coplin jar to be used for total binding. Then add competing ligand to the remaining radioligand solution to produce a final concentration of 1 μM.

[^{3}H]SCH 23390: (+)-Butaclamol.

[^{125}I]Iodosulpiride: (+)-butaclamol.

[^{3}H]PD 128907: Spiperone.

Transfer this solution into the Coplin jar to be used for nonspecific binding.

3. Count an aliquot of the radioligand solution to allow determination of the exact concentration.
4. Place slides in the respective Coplin jars and incubate at room temperature for 2 h. This is sufficient time for the system to reach steady state.

    If large numbers of slides are to be assayed, multiple Coplin jars may be used and/or multiple batches of slides may be run. If multiple batches of slides are assayed, it is wise to count the radioligand binding solution between each batch to ascertain that the concentration of radioligand is not depleted after multiple incubations.
5. Terminate the binding reaction by dipping the slides in ice-cold assay buffer. This is followed by two wash incubations in ice-cold assay buffer followed by a dip in ice-cold dH_2O. Washes can be accomplished by either transferring slides from one wash vessel to the next or changing the buffer in the Coplin jar containing the slides. Wash times for the two wash incubations for each radioligand are as follows (*see* **Note 18**):

    [^{3}H]SCH 23390: 2 × 5 min.
    [^{125}I]Iodosulpiride: 2 × 5 min.
    [^{3}H]PD 128907: 2 × 2 min.
6. Blot slides and allow them to dry. Drying can be accomplished by standing slides in the dividers that come in boxes of 20-mL scintillation vials set on a piece of absorbent bench paper. Alternatively, slides can be aspirated using a Pasteur pipet attached to a vacuum source. Once dry, slides can be placed in slide boxes and allowed to dry thoroughly (at least overnight). Drying can be augmented by blowing slides with cool air using a fan, blower etc.
7. Once dry, arrange slides in X-ray cassettes. Include appropriate methylmethacrylate radioactivity standards on *each* film. In the dark room, apply ^{3}H-Hyperfilm to the sections with the emulsion (dull) side directly against the sections. Exposure of autoradiograms with each radioligand will take roughly the following amount of time:

    [^{3}H]SCH 23390: 3 wk.
    [^{125}I]Iodosulpiride: 48 h.
    [^{3}H]PD 128907: 12 wk.

    This time may vary depending on variations in specific activity, actual radioligand concentration, tissue handling, experimental treatment, brain region to be examined, and so forth. Accordingly, it is often useful to also expose several "test films," which expose a pair of slides and can be developed at various times, to determine the adequacy of exposure before developing the films with the actual experimental sections. It is equally important not to overexpose the films (*see* **Note 19**).
8. When exposure is adequate, develop films according to the manufacturer's instructions. Note that because the emulsion on ^{3}H-Hyperfilm has no coating, the film must be developed by hand. It will be damaged by automated X-ray film processing machines. It also scratches very easily. Therefore, it is essential that it be developed with the emulsion side up. Care should also be exercised

when handling the film. After drying, store films in plastic sheet protectors to prevent scratching.

9. Capture images using a video-based densitometry image analysis system such as NIH Image, MCID, and so forth. To quantify autoradiograms, use the radioactivity standards to generate a standard curve to describe the relationship between optical density (OD) and radioactivity (using the values supplied by the manufacturer for nanoCuries per milligram of tissue equivalent). Images from each film can then be calibrated according to the standards included on that film. Sample the brain regions of interest from the total binding sections and the adjacent sections used for nonspecific binding. Specific binding in that brain area can then be calculated using the following equation:

$$\text{Total binding} - \text{nonspecific binding} = \text{specific binding}$$

The density of ligand binding can then be calculated:

$$\frac{\text{Specific binding (nCi/mg tissue equivalent)}}{\text{Specific activity (Ci/mmol)}} = \text{Specific binding (pmol/mg tissue equiv.)}$$

4. Notes

1. Saturation analysis represents only one type of radioligand binding experiment. It is used to determine receptor affinity and density. For detailed descriptions of a variety of types of receptor binding experiments see Bylund and Yamamura *(10)* and Levant *(11)*.

2. Although [^{3}H]SCH 23390 exhibits high selectivity for D_1 receptors over D_2, this ligand also possesses some affinity for the 5-hydroxytryptamine$_2$ (5-HT$_2$) receptor. Because caudate expresses very low density of 5-HT$_2$ receptors, the contribution of the sites is negligible in membranes prepared from that brain area. If other brain areas or tissues are used, 5-HT$_2$ receptors may be blocked in the presence of 40 µM ketanserin.

3. Although the [^{3}H]spiperone binding is the most widely used assay for the D_2-like sites, this assay is somewhat limited by the fact that it utilizes a radioligand with high affinity but low specific activity. In striatal membranes, the ligand also exhibits relatively high nonspecific binding. As such, assays using [^{3}H]spiperone must use relatively low concentrations of membrane homogenate to avoid ligand depletion. As a result, the magnitude of specific binding in terms of dpm obtained with [^{3}H]spiperone is striatal membranes is relatively low compared to many other radioligand binding assays.

   It should also be noted that while [^{3}H]spiperone exhibits high selectivity for D_2 receptors over D_1, this ligand also possesses relatively high affinity for the 5-HT$_2$ receptor. As striatum expresses very low density of 5-HT$_2$ receptors, the contribution of the sites is negligible in membranes prepared from that brain area. If other brain areas or tissues are used, 5-HT$_2$ receptors may be blocked in the presence of 40 µM ketanserin. Interaction of [^{3}H]spiperone with an acceptor site, the spirodecanone site, has also been reported *(12)*.

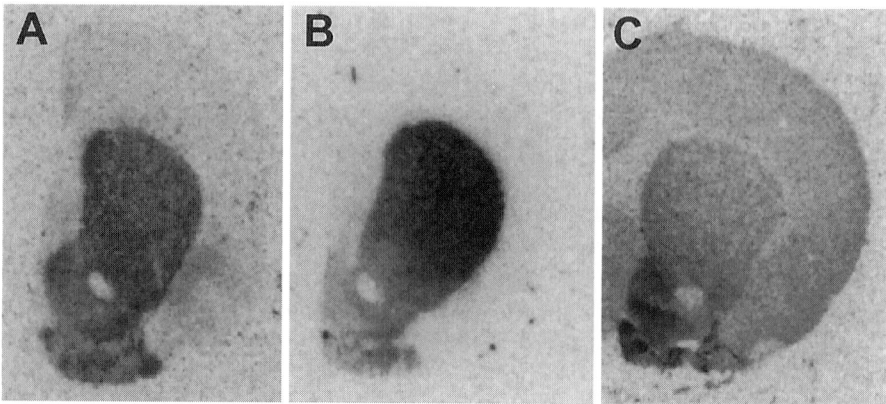

Fig. 4. Representative autoradiograms. Sections shown represent (A) [^{3}H]SCH 23390 binding, (B) [^{125}I]iodosulpiride binding, and (C) [^{3}H]128907 binding in coronal sections of rat brain.

4. Obtaining selective labeling of the D_3 site with this, or other, radioligands, appears to be dependent on the use of assay conditions that disfavor agonist binding at the D_2 site. The greatest D_2/D_3 selectivity for these ligands has been obtained in the absence of Mg^{2+} and the presence of EDTA *(7,9)* in concordance with previous studies indicating that the high affinity agonist state of D_2-like receptors is not favored in the absence of Mg^{2+} *(13)*. Because selective visualization of the putative D_3 receptor appears to require in vitro assay conditions that disfavor D_2 binding, interpretation of these results must consider that these conditions may also affect the binding or functional properties of the D_3 site.
5. Other types of filtering apparatus may be used. However, for best results, it is desirable that all binding reactions be subjected to uniform filtering conditions. Accordingly, a cell harvester produces the most satisfactory results.
6. Buy precleaned slides or wash slides in soap and water followed by an ethanol dip to remove all dust and oil. Place in slide racks. To sub (gelatin-coat) slides, prepare a subbing solution containing 5 g of porcine gelatin (300 bloom) and 0.5 g of Chrome-Alum (chromium potassium sulfate) in 1 L of dH_2O. The actual amount of subbing solution required will depend on the size of the histology dishes to be used for treating the slides. Heat while stirring until gelatin is dissolved. Do not boil. Filter and allow to cool. Dip slides into subbing solution several times to ensure complete coating. Drain excess. Allow coating to dry. Protect slides from dust. Store at room temperature. Alternatively, charged slides (Fisher Plus) or poly-L-lysine-coated slides may be used.
7. After the animal is euthanized, remove the brain carefully. To maintain the shape and structural integrity of the brain, snap-freeze the brain by swirling it in an isopentane (2-methylbutane), dry-ice bath. When the brain is frozen, it will be opaque white in appearance. Some cracking of the brain along the longitudinal

fissure may occur if the isopentane is too cold. If severe, the brain may split in half but will still be usable. Such cracking can be avoided with practice and experience. Liquid nitrogen is not recommended because it will tend to cause the brain to shatter.

8. [^{125}I]Iodosulpiride produces selective labeling of D_2-like receptors *(14)*. Iodinated ligands have the advantage of producing relatively high-resolution autoradiograms in a short period of time. If a tritiated ligand is desired, 0.7 nM [^{3}H]spiperone may be used in the presence of 40 μM ketanserin *(15)*. The agonist [^{3}H]quinpirole is also useful for autoradiographic localization of D_2-like sites *(16)*.

9. Although some films, such as Kodak Biomax MS, are sensitive to tritium, in our experience, ^{3}H-Hyperfilm is the only film that produces satisfactory autoradiograms with low-energy β-emitters for use in quantitative analysis of receptor binding in specific brain areas.

10. Brain regions may be identified using any of a variety of brain atlases such as the rat brain atlas of Paxinos and Watson *(17)*.

11. Using a Polytron, homogenization at a setting of 6 out of 10 is sufficient.

12. Before counting, allow sufficient time for the cocktail to saturate the filters and for the mixture of radiolabeled membranes and scintillation cocktail to reach steady state. The amount of time required will depend on the cocktail used and can be determined experimentally.

13. In striatal membranes, saturation analysis with [^{3}H]SCH 23390 should yield a K_D value of 0.1–0.5 nM and a B_{max} value of 700–1000 fmol/mg of protein *(18–21)*. K_D values for [^{3}H]spiperone in striatal membranes are most commonly reported to be 0.05–0.3 nM *(21)*. A B_{max} value of approx 300–500 fmol/mg of protein should be anticipated *(22,23)*. K_D values for [^{3}H]PD 128907 in ventral striatal membranes are approx 0.3 nM. A B_{max} value of approx 20–40 fmol/mg of protein should be anticipated *(24)*.

14. Computer packages for analysis of radioligand binding data provide unbiased analysis of untransformed data. Accordingly, use of such a package is highly recommended. However, it is often useful to analyze the data manually as well, thus allowing for verification of the computer output.

15. Binding in a saturation analysis is most appropriately expressed as a function of free radioligand. However, because binding in these assays at the K_D represents <10% of total counts added, depletion of free radioligand is negligible. Hence, for simplicity of calculation, the use of total radioligand, rather than free radioligand will generate a reasonable estimate of K_D and B_{max} values in manual calculations.

16. Temperatures for cutting brain tissue will vary depending on the instrument used. Typically, temperatures in the range of –11 to –20°C work well. Key points for successful cutting of brain tissue include having the tissue at the proper temperature. This will require moving the brain from storage at –70°C to the cryostat about 30 min before cutting. A sharp microtome knife is essential. Proper anti-roll plate adjustment is also important. Sections may be collected on either chilled or room temperature slide as preferred by the individual.

17. It is critical that the sections dry thoroughly or pieces of the sections will come off the slides during incubation.
18. Wash times were determined experimentally to maximize specific binding and minimize nonspecific binding.
19. The greater the number of counts bound and with high-energy isotopes, the greater the risk of overexposure. Accordingly, particular care should be taken with [^{125}I]iodosulpiride to avoid overexposure. On the other risk of overexposure is relatively low with tritiated ligands, particularly [^{3}H]PD 128907. As such, one might err on extending the duration of exposure to ensure that sufficient signal is generated.

References

1. Fibiger, H. C. and Phillips, A. G. (1988) Mesocorticolimbic dopamine systems and reward. *Ann. NY Acad. Sci.* **537,** 206–215.
2. Seeman, P. and Grigoriadis, D. (1987) Dopamine receptors in brain and periphery. *Neurochem. Int.* **10,** 1–25.
3. Levant, B. (1996) Distribution of dopamine receptor subtypes in the CNS, in *CNS Neurotransmitters and Neuromodulators. Dopamine* (Stone, T. W., ed.), CRC Press, Boca Raton, FL, pp. 77–87.
4. Sokoloff, P., Giros, B., Martres, M. P., Bouthenet, M. L., and Schwartz, J. C. (1990) Molecular cloning and characterization of a novel dopamine receptor (D3) as a target for neuroleptics. *Nature* **347,** 146–151.
5. Sunahara, R. K., Guan, H.-C., O'Dowd, B. F., et al. (1991) Cloning of the gene for a human D_5 receptor with higher affinity for dopamine than D_1. *Nature* **350,** 614–619.
6. Van Tol, H. M. M., Bunzow, J. R., Guan, H.-C., et al. (1991) Cloning the gene for a human dopamine D_4 receptor with high affinity for the antipsychotic clozapine. *Nature* **350,** 610–614.
7. Akunne, H. C., Towers, P., Ellis, G. J., et al. (1995) Characterization of binding of [^{3}H]PD 128907, a selective dopamine D3 receptor agonist ligand, to CHO-K1 cells. *Life Sci* **57,** 1401–1410.
8. Burris, K. D., Filtz, T. M., Chumpradit, S., et al. (1994) Characterization of [^{125}I](*R*)-*trans*-7-hydroxy-2-[*N*-propyl-*N*-(3′-iodo-2′-propenyl)amino] tetralin binding to dopamine D_3 receptors in rat olfactory tubercle. *J. Pharmacol. Exp. Ther.* **268,** 935–942.
9. Lévesque, D., Diaz, J., Pilon, C., et al. (1992) Identification, characterization, and localization of the dopamine D_3 receptor in rat brain using 7-[^{3}H]hydroxy-*N,N*-di-*n*-propyl-2-aminotetralin. *Proc. Natl. Acad. Sci. USA* **89,** 8155–8159.
10. Bylund, D. B. and Yamamura, H. I. (1990) Methods for receptor binding, in *Methods in Neurotransmitter Receptor Analysis* (Yamamura, H., Enna, S., and Kuhar, M., eds.), Raven Press, New York, pp. 1–36.
11. Levant, B. (1998) Dopamine receptors, in *Current Protocols in Pharmacology* (Ferkany, J. and Enna, S. J., eds.), John Wiley & Sons, New York, pp. 1.6.1–1.6.16.

12. Howlett, D. R., Morris, H., and Nahorski, S. R. (1979) Anomalous properties of [^{3}H]-spiperone binding sites in various areas of the rat limbic system. *Mol. Pharmacol.* **15,** 506–514.
13. Sibley, D. R. and Creese, I. (1983) Regulation of ligand binding to pituitary D-2 dopaminergic receptors. *J. Biol. Chem.* **258,** 4957–4965.
14. Levant, B., Grigoriadis, D. E., and DeSouza, E. B. (1995) Relative affinities of dopaminergic drugs at D_2 and D_3 dopamine receptors. *Eur. J. Pharmacol.* **278,** 243–247.
15. Joyce, J. N. and Marshall, J. F. (1987) Quantitative autoradiography of dopamine D2 sites in rat caudate–putamen: localization to intrinsic neurons and not to neocortical afferents. *Neuroscience* **20,** 773–795.
16. Levant, B., Grigoriadis, D. E., and DeSouza, E. B. (1993) [^{3}H]Quinpirole binding to putative D_2 and D_3 dopamine receptors in rat brain and pituitary gland: a quantitative autoradiographic study. *J. Pharmacol. Exp. Ther.* **264,** 991–1001.
17. Paxinos, G. and Watson, C. (1986) *The Rat Brain in Stereotaxic Coordinates.* Academic Press, Sydney.
18. Billard, W., Ruperto, V., Crosby, G., Iorio, L. C., and Barnett, A. (1984) Characterization of the binding of ^{3}H-SCH 23390, a selective D-1 receptor antagonist ligand, in rat striatum. *Life Sci.* **35,** 1885–1893.
19. Briere, R., Diop, L., Gottberg, E., Grondin, L., and Reader, T. A. (1987) Stereospecific binding of a new benzazepine, [^{3}H]SCH23390, in cortex and neostriatum. *Can. J. Physiol. Pharmacol.* **65,** 1507–1511.
20. Kilpatrick, G. J., Jenner, P., and Marsden, C. D. (1986) [^{3}H]SCH 23390 identifies D-1 binding sites in rat striatum and other brain areas. *J. Pharm. Pharmacol.* **38,** 907–912.
21. Seeman, P. (1993) *Receptor Tables*, Vol. 2: *Drug Dissociation Constants for Neuroreceptors and Transporters*, SZ Research, Toronto.
22. Grigoriadis, D. E. and Seeman, P. (1985) Complete conversion of brain D_2 dopamine receptors from the high- to low-affinity state for dopamine agonists using sodium ions and guanine nucleotides. *J. Neurochem.* **44,** 1925–1935.
23. Richfield, E. K., Penney, J. B., and Young, A. B. (1989) Anatomical and affinity state comparisons between dopamine D_1 and D_2 receptors in the rat central nervous system. *Neuroscience* **30,** 767–777.
24. Bancroft, G. N., Morgan, K. A., Flietstra, R. J., and Levant, B. (1998) Binding of [^{3}H]PD 128907, a putatively selective ligand for the D_3 dopamine receptor, in rat brain: a receptor binding and quantitative autoradiographic study. *Neuropsychopharmacology* **18,** 305–316.

21

Analysis of DNA-Binding Activity in Neuronal Tissue with the Electrophoretic Mobility-Shift Assay

Christine L. Konradi

1. Introduction

DNA is bound by many proteins in a sequence specific manner. Electrophoretic mobility-shift assays (EMSAs), also known as gel shift assays, are designed to determine the amount and identity of proteins from a particular sample that bind to a specific DNA sequence, usually an enhancer element. The defined DNA sequence is presented as a radiolabeled, synthetic oligonucleotide to a protein extract, which contains DNA-binding proteins. These proteins can then bind to the oligonucleotide. The sample is electrophoresed through a nondenaturing acrylamide gel in a vertical electrophoresis system; the gel is dried and exposed to X-ray film. Whereas the oligonucleotide will move quickly through the gel, oligonucleotides that are bound by protein will move slower and thus be shifted. The more DNA-binding proteins bind to oligonucleotides, the stronger the radioactive signal of the shifted band.

Often, the analysis focuses on how particular treatments change protein binding to a specific oligonucleotide. It is assumed that the same protein binds to the native enhancer element and regulates the expression of genes in vivo. Specific proteins binding to the oligonucleotide can be characterized with antibodies that, when interacting with the protein, further shift the band. Specificity of the gel shift assay can be confirmed with synthetic proteins, and in studies where various amounts and types of unlabeled oligonucleotides are used to compete for binding to the labeled enhancer element.

2. Materials
2.1. Annealing of Oligonucleotides

1. Annealing buffer: 10 mM Tris-HCl, pH 7.5–8.0, 50 mM NaCl, and 1 mM EDTA.

2.2. Radiolabeling of the Enhancer

1. 100 mM Dithiothreitol (DTT) is prepared in distilled water and stored at –20°C.
2. NucTrap Push columns are available from Stratagene, and can be substituted with any other type of prepacked size-exclusion chromatography column that retains nucleotides and short oligonucleotides to separate them from larger pieces of DNA or RNA.
3. 1X STE buffer: 100 mM sodium chloride; 20 mM Tris-HCl, pH 7.6; and 10 mM EDTA in distilled water.

2.3. Preparation of Protein Extracts

1. Sonication buffer for quick and easy protocol (100 mL): 2 mL of 1 M N-(2-hydroxyethyl) piperazine-N'-(2-ethanesulfonic acid) (HEPES)–KOH buffer, pH 7.9 (20 mM), 25 mL of 100% glycerol (25%), 25 mL of 2 M KCl (0.5 M), 150 µL of 1 M MgCl$_2$ (1.5 mM), and 40 µL of 0.5 M sodium EDTA, pH 8.0 (0.2 mM). Bring to 100 mL with distilled water. Additives: Before use add to an aliquot of sonication buffer (e.g., 1 mL): 1 µL of 1 M DTT/mL (1 mM) and 1X mammalian protease inhibitor cocktail (commercially available from Calbiochem; cat. no. 539134) (*see* **Note 1**). If state of phosphorylation is important, add 1 µL of 1 M NaF/mL (1 mM), 1 µL of 5 µM okadaic acid/mL (5 nM), and 1 µL of 100 mM Na$_3$VO$_4$/mL (100 µM).
2. Buffer A for crude nuclear extract (100 mL): 1 mL of 1 M HEPES-KOH, pH 7.9 (10 mM), 150 µL of 1 M MgCl$_2$ (1.5 mM), 500 µL of 2 M KCl (10 mM), 200 µL of 0.5 M sodium EDTA, pH 8.0 (1 mM), and 25 mL of 100% glycerol (25%). Bring to 100 mL with distilled water. Additives: Before use add to an aliquot: 1 µL of 1 M DTT/mL (1 mM) and the 1X mammalian protease inhibitor cocktail (*see* **Note 2**).

2.4. Preparation of the Gel

1. 10X TBE buffer (Tris–borate–EDTA): 108 g of 0.89 M Tris base, 55 g of 0.89 M boric acid, and 7.4 g of 20 mM EDTA disodium salt in 1 L of distilled water; adjust pH to 8.3.
2. 10% Ammonium persulfate (APS) in distilled water: 1 g of APS in 10 mL of distilled water. Lasts approx 1 mo at 4°C.
3. 30% Acrylamide–*bis*-acrylamide solution in distilled water (37.5:1): 29.2 g of acrylamide and 0.8 g of *bis*-acrylamide in 100 mL of distilled water.

2.5. EMSA

1. 5X EMSA buffer: 50 mM HEPES-KOH, pH 7.9, 50% glycerol, and 0.5 mM EDTA (from 0.5 M sodium EDTA, pH 8.0) (*see* **Note 3**).

2. Poly (dI-dC): make stock of 1 μg/μL in TE (Pharmacia, cat. no. 27-7880-02). Heat to 70°C for 10 min, let slowly cool down to room temperature, and store at −20°C in aliquots.
3. 1 M DTT stock. Store at −20°C.
4. Phenylmethylsulfonyl fluoride (PMSF): 100 mM stock solution in ethanol at −20°C.
5. 1X Dye: 5% glycerol in distilled water, 0.5 mg/mL of bromophenol blue, and 0.5 mg/mL of xylene xyanol.

3. Methods

The EMSA method contains five parts: (1) the design and annealing of both DNA strands of the enhancer element, (2) the radiolabeling of the enhancer, (3) the preparation of the protein extract, (4) the preparation of the gel, and (5) the actual electrophoretic mobility shift assay (gel-shift assay).

3.1. Design and Annealing of Both DNA Strands of the Enhancer Element

3.1.1. Design of Oligonucleotides

For the protocol provided, dG overhangs are needed for the incorporation of radiolabeled dCTP; for example, dGdG + sense oligonucleotide, dGdG + antisense oligonucleotide (**Fig. 1**). The overhangs are filled in with radiolabeled dCTP. The entire enhancer can thus incorporate up to four labels.

3.1.2. Annealing of Oligonucleotides

Make oligonucleotides in oligosynthesizer or order from supplier.

1. Dry oligonucleotides down.
2. Resuspend in 200 μL of annealing buffer.
3. Centrifuge for 10 min to pellet impurities.
4. Take supernatant into a fresh tube.
5. Determine the optical density of the supernatant at 260 nm to obtain the concentration. The concentration of oligonucleotides in the sample can be calculated from the spectrophotometric reading at 260 nm in a quartz cuvette. An OD of 1.0 corresponds to approx 30 μg/mL of oligonucleotides (*see* **Note 4**).
6. Combine equal molar amounts of both oligonucleotides (the higher, the better).
7. Boil for 5 min.
8. Incubate at 42°C for 3 h.
9. Leave at room temperature for a couple of hours.
10. Determine the optical density at 260 nm to get concentration. Dilute to 100 ng/μL.
11. Freeze in aliquots at −20°C.

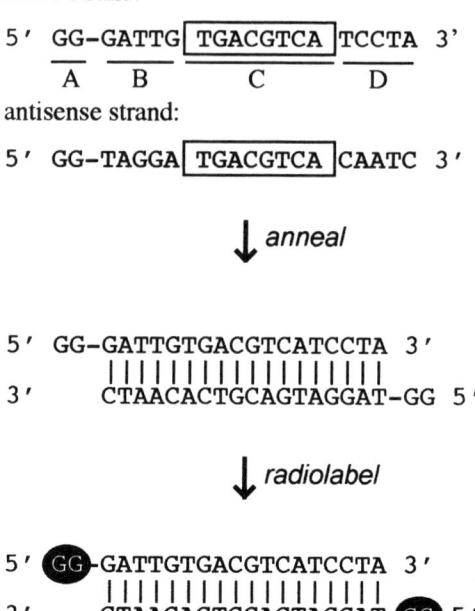

Fig. 1. Synthesis and preparation of the oligonucleotides for the labeled enhancer element ("probe"). The sense strand contains the following elements: A, Two deoxyguanosines which are needed for labeling the double stranded oligonucleotide; B, five to seven nucleotides immediately preceding the enhancer element (as present in the gene of interest); C, the actual enhancer sequence; and D, five to seven nucleotides 3′ of the gene of interest. The antisense strand needs to be complementary to the sense strand; in addition, it has two deoxyguanosines overhanging at the 5′ end. After annealing, the four deoxyguanosines serve as templates for the incorporation of the radiolabeled deoxycytidine triphosphates.

3.2. Radiolabeling of the Enhancer

1. Mix:

Double-stranded enhancer element (at 100 ng/μL)	1 μL
5X Reverse transcription buffer (provided with the reverse transcriptase by the company)	4 μL
[α-^{32}P]dCTP (800 Ci/mmol; 10 μCi/μL) (*see* **Note 5**)	5 μL
Distilled H$_2$O	7 μL
100 m*M* DTT	2 μL
Reverse transcriptase enzyme (superscript reverse transcriptase; Invitrogen/GIBCO)	1 μL
Final volume	20 μL

2. Incubate at 37°C for 30 min to 1 h.
3. Separate unincorporated nucleotides by size-exclusion chromatography (e.g., Nuctrap Push Columns, Stratagene).
4. Nuctrap push columns: Equilibrate the column with 80 μL of 1X STE buffer. Bring sample volume to 80 μL with STE buffer, and then load onto the resin. Elute sample with 80 μL of STE buffer, and bring eluate to a final concentration of 100 μL, which equals 1 ng of labeled oligonucleotide/μL. Measure radioactivity of 1 μL of eluate in a scintillation counter.
5. Freeze labeled, double-stranded oligonucleotides at −20°C in a Plexiglas container.

3.3. Preparation of Protein Extracts

3.3.1. Quick and Easy Whole Cell Protocol

This is a rapid protocol, which reduces the likelihood of post-harvest protein modifications that may affect the interaction between binding proteins and DNA. It is particularly recommended with samples in which the state of phosphorylation of proteins influences the binding properties. Phosphorylation of DNA-binding proteins is fairly common and should be taken into consideration when designing EMSA experiments. However, whole cell extracts are not appropriate for all DNA-binding proteins. Some DNA-binding proteins are regulated by compartmentalization, for example, they are kept outside the nucleus and translocated to the nucleus to activate transcription whenever necessary. In these cases it is recommended to use nuclear extracts for EMSA (*see* **Subheading 3.3.2.** and **Note 6**).

3.3.1.1. NEURONAL CULTURE

1. Scrape cells in 100 μL of sonication buffer/1 × 10^6 neurons (1 well of a 12-well plate).
2. Sonicate for 10 s with an ultrasonic dismembrator at medium setting. Some proteins may be sensitive to sonication. In these cases, homogenize with a tight-fitting glass or Teflon-pestle homogenizer.
3. Centrifuge at 16,000*g* for 10 min at 4°C.
4. Take supernatant for EMSA. Freeze aliquots of supernatant in liquid nitrogen and store at −80°C. All samples should have the same protein concentration if you start out with the same number of cells. Measure protein concentration in some selected experiments to confirm that protein measurements are not needed.

3.3.1.2. BRAIN TISSUE

1. Dissect tissue, quick freeze, and weigh frozen tissue.
2. Add 1 μL of sonication buffer to each 20 μg of frozen tissue.

3. Sonicate for 10 s with an ultrasonic dismembrator at medium setting, and make sure no tissue clumps remain. Some proteins may be sensitive to sonication. In these cases, homogenize with a tight-fitting glass or Teflon-pestle homogenizer.
4. Centrifuge at 16,000g for 10 min at 4°C.
5. Take supernatant, measure protein, and equalize protein concentration between samples by adding sonication buffer with all additives to a final concentration of 2 µg of protein/µL.
6. Proceed to EMSA or quickly freeze aliquots in liquid nitrogen and store at –80°C. Repeated freeze–thaw can change the binding properties of DNA-binding proteins. Freezing multiple aliquots of each sample is strongly recommended.

3.3.2. Crude Nuclear Extracts

This protocol is recommended if nuclear proteins need to be separated from cytosolic proteins. It is based on **refs. *1–4***. Tissue must be fresh and not previously frozen. In the first step, cells are swollen in low-salt buffer and disrupted in a homogenizer. Nuclei should stay intact. Nuclei are then pelleted, separated from cytoplasmic proteins, and nuclear proteins are extracted in a high-salt buffer. Membranes are pelleted and discarded. All solutions need to be ice cold.

3.3.2.1. NEURONAL CULTURE *(5)*

1. Harvest fresh cultured cells in 500 µ of buffer A.
2. Incubate on wet ice for 10 min to swell cells.
3. Homogenize with a loose-fitting (dounce B) homogenizer, 15 strokes up and down.
4. Incubate on wet ice for 5 min.
5. Proceed to **step 6** of the brain tissue protocol.

3.3.2.2. BRAIN TISSUE *(6)*

1. Dissect brain area of interest.
2. Chop tissue into small pieces and suspend in 1 mL of buffer A; pipet up and down with a 1-mL pipet to suspend tissue further.
3. Incubate on wet ice for 10 min.
4. Homogenize with a dounce B homogenizer, 15 strokes up and down.
5. Incubate on wet ice for 5 min.
6. Centrifuge at 4°C at 800g for 10 min.
7. Discard supernatant (or keep if you want to analyze the cytoplasmic fraction).
8. Resuspend pellet in equal volume of sonication buffer plus additives, and carefully resuspend with a 1-mL pipet.
9. Incubate for 30 min on wet ice.
10. Centrifuge at 16,000g, 10 min, 4°C.
11. Take supernatant and measure protein, and discard the pellet.
12. Dilute protein with sonication buffer to 1 µg/1 µL, and mix well.

Electrophoretic Mobility-Shift Assay

13. Proceed to EMSA or quickly freeze aliquots in liquid nitrogen and store at –80°C.

3.4. Preparation of the Gel

A nondenaturing, 4% acrylamide–*bis*-acrylamide gel (37.5:1) is used with vertical electrophoresis for EMSA. TBE buffer is used as running buffer at 0.25X.

1. Assemble the glass plates and spacers. Both plates and spacers need to be washed with water and rinsed with ethanol. The plates need to be spotless, but soaps should be avoided because they may interact with the gel matrix during the run and degrade the gel. We use 1.5-mm spacers for EMSAs.
2. Mix for a 30-mL solution: 0.75 mL of 10X TBE, 1 mL of 100% glycerol, 4 mL of 30% acrylamide–*bis*-acrylamide solution (37.5γ1), and 24 mL of distilled water. Mix well and make sure glycerol is entirely dissolved.
3. To start polymerization, add 300 µL of 10% APS and 30 µL of N,N,N',N'-tetramethylethylenediamine (TEMED).
4. Quickly pour solution between glass plates, insert comb, and let rest for at least 30 min to polymerize.
5. Prerun gel in the cold room (4°C) for 30 min at 200 V with 0.25X TBE.

3.5. EMSA

1. To make 2.5X EMSA buffer, mix on ice 50 µL of 5X EMSA buffer, 45.25 µL of distilled water, 2.5 µL of poly (dI-dC), 1.25 µL of $MgCl_2$ (stock: 1 M), 0.5 µL of DTT (stock: 1 M), and 0.5 µL of PMSF (stock: 100 mM) (*see* **Note 7**).
2. Take 6 µL of 2.5X EMSA buffer/sample.
3. Add 4 µL of protein solution (or 4 µL of sonication buffer for the control lane without protein).
4. Incubate 10 min on wet ice for nonspecific binding to poly (dI-dC).
5. Make a stock solution of 4 µL of distilled water and 1 µL of labeled oligonucleotide for $n + 1$ samples.
6. Add 5 µL of water and labeled oligonucleotide stock solution to each sample (= 1 ng labeled oligonucleotide per sample).
7. Mix by pipetting.
8. Incubate at room temperature for 10 min for protein–DNA binding.
9. Load 10 µL onto a prerun gel in the cold room, and start with control sample without protein. Flush out the wells of the gel with running buffer before loading the samples.
10. Load dye in adjacent wells. Do not load dye with sample because it interferes with the electrophoresis (*see* **Note 8**).
11. Electrophorese for 1 h to 90 min at 200 V in a cold room. Bromophenol blue should migrate two thirds through the gel. In a 4% gel, the bromophenol blue migrates at approx 40 bases. **Figure 2** shows some examples of EMSAs.

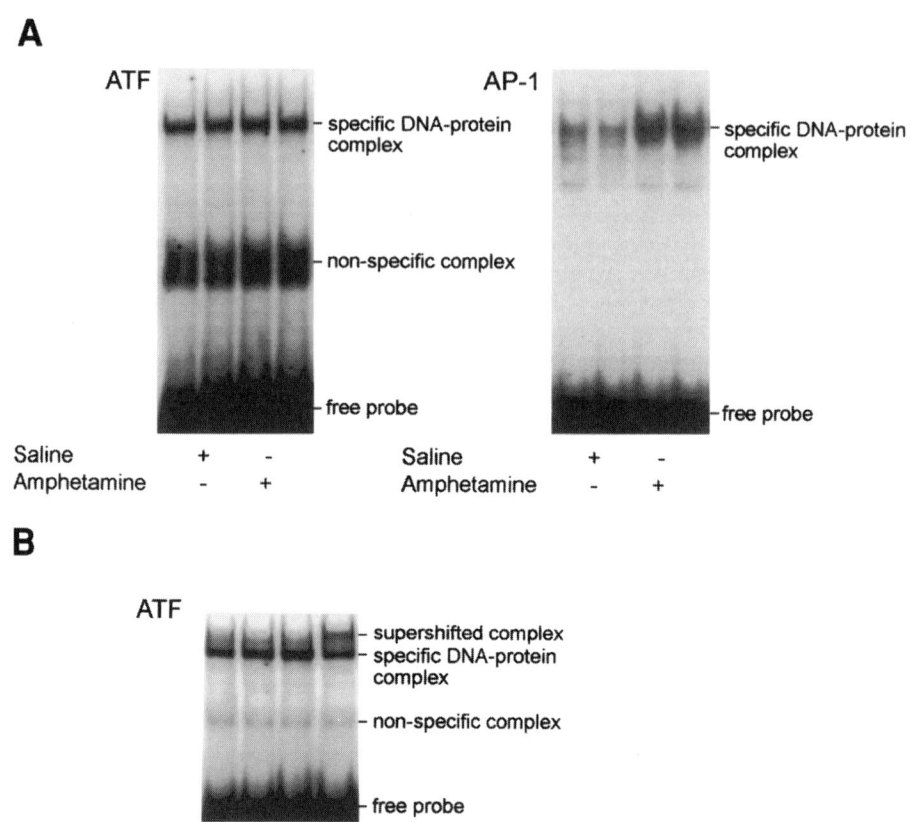

Fig. 2. EMSAs in primary striatal culture and in rat striatum. (**A**) Striatal extracts from two rats treated with saline and two rats treated with amphetamine, i.p. Levels of the protein(s) binding to the ATF enhancer are unchanged (*see also* **ref. 7** for further experimental details), whereas levels of proteins binding to the AP-1 enhancer are induced by amphetamine. (**B**) In primary striatal culture, treatment with dopamine also does not affect binding to the ATF site. However, the addition of a phosphorylation-specific antibody, phosphorylated cAMP response element binding protein (pCREB), reveals that the state of phosphorylation of the protein bound at the ATF site, CREB, is altered by treatment with dopamine. Indeed, the activity of the protein binding to the ATF site is regulated by phosphorylation, whereas the activity of the protein(s) binding to the AP-1 site is regulated by protein concentration. The specificity of the binding was established with various antibodies in supershift experiments, and with a panel of unlabeled enhancer elements in competition experiments *(7)*.

3.5.1. Cold Competition (7)

1. EMSA buffer–protein solution stock: Mix 6 µL of 2.5X EMSA buffer for $n + 1$ samples with 4 µL of protein solution for $n + 1$ samples in an Eppendorf tube.
2. Incubate this stock solution for 10 min on wet ice for nonspecific binding to poly (dI-dC).
3. Make a stock solution of 3 µL of distilled water and 1 µL of labeled oligonucleotide for $n + 1$ samples.
4. Into individual polymerase chain reaction (PCR) tubes mix labeled and unlabeled oligonucleotides for each competition sample; 1 µL of unlabeled oligonucleotide at the appropriate concentration with 4 µL of water and labeled oligonucleotide stock. For a fivefold competition, add 5 ng of unlabeled oligonucleotide in 1 µL of distilled water, and so forth. Do not forget to add a tube with no competitor (1 µL of distilled water instead of unlabeled oligonucleotide).
5. Add 10 µL of protein-EMSA buffer stock to each oligonucleotide mixture, and mix by pipetting.
6. Incubate at room temperature for 10 min for protein–DNA binding.
7. Load 10 µL onto prerun gel in cold room. Flush out the wells of the gel with running buffer before loading the samples.
8. Load dye in adjacent wells.
9. Electrophorese for 1 h to 90 min at 200 V in cold room. **Figure 3** shows the mechanism behind competition in EMSAs.

3.5.2. Antibody Supershift (5,6)

1. Take 6 µL of 2.5X EMSA buffer/sample.
2. Add 1 µL of antibody to each sample tube in a preliminary experiment, and find out the optimal concentration of antibody needed.
3. Add 4 µL of protein solution to each sample, and mix by pipetting.
4. Incubate 15 min at room temperature for nonspecific binding to poly (dI-dC) and for binding of antibody to protein.
5. Make a stock solution of 3 µL of distilled water and 1 µL of labeled oligonucleotide for all +1 samples.
6. Add 4 µL of water and labeled oligonucleotide stock solution to each sample. Mix by pipetting.
7. Incubate at room temperature for 10 min for protein–DNA binding.
8. Load 10 µL onto the gel in the cold room. Flush out the wells of the gel with running buffer in a syringe before loading the samples.
9. Load dye in adjacent wells.
10. Electrophorese for 1 h to 90 min at 200 V in the cold room. **Figure 4** shows the mechanism behind antibody supershifts in EMSAs (*see* **Note 9**).

3.6. Disassembling the Gel and Exposing It to X-ray Film

1. Turn off the power supply, disconnect the cables, remove the lid, and take out the gel.

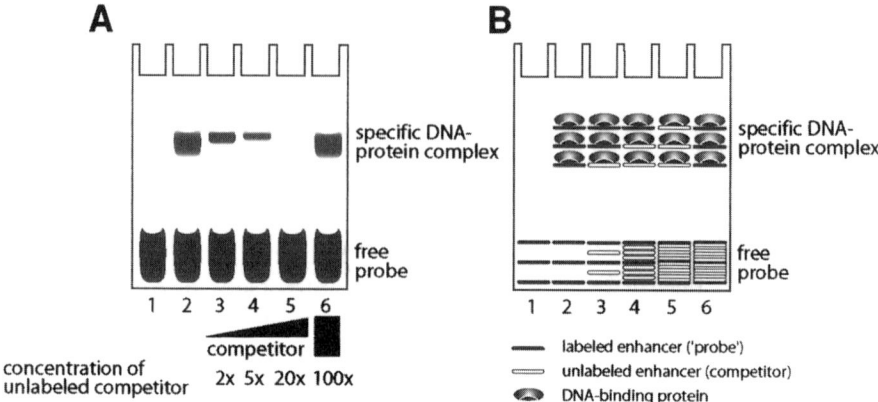

Fig. 3. Theory behind EMSAs and gel competition assays. (**A**) *Lane 1*: Labeled enhancer without protein added. *lane 2*: Labeled enhancer with protein sample added and protein binding to the enhancer. *Lanes 3–5*: Increasing concentrations of unlabeled competitor displace the labeled enhancer, causing a disappearance of the specific band. *lane 6*: High concentrations of an unrelated enhancer have no effect on the specific band, as they will not compete for the protein bound to the labeled enhancer. The concentrations below the gel are approximated from our experience with competition studies. (**B**) Illustration of the interaction between labeled enhancer, unlabeled competitor, unrelated enhancer, and protein binding to the labeled enhancer. *Lane 1*: In the absence of protein extract, all labeled enhancer migrates to the bottom of the gel as free probe. *Lane 2*: In the presence of DNA-binding protein, some of the labeled enhancer is shifted due to the higher molecular weight of the DNA–protein complex. *Lanes 3–5*: In the presence of unlabeled competitor, protein binding is proportionally distributed between protein bound to the labeled enhancer, and protein bound to the competitor. Only protein bound to the labeled enhancer is visible on the gel. *Lane 6*: The unrelated enhancer does not compete with the labeled enhancer for the protein specifically bound to the labeled enhancer. However, if the protein binding to the labeled enhancer is nonspecific, the unrelated enhancer will compete. *Note:* The amount of labeled enhancer is in vast excess of DNA-binding protein, such that the enhancer held back in the gel is not detectable in the levels of free probe.

2. Carefully remove one glass plate by prying out the spacers and using a spatula to separate the plates. Work behind a radiation shield.
3. Put a piece of dry blotting paper, slightly bigger than the gel, onto the gel. Gel should stick to the blotting paper. Mark the orientation of the gel wells on the paper.
4. Carefully lift gel-blotting paper sandwich from glass plate.
5. Cover the gel-blotting paper sandwich with plastic wrap.
6. Put two layers of filter paper (3 MM paper; Whatman) on a gel dryer that is preheated to 80°C. Place blotting paper–gel–plastic wrap sandwich on top of the

Electrophoretic Mobility-Shift Assay

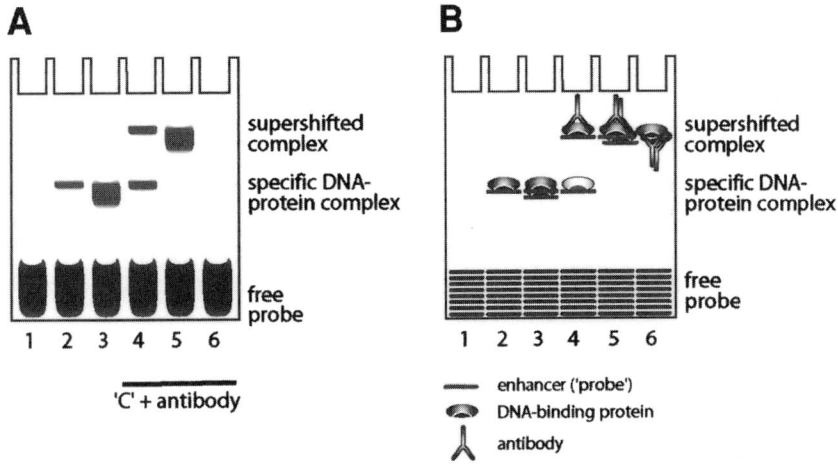

Fig. 4. The role of induction of DNA-binding proteins, and the role of antibodies in EMSA. (**A**) *Lane 1*: Labeled enhancer element (probe) without protein bound. *Lane 2:* A control protein sample has low levels of DNA-binding protein(s) that interact specifically with the labeled enhancer. *Lane 3*: A protein sample with induced DNA-binding protein(s) that interacts with the labeled enhancer. *Lane 4:* An antibody interacts with one protein bound to the labeled enhancer and supershifts part of the band, while a second, different DNA-binding protein bound to the enhancer is not supershifted. *Lane 5:* An antibody interacts with the DNA-binding protein and supershifts the entire band. However, this protein may still not be the only protein binding to the enhancer element, as it could be forming a heterodimer with another DNA-binding protein. Because the entire band is shifted, each protein complex bound must contain the protein the antibody is raised against. To characterize potential partners in heterodimers, additional antibodies need to be used in the preparation. *Lane 6:* The antibody binds the DNA-binding protein and blocks the interaction of the DNA-binding protein with the labeled enhancer, causing a disappearance of the specific band. *Lanes 4–6* use the protein sample shown in *lane 3*. (**B**) Diagram of the interaction between enhancer, DNA-binding protein and antibody. Lanes are the same as in **A**.

two layers with the plastic wrap on top, close gel dryer, and make sure plastic wrap is on top only and does not wrap around the gel-blotting paper sandwich. The gel dryer may become radioactive. Mark the gel dryer with radioactive tape and test the gel dryer with a Geiger counter after each use.

7. Turn on the vacuum pump and let gel dry for 1 h.
8. Gel will dry into blotting paper. When gel is totally dry, it should be no thicker than the blotting paper and it should not be sticky. Remove the two lower layers of filter paper and discard as radioactive waste. Expose the blotting paper–gel–plastic wrap to X-ray film in a light-tight cassette with intensifying screen. Store cassette at −80°C for signal amplification.

4. Notes

1. The following mixture of reagents can be used instead of the protease inhibitor cocktail: 1 mM PMSF, 1 μg/mL of pepstatin A, 10 μg/mL of leupeptin, and 10 μg/mL of aprotinin.
2. Some DNA-binding proteins may change their affinities depending on phosphorylation. It is therefore advisable to add phosphatase inhibitors to buffer A and the sonication buffer. However, preserving states of phosphorylation in in vivo experiments is not always possible, despite the preventive measures. Primary neuronal culture can be useful for the characterization of the role of phosphorylation in DNA–protein interactions.
3. Spermidine may improve protein binding to DNA. If binding of protein to DNA is low, try adding spermidine to the 2.5× EMSA buffer to a final concentration of 1 mM.
4. Molar concentrations of oligonucleotides can also be calculated with programs available as shareware on the internet.
5. The specific activity of [α-^{32}P]dCTP can be increased to 3000 Ci/mmol.
6. EMSAs with purified transcription factors: purified transcription factors are commercially available. They can also be synthesized from DNA vectors using rabbit reticulocyte lysate systems (commercially available) *(6)*. Purified transcription factors are helpful to optimize gel-shift assays, to test the performance of specific antibodies in the assay, and to characterize which proteins can potentially bind to a novel enhancer element.
7. The binding of protein to DNA is influenced by the concentration of $MgCl_2$, KCl, and DTT. Optimal concentrations of these reagents should be determined for binding to each enhancer element.
8. In antibody supershift experiments, a band may appear that is due to a nonspecific interaction of the serum background with the enhancer. This could be easily mistaken as a supershifted band. If available, control lanes should be run with protein exposed to preimmune serum, for example, serum obtained from the host before inoculation with the antigen. If preimmune serum is not available, normal serum from the host species should be used.
9. Because the samples do not contain dye, it can be very difficult to see the wells or to keep track of the loading. We mark the bottoms of the wells with marker on the glass plates before we remove the comb. As an alternative, a template of the wells can be drawn on a transparency, cut out, and affixed to the glass plate with water, making use of capillary forces. After loading, the template is removed. The template can be reused.

References

1. Schreiber, E., Matthias, P., Muller, M. M., and Schaffner, W. (1989) Rapid detection of octamer binding proteins with 'mini-extracts', prepared from a small number of cells. *Nucleic Acids Res.* **17,** 6419.

2. Dale, T.C., Imam, A. M., Kerr, I. M., and Stark, G. R. (1989) Rapid activation by interferon alpha of a latent DNA-binding protein present in the cytoplasm of untreated cells. *Proc. Natl. Acad. Sci. USA* **86,** 1203–1207.
3. Dignam, J. D., Lebovitz, R. M., and Roeder, R. G. (1983) Accurate transcription initiation by RNA polymerase II in a soluble extract from isolated mammalian nuclei. *Nucleic Acids Res.* **11,** 1475–1489.
4. Lewis, S. E. and Konradi, C. (1996) Analysis of DNA–protein interactions in the nervous system using the electrophoretic mobility shift assay. *Methods* **10,** 301–311.
5. Konradi, C., Cole, R. L., Green, D., et al. (1995) Analysis of the proenkephalin second messenger-inducible enhancer in rat striatal cultures. *J. Neurochem.* **65,** 1007–1015.
6. Konradi, C., Kobierski, L. A., Nguyen, T. V., Heckers, S., and Hyman, S. E. (1993) The cAMP-response-element-binding protein interacts, but Fos protein does not interact, with the proenkephalin enhancer in rat striatum. *Proc. Natl. Acad. Sci. USA* **90,** 7005–7009.
7. Konradi, C., Cole, R. L., Heckers, S., and Hyman, S. E. (1994) Amphetamine regulates gene expression in rat striatum via transcription factor CREB. *J. Neurosci.* **14,** 5623–5634.

IV

GENE FUNCTION ANALYSIS

22

Viral-Mediated Gene Transfer to Study the Behavioral Correlates of CREB Function in the Nucleus Accumbens of Rats

William A. Carlezon, Jr. and Rachael L. Neve

1. Introduction

There is an enormous initiative to establish causal relationships between brain biology (including patterns of gene expression) and behavior. Unfortunately, genetic intervention is not accomplished easily in the brain. One strategy is to engineer and deliver to the brain specialized viral vectors that carry a gene (or genes) of interest, thereby exploiting the natural ability of viruses to insert genetic material into cells. When delivered to the brain, these vectors cause infected cells to increase expression of the genes of interest. Viral vectors are particularly useful when the goal is to manipulate expression of a single gene (1) in a specific brain region (2) at a specific time (3) in animals that developed normally. As such, this technology has the potential to offer new insights into the etiology of a wide variety of neuropsychiatric disorders; ultimately it may usher in a new generation of "smart" pharmacotherapies that are designed to negate or reverse alterations in the molecular structure of the brain that lead to pathophysiological changes in behavior.

The use of any viral vector system for gene transfer studies in brain depends on the dynamic interaction of a number of factors, ranging from the hypothesis that is being tested, the brain region targeted for study, the viral backbone that is used as the vector, the titer of the vector, and the volume of the vector that is injected into the brain. Unfortunately, there is little information in the literature about how one, once in possession of a vector of interest, would embark on a gene transfer study. This chapter describes some of the protocols that we developed for viral-mediated gene transfer studies in which we used herpes simplex virus (HSV) vectors to study the molecular biology of cocaine addiction. Included are time-efficient steps to ensure that the viral vector increases expression of the target gene, that it is not toxic in the brain, and that it causes meaningful changes in behavior. The steps in engineering and constructing a viral vector are beyond the scope of this chapter; we assume that the reader has obtained the vectors that will be used, although we make reference to virtually all of the steps involved in performing a viral vector study and we provide citations for those procedures that are not covered in depth. As a template, we describe here some experiments that we conducted to examine how drug-induced activation of the transcription factor cAMP response element binding protein (CREB) in the basal forebrain (nucleus accumbens) [NAc] affects behavior *(1)*.

2. Materials
2.1. Viral Packaging

For an overview of the packaging procedure, see Carlezon et al. *(2)*.

1. Laminar flow hood and 37°C humidified incubator with 10% CO_2 atmosphere.
2. Tissue culture microscope and hemocytometer.
3. 60-mm, 100-mm, and 150-mm tissue culture dishes.
4. 5*dl*1.2 helper virus derived from HSV-1 strain KOS by deletion of most of the *IE2* (or *ICP27*) gene.
5. 2–2 cell line derived from the African green monkey VERO kidney epithelium cell line by transfection with a plasmid expressing the *IE2* gene.
6. 0.05% Trypsin–0.02% EDTA (Sigma T3924).
7. Dulbecco's minimum essential medium (DMEM): 1% penicillin–streptomycin and 4 m*M* glutamine (50× glutamine in water can be stored at –20°C) with either 5% or 10% fetal bovine serum (FBS; Hyclone) as indicated.
8. Dulbecco's phosphate-buffered salt solution (D-PBS): both Ca^{2+}, Mg^{2+}-free D-PBS (Fisher MT-21-031-CV) and D-PBS + Ca^{2+}, Mg^{2+} (Fisher MT-21-30-CV).
9. LipofectAMINE (Life Technologies).
10. OptiMEM (Life Technologies).
11. Cell lifters.
12. Dry ice–ethanol bath, 37°C water bath.
13. 15- and 50-mL conical tubes, polypropylene, with plug seals.
14. 50-mL conical tubes, polystyrene.
15. Cup-type sonicator.
16. 10%, 30%, and 60% sucrose solutions in D-PBS with Ca^{2+}, Mg^{2+}.
17. Beckman Ultra-Clear 25 × 89-mm (SW28), and 14 × 89 (SW41)-mm tubes.
18. Ultracentrifuge, Beckman SW28 and SW41 (or SW40) rotors.
19. 3-mL syringes and 18-gauge needles.
20. Screw-capped vials for virus storage.

2.2. Titering the Vector

1. Laminar flow hood and 37°C humidified incubator with 10% CO_2 atmosphere.
2. Tissue culture microscope and hemocytometer.
3. 24-Well tissue culture dishes.
4. PC12 cell line derived from rat pheochromocytoma (adrenal cells with nerve growth factor [NGF] reponsiveness).
5. DMEM + 10% horse serum (HS) + 5% FBS.
6. Phosphate-buffered saline (PBS): prepare a 10X PBS stock by mixing 1 g of KH_2PO_4, 10.8 g of $Na_2HPO_4 \cdot 7H_2O$, 1 g of KCl, 40 g of NaCl, and water to 500 mL final volume. The pH should be approx 7.0. Filter-sterilize, and store at room temperature (RT). Dilute 10-fold before use.
7. Poly-D-lysine (PDL): Prepare 1 mg/mL of PDL (PDL, mol wt 70,000–150,000 Sigma P6407) in water, filter sterilize, and store at –20°C. This is a 50X stock. Dilute to 20 µg/mL in water immediately prior to use.
8. Tris-buffered saline (TBS): 100 mM Tris-HCl, pH 7.5, and 150 mM NaCl. Store at RT.
9. TBS–serum–Triton (TST): 1% goat serum and 0.1% Triton X-100 in TBS. Store at 4°C.
10. Fe solution: Mix 200 mL of PBS, 332 mg of potassium ferricyanide, 424 mg of potassium ferrocyanide, 0.2 mL of 1 M $MgCl_2$, 0.2 mL of 20% Nonidet P-40, and 0.2 mL of 10% sodium deoxycholate. Store at 4°C in a darkened container, as this reagent is light sensitive.
11. X-gal (5-bromo-4-chloro-3-indolyl-β-D-galactoside; a 50 mg/mL stock in dimethyl sulfoxide can be stored in aliquots at –20°C).
12. 4% Paraformaldehyde: Add 20 g of paraformaldehyde to 300 mL of H_2O and heat to 55–60°C. Slowly add 1 M NaOH dropwise over about 10 min until the solution becomes clear. Cool the solution to room temperature. Use pH paper to check that the pH is 7.0–7.5 (add more NaOH if necessary). Add 100 mL of 0.5 M sodium phosphate buffer, pH 7.0, and then water to a final volume of 500 mL. The final pH should be 7.0–7.5. Store at 4°C.
13. Alkaline phosphatase (AP) buffer: 100 mM Tris-HCl, pH 8.5, 150 mM NaCl, and 5 mM $MgCl_2$.
14. AP substrate: Prepare 5-bromo-4-chloro-3-indolyl phosphate (BCIP) stock solution by dissolving one tablet (Sigma B-0274) in 0.5 mL of 100% dimethyl formamide. Prepare nitroblue tetrazolium (NBT) stock solution by dissolving one tablet (Sigma N-5514) in 1 mL of water. Make substrate solution by adding 330 µL of NBT stock to 10 mL of AP buffer. Mix, and then add 33 µL of BCIP stock. Store at –70°C in aliquots.

2.3. Cell Culture

1. Primary neuronal cultures.
2. Western immunoblotting: Rabbit anti-CREB antibody, 1:4000 (Cell Signaling), goat peroxidase labeled antibody to rabbit IgG, 1:5000 (Vector Laboratories), and chemiluminescence (Perkin Elmer).

2.4. Surgery

1. Rats, male, 300–325 g at the time of surgery.
2. Stereotaxic instrument (two arms: left and right).
3. Two Hamilton syringes (5 µL) with 26-gauge, blunt-tipped needles.
4. Sterile field (if aseptic procedures required).
5. Distilled water, and 95% ethanol.
6. 0.9% Saline.
7. Isopropyl alcohol and Betadine wipes.
8. Standard surgical instruments (scalpel, hemostats, autoclip applier, etc.).
9. Anesthetic (if pentobarbital, then also atropine).
10. Drill.
11. Timer.

2.5. Histology

2.5.1. Fixed Tissue

1. Peristaltic perfusion pump.
2. Physiological saline (~500 mL per rat).
3. 4% Paraformaldehyde (~500 mL per rat).
4. Anesthetic.
5. Standard surgical instruments.
6. Decapitator.

2.5.2. Fresh Tissue

1. Decapitator.
2. Artificial cerebrospinal fluid (aCSF) buffer: 126 mM NaCl, 5 mM KCl, 1.25 mM NaH$_2$PO$_4$, 25 mM NaHCO$_3$, 2 mM CaCl$_2$, 2 mM MgCl$_2$, and 10 mM D-glucose, at pH 7.4.
3. Tissue slicer (Stoelting, 51425).
4. Dry ice.
5. Tissue punches.
6. Microcentrifuge tubes for sample collection.

2.6. Immunohistochemistry

1. Freezing microtome.
2. Normal goat serum (NGS).
3. Bovine serum albumin (BSA).
4. Rabbit polyclonal antibody to CREB, 1:1000 (Cell Signaling).
5. Goat biotinylated antibody to rabbit IgG, 1:200 (Vector Laboratories).
6. Avidin–biotin (ABC Elite, Vector).
7. 3,3′-Diaminobenzidine tetrahydrochloride (DAB).
8. Slides (Superfrost Plus, Fisher).
9. Dehydration/staining kit.

2.7. Behavioral Assessment

1. "Unbiased" place conditioning apparatus (Med Associates): different wall cues (e.g., white/black, horizontal stripes/vertical stripes), different floor textures (e.g., rods, mesh, solid), and adjustable ceiling lights (e.g., bright/dark).
2. Physiological saline (or appropriate vehicle).
3. Drug (e.g., cocaine).
4. Syringes and needles.
5. Isopropyl alcohol wipes.

3. Methods

Theoretically, it is possible to engineer viral vectors that express any biological entity with a known sequence. Vectors have been used that encode enzymes (e.g., tyrosine hydroxylase, cre recombinase), transcription factors (e.g., CREB), receptors (e.g., dopamine D1), receptor subunits (e.g., glutamate receptor sununits including GluR1, GluR2), growth-associated proteins (e.g., GAP43), and reporter proteins (β-galactosidase, green fluorescent protein [GFP]). Moreover, vectors encoding the antisense sequences or dominant negative mutations (e.g., dominant negative mutant CREB [mCREB], which acts as an antagonist of endogenous CREB) of any of these entities can be engineered. There are some practical limitations, including the size of the entity and the ease with which it can be subcloned into the viral backbone (e.g., HSV-PrpUC; **Fig. 1**). The larger the construct, the less efficient the amplification will be during replication. This will lead to less favorable amplicon-to-helper virus ratios, and lower titers of the vector. Assuming that the sequence of the construct is known, the cDNA itself is available, and that no untoward complications are encountered in subcloning, new HSV vectors can be engineered and generated usually within a period of several weeks, and sometimes within days.

3.1. Viral Packaging

To achieve a favorable ratio of recombinant vector to helper virus, the stocks derived from transfection of the packaging cells followed by superinfection with helper virus are repeatedly passaged (infected, grown, and harvested) on the permissive host (2–2 cells). During replication, the recombinant vector is packaged as long, repeated sequences of the entire plasmid (concatemers). Because each recombinant virus particle contains multiple origins of replication (e.g., in the case of HSV-CREB, repeated 5-kb plasmids up to 150-kb total length = 30 $HSV_{ori}s$), the vector-containing virus has a large replicative advantage compared to the helper virus (which contains only one HSV_{ori} for its entire 150-kb length). The efficiency of the initial transfection of vector DNA into the packaging line is therefore critical to the success of the packaging.

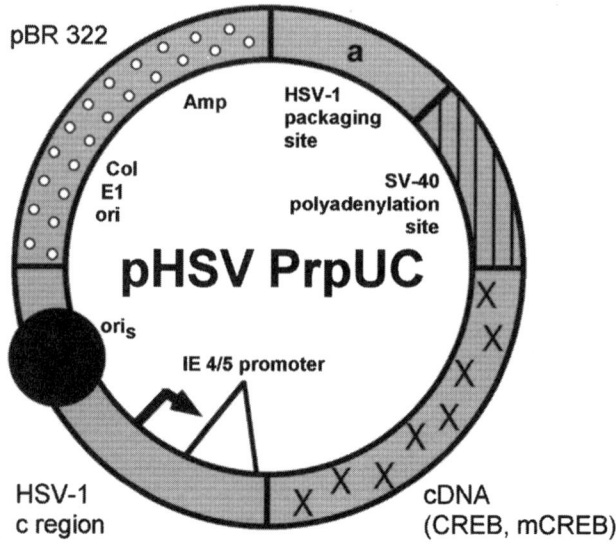

Fig. 1. The vector plasmid (p) HSV-PrpUC, which serves as the viral backbone for all of our viral-mediated gene transfer studies.

Lim et al. *(3)* showed that the transfection of the vector DNA at the start of the packaging procedure was significantly more efficient using lipofection than using calcium phosphate, and thereby achieved a favorable ratio (>1) of vector to helper. Since then, using a total of three passages (infections/harvests) in 2–2 cells, we have achieved vector/helper ratios greater than 100:1. Surprisingly, this ratio becomes less favorable, rather than more favorable, when more than three passages are used. An additional improvement to the packaging procedure, the banding of the virus on a sucrose step gradient, followed by a high-speed centrifugation to pellet the virus, has reduced further the cytotoxicity of the virus preparations. This step simultaneously removes toxic factors present in the crude cell lysates and enables concentration of the vector to titers exceeding 10^8 infectious units (iu)/mL.

3.1.1. Transfect 2–2 Cells

1. Maintain 2–2 cells at 37°C in a humidified, 10% CO_2 incubator in DMEM + 10% FBS. When a fresh aliquot of cells is thawed, it should be passaged at least two times before the cells are plated for the transfection.
2. Two days before transfection, plate 2–2 cells at 2.5×10^5 per 60-mm dish in 5 mL of DMEM.
3. Dilute 2 μg of DNA (purified by QIAGEN column) with 250 μL of OptiMEM in a sterile 1.5-mL microcentrifuge tube. Dilute 12 μL of LipofectAMINE with 250 μL

of OptiMEM in a second tube, and then add it to the 250-μL DNA mixture in the first tube. Leave the DNA-LipofectAMINE mixture at RT for 20–45 min, for liposomes to form.
4. Remove the medium from the plates and wash them once with 2 mL of OptiMEM, remove, and replace with 2 mL of OptiMEM.
5. Add the DNA–LipofectAMINE mix to the cells dropwise evenly over the whole plate. Incubate at 37°C for 5–7 h.
6. Prewarm D-PBS + Ca^{2+}, Mg^{2+} and DMEM + 10% FBS to 37°C (about 20 min).
7. Wash the cells three times with 2 mL of D-PBS + Ca^{2+}, Mg^{2+}. Replace medium with 3 mL DMEM + 10% FBS. Incubate cells overnight at 37°C.

3.1.2. Superinfect Transfected Cells—P0

1. Allow the cells to recover at least 20 h from the time of the last wash after the transfection. Prewarm DMEM + 5% FBS to 37°C (about 20 min).
2. Remove the medium from the plates and add 5 mL of DMEM + 5% FBS. Add approx 6×10^5 pfu of helper virus and incubate at 37°C until 95% of the cells show cytopathic effects (CPE) (have rounded up). This will take 30–40 h.

3.1.3. Harvest Cells—P0

1. Harvest the cells by pipetting the medium onto them until they have all detached from the plate or scraping them up with a cell lifter, and transfer cells and medium to 15-mL polypropylene conical tubes with plug seals.
2. Freeze–thaw the cells three times using a dry ice–ethanol bath and a 37°C water bath. Minimize the amount of time that they are thawed in the 37°C bath.
3. Transfer the cells to 15-mL polystyrene tubes and sonicate them for 2 min in a cup sonicator (power setting 6, 50% duty cycle, 1-s cycles). Centrifuge the cells for 5 min at low speed (1350g for 5 min) to pellet cell debris but not virus particles. Plate the supernatant on cells for the P1 passage or store it at –70°C.

3.1.4. Amplify Virus Stock—P1

1. Plate fresh 2–2 cells: 4×10^5 cells per 60-mm dish in 5 mL of DMEM + 10% FBS. Incubate at 37°C for 2 d.
2. Replace the medium with 4 mL of DMEM + 2% FBS and add 4 mL of the P0 supernatant.
3. Harvest and process the cells as in **Subheading 3.1.3.** when they show 95% CPE (about 24 h).

3.1.5. Amplify Virus Stock—P2

1. Plate fresh 2–2 cells: 1×10^6 per 100-mm dish in 10 mL of DMEM + 10% FBS (two dishes per sample).
2. Two days later, replace the medium in each dish with 6.0 mL of DMEM + 5% FBS and add 4.0 mL of P1 supernatant per dish.
3. Incubate the cells at 37°C overnight, and process them as described in **Subheading 3.1.3.** (but using 50-mL instead of 15-mL conical tubes) when they show 95% CPE (about 24 h).

3.1.6. Amplify Virus Stock—P3

1. Plate fresh 2-2 cells: 1.2×10^6 per 100-mm dish in 10 mL of DMEM + 10% FBS (six dishes per sample).
2. Two days later, replace the medium in each dish with 6 mL of DMEM + 5% FBS and add 4.0 mL of P2 supernatant per dish.
3. Incubate the cells at 37°C overnight, and process them as described in **Subheading 3.1.3.** (but using 50- instead of 15-mL conical tubes, two per virus sample from six dishes) when they show 95% CPE (about 24 h).
4. Prepare sucrose step gradients in 40-mL (25×89 mm) Ultra-Clear Beckman SW28 ultracentrifuge tubes (or their equivalent) at RT, by layering the following sucrose solutions into the tube: 7 mL of 60% sucrose in D-PBS + Ca^{2+}, Mg^{2+}, 6 mL of 30% sucrose in D-PBS + Ca^{2+}, Mg^{2+}, and 3 mL of 10% sucrose in D-PBS + Ca^{2+}, Mg^{2+}. Three tubes are needed for each virus preparation.
5. Load 20 mL of crude virus onto the gradient and centrifuged 1 h at 125,000g (SW28 rotor, 112,000g at 18°C).
6. Immobilize each tube in a clamp on a ring stand, and place it in front of a black background. The virus will appear as a sharp, very thin band at the 30%/60% interface, while contaminants will form a diffuse band close to the 10%/30% interface.
7. Using a 5-mL syringe with an 18-gauge needle attached, pierce the tube underneath the band, with the beveled edge of the needle pointing upwards. Slowly pull the band into the syringe, in a volume of 2 mL. Transfer this to an 11.5-mL (14×89 mm) Ultra-Clear Beckman SW41 tube and discard the remainder.
8. Dilute the virus in each tube with 9.5 mL of D-PBS + Ca^{2+}, Mg^{2+} and gently mix the contents of each tube by pipetting it up and down.
9. Centrifuge at 125,000g (SW41 or SW40, 26K, 18°C) for 1 h 15 min. Carefully aspirate the supernatant.
10. Add 200 µL of 10% sucrose in D-PBS + Ca^{2+}, Mg^{2+} to the pellet in each tube, and resuspend it by shaking it in a rack on a platform shaker at 4°C overnight. Very briefly triturate the virus the following day before dispensing it into aliquots and storing it at –70°C (*see* **Note 1**).

3.2. Titering the Vector

For any viral-mediated gene transfer study, titering the vectors is critical. Determining the titer of the vector confirms that the packaging has produced a viable vector, and it may provide an early indication that the vector has inadvertent cytotoxic effects. Also, it is critical to be aware of titers if vectors from different batches are to be used in a single study, although *post hoc* dilutions of the vector preparations is not ideal. In general, we try to make in a single batch enough vector to complete an entire study. PC12 cells are used for vector assays because they are round and easy to distinguish as single cells when positive for expression, which is usually quite high in this cell line. The following protocol is optimized for PC12 cells. If you use a different

Viral-Mediated Gene Transfer

cell line, plate it as you do normally to get a monolayer that will be confluent within 24 h.

1. Coat 24-well plates with PDL to get the PC12 cells to adhere properly to the bottom of the plates. Incubate the plates in 500 µL/well of a 20 µg/mL solution (1 mg/mL PDL in water = 50X stock—store at –20°C), at least 5 min at RT. Aspirate completely before using.
2. Pass the PC12 cells through a 21-gauge needle to dissociate aggregates. Count cells with a hemocytometer, and seed them at a density of 3×10^5/well in a volume of 500 µL of DMEM + 10% HS + 5% FBS.
3. The next day, change the medium. Add virus (sample volumes up to 100 µL). Incubate the infected cells at 37°C overnight. Always include an uninfected well as a negative control for background staining.
4. The next day, wash the cells with PBS. Fix them with 0.5 mL of 4% paraformaldehyde in 0.1 M phosphate, pH 7, for at least 15 min at RT. If the paraformaldehyde is old, or if you do not fix long enough, the cells will collapse and shrink, and background staining problems occur.
5. Wash the cells once with 500 µL of PBS or TBS. Phosphate inhibits AP beware if you use the antibody–AP staining procedure (*see* **steps 8–12**).
6. For X-gal staining, warm in a 37°C water bath 500 µL of Fe solution per well of cells to be stained. Warm the X-gal stock and add it to the warm Fe solution, to a final concentration of 1 mg/mL (dilute 50X). If the solutions are not warmed completely to 37°C before they are combined, a precipitate forms.
7. Remove the PBS or TBS from the cells and add the X-gal/Fe solution (500 µL per well). Incubate at 37°C for 30 min to overnight. Stop the reaction by washing twice with PBS. If crystals of X-gal are seen after the staining (due to low ambient temperature), these may be removed by incubating the sample in 50% dimethyl sulfoxide (DMSO)/H_2O for 5 min. For immunocytochemical staining, follow **steps 8–12**.
8. Use primary antibody diluted in TST (e.g., rabbit anti-HSV at 1:10,000). Leave overnight at 4°C.
9. Wash the cells with TBS twice, then leave them at RT for 10 min.
10. Incubate the cells with AP–anti-rabbit conjugate (or anti-mouse if appropriate) at 1:2000 in TST for about 1 h at room temperature.
11. Wash the cells with TBS twice, then wash with AP buffer twice. Add substrate solution and incubate in dimly lit area. Check with microscope for color development.
12. Stop color development by washing twice with PBS. Leave the cells in PBS + 10 mM EDTA to prevent further darkening.

3.3. Cell Culture

Before we initiate in vivo studies, we confirm that the vector causes elevated transgene expression in vitro. To accomplish this, we infect cell cultures with the vector of interest (e.g., HSV-CREB). For comparison, we use a vector that encodes a control protein (e.g., *E. coli* β-galactosidase, encoded by the *LacZ* gene) and, typically, a "mock" infection (10% sucrose, the vehicle in which the vector is resuspended after packaging). We do not proceed to in vivo studies unless it is clear that the vector causes substantial up-regulation of the protein encoded by the transgene. General procedures for infection of primary neuronal cultures with HSV vectors and subsequent processing of the cultures for immunoblots have been described *(4)*. The following are details specific to our Western blotting procedures for CREB (or mCREB, expression of which is indistinguishable from that of CREB using the antibodies described here):

1. Load 20 µg of protein (in homogenate containing 1% sodium dodecyl sulfate [SDS]) per lane.
2. Block nonspecific protein binding to membranes for 2 h at room temperature with a 5% milk in PBS containing 0.1% Tween 20 (PBS-T) blocking solution.
3. Incubate membrane in primary antibody (anti-CREB, 1:4000 in PBS-T) for 2 h at RT.
4. Wash membranes three times for 15 min each in PBS-T at room temperature.
5. Incubate membrane in secondary antibody (anti-rabbit, 1:5000 in 5% milk in PBS-T) for 2 h at room temperature.
6. Wash membranes three times for 15 min each in PBS-T at room temperature.
7. Detect CREB protein on the membrane using chemiluminescence (**Fig. 2**).

3.4. Surgery

The final step before initiating behavioral studies is to confirm that the vector expresses in vivo. We unilaterally microinject the vector at full strength titer into the NAc, perfuse the rat on day 3 (when we normally begin the behavioral studies, a time at which transgene expression is maximal), and perform within-animal immunohistochemical comparisons (gene expression within the injected vs noninjected hemisphere) to ensure protein overexpression (*see* **Notes 2–4**).

1. Establish an aseptic surgical field if necessary. Attach the Hamilton syringes to the stereotaxic apparatus, and angle the arms at 10° from the midline (*see* **Note 5**). Thoroughly clean the Hamilton syringes with alternating washes of 95% ethanol and distilled water. The final wash should be distilled water, such that the vector does not come into contact with ethanol.
2. Anesthetize each rat with intraperitoneal (i.p.) pentobarbital (65 mg/kg sodium pentobarbital) accompanied by subcutaneous (s.c.) atropine (0.25 mg/kg) to minimize bronchial secretions (*see* **Note 6**). Once the rat is anesthetized, shave the scalp with an electric trimmer.
3. Place the rat (300–325 g) in a stereotaxic instrument. Use a scalpel incision to expose the skull, and clean the area with sterile saline.

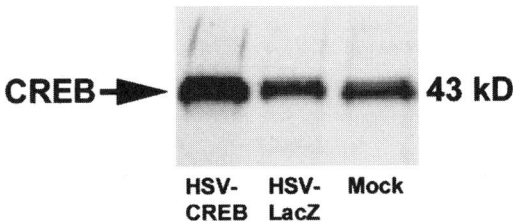

Fig. 2. Western immunoblot confirming that HSV-CREB elevates expression of CREB in cell culture. For comparison, HSV-LacZ (encoding *E. coli* β-galactosidase) was used as a viral vector control, and mock (10% sucrose) was used as a control for the vehicle into which the vector constructs are resuspended after packaging.

4. Set the stereotaxic arms to perpendicular (90°) at their base. Once set, the arms should not be pivoted, or all measurements should be reestablished. Ensure that the skull is flat by using one of the Hamilton syringes to obtain measurements for the height of the skull surface within the dorsal–ventral plane at lambda and bregma. If the measurements are not within 0.1 mm, adjust the height of the head at the bite-bar on the stereotaxic instrument. After any adjustment, always re-measure both lambda and bregma.
5. The stereotaxic coordinates that we use for the NAc are 1.7 mm anterior to bregma and ± 2.3 mm from the midline *(5)*. Using a 10° angle from the midline for the Hamilton syringes, bilateral microinjections can proceed in parallel. Raise the stereotaxic arms to their maximal height, and mark the appropriate points on the skull with a pen if necessary. Do not pivot the arms, or the stereotaxic coordinates should be reestablished.
6. Drill burr holes (1 mm in diameter) into the skull, to the level of dura. We aim our microinjections into the NAc at 6.5 mm below dura. If dura is damaged during drilling, assume that the skull of a 300–325 g rat is 0.5 mm thick, and subtract this thickness from the dorsal–ventral measurements used to ensure that the skull was flat. Without rupturing dura (or penetrating the brain if the dura was damaged), ensure that the trajectory of each injection needle will not be affected by the size or the shape of the burr holes.
7. Load the Hamilton syringes with vector. The vectors should be thawed as close to the time of microinjections as possible (*see* **Subheading 3.1.**). Lower the injection syringes to dura, but use a sterile 26-gauge needle to puncture dura. Slowly lower the injection needle to the appropriate depth, over 20–30 s. Once at the desired depth, retract the needle 0.1 mm.
8. Inject the vectors over a period of 10 min. We perform the microinjections manually, at a rate of 0.1 μL/30 s. After the first microinjection, we start a timer; as such, the last microinjection is performed at 9 min 30 s. We then wait an additional 5 min before removing the injection needle from the brain, to allow

maximum diffusion from the microinjector tip and absorption into the tissue. We then remove the needle slowly, over 20–30 s.
9. Clean the incision and ensure that the area is not bleeding actively before suturing. We use wound clips (three per incision). Cover the wound with antibiotic ointment.
10. Allow the rat to recover completely from the anesthetic before replacing it in its home cage. This is particularly important if the home cage contains wood shavings. We consider it safe to place the rats in their home cages once they begin drinking or grooming.

3.5. Histology

To confirm viral-mediated transgene expression in vivo, we regularly use immunohistochemistry or immunoblotting. Unfortunately, these techniques require vastly different (and incompatible) methods of tissue preparation. Generally we use immunohistochemistry because it involves fixed tissue and, as such, it offers the ability to localize injection placements, estimate the size of the sphere of infected cells, and identify rare occasions of untoward effects within the targeted region (excessive damage, hemorrhage, poor injection placements, diffusion of the vector up the injector track or into the ventricles). Because we section each fixed brain into six-well containers, we can perform up to six types of analyses, or replicates of particularly critical analyses. We use at least one well for Nissl staining, which provides the best index of the localization of the injection sites. A distinct advantage of this methodology is that if the sections are analyzed by an investigator who is unaware of the experimental conditions, rats for which the injections sites are not within the targeted region can be eliminated from the study. Conversely, for immunoblotting we obtain 14-gauge punches from fresh tissue, which eliminates the ability to ensure that there were no untoward effects at the injection site. Immunoblotting offers the ability to estimate relative increases in overall transgene expression in homogenized tissue, rather than increases in the number of transgene-expressing cells.

When characterizing the effects of new vectors, we perfuse the rats on day 3 after viral-mediated gene transfer. We usually begin our behavioral studies on day 3, so we want to ensure that we can expect transgene expression to be high at this time. In rats used in behavioral studies, we perfuse the animal immediately after testing to preserve transgene expression, which typically begins to wane after days 4–5, as much as possible (**Fig. 3**).

3.5.1. Brain Fixation for Immunohistochemistry
1. Deeply anesthetize the rat with 130 mg/kg (i.p.) pentobarbital.
2. Exsanguinate the rat by transcardial perfusion with ice-cold 0.9% saline. If perfusing with a peristaltic pump, use a low flow rate. Generally, exsanguination requires 300–500 mL of saline.

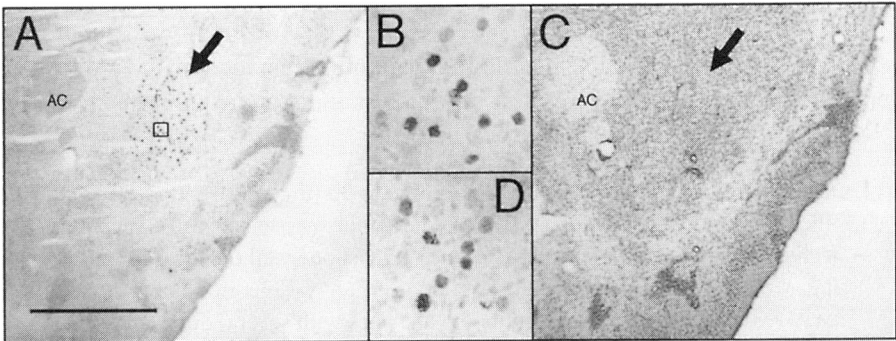

Fig. 3. Histological examination of NAc after gene transfer. (**A**) Expression of CREB at the completion of a place conditioning study, 5 d after microinjection of HSV-CREB into the left NAc shell (×40). *Arrow* indicates injection site; scale bar = 1 mm. (**B**) Higher magnification (×200) of the area indicated by the box in (**A**), confirming nuclear localization of CREB expression. (**C**) An adjacent, Nissl-stained slice from the same brain, showing minimal evidence of toxicity or damage despite strong transgene expression. (**D**) Expression of mCREB 5 d after microinjection of HSV-mCREB into the NAc. Expression of mCREB, which contains a point mutation at Ser133 that prevents phosphorylation, is indistinguishable from that of CREB with the antibodies used. AC, anterior commissure.

3. Fix the rat with ice-cold 4% paraformaldehyde (for detection of β-galactosidase, 2% paraformaldehyde works best) using a low flow rate. Generally, fixation requires 300–500 mL of paraformaldehyde.
4. Postfix the brain for 1–2 h in paraformadehyde solution.
5. Cryoprotect the brain overnight in 20% glycerol–50 mM PBS. When first placing the brain in this solution, it should float. Do not section the brain unless it has sunk into the solution (usually within 12 h). For best results, section the brain within 48 h of fixation.

3.5.2. Harvest Brain Tissue for Immunoblot

1. Remove the brain as rapidly as possible, and momentarily cool the whole brain by placing it into ice-cold aCSF buffer.
2. Use a tissue slicer to obtain a 1-mm thick section of brain that includes the NAc (i.e., 1.0–2.0 mm anterior to bregma; *see* **ref. 5**).
3. Place the 1-mm thick section into a shallow Petri dish filled with ice-cold aCSF. Use a 14-gauge tissue punch to obtain the NAc.
4. Place the tissue punches in prelabeled microcentifuge tubes. Flash freeze the tissue by placing the tubes in ground dry ice.
5. Perform Western immunoblotting analyses as described in **Subheading 3.3.**).

3.6. Immunohistochemistry

The following are details specific to our immunohistochemistry procedures for CREB (or mCREB, expression of which is indistinguishable from that of CREB using the antibodies described here):

1. Section the brains on a freezing microtome at 20 µm. We use 18-well containers filled with 0.005% sodium azide. For each brain we use six wells, such that each well contains every sixth section. We use at least one well of tissue for Nissl staining, whereas the other five wells can be used for various immunohistochemistry analyses (replicates with one antibody, analyses with several different antibodies).
2. We use a blocking buffer containing 2% NGS and 1% BSA, to maximize suppression of background CREB expression.
3. We incubate overnight at 4°C in rabbit polyclonal antibody to CREB (1:1000) within the NGS–BSA blocking solution.
4. We incubate for 2 h at room temperature in goat biotinylated anti-rabbit IgG (1:200).
5. We mix the avidin–biotin complex at least 30 min before incubating the sections in it for 1 h at room temperature.

3.7. Behavioral Assessment

The most critical element in our place conditioning procedures is the use of an "unbiased" place conditioning apparatus. An unbiased apparatus is one in which the rats have no reliable *a priori* preference for any of the compartments, although it is important that each compartment have qualities that allow the animals to discriminate among environments. Our apparatus has three types of distinguishing features: wall color (white and black), floor texture (rods and screen), and most importantly, light intensity (dim to bright) (**Fig. 4**; *see* **Note 7**).

1. Place the apparatus in a quiet room. Minimize personnel entry into the room during place conditioning sessions, because disruptions may change the behavior of the rats. All portions of the place conditioning procedure should be conducted during the light portion of the activity cycle with the lights off in the test room. These conditions maximize exploratory behavior in the apparatus.
2. Before each screening session, ensure that all chambers in the apparatus are accessible (i.e., doors open) and clean. To minimize olfactory cues associated with other rats, clean the entire apparatus with isopropyl alcohol wipes if necessary (*see* **Note 8**). Leave the tops open until ready for testing to minimize trapping odors within the apparatus.
3. *Day 0*. Close the tops of the side compartments. Gently place a naive rat into the center compartment of each apparatus, and close the top. Turn the lights off in the test room.

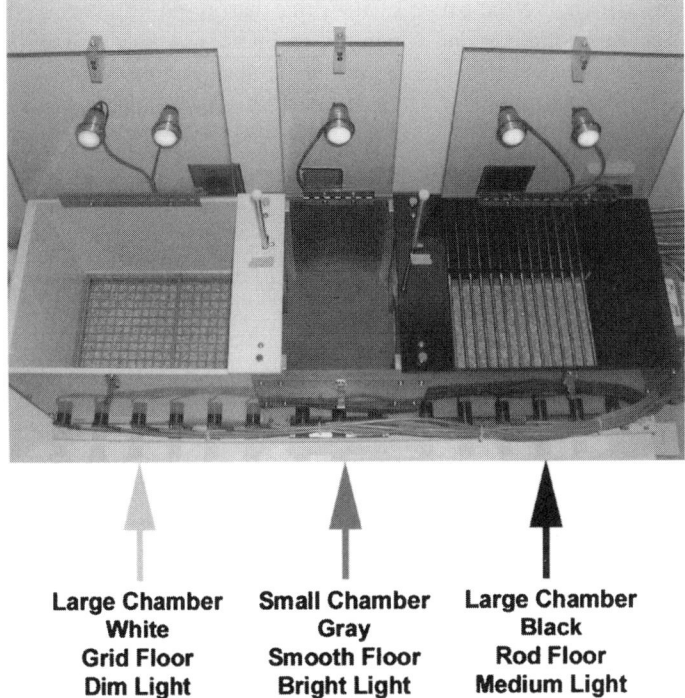

Fig. 4. "Unbiased" place conditioning apparatus. (Med Associates, St. Albans VT.)

4. Screen the rats for 30 min. Remove the rats from the apparatus immediately after the testing session ends.
5. Counterbalance the rats such that each of the four conditions (*see* **Note 9**) is represented within each experimental condition.
6. Perform stereotaxic microinjections of viral vectors (as **Subheading 3.4.**).
7. *Day 1–2.* Allow the rats to recover from the surgery.
8. *Day 3.* Begin "compressed" conditioning paradigm (two sessions per day; *see* **Note 10**). Close the doors in the dividers that separate the compartments of the apparatus. In the morning, administer saline (1 mL/kg) and place each rat in the "nondrug"-assigned compartment. Turn the lights off in the test room, and condition the rats for 1 h. After the conditioning session, return the rats to their home cages and the vivarium. Clean the entire apparatus with isopropyl alcohol wipes, and leave the tops open to minimize trapped odors. In the afternoon (at least 3 h later), administer the drug under study (e.g., cocaine) and place each rat in the "drug" assigned compartment. Turn the lights off in the test room, and condition the rats for 1 h. After the conditioning session, return the rats to their home cages and the vivarium. It is particularly important to remove the rats from the apparatus immediately at the end of the cocaine conditioning sessions,

to avoid associating the offset of the drug that might be aversive (*6*) with the drug-assigned compartment. Clean the entire apparatus with isopropyl alcohol wipes, and leave the tops open to minimize trapped odors.
9. *Day 4.* Repeat exactly the procedures used on d 3. Do not alternate the order of nondrug and drug pairings (*see* **Note 10**).
10. *Day 5.* Test the animals under the conditions used for screening (**items 2–4**).
11. Immediately after testing, collect brain tissue (fresh or perfused, as in **Subheading 3.6.**) for histological or molecular analyses. In some cases, only one will be possible (i.e., if Western immunoblotting studies will be conducted, it is necessary to collect tissue from the area of the microinjection). Generally, we favor procedures that will allow detailed examination of the microinjection placements, because rats in which the gene transfer was not targeted to the appropriate region (i.e., NAc) should be eliminated from statistical analyses.
12. Perform statistical analyses. To maximize statistical power, we typically use analyses of variance (ANOVA) with repeated measures: we compare the net differences in time spent in the drug side (i.e., time spent in drug side minus time spent in saline side, in seconds) before and after treatment. Significant effects are analyzed further with *post hoc* tests (Fisher's *t*-tests).

4. Notes

1. The viruses should be divided into the smallest aliquots that are practical. Repeated freeze–thaw cycles can damage the virus particles, effectively decreasing titer. Because we typically use bilateral microinjections of 2.0 µL of vector for each rat, we freeze the vectors in 5.0-µL aliquots. Whenever freezing or thawing of aliquots is necessary, it should be performed as quickly as possible (i.e., using a dry ice–ethanol slurry or a 37°C water bath, respectively).
2. Presently, the full-strength titer of HSV vectors is on the order of 10^8 infectious units (iu)/mL. In our early studies in brain, we examined gene transfer in the NAc after various combinations of viral titers and injection volumes. We started with 2.0-µL microinjections (within a range of volumes often used for intra-NAc microinjections of drugs, for example) and full-strength titer. For comparison, we also used 1.0-µL microinjections of full-strength titer, and 2.0-µL or 1.0-µL microinjections of titer diluted by 50% (in 10% sucrose). We found that the most important factor for transgene expression in the VTA was injection volume: regardless of the vector titer, the sphere of transgene-expressing neurons in the VTA was substantially larger with 2.0-µL microinjections than it was with 1.0-µL microinjections. We proceeded with behavioral studies using full-strength titers because we found more transgene-expressing neurons in rats that received the undiluted vector than in rats that received the 50% diluted vector, but overall, the effects attributable to titer were less striking than the effects attributable to injection volumes. These findings suggest that, at some critical point specific for each type of vector, brain microinjections of higher titers simply cause more infections per cell rather than more cells infected. Thus the main advantage of high-titer vectors may be that, eventually, it will be possible to use diluted titer to

minimize the delivery of any nonvector material (helper virus, cell debris), while maintaining an injection volume that causes the desired sphere of transgene expression within the targeted brain region.

3. Another consideration is the targeted brain region itself. One of the most important current limitations of viral-mediated gene transfer is that it is difficult to design vectors that target specific subtypes of neurons (e.g., dopamine- or γ-aminobutyric acid [GABA]-containing neurons). This current limitation is less of a concern when the targeted brain regions are relatively homogeneous in neuronal composition (e.g., the NAc) or when the transgenes are expressed ubiquitously in neurons (e.g., CREB), but eventually it will be important to target specific neuronal populations in brain nuclei with more heterogeneous compositions. Ectopic transgene expression—expression of a transgene in a cell that normally does not express the gene—is a potential experimental confound in any studies in which only a certain percentage of the cells within a target nucleus express the gene under study. However, this limitation is not unlike that often encountered with transgenic mice, in which transgene expression can also be ectopic. As with any technology, advantages (the ability to alter expression of a single gene within a specific brain region in an adult animal that developed normally) must be weighed against disadvantages (limited neuronal specificity).

4. HSV-based vectors are Biohazard level 2 (BL2) agents. They do not spread through air and, as such, they are associated with minimal health hazards if handled with good laboratory practices. Poor handling may make the vectors—and many other reagents used routinely in molecular biology laboratories—pose health risks. Researchers performing surgery to microinject the vectors into the brain should wear disposable protective clothing, including a lab coat, gloves, shoe covering, hair covering, a mask, and protective eyewear. Surgical procedures should be performed behind a shield (e.g., within a hood with a sash). Items that come into contact with the vectors or animals transfected with the vectors should be placed in special biohazard bags that can be autoclaved or disinfected before discarding. The area in which the surgery was performed should be cleaned with a commercial disinfectant. Cages that contain animals used for viral-mediated gene transfer studies should be identified with biohazard stickers. Any apparatus used for behavioral studies should be cleaned thoroughly between animals and at the end of each day. Above all, all procedures used must be approved in advance by the institutional safety and animal care committees. This list of procedures should not be considered to be complete, or an endorsement of minimally acceptable steps to ensure safety.

5. For the microinjections of vectors into the NAc, we angle the injection syringes at 10° from the midline to avoid penetrating the ventricles. If penetrated, the ventricles can act as a pressure sink, essentially pulling the vector out of the targeted region and into the ventricles. We use blunted needles on the injection syringes, because beveled needles can direct the flow of the vector and affect the shape of the sphere of transfected neurons. Finally, after lowering the injection needle to the appropriate depth, we retract it 0.1 mm. This manipulation

provides a small "pocket" of fresh tissue into which the vector is injected, and minimizes the flow of the vector up the injector track into more dorsal regions. The stereotaxic coordinates that we use (*see* **Subheading 3.4.**) were derived empirically, through repeated trials, and should be used only as a guide. Each individual researcher should ensure that his or her coordinates are appropriate and reliable before initiating behavioral studies.
6. We use pentobarbital anesthetic for viral-mediated gene transfer surgery. In our experience, a major limitation of pentobarbital anesthesia is that its duration of action can be unpredictable and difficult to titrate in rats. We accompany pentobarbital anesthetic with atropine to minimize bronchial secretions. During surgery, we use a toe pinch at regular intervals to ensure that the rats are sufficiently anesthetized. Supplemental anesthetic is administered if necessary. We allow the rats to awaken from the anesthetic in a clean and empty enclosure (i.e., a Plexiglas cage) before replacing them in their home cage. In particular, we do not place partially anesthetized rats in wood-based bedding: the rats will often attempt to eat the bedding as they emerge from the anesthetic, but will be unable to swallow because the atropine minimizes salivation.
7. To first characterize our place conditioning apparatus, we used a number of rats that were not part of the viral vector studies; in some cases the rats had been in other types of behavioral studies, although none had experience in the place conditioning apparatus. We found in 30-min screening sessions that, in general, rats prefer the compartment with the black walls more than the compartment with the white walls, and the screen floor more than the rod floor. Accordingly, we exploited the modular capabilities of our apparatus by placing the screen floor in the white compartment and the rod floor in the black compartment. In our experience, this combination still favors the black compartment, so we increased the intensity of the lights in the black compartments and decreased the intensity in the white compartments. The lights are always most intense in the middle compartment to discourage the rats from spending large amounts of time in this area, as it is never associated with either drug or nondrug conditions; the middle compartment is smaller than the side compartments, and it serves only as the starting point for test sessions and a connector between the two large side compartments. We continued to adjust the lighting until there were no systematic preferences for either of the large side compartments, as indicated by similar overall averages for the time spent in the white compartment and black compartment across a group of rats (e.g., >10 rats). After these conditions were established, we avoided further adjustments to the lighting.
8. We thoroughly clean the apparatus between each session, to minimize possible confounds associated with olfactory stimuli. We clean the walls, tops, and floors with isopropyl alcohol wipes, and we replace the absorbent material (wood chips) in the waste trays located below the floors.
9. Individual rats often show nominal preferences for one of the large side compartments. We counterbalance the compartment assignments such that some rats will receive drug in their preferred environment and some will receive it in their

nonpreferred environment. We also counterbalance each treatment group such that approximately the same number of rats receive drug associated with the white compartment as in the black compartment. Accordingly, there are four possible types of drug–environment pairings: the drug paired with the white/preferred environment; drug with the white/nonpreferred environment; drug with the black/preferred environment; and drug with the black/nonpreferred environment. Ideally, each possibility should be represented equally within each group in the study. Occasionally, rats show strong preferences for one of the environment. We screen the rats before surgery so that we can eliminate those with large *a priori* preferences for a compartment without performing the most labor-intensive portions of the study (e.g., viral-mediated gene transfer and conditioning).

10. The goal of viral-mediated gene transfer studies is to establish causal relations between genes and behavior. Accordingly, drug treatments should be given in some rats when transgene expression is maximal, and in other rats after transgene expression has waned. Because of the transient nature of viral-mediated transgene expression—in this case, HSV-mediated transgene expression begins to wane only a few days after gene transfer—it can be challenging to design behavioral studies in which repeated drug treatments correspond with maximal transgene expression. To test the rats while transgene expression is maximal, we developed a "compressed" place conditioning protocol in which the conditioning portion of our studies is conducted over 2 d. On each day, the rats receive two conditioning sessions: nondrug (i.e., vehicle) in the morning, and drug in the afternoon. In this compressed protocol, it is important that the nondrug pairings always precede drug pairings; if the rats receive drug in the morning conditioning session, it is possible that they could associate symptoms of dysphoria (e.g., acute drug withdrawal) with the nondrug environment during the afternoon session. This "compressed" conditioning paradigm may generate dose–effect functions for each drug that are different than those reported in articles in which more extended conditioning protocols are used. Accordingly, it is advisable to identify the appropriate dose ranges for each drug in naive animals before initiating labor-intensive gene transfer studies.

Acknowledgments

Our work with viral vectors is sponsored by NIH (DA12736 and MH63266 to W. C., AG12954 to R. L. N.) and an unrestricted research grant from Johnson & Johnson (to W. C.).

References

1. Carlezon, W. A., Jr., Thome, J., Olson, V., et al. (1998) Regulation of cocaine reward by CREB. *Science* **282,** 2272–2275.
2. Carlezon, W. A., Jr., Nestler, E. J., and Neve, R. L. (2000) Herpes simplex virus-mediated gene transfer as a tool for neuropsychiatric research. *Crit. Rev. Neurobiol.* **14,** 47–68.

3. Lim, F., Hartley, D., Starr, P., et al. (1996) Generation of high-titer defective HSV-1 vectors using an IE 2 deletion mutant and quantitative study of expression in cultured cortical cells. *Biotechniques* **20,** 460–469.
4. McPhie, D. L., Lee, R. K., Eckman, C. B., et al. (1997) Neuronal expression of beta-amyloid precursor protein Alzheimer mutations causes intracellular accumulation of a C-terminal fragment containing both the amyloid beta and cytoplasmic domains. *J. Biol. Chem.* **272,** 24,743–24,746.
5. Paxinos, G. and Watson, C. (1997) *The Rat Brain in Stereotaxic Coordinates*, Compact 3rd edit., Academic Press, San Diego.
6. Pliakas, A. M., Carlson, R., Neve, R. L., Konradi, C., Nestler, E. J., and Carlezon, W. A., Jr. (2001) Altered responsiveness to cocaine and increased immobility in the forced swim test associated with elevated cAMP response element binding protein expression in nucleus accumbens. *J. Neurosci.* **21,** 7397–7403.

23

Generating Gene Knockout Mice for Studying Mechanisms Underlying Drug Addiction

Jianhua Zhang and Ming Xu

1. Introduction

Drug addiction is a chronic relapsing disease with psychological and social factors *(1)*. Compulsive drug-taking is a central feature of drug addiction *(2–4)*. A major goal of drug abuse research is to understand the cellular and molecular mechanisms underlying the development of the loss of control over drug-taking *(2–4)*. Because the neurobiological mechanisms underlying the development of uncontrolled drug-taking behaviors are associated with the brain dopaminergic and glutamatergic systems *(5–8)*, animal models are obviously needed for the various investigations. With the development of the gene targeting approach *(9–13)*, genetically engineered mouse models have become increasingly useful for studying molecular mechanisms underlying drug addiction. Gene targeting allows a direct assessment of the contribution of individual genes to specific behaviors in mice. This technology provides a very useful alternative as opposed to pharmacological approach to dissect complex biological mechanisms, including how the dopamine (DA) receptors and the DA transporter function in vivo *(14–21)*.

Here we summarize the gene targeting procedures we have used to generate the DA D1 and D3 receptor knockout mice *(14,19)*. There are three major steps in the procedure including the generation of a DNA targeting construct, homologous recombinants in mouse embryonic cells (ES) and knockout mice *(22–25)*. Conditional gene targeting including the generation of region-specific and inducible knockout mice can overcome some of the limitations associated with the straightforward gene targeting method. Although the procedures for such advanced gene targeting are not discussed here, they are also based on the basic protocols summarized below.

2. Materials

1. Mouse 129 λ phage genomic DNA library (Stratagene, CA).
2. XL1-Blue bacterial cells (GIBCO-BRL, NY).
3. pBluescript II KS-cloning vector (Stratagene).
4. Restriction enzymes, T4 DNA ligases (Stratagene).
5. Agarose gel electrophoresis apparatus and power supply (Bio-Rad, CA).
6. [^{32}P]dCTP (Perkin Elmer, MA): This is a radiation hazard. Use according to institutional guidelines.
7. Hybridization buffer: 5% sodium dodecyl sulfate (SDS), 6X sodium chloride/sodium citrate (SSC), 10% dextran sulfate, and 100 µg/mL of sonicated salmon sperm DNA.
8. Zeta-probe membrane (Bio-Rad).
9. Standard water baths and incubators.
10. Mitomycin C-treated mouse embryonic fibroblasts isolated from d 13.5 C57BL/6J×129/Sv embryos.
11. 129/Sv mouse embryonic stem cells.
12. Dulbecco's modified Eagle's medium (DMEM), glutamine, nonessential amino acids, and pen-strep (GIBCO-BRL).
13. Fetal calf serum (Hyclone, UT): Heat-inactivate at 56°C, aliquot, and store at –20°C as suggested by the vender.
14. β-Mercaptoethanol (Bio-Rad): Dilute to 10 mM in phosphate-buffered saline (PBS), filter sterilize, and store at 4°C. Make fresh solution every 2 wk.
15. Leukemia inhibitory factor (also called ESGRO™, GIBCO-BRL), store at 4°C.
16. 100X Nucleosides: Dissolve 80 mg of adenosine, 85 mg of guanosine, 73 mg of cytindine, 73 mg of uridine, and 24 mg of thymidine (Sigma, MO) in 100 mL of water, filter sterilize, and store at –20°C.
17. G418 sulfate (also called Geneticin, GIBCO-BRL): Make a 15 mg/kg stock, filter sterilize, and store at –20°C.
18. Standard tissue culture plasticware, serological pipets, and filter units.
19. Standard tissue culture incubators, tissue culture hoods, and light microscope.
20. C57BL/6J male and female mice, F1(C57BL/6J×DBA2) female mice and vasectomized F1(C57BL/6J×DBA2) male mice (The Jackson Laboratory, ME).
21. Standard surgical equipment, including scissors, forceps, and a dissecting microscope.
22. 100% Avertin: Mix 10 g of tribromoethyl alcohol with 10 mL of tertiary amyl alcohol. Dilute to 2.5% in water, wrap in foil, and store at 4°C.
23. Pipet puller (Sutter Instrument Co., CA).
24. Borosilicate glass pipets for making holding, injection and transfer pipets, 0.8 mm inner diameter (i.d.) and 1.0 mm outer diameter (o.d.) (Corning, NY).
25. Wound clips (Arthur H. Thomas Co., PA)

Generating Knockout Mice

26. An injection setup including two Leitz micromanipulators (Leica, IL), an inverted fixed-stage microscope with a cooling stage (Nikon, Japan) and an air table.
27. 20X SSC: 175.3 g of NaCl, 88.2 g of sodium citrate in 1 L of water, pH 7.0.
28. Tris-EDTA (TE): 10 mM Tris-HCl, 1 mM EDTA, pH 7.0.
29. SM buffer: 5.8 g of NaCl, 2 g of $MgSO_4 \cdot 7H_2O$, 50 mL of 1 M Tris-HCl, pH 7.5, and 5 mL of 2% gelatin in 1 L of water.
30. EF culture medium: 10% Fetal calf serum, 1X pen-strep, and 1X glutamine in DMEM medium, filter sterilized and stored at 4°C.
31. ES medium: 15% fetal calf serum, 0.5X pen-strep, 1X glutamine, 1X nonessential amino acids, 1X nucleosides, 10 mM β-mercaptoethanol, and leukemia inhibitory factor, at 100 U/mL, in DMEM medium, filter sterilized and stored at 4°C.
32. Gene Pulser Electroporator (Bio-Rad).
33. 2X Frozen medium: 80% fetal calf serum and 20% dimethyl sulfoxide, filter sterilized.
34. DNA digestion buffer: 100 mM Tris-HCl, pH 8.0, 5 mM EDTA, 0.2% SDS, 200 mM NaCl with 100 μg/mL of pronase K.

3. Methods

The process of making a knockout mouse includes the generation of a DNA targeting construct, homologous recombinants in ES cells, and knockout mice. We assume that interested readers have some knowledge about the techniques in basic molecular biology and cell culture.

3.1. Targeting Construct

The goal for this step is to clone genomic DNA sequences covering the gene of interest to make a targeting construct, and to clone hybridization probes for screening homologous recombinants.

3.1.1. Gene Cloning

Once a target gene is chosen, the first step is to clone and map in detail a segment of the DNA covering parts of the gene (*see* **Note 1**).

1. Obtaining a gene-specific probe: Use a cDNA or DNA sequences covering exons of the gene of interest if available. Alternatively, perform a polymerase chain reaction (PCR) to obtain a segment of the gene as the probe if the sequence information for the gene of interest is known. We prefer probes covering the first one or two exons of the gene. Check the probe by restriction enzyme digestion followed by gel electrophoresis to ensure its validity. Alternatively, sequence the probe for verification.
2. Screening the genomic library: Plate out a mouse 129 λ phage genomic library of 1.8 million plaque forming units on 30 150-mm plates. Use XL1-Blue cells as the infection host. Lift two sets of filters from the 30 plates. Use a ^{32}P-labeled

gene-specific DNA probe to hybridize with the two sets of filters separately at 65°C for 16–20 h. Wash the filters at 65°C in 0.1X SSC containing 0.5% SDS for four times with 30 min each and expose the filters. After autoradiography, pick out plaques that produce common positive hybridization signals on both sets of filters. Plate out these plaques in duplicate at 500 and 2500 plaque forming units per 150-mm plate and screen again until each positive plaque is free of contaminating plaques.
3. Purifying the genomic clones: Amplify three or four individual positive plaques by plating out on 10 150-mm plates at 300,000 plaque forming units per plate. Collect the plaques by incubating in an SM buffer for 3 h, incubate at room temperature for 30 min with 1 µg/mL of pancratic DNase and RNase, precipitate in polyethylene glycol and 1 M NaCl at 4°C for 1 h, centrifuge at 11,000g, and resuspend the phage particles in TE containing 0.5% SDS. Digest the phages with 100 µg/mL of pronase K for 2 h at 56°C, precipitate the phage DNA in ethanol, and resuspend in 0.5 mL of TE, pH7.0.
4. Restriction mapping the genomic clones: Perform single and double restriction enzyme digestions of individual genomic clones followed by gel electrophoresis. Construct a relatively detailed restriction map based on the digestion results. The analysis of multiple genomic clones usually gives a restriction map for 15–20 kilobase pair sequences, which will be usually sufficient for making both the targeting construct and the hybridization probes.
5. Gene cloning by direct PCR: Given a significant portion of the mouse genomic sequence is known to the public, one can refer to the sequence information and perform PCR to clone the sequences required to make both the targeting construct and the hybridization probes.

3.1.2. The Targeting Construct

The design is to make a construct in which a selectable maker gene is flanked by DNA sequences upstream (5′ arm) and downstream (3′ arm) of the region that will be deleted on homologous recombination (**Fig. 1**). In general, the larger the total length of DNA sequences used in the targeting construct, the higher the homologous recombination rate will be. We routinely use 8–12 kilobase pairs of total DNA sequence for making the targeting constructs (*see* **Note 2**).

1. Preparation of 5′ and 3′ arms: Digest 50 µg of the genomic clones by appropriate restriction enzymes, and separate the DNA segments by gel electrophoresis. Purify the relevant DNA segments using a gene clean kit (BIO 101, CA) and resuspend in TE. Alternatively, obtain the DNA segments by PCR.
2. Cloning vector preparation: We have used a cloning vector based on the KS-plasmid which also contains a selectable marker, the *neo* gene driven by a phosphoglycerate kinase gene promoter. This cloning vector contains an array of cloning sites that are convenient for making the targeting construct. Digest this vector with relevant restriction enzymes that match the ends in the 5′ and

Generating Knockout Mice

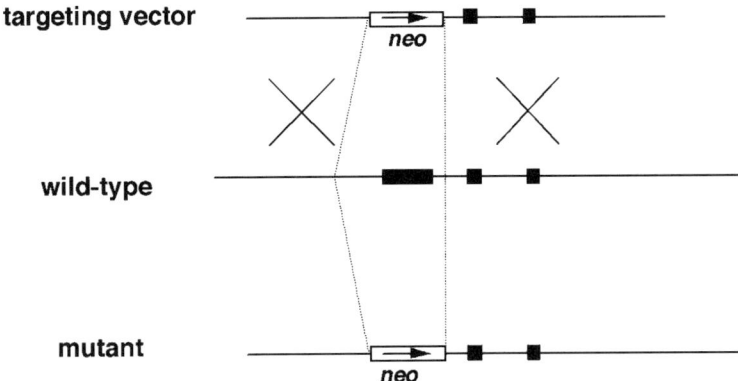

Fig. 1. Strategy for generating a knockout mouse. Genomic DNA locus of a hypothetical wild-type gene with three exons (*black boxes*); the targeting vector and the mutant gene locus are indicated. In this particular design, the first exon of the gene is replaced by a selectable marker, the *neo* gene, on proper homologous recombination.

3′ arms such that they can be ligated appropriately to generate the targeting construct, perform gene clean procedure to purify the cloning vector DNA, and resuspend in TE.

3. Targeting vector construction: Ligate the 5′ and 3′ arms with the cloning vector. Depending on the ligation strategy, perform either stepwise ligation, which is usually easier, or four-piece ligation. Transform the ligation mix into the XL1-Blue bacteria and plate out on LB plates containing ampicillin. Grow individual colonies at 37°C and isolate the plasmid individually. Digest the plasmid DNA and perform gel electrophoresis to screen for the targeting construct (*see* **Note 3**).

3.1.3. Quantification and Verification of the Targeting Construct

Once the targeting construct is made, grow 0.5 L of bacteria harboring the construct. Isolate and purify the plasmid twice by CsCl ultracentrifugation. Verify the plasmid by restriction enzyme digestion and gel electrophoresis. Quantify the plasmid by OD_{260} reading.

3.1.4. Hybridization Probes

The 5′ and 3′ probes should ideally be located outside of the 5′ and 3′ arms used in making the targeting construct, detect DNA size changes predicted by the desired homologous recombination, and detect single-copy DNA sequences in the genome. Isolate candidate probe DNA fragments and perform genomic Southern blotting to verify the probes. Alternatively, perform PCR to obtain the probes and verify by Southern blotting (*see* **Notes 1** and **4**).

3.2. Homologous Recombinants

The goal for this step is to obtain multiple ES cell clones carrying a homologous recombination in the desired gene locus. For the best results, the whole process should be performed in a designated tissue culture room (*see* **Note 5**).

3.2.1. Mouse Embryonic Fibroblasts (EF) and ES Cell Culture and Targeting Construct Preparation

1. EF cell culture: Treat a T25 and a T175 tissue culture flask with 5 and 30 mL of 0.1% gelatin for 20 min at room temperature. Thaw 3×10^6 mitomycin C-treated mouse EF feeder cells at 37°C, mix with 9 mL of EF culture medium, and centrifuge at 2000g for 5 min. Remove the supernatant, resuspend cells in 30 mL of EF medium, and distribute 5 mL and 25 mL of cells to the two tissue culture flasks, respectively. Culture the EF feeder cells at 37°C with humidity and 5% CO_2.
2. ES cell culture: Four hours after plating out the EF cells, thaw 3×10^6 ES cells and plate out in one T25 flask in 5 mL of ES medium. Change medium daily and take extreme care not to overgrow the cells. When the cells become 50% confluent, remove supernatant, add 2 mL of trypsin-EDTA, incubate at 37°C for 5 min, pipet the cells up and down to make a single cell suspension, and add to the T175 flask that contains the EF feeder cells (*see* **Note 6**).
3. Targeting construct preparation: Linearize 70 µg of the targeting construct either at the 5′ end of the 5′ arm or at the 3′ end of the 3′ arm by restriction enzyme digestion, check the digest by gel electrophoresis, and precipitate the DNA by ethanol. Take the linearized targeting construct into the tissue culture room, wash with 70% ethanol, air-dry for 15 min, and resuspend in 600 µL of PBS, pH 7.4.

3.2.2. ES Cell Transfection and G418 Selection

1. Harvesting ES cells: Plate out 9×10^6 EF feeder cells in 10 100-mm tissue culture dishes at least 4 h before the transfection. The ES cells are ready for transfection when they become 50% confluent in the T175 flask. Remove medium from the T175 flask and treat the ES cells using 5 mL of trypsin–EDTA for 5 min. Pipet the cells up and down to make a single cell suspension, add 5 mL of ES medium, and collect ES cells by centrifuging at 2000g. Remove the supernatant and tap the remaining cell pellet to resuspend.
2. Electroporation: Combine the ES cells with the linearized DNA and transfer the mixture into a 0.4-cm electroporation cuvette. Use a Gene Pulser Electroporator at the following settings: 800 V and 3.0 µF with no resistance and no capacitance extender. The time constant after the electroporation is typically 0.1 ms. Incubate the cell–DNA mixture on ice for 10 additional minutes and resuspend the mixture in 100 mL of ES cell medium. Distribute the cells into 10 100-mm culture dishes and incubate at 37°C with humidity and 5% CO_2.

Generating Knockout Mice

3. G418 selection: At 16–24 h after the transfection, change medium and add G418 at 200 µg/mL. Change medium daily with G418 and most of cell death is observed around d 3–4 of G418 selection.

3.2.3. Picking Colonies

1. Preparation: Around d 7 of G418 selection, individual colonies representing individual transfectants become visible and they are ready to be picked. Plate out EF feeders on 96-well flat bottom plates at the same concentration as before. Change to ES medium just before colony picking.
2. Picking colonies: Change medium 2 h before colony picking. Pick out individual colonies under the microscope using a pipetman under a tissue culture hood. Care should be taken to pick only undifferentiated colonies. Trypsinize the colonies for 5 min, resuspend in 200 µL of ES medium, and place in 96-well plates. Three hundred or more colonies should be picked per transfection.
3. Passage and freezing of ES colonies: Change medium daily. When the cells are 50% confluent, trypsinize and split the individual ES clones into two sets of 24-well dishes that also contain EF feeder cells. One set will be for keeping a source of ES cells for future amplification. When the cells are 50% confluent, freeze the cells by adding 0.5 mL of 2X frozen medium. Store the plates at –80°C.

3.2.4. Identification of Homologous Recombinants

1. Genomic DNA isolation: Another set of 24-well plates will be for genomic DNA isolation. To obtain enough DNA, culture the cells till they are overgrown. Remove the culture medium and add 0.5 mL of DNA digestion buffer. Incubate the plates at 56°C overnight. Mix the digests with equal volume of isopropanol very well. Fish out the genomic DNA and resuspend the DNA in 50 µL of TE.
2. Southern blotting analysis of DNA: Digest all DNA samples with proper restriction enzymes, separate by gel electrophoresis, and transfer to membranes. Hybridize the filters with either ^{32}P-labeled 5' or 3' probes. Verify potential homologous recombinants using the other probe. Alternatively, perform PCR for screening potential homologous recombinants.

3.2.5. ES Cell Amplification and Verification

1. ES cell amplification: Once homologous recombinant clones are identified, thaw ES cells in the 24-well frozen dishes and plate out in 24-well dishes with EF feeders. Change medium with G418 daily and do not overgrow cells. Once 50% confluent, trypsinize and split ES cells into 12 wells, then to six wells, and eventually to two T25 flasks sequentially. When cells are 50% confluent, freeze them into 10 vials and store them in liquid nitrogen. Three or more independent ES cell clones should be amplified.
2. ES cell verification: Isolate genomic DNA from the amplified ES cell clones and perform Southern blotting with both the 5' and 3' probes as before. Alternatively,

perform PCR for such verification. Once verified, ES cells are ready for injection (*see* **Notes 7** and **8**).

3.3. Knockout Mice

The goal for this step is to introduce the ES cells harboring proper homologous recombination into mouse blastocysts to generate male chimeric mice. If the injected ES cells, which are of 129 origin, contribute to the development of a mouse, the coat color of the mouse may be chimeric which can be easily visualized. Each injection process takes 5 d. Then, the chimeras are bred repeatedly with C57BL/6J mice to obtain mice homozygous for the desired gene mutation.

3.3.1. Establishing a Mouse Colony

Two groups of mice are required. One group is for blastocyst production. For traditional reasons, C57BL/6J inbred strain of mouse is used. Forty male stud mice 6–16 wk of age and 160 female mice 4–8 wk of age are needed. Another group is to provide foster mothers. Twenty vasectomized C57BL/6J×DBA2 F1 males 6–16 wk of age and 80 normal C57BL/6J×DBA2 F1 females 10–18 wk of age are needed. This steady-state colony size is sufficient to support three injections per week.

3.3.2. Blastocysts and ES Cell Preparation

1. Timed breeding: On d 1, select 20 C57BL/6J female mice that are in estrous cycle and breed with C57BL/6J male mice individually between 4:00 and 6:00 P.M. On d 2, check the female C57BL/6J mice for vaginal plugs before 11:00 A.M. On the same day, select 10 C57BL/6J×DBA2 F1 female mice that are in estrous cycle and breed with individual vasectomized C57BL/6J×DBA2 F1 male mice between 4:00 and 6:00 P.M. On d 3, check plugs for the C57BL/6J×DBA2 F1 females before 11:00 A.M. More than 50% of the females are plugged on the average. The plugged C57BL/6J females and C57BL/6J×DBA2 F1 females can be group-housed separately. If a higher yield of blastocysts is desired, superovulation can be performed.
2. ES cell preparation: Treat three wells in a 12-well tissue culture dish with 0.1% gelatin for 20 min at room temperature. Thaw one vial of amplified ES cells from **Subheading 3.2.5.**, resuspend in 9 mL of ES cell medium, centrifuge at 2000*g*, resuspend in 3 mL of ES cell medium, plate out in three wells in the 12-well dish, and incubate at 37°C in 5% CO_2. No EF feeder cells are needed. On d 4, change medium for the ES cells. On d 5, change medium again and ES cells should be about 30–60% confluent. Just before the injection, remove culture medium, trypsinize for 30 s and resuspend in 1 mL of ES cell medium one well at a time.
3. Blastocyst isolation: On d 5, set up a 25-gauge needle with a 10-mL syringe and a 50-mm Petri dish with two drops of ES medium covered with dimethylpolysi-

loxane. Kill all plugged female C57BL/6J mice by cervical dislocation. Dissect out uteri and flush into a 50-mm Petri dish with ES medium using the syringe. Collect blastocysts under the microscope using a transfer pipet connected to a mouthpiece. Transfer all blastocysts into the above culture dish and leave the dish in the tissue culture incubator. Although the yield may vary, 50 blastocysts typically can be obtained.

3.3.3. Blastocyst Injection

1. Setting up an injection dish: In the center of a 50-mm Petri dish, place 0.5 mL of ES medium in a rectangular way. Cover the medium with dimethylpolysiloxane. Transfer about 10 blastocysts and a few hundred ES cells into the injection dish by mouth pipeting. Adjust both the holding pipet and the injection pipet so that they are in the same plane.
2. Blastocyst injection: Use the injection pipet to collect undifferentiated ES cells with minimum volume of medium under the microscope. Avoid air bubbles. Find and hold a healthy blastocyst from the side with the inner cell mass using the holding pipet. Move the tip of the injection pipet close to the blastocyst and penetrate the wall of the blastocyst. Gently release 8–10 ES cells into the blastocyst. Gently move the injection pipet when exiting the injected blastocyst to prevent the ES cells from being expelled out. Repeat the injection process (*see* **Note 9**).
3. Blastocyst culture: Transfer the injected blastocysts to another Petri dish containing the ES medium covered with dimethylpolysiloxane. Culture the injected blastocysts in the incubator for at least 2 h before returning them to foster mothers.
4. Additional injections: Each ES clone should be injected at least twice. Moreover, three or four independent ES clones should be injected (*see* **Note 10**).

3.3.4. Transfer of the Injected Blastocysts Back into Foster Mothers

1. Anesthetize the foster mothers by an intraperitoneal injection of 2.5% avertin at 0.016 mL/g of body weight.
2. Make a small incision on one side of the mouse and avoid lesioning blood vessels. Pull out one side of the uterus through the associated fat using a pair of forceps.
3. Collect about six to eight injected and two noninjected blastocysts using a transfer pipet connected to a mouthpiece. Wash the blastocysts twice in fresh ES cell medium.
4. Gently punch the uterine wall from the oviduct end of the foster mother with a 25-gauge needle. Then, put the tip of the transfer pipet in and gently blow the blastocysts into the uterus. Close the incision with a wound clip.
5. Leave the foster mothers in a cage on top of a 37°C warm plate till waking up. Return foster mothers to the mouse colony. Single house foster mothers when they are ready to give birth.

3.3.5. Identification of Mice Carrying the Desired Gene Mutation and Breeding

1. Chimeric mice: At 17–18 d after the blastocyst return, the foster mothers will give birth to pups. About 10 d after birth, the percent chimerism for individual pups can be determined. When reaching 6 wk of age, male chimeric mice with a high percentage of chimerism are bred with C57BL/6J female mice.
2. Germ-line transmission of the desired gene mutation: When the C57BL/6J females give birth after breeding with the male chimeric mice, the appearance of the agouti color pups is the first indication that the injected ES cells have contributed to the germ cell development. To verify this, isolate genomic DNA from the tail biopsies followed by either Southern blotting or PCR as described previously.
3. Homozygous knockout mice: Breed heterozygous mice to obtain mice homozygous for the desired deletion. Identify the knockout mice by either Southern blotting or PCR as described previously (*see* **Notes 11–15**).

4. Notes

1. Care should be taken that the right gene has been cloned particularly when dealing with a gene family. Preferably, DNA sequencing is performed or at least detailed restriction mapping information is obtained before making the targeting construct. Moreover, the hybridization probes should be specific for the gene of interest and should be similarly verified.
2. When designing the targeting construct, we prefer to eliminate the translation start site rather than to insert a sequence in an exon, particularly in the downstream exons. This helps to eliminate the potential of generating a part of the gene product or a hybrid gene product in the knockout mice.
3. Other researchers have used a thymidine kinase (*TK*) gene together with the *neo* gene in the targeting construct to increase the homologous recombination rate. To minimize the exposure of ES cells to chemicals, we have not used the *TK* gene for selection. In our experience, we get a reasonable recombination rate with a 5–20% range.
4. If ES cells from the 129 mouse strain will be used to generate the knockout mice, for maximum homologous recombination rate, genomic library constructed from DNA isolated from the same mouse strain should be used.
5. For high-quality cell culture work, use sterile and disposable plasticware. Moreover, use the highest grade chemicals when possible. If making up own cell culture medium, use double distilled water.
6. Extreme care should be taken to limit potential differentiation of ES cells. Leukemia inhibitory factor should be added to the ES cell medium and ES cells should not be overgrown.
7. After the amplification of ES cells, genomic DNA should be analyzed again by either Southern blotting or PCR to ensure the amplified cells are relatively homogeneous with respect to the gene mutation.

8. Before the blastocyst injection, a mycoplasma test is recommended to ensure that the ES cells are not contaminated, which will prevent germ-line transmission.
9. During the blastocyst injection, choose the best ES cells that are not differentiated and are with a round morphology and sharp edges. Do not choose those that have a flat morphology and fussy edges.
10. It is difficult to predict which ES cell clone will give rise to mice with a high percentage of chimerism and germ-line transmission. To increase the chance of germ-line transmission, multiple ES clones should be injected.
11. Maintain a pathogen-free mouse colony.
12. Mice with different genetic background can perform differently, particularly in behavioral studies. Depending on the goal of the analysis, littermate controls should be used. Preferably and eventually, one will need to breed the knockout mice back into a pure genetic background of choice.
13. Because developmental compensations can occur during mouse development, care should be taken to check for any detectable deficiencies in the knockout mice. Moreover, caution should be exercised in data interpretation. Conditional gene targeting may help to overcome such compensations.
14. Timetable: Targeting construct: 1–3 mo. Homologous recombinants: 2 mo. Knockout mice: 6–7 mo.
15. After obtaining the knockout mice, one will need to compare the expression of gene of interest in the knockout mouse and normal control mouse. Moreover, we recommend performing a general development and behavioral survey of the knockout mouse. Then, depending on the scientific questions asked, one can combine behavioral, electrophysiological, neuroanatomical, and molecular biological approaches to analyze the knockout mice. The combined use of genetically engineered mouse models with proper analyses has tremendous potential to provide novel insights into mechanisms underlying drug addiction.

Acknowledgment

J. Z. and M. X. were supported by grants from the NIH, the U.S. Army, and the Epilepsy Foundation of America.

References

1. Leshner, A. I. (1997) Addiction is a brain disease, and it matters. *Science* **278,** 45–47.
2. Koob, G. F., Sanna, P. P., and Bloom, F. E. (1998) Neuroscience of addiction. *Neuron* **21,** 467–476.
3. Berke, J. D. and Hyman, S. E. (2000) Addiction, dopamine, and the molecular mechanisms of memory. *Neuron* **25,** 515–532.
4. Nestler, E. J. and Aghajanian, G. K. (1997) Molecular and cellular basis of addiction. *Science* **278,** 58–63.
5. Le Moal, M. and Simon, H. (1991) Mesocorticolimbic dopaminergic network: functional and regulatory roles. *Physiol. Rev.* **71,** 155–234.

6. Koob, G. F. (1992) Drugs of abuse: anatomy, pharmacology and function of reward pathways. *Trends Pharmacol. Sci.* **13,** 177–184.
7. Wolf, M. E. (1998) The role of excitatory amino acids in behavioral sensitization to psychomotor stimulants. *Prog. Neurobiol.* **54,** 679–720.
8. Vanderschuren, L. J. and Kalivas, P. W. (2000) Alterations in dopaminergic and glutamatergic transmission in the induction and expression of behavioral sensitization: a critical review of preclinical studies. *Psychopharmacology* **151,** 99–120.
9. Doetschman, T., Gregg, R. G., Maeda, N., Hooper, M. L., Melton, D. W., Thompson, S., and Smithies, O. (1987) Targeted correction of a mutant HPRT gene in mouse embryonic stem cells. *Nature* **330,** 576–578.
10. Thomas, K. R. and Capecchi, M. R. (1987) Site-directed mutagenesis by gene targeting in mouse embryo-derived stem cells. *Cell* **51,** 503–512.
11. Evans, M. J. (2001) The cultural mouse. *Nat. Med.* **7,** 1081–1083.
12. Smithies, O. (2001) Forty years with homologous recombination. *Nat. Med.* **7,** 1083–1086.
13. Capecchi, M. R. (2001) Generating mice with targeted mutations. *Nat. Med.* **7,** 1086–1090.
14. Xu, M., Moratalla, R., Gold, L. H., et al. (1994) Dopamine D1 receptor mutant mice are deficient in striatal expression of dynorphin and in dopamine-mediated behavioral responses. *Cell* **79,** 729–742.
15. Drago, J., Gerfen, C. R., Lachowicz, J. E., et al. (1994) Altered striatal function in a mutant mouse lacking $D1_A$ dopamine receptors. *Proc. Natl. Acad. Sci. USA* **91,** 12,564–12,568.
16. Baik, J., Picetti, R., Saiardi, A., et al. (1995) Parkinsonian-like locomotor impairment in mice lacking dopamine D2 receptors. *Nature* **377,** 424–428.
17. Giros, B., Jaber, M., Jones, S. R., Wrightman, R. M., and Caron, M. G. (1996) Hyperlocomotion and indifference to cocaine and amphetamine in mice lacking the dopamine transporter. *Nature* **379,** 606–612.
18. Accili, D., Fishburn, C. S., Drago, J., et al. (1996) A targeted mutation of the D_3 dopamine receptor gene is associated with hyperactivity in mice. *Proc. Natl. Acad. Sci. USA* **93,** 1945–1949.
19. Xu, M., Tirado, G., Moratalla, R., White, N. M., Graybiel, A. M., and Tonegawa, S. (1997) Dopamine D3 receptor mutant mice exhibit increased behavioral sensitivity to concurrent stimulation of D1 and D2 receptors. *Neuron* **19,** 837–848.
20. Rubinstein, M., Phillips, T. J., Bunzow, J. R., et al. (1997) Mice lacking dopamine D4 receptors are supersensitive to ethanol, cocaine, and methamphetamine. *Cell* **90,** 991–1001.
21. Kelly, M. A., Rubinstein, M., Phillips, T. J., et al. (1998) Locomotor activity in D2 dopamine receptor-deficient mice is determined by gene dosage, genetic background, and developmental adaptations. *J. Neurosci.* **18,** 3470–3479.
22. Sambrook, J., Fritsch, E. F., and Maniatis, T., eds. (1989) *Molecular Cloning, A Laboratory Manual*, 2nd edit. Cold Spring Harbor Laboratory Press, Cold Spring Harbor, NY.

23. Hogan, B., Costantini, F., and Lacy, E., eds. (1986) *Manipulating the Mouse Embryo: A Laboratory Manual.* Cold Spring Harbor Laboratory Press, Cold Spring Harbor, NY.
24. Wassarman, P. M. and DePamphilis, M. L., eds. (1993) *Methods in Enzymology: Guide to Techniques in Mouse Development*, Vol. 225. Academic Press, San Diego.
25. Doetschman, T. (1994) Gene transfer in embryonic stem cells, in *Transgenic Animal Technology: A Laboratory Handbook* (Pinkert, C. A., ed.), Academic Press, San Diego, pp. 115–146.

24

Antisense Approaches in Analyzing the Functional Role of Proteins in the Central Nervous System

John Q. Wang and Limin Mao

1. Introduction

Along with the emergence and development of the new field of "proteomics," neuroscientists together with other life scientists have been offered the greater opportunity to clone and sequence new biologically active proteins and to assess alterations in protein expression at both mRNA and protein levels in response to a variety of experimental manipulations (1). As proteins serve as important elements and molecules in cellular structures, receptors, transporters, factors, enzymes, and so on, functional roles of these various proteins largely determine cellular physiology. To assess the role of proteins, especially receptors and enzymes, pharmacological antagonists/inhibitors have been used as effective tools for decades. However, although these pharmacological tools possess potent antagonistic actions, the selectivity against their specific targets is relative and sometimes far from satisfactory. Moreover, the antagonistic properties of these tools, which are usually characterized in vitro, may not be reflected in vivo. Lack of selective agents at all is another dilemma for analyzing the function of some proteins, such as structural proteins, transcription factors, and so forth.

Recent advances in molecular biology have led to the development of an entirely new type of agents for analyzing functions of biological proteins. This agent is a short, single-stranded DNA sequence, for example, antisense oligodeoxynucleotide (2,3), which is complementary to a selected region of an endogenous mRNA encoding a specific protein. By binding to selected areas of the target mRNAs in a sequence-specific manner, agents interfere with the mRNA-directed translation and, as a result, prevent the synthesis of proteins.

The reduced protein production may lead to a deficit of protein-mediated functions that could be detected experimentally after antisense application. Apparently, antisense oligonucleotides provide a new and promising approach to alter selectively the synthesis of any proteins with known sequences and subsequently to analyze their functional roles. Indeed, the antisense approach has been utilized frequently and successfully in recent years to evaluate the functional role of various proteins in brain cells both in vivo and in vitro *(2,3)*. In the basal ganglia, this technology has also demonstrated its valid efficacy and selectivity in reducing protein levels and therefore functions, including receptors *(4–7)*, transporters *(8,9)*, kinases *(10)*, neuropeptides *(11)*, neurotrophic factors *(12,13)*, or transcription factors *(14,15)*.

The principle of antisense technology is to reduce the synthesis of proteins by preventing their translation at the mRNA level. The precise mechanisms underlying the antisense attenuation are not fully understood. However, two major mechanisms have been proposed in this regard. As illustrated in **Fig. 1**, the antisense oligonucleotide is a short sequence of deoxynucleotides complementary to a portion of the target mRNA. This antisense oligo can hybridize with the mRNA and physically prevent it from being translated into a protein, i.e., translational arrest. Besides the translational arrest mechanism, enhanced degradation of antisense–mRNA hybrids by the enzyme RNase H is believed to be another major mechanism *(2,3)*. In addition to the two major mechanisms, other possible mechanisms have also been suggested. For instance, heterologous nuclear RNA (hnRNA) is converted into the matured template mRNA by removing the introns (noncoding segments between exons) and adding a cap and poly(A) tail. If designed to hybridize at a splice (intron–exon) junction, a cap site, or a poly(A) tail site, the antisense oligos may arrest the process that converts hnRNA into the template mRNA *(16)*. Antisense may also hybridize with DNA in the nucleus to block transcription *(2,17)*.

There are obvious advantages of using antisense oligos *(2)*. Among the main advantages are its high selectivity and specificity in inhibiting target proteins. Because antisense oligos bind to a given mRNA molecule in a sequence-specific fashion, these oligos are thought to achieve higher selectivity than pharmacological antagonists in interfering protein production and functions. Another main advantage is the ability of oligos to reduce the expression of proteins for which no pharmacological agents are available, or where the available agents show poor specificity. This advantage seems very useful for probing the functional role of intracellular and intranuclear proteins, such as organelle structural proteins, enzymes, kinases, transcription factors, and oncogenes, given the fact that the development of pharmacological antagonists for these proteins is difficult. In addition, blocking effects of pharmacological agents at receptors may result in the compensatory up-regulation of receptor

Antisense Pharmacology

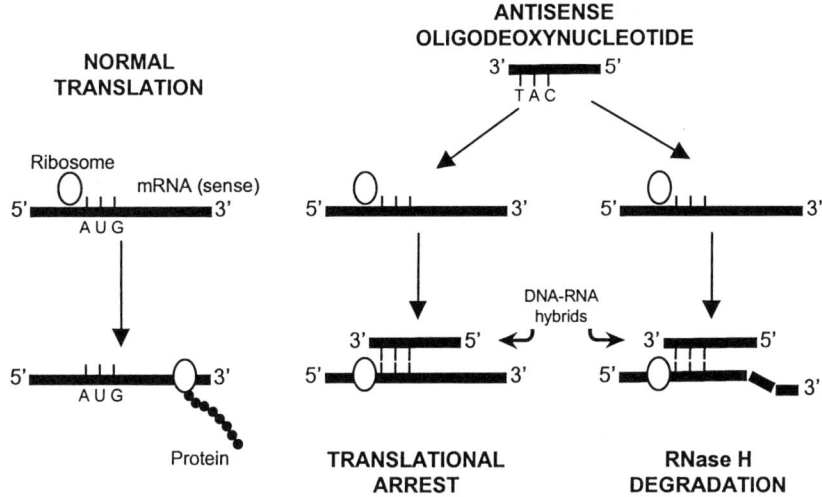

Fig. 1. A schematic diagram illustrating the process of normal protein translation and two principal mechanisms responsible for the suppression of protein translation by antisense oligodeoxynucleotides. DNA has a sense and antisense strands, and mRNA is copied from the antisense strand so that it has the same sequence of the sense DNA (except for base T in DNA and base U in RNA). Antisense oligodeoxynucleotides have the same antisense DNA sequence complementary to a portion of endogenous mRNAs, usually around the AUG initiation codon. Thus, they bind to the target mRNAs in a sequence-specific fashion to form DNA–RNA hybrids, preventing ribosome-catalyzed protein translation (translation arrest). Alternatively, mRNA and antisense DNA hybrids might be more readily or rapidly degraded by RNase H.

gene expression and functions via a negative feedback mechanism. The antisense blockade of receptor production is less likely accompanied by such a compensatory up-regulation. Taken together, antisense technology, because of its remarkable advantages, represents a useful tool for analyzing protein functions both in vivo and in vitro.

Antisense technology also has its own disadvantages. First, it can only be used to reduce the synthesis of a protein whose gene sequence is known. Second, complete suppression of protein production is rarely, if ever, obtained as a result of antisense application. The incomplete suppression of protein synthesis is particularly likely when constitutive expression of a protein is high and turnover of the protein is low. The incomplete blockade may not result in any physiological consequences if remaining proteins can still carry out their functions. Finally, in vivo use of antisense oligonucleotides in the central nervous system may be toxic to the cells exposed to antisense oligos. The reduction of target protein content at a tissue level may result from cell loss instead of protein synthesis inhibition in intact cells of tissue. Thus, careful designing and modulation of antisense molecules are required to minimize neurotoxicity. In general, successful use of an antisense strategy is not always guaranteed because of the disadvantages described above. The appropriate and effective use of this method can be better served by acknowledging and minimizing its limitations.

In this chapter, an example of the in vivo experiment that was recently conducted in this laboratory with antisense oligos against metabotropic glutamate receptor subtype 5 (mGluR5) is presented to illustrate different perspectives of antisense methodology. The mGluR5 antisense oligonucleotides that were directly infused into the rat dorsal striatum showed a successful knockdown of mGluR5 expression in striatal neurons and affected cocaine- and amphetamine-stimulated behaviors and striatal neuropeptide gene expression. In addition to the mGluR5 receptor, the following protocol and others *(18)* are also largely applicable for knocking down many other surface receptors and intracellular signaling proteins in striatal neurons either in vivo or in vitro.

2. Materials

1. Male young adult rats averaging 200–225 g at time of surgery.
2. Stereotaxic surgical materials on small animals:
   a. Stereotaxic apparatus.
   b. Dual-pipe fiberoptic illuminators.
   c. Heating pad.
   d. Small animal clippers.
   e. Syringes (5 mL) with 22-gauge needles.
   f. Instruments (e.g., scalpel, scissors, spatula, eye forceps, hemostat, autoclip wound clips, and antoclip applier).
   g. High-speed microdrill and steel burrs (Fine Science Tools).
   h. Dental carboxylate cement (Durelon).
   i. Screwdriver and stainless steel machine screws (Small Parts).
   j. Sterile towel/drape, Povidone–iodine swabsticks and cotton swabs.
   k. The rat brain atlas by Paxinos and Watson *(19)*.
3. Hot glass bead sterilizer.
4. Anesthetic solution (4% chloral hydrate).
5. Intrastriatal infusion system:
   a. 24-Gauge stainless steel guide cannula cut 10 mm (Small Parts).
   b. 30-Gauge stainless steel wire (stylet) cut 10 mm (Small Parts).
   c. 30-Gauge stainless steel cannula (injector) 2.5 mm longer than guide cannula (Small Parts).
   d. Standard PE10 catheter tubing and 10-µL Hamilton syringe.
   e. Microliter syringe pump (Harvard Apparatus).

6. Antisense oligodeoxynucleotides and controls (custom-synthesized, 18-mer phosphorothioate-modified and HPLC-purified; GIBCO/BRL) (*see* **Note 1**):
   MGluR5 antisense oligonucleotide complementary to nucleotide –6 through +12 of the rat mGluR5 mRNA *(20)*: 5'-CAGAAGGACCATTTTAGG-3'. Position +1 corresponds to the A nucleotide of the initiation AUG codon on mGluR5 mRNA (ATG on DNA) (*see* **Note 2**).
   Three control oligodeoxynucleotides (*see* **Note 3**):
   a. Sense oligonucleotide: 5'-CCTAAA<u>ATG</u>GTCCTTCTG-3'.
   b. Mismatched oligonucleotide (3 mismatched bases in the mGluR5 antisense oligonucleotide): 5'-CAGAAG<u>C</u>ACCATA<u>T</u>TA<u>C</u>G-3'.
   c. Scrambled oligonucleotide (scrambled bases in the mGluR5 antisense oligonucleotide): 5'-GCTGATACAGAGATGACT-3'.
7. Sterile artificial cerebrospinal fluid (aCSF): 123 mM NaCl, 0.86 mM CaCl2, 3.0 mM KCl, 0.89 mM MgCl$_2$, 25 mM NaHCO$_3$, 0.50 mM NaH$_2$PO$_4$, and 0.25 mM Na$_2$HPO$_4$ aerated with 95% O$_2$–5% CO$_2$, pH 7.2–7.4.

Use sterile technique during preparation of all solutions, including the oligodeoxynucleotide solutions. All surgical instruments must be sterile, and surgical and oligo infusion procedures should be performed in a sterile manner.

3. Methods

The methods described below outline (1) implantation of guide cannula in the striatum, (2) intrastriatal infusion of oligonucleotides, and (3) assessment of efficacy and selectivity of antisense oligonucleotides.

3.1. Implantation of Guide Cannula

Before surgery, all surgical instruments need to be sterilized by placing them in the hot glass bead sterilizer. The sterilizer, by allowing rapid sterilization of instruments, is better adapted than an autoclave for surgery involving multiple animals.

1. Anesthetize rats by injecting 4% chloral hydrate intraperitoneally at a dose of 400 mg/kg using a 5-mL syringe with a 22-gauge needle.
2. Shave the top of head and place the head in the stereotaxic apparatus. Wipe the top of head with Povidone–iodine swabsticks before making a midline sagittal incision with a scalpel and exposing the skull.
3. Scrape the exposed bone area gently with cotton swabs to remove the periosteal connective tissue and stop any bleeding.
4. Locate the bregma and determine the location for cannula placement in the left or right hemisphere using the coordinates: 1.0 mm rostral to bregma and 2.0–2.5 lateral to midline (5 mm below the skull surface for targeting the central part of the dorsal striatum) according to the atlas of Paxinos and Watson *(19)*. Drill a

small hole using a high-speed microdrill at the location of cannula placement. Be careful not to drill through the dura and brain tissue, which causes bleeding.
5. Drill three small holes surrounding the cannula hole. Insert the three screws through the holes partially. The exposed screw heads act as anchors to secure the dental cement.
6. Insert a 30-gauge stylet (12 mm in length) into a 24-gauge 10 mm guide cannula. Bend the stylet at 10 mm to have 2 mm of it stay outside of the top of the guide cannula. Attach the guide cannula to the arm of the stereotaxic apparatus. Lower the cannula through the hole to a depth of 2.5 mm below the skull surface.
7. Mix the powdered dental cement with the solvent provided in the same kit. Cover the entire exposed bone, the three screws, and the guide cannula with dental cement. It is essential to keep dry of the covering area for tight dental cement adhesion.
8. After the cement has hardened, remove the arm of the stereotaxic apparatus from the guide cannula and remove the head of animal from the stereotaxic apparatus. The animal will be allowed to recover from the surgery for 5–7 d. During the recovery, the animal will be closely watched for incision infection and any other abnormal activities.

3.2. Infusion of Oligonucleotides

Generally, oligonucleotides need to be administered repeatedly or continuously to maintain the appropriate oligonucleotide concentration at the local sites for days. This is especially critical to knock down proteins with slow turnover. Thus, in this study, mGluR5 antisense oligos and their controls are given in a repeated infusion paradigm. Intrastriatal infusions are made twice a day at 8 A.M. and 6 P.M. (2 nmol/1 µL each time) for consecutive three days (*see* **Note 4**).

1. Dissolve bulk oligodeoxynucleotides (antisense or controls) in aCSF, pH 7.0, to a concentration of 2 nmol/µL. If the pH value is too acidic or basic, use 0.1 *N* NaOH or 0.1 *N* HCl to adjust, respectively.
2. Slightly bend the 30-gauge injector with a length of 15 mm at 12.5 mm. This will allow the injector to protrude 2.5 mm beyond the tip of the 10-mm guild cannula after being inserted into the guide cannula. Connect the bent end of the injector to the PE10 tubing.
3. Connect the Hamilton syringe filled with sterile distilled water to the other end of PE10 tubing. Slowly fill the tubing and injector until they are completely filled with sterile distilled water. Fill the Hamilton syringe the last time with sterile distilled water and connect to the tubing (make sure that no air bubbles are trapped in the whole system). Insert the syringe into the syringe pump.
4. Completely empty the syringe, and then withdraw the plunger slightly to trap a tiny air bubble in the injector. Immerse the injector in the oligodeoxynucleotide solution at a concentration of 2 nmol/µL, then draw the solution into the

injector/tubing system by slowly pulling the plunger of the syringe. You should be able to visualize the moving of a small air bubble in the PE10 tubing when drawing the solution.
5. Remove the stylet and insert the injector cannula. This step can be normally done without anesthesia. However, prior handling of the animals is recommended to allow the animals to be accustomed to the experimenter.
6. Start infusion of the oligonucleotide solution at a rate of 0.1 µL/min. A total dose/volume is 2 nmol/1 µL for a 10-min infusion. Progress of infusion can be monitored through observing movement of the small air bubble through the PE-10 tubing.
7. After the termination of the infusion, the injector will be left in the place for an additional 5 min to reduce any possible backflow of the solution along the injection track.

3.3. Assessment of Efficacy and Selectivity

Effective and selective properties of antisense oligonucleotides in reducing target gene expression need to be evaluated before any functional study is conducted. The mGluR5 mRNA and protein are expressed in both striatonigral and striatopallidal projection neurons of rat striatum *(21–23)*. Among all eight subtypes of mGluRs, mGluR5 represents one of subtypes with the highest level of expression in postsynaptic striatal neurons. The presynaptic distribution of mGluR5-immunoreactivity is, however, minimal *(24)*. These distribution features make mGluR5 a fitting target for the antisense inhibition.

3.3.1. Assessment of Protein Levels

The efficacy and selectivity can be first evaluated at anatomical levels. This is to detect alterations in basal mGluR5 protein levels following repeated treatments with mGluR5 antisense and their controls. With Western immunoblot, we found that basal mGluR5 expression in the ipsilateral dorsal striatum was reduced by 50–60%. The reduction of mGluR5-like immunoreactivity at the same extent was also seen in the dorsal striatum using immunohistochemistry. MGluR5 expression returned to normal levels 24–48 h after the termination of antisense treatment, indicating mGluR5 reduction is reversible and not due to cell loss (*see* **Note 5**). In contrast to mGluR5, other subtypes of mGluRs surveyed, including mGluR1, mGluR2/3, and mGluR7, did not show significant changes in their basal expression in the injected striatum. Repeated treatments with aCSF or any of the three control oligos, i.e., sense, mismatched, and scrambled oligos, had no significant effects on expression of mGluR5 as well as other subtypes. Evidently, the mGluR5 antisense oligonucleotides used in our study possess the ability to suppress effectively and selectively constitutive mGluR5 synthesis.

3.3.2. Assessment of Protein Functions

Even though basal protein levels are reduced after antisense application, protein-mediated functions may not be affected if remaining proteins are adequate for carrying out their functions. Thus, in addition to the assessment of mGluR5 expression levels, mGluR5 binding properties and mGluR5-mediated cellular activity need to be evaluated after mGluR5 antisense administration. Using the radioactively labeled mGluR5 agonist or antagonist, quantitative receptor binding assay can be performed to assess antisense-induced changes in mGluR5 binding affinity and the number of binding sites. MGluR5 is positively coupled to phospholipase C through G-proteins *(25)*. Activation of mGluR5 increases phosphoinositide turnover and thus hydrolyzes phosphoinositol-4, 5-biphosphate into 1,4,5-triphosphate (IP3) and diacylglycerol. By measuring the production of IP3 or diacylglycerol in striatal tissues after mGluR5 stimulation with the selective agonist (RS)-2-chloro-5-hydroxy-phenylglycine (CHPG), mGluR5 capacity in producing these specific intracellular products can be evaluated after antisense treatments (*see* **Notes 6** and **7**).

4. Notes

1. Unmodified DNA oligonucleotides are believed to be degraded rapidly by various enzymes. This may result in an inadequate accumulation of high concentrations of oligos in neurons to interfere with mRNA–protein translation. Thus, the oligonucleotides for in vivo infusion need to be modified to increase their stability. Among many various modification techniques *(2)*, phosphothioate addition is the most popular one. By constructing the oligonucleotides with a phosphothioate "backbone," resistance of the modified oligos to cellular nucleases can be dramatically enhanced *(26,27)*. The other modifications are available to improve uptake and prevent adverse effects.
2. A number of factors determine the tactics in designing which part of the mRNA to target and how long the antisense ought to be. This issue has been discussed in early excellent reviews *(2,3,28)*. In brief, using the same sequence that has demonstrated efficacy and selectivity in others' studies would be the first and easier choice. If this is not available, you can design it yourself with an aid from a molecular biologist. First, an oligonucleotide should be at least 15 bases long to ensure specific hybridization *(17)*. On the other hand, it is desirable to limit the length to a maximum of 20–25 bases to optimize cellular uptake *(28)*. Second, self-hybridization of antisense oligos should be avoided, which happens when the parts of oligos are complementary each other. With regard to which part of the gene to target, it is a common strategy to target the region around the AUG initiation codon. This has been practiced successfully in most of the experiments. It should be avoid choosing a sequence that might be complementary to more than one gene. It is very useful to check the percent match of your antisense oligonucleotide against all known mRNAs in the NIH gene databases

at www.ncbi.nlm.nih.gov/BLAST by selecting standard nucleotide-nucleotide BLAST (blastn).
3. In addition to aCSF as controls, three different types of oligonucleotides were used in our study, which is considered to be more stringent than just using one of them. The sense oligonucleotide that has a different base composition to the antisense oligonucleotide could be potentially problematic because the former may cause different base concentrations extracellularly and, especially, intracellularly after the degradation. Different base concentrations may provoke different metabolic effects on cellular functions. Accordingly, mismatched and scrambled oligonucleotides that have the same base composition to the antisense oligos seem to be better than sense oligos.
4. Frequency and dose of antisense delivery are determined by expression patterns of genes of interest and can be best worked out by trial and error. If the gene has a low basal expression, and turns over rapidly, one injection 6–8 h prior to study may be sufficient, considering that exogenous oligonucleotides usually reach a maximal intracellular concentration after 2–16 h in most cases *(2)*. If the basal expression is high with a low turnover, longer and repeated pretreatments are needed *(29)*. Empirically, 2–3 d are sufficient for a substantial effect on receptors, and injections twice daily seem to work well *(2,28; this study)*. An alternative delivery of antisense is the use of osmotic minipumps loaded with antisense oligonucleotides. The implanted pump allows a continuous infusion for at least 7 d, which may be particularly preferable for inhibiting certain proteins *(30)*. A detailed protocol for delivering antisense oligonucleotides with an osmotic pump was described elsewhere *(18)*.
5. High doses of oligonucleotides may cause nonspecific effects and loss of neural cells surrounding the infusion site at varying extents. Sequence containing four consecutive G bases seems more toxic than other sequences *(31)*. Thus, a conventional histological examination of infusion site after the end of the study is required to evaluate extent of possible neurotoxicity. Reversibility of antisense effect will also be a valid reassurance of lack of toxicity.
6. Mechanisms underlying the uptake of oligonucleotides into cells are not explicitly understood. It may be processed by a simple event similar to fluid-phase endocytosis (pinocytosis) given a water-soluble nature of oligonucleotides. However, the uptake of oligos into the cells appears to be cell-type specific. Thus, surface receptors or channels may likely participate in the uptake.
7. There is a simple way to assess the uptake and intracellular concentration of the oligonucleotide. Application of the oligonucleotide conjugated with fluorochromes (fluoroscein isothiocyanate [FITC] or tetramethylrhodamine isothiocyanate [TRITC]) will allow detection of kinetics and intracellular distribution of the oligo and diffusion area of the labeled oligo in the injected tissue. However, the intracellular level of oligos with conjugated flurochromes may not necessarily represent the real level of intact oligos because oligos may undergo rapid degradation after uptaken into the cells. Intracellular accumulation of fluorescent oligos may genuinely mean an effective uptaking process.

Acknowledgments

This work was supported by the NIH Grants DA10355 and MH61469 (to J. Q. Wang).

References

1. Pennington, S. R. and Dunn, M. J., eds. (2001) *Proteomics: From Protein Sequence to Function.* Springer-Verlag, New York.
2. Pilowsky, P. M., Suzuki, S., and Minson, J. B. (1994) Antisense oligonucleotides: a new tool in neuroscience. *Clin. Exp. Pharmacol. Physiol.* **21,** 935–944.
3. Weiss, B., Davidkova, G., and Zhang, S. P. (1997) Antisense strategies in neurobiology. *Neurochem. Int.* **31,** 321–348.
4. Sibley, D. R. (1999) New insights into dopaminergic receptor function using antisense and genetically altered animals. *Annu. Rev. Pharmacol. Toxicol.* **39,** 313–341.
5. Weiss, B., Zhang, S. P., and Zhou, L. W. (1997) Antisense strategies in dopamine receptor pharmacology. *Life Sci.* **60,** 433–455.
6. Standaert, D. G., Testa, C. M., Rudolf, G. D., and Hollingsworth, Z. R. (1996) Inhibition of *N*-methyl-D-aspartate glutamate receptor subunit expression by antisense oligonucleotides reveals their role in striatal motor regulation. *J. Pharmacol. Exp. Ther.* **276,** 342–352.
7. Mukhin, A., Fan, L., and Faden, A. I. (1996) Activation of metabotropic glutamate receptor subtype mGluR1 contributions to post-traumatic neuronal injury. *J. Neurosci.* **16,** 6012–6020.
8. Silvia, C. P., Jaber, M., King, G. R., Ellinwood, E. H., and Caron, M. G. (1997) Cocaine and amphetamine elicit differential effects in rats with a unilateral injection of dopamine transporter antisense oligodeoxynucleotides. *Neuroscience* **76,** 737–747.
9. Rothstein, J. D., Dykes-Hoberg, M., Pardo, C. A., et al. (1996) Knockout of glutamate transporters reveals a major role for astroglial transport in excitotoxicity and clearance of glutamate. *Neuron* **16,** 675–686.
10. Cai, G., Zhen, X., Uryu, K., and Friedman, E. (2000) Activation of extracellular signal-regulated protein kinases is associated with a sensitized locomotor response to D(2) dopamine receptor stimulation in unilateral 6-hydroxydopamine-lesioned rats. *J. Neurosci.* **20,** 1849–1857.
11. Broberger, C., Nylander, I., Geijer, T., Terenius, L., Hokfelt, T., and Georgieva, J. (2000) Differential effects of intrastriatally infused fully and endcap phosphorothioate antisense oligonucleotides on morphology, histochemistry and prodynorphin expression in rat brain. *Mol. Brain Res.* **75,** 25–45.
12. Lau, Y. S., Hao, R., Fung, Y. K., et al. (1998) Modulation of nigrostriatal dopaminergic transmission by antisense oligodeoxynucleotide against brain-derived neurotrophic factor. *Neurochem. Res.* **23,** 525–532.
13. Batchelor, P. E., Liberatore, G. T., Porritt, M. J., Donnan, G. A., and Howells, D. W. (2000) Inhibition of brain-derived neurotrophic factor and glial cell line-

derived neurotrophic factor expression reduces dopaminergic sprouting in the injured striatum. *Eur. J. Neurosci.* **12,** 3462–3468.
14. Sommer, W. and Fuxe, K. (1997) On the role of c-*fos* expression in striatal transmission. The antisense oligonucleotide approach. *Neurochem. Int.* **31,** 425–436.
15. Konradi, C., Cole, R. L., Heckers, S., and Hyman, S. E. (1994) Amphetamine regulates gene expression in rat striatum via transcription factor CREB. *J. Neurosci.* **14,** 5623–5634.
16. Smith, C. A., Aurelian, L., Reddy, M. P., Miller, P. S., and Tso, P. O. P. (1986) Antiviral effect of an oligo (nucleotide methylphosphnate) complementary to the splice junction of herpes simplex virus type 1 immediate early pre-mRNAs 4 and 5. *Proc. Natl. Acad. Sci. USA* **83,** 2787–2791.
17. Crooke, S. T. (1993) Progress toward oligonucleotide therapeutics: pharmacodynamic properties. *FASEB J.* **7,** 533–539.
18. Ouagazzal, A. M., Tepper, J. M., and Creese, I. (1997) Reducing gene expression in the brain via antisense methods, in *Current Protocols in Neuroscience*, John Wiley & Sons, Unit 5.4.1.–5.4.15.
19. Paxinos, G. and Watson, C., eds. (1986) *The Rat Brain in Stereotaxic Coordinates*. Academic Press, Sydney.
20. Abe, T., Sugihara, H., Nawa, H., Shigemoto, R., Mizuno, N., and Nakanishi, S. (1992) Molecular characterization of a novel metabotropic glutamate receptor mGluR5 coupled to inositol phosphate/Ca^{2+} signal transduction. *J. Biol. Chem.* **267,** 13,361–13,368.
21. Kerner, J. A., Standaert, D. G., Penney, J. B., Young, A. B., Jr., and Landwehrmeyer, G. B. (1997) Expression of group one metabotropic glutamate receptors subunit mRNAs in neurochemically identified neurons in the rat neostriatum, neocortex, and hippocampus. *Mol. Brain Res.* **48,** 259–269.
22. Tallaksen-Greene, S. J., Kaatz, K.W., Romano, C., and Albin, R.L. (1998) Localization of mGluR1a-like immunoreactivity and mGluR5-like immunoreactivity in identified population of striatal neurons. *Brain Res.* **780,** 210–217.
23. Testa, C. M., Standaert, D. G., Landwehrmeyer, G. B., Penney, J. B., Jr., and Young, A. B. (1995) Differential expression of mGluR5 metabotropic glutamate receptor mRNA by rat striatal neurons. *J. Comp. Neurol.* **354,** 241–252.
24. Wang, J. Q., Mao, L., and Lau, Y. S. (2002) Glutamate cascade from metabotropic glutamate receptors to gene expression in striatal neurons: implications for psychostimulant dependence and medication, in *Glutamate and Addiction* (Herman, B. R., ed.), Humana Press, Totowa, NJ, pp. 157–170.
25. Conn, P. J. and Pin, J. P. (1997) Pharmacology and function of metabotropic glutamate receptors. *Annu. Rev. Pharmacol. Toxicol.* **37,** 205–237.
26. Milligan, J. F., Matteucci, M. D., and Martin, J. C. (1993) Current concepts in antisense drug design. *J. Med. Chem.* **36,** 1923–1937.
27. Sommer, W., Hebb, M. O., and Heilig, M. (2000) Pharmacokinetic properties of oligonucleotides in brain. *Methods Enzymol.* **314,** 261–275.
28. Wahlestedt, C. (1994) Antisense oligonucleotide strategies in neuropharmacology. *TiPS* **15,** 42–46.

29. Ramanathan, M., Macgregor, R. D., and Hunt, C. A. (1993) Predictions of effect for intracellular antisense oligodeoxyribonucleotides from a kinetic model. *Antisense Res. Dev.* **3,** 3–18.
30. Whitesell, L., Geselowitz, D., Chavacy, C., et al. (1993) Stability, clearance, and disposition of intraventricularly administered oligodeoxynucleotides: implications for therapeutic application within the central nervous system. *Proc. Natl. Acd. Sci. USA* **90,** 4665–4669.
31. Yaswen, P., Stampfer, M. R., Ghosh, K., and Cohen, J. S. (1993) Effects of sequence of thioated oligonucleotides on cultured human mammary epithelial cells. *Antisense Res. Dev.* **3,** 67–77.

V

STRIATAL CULTURE PREPARATION

25

Primary Striatal Neuronal Culture

Limin Mao and John Q. Wang

1. Introduction

The incredible in vivo complexity of nervous system and activity oftentimes prevents the neuroscientist from attempting to understand how individual neural cells or a specific phenotype of neural cells contribute to a given function of the nervous system. To elucidate intercellular and intracellular mechanisms, the complexity therefore needs to be reduced experimentally. Primary neural culture is an in vitro approach attempting to reduce this complexity, and establishes a relatively purified model for studying the nervous system at a single cell level. Because the culture environment can be readily controlled, neural cultures can be set up to possess biological properties similar or distinct to the original tissue. Research data from the culture study can therefore be valuable to predict the behavior of specific neural cells in vivo.

Primary cell culture from striatal tissues is a useful tool for studying anatomical characteristics, developmental profiles, electrophysiological properties, and molecular expression repertoires of striatal neurons in vitro *(1)*. The utilization of this model has been increased substantially recently owing to an increasing need to study intracellular activity and signal transduction pathways, which are difficult to be pursued in vivo. We have recently established primary neuronal cultures from rat striatal tissues to dissect the signaling pathways bridging surface glutamate receptor stimulation to nuclear gene expression *(2)*. In this culture preperation, E19 or E20 rat embryos are used and striatal cells are dissociated mechanically with an aid of a protease. The dissociated cells are then plating on poly-D-lysine-coated glass slides or plastic dishes. Cells are cultured in the medium containing a mitotic inhibitor. After 12–14 d in culture, a viable, predominantly neuronal culture can be obtained for the further evaluation of a variety of activities in striatal neurons *(2)*. This chapter summarizes step-by-step procedures of the striatal neuronal culture preparation validated in this laboratory and others. However, it should be pointed out that there is no universally adopted technique, although the methods described in this chapter are used successfully in several laboratories and routinely produce reliable results.

2. Materials
2.1. Animals and Equipment

1. An E19 or E20 pregnant rat (usually has litters of 10–12 with a pregnant period of 19–21 d) (*see* **Note 1**). The embryonic age of the litter can be known by using a time-mated rat. However, true age can be determined only after the pregnant mother has been killed. There were occasional unsatisfactory experiences when timed-pregnant rats were obtained from commercial vendors.
2. Jar with cotton wool at base with screw-lid for gas anesthesia.
3. Surgical tools for dissection of striatal tissues, such as scissors (small and microdissecting scissors) and fine forceps.
4. Petri dish (6 cm in diameter) half-filled with 1X phosphate-buffered saline (PBS).
5. 15-mL Plastic centrifuge tube with screw-cap.
6. Eppendorf pipet filler/dispenser and 10-mL pipet.
7. Pasteur pipets.
8. Bright-line hemocytometer.
9. Eight-chamber removable glass slides (Nunc) for *in situ* hybridization measurement of mRNAs or for immunocytochemical detection of proteins, eight-chamber removable glass coverslips (Nunc) for immunofluorescent labeling or fluorescent Ca^{2+} imaging, or 24-well plastic dishes for other purposes.
10. Water bath.
11. Refrigerated centrifuge.
12. Dissecting microscope.

2.2. Solutions and Reagents

1. Poly-D-lysine solution: Made up into a 0.01% solution with distilled water and filtered, and used as a substrate on which the cells can adhere and extend processes.
2. 1X (~0.15 *M*) PBS.
3. Trypsin (1:250, 2.5%) for proteolytic digestion of striatal tissues.
4. Trituration solution: 50 mg of bovine serum albumin (Sigma), 500 µg of DNase I (Sigma), and 25 mg of soybean trypsin inhibitor (Sigma) in 1X PBS to 50 mL and filter. Store at 4°C.
5. Dulbecco's modified Eagle medium (DMEM): To make approx 100 mL of DMEM culture medium, dissolve 10 mL of 100% fetal bovine serum (GIBCO), 2 mL of 50X B-27 (GIBCO), 10 mL of glucose, 100 µL of gentamicin, and 100 µL

of penicillin–streptomycin into 78 mL of DMEM–F12 (GIBCO). Store at 4°C and use within 1 wk.
6. Neurobasal (NB) culture medium: 2 mL of 50X B-27 and 250 µL of L-glutamate into 97.75 mL of NB (GIBCO). Store at 4°C and use within 1 wk. B-27 suppresses glial growth *(3)*.
7. 10 µ*M* Cytosine β-D-arabino-furenoside (Ara-c) in a solution containing 70% DMEM and 30% NB.
8. Trypan blue.

3. Methods
3.1. Coating Glass Slides and Plastic Dishes

1. In the tissue culture hood, aliquot 0.2–0.4 mL of 0.01% poly-D-lysine solution into each well of the eight-well chamber glass slide (well volume = 0.7 mL) or 1 mL into each well of the 24-well plastic culture dish (well volume = 3.5 mL). Leave in the hood at room temperature overnight (*see* **Note 2**).
2. Remove poly-D-lysine solution and rinse each well three times with sterile 1X PBS. Rinse with DMEM medium one time before use. Use immediately or store in sterile 1X PBS at 4°C covered with parafilm.

3.2. Dissecting out the Striatum

1. Asphyxiate an E19 or E20 pregnant rat with CO_2. Intraperitoneal injection of anesthetic agents is not recommended because it may cause damage to embryos.
2. After wiping the abdomen with 70% ethanol, make a vertical incision on abdomen with scissors to expose abdominal contents. Lift up the string of embryos with forceps to visualize individual embryo in embryonic sac. Incise each embryonic sac with small scissors to allow the embryo to come out through the incision. Remove each embryo carefully into a Petri dish half-filled with 1X PBS.
3. The rat is decapitated. A whole brain is removed and placed into another clean Petri dish, which is half-filled with 1X PBS and placed on the ice.
4. Place the dish under a dissecting microscope at 10X–15X magnification. Lie the brain so that the dorsal surface of hemispheres faces you.
5. With fine iridectomy scissors and Dumont-style forceps, make an incision through midline along the dorsal surface of each hemisphere *(4)*. Make sure not to incise too deep so as not to damage striatal structures.
6. Splay open each hemisphere through cut surface. The incision should lie along the complete anteroposterior axis of the lateral ventricles. On the lateral wall of each ventricle, the striatal primordia consisting of a bifid ridge is visualized.
7. Carefully slice this ridge off adjacent tissue with scissors from both side of hemisphere.
8. Repeat with the all brains and place all striata obtained into a new dish filled with approx 8 mL of 1X PBS. (If one pregnant rat with 10–12 litters is used, all 20–24 striata obtained can be pooled into the new dish for the following procedures.)

3.3. Cell Trituration, Suspension, and Plating

1. Remove the dish to culture hood. Cut striata into approx 1-mm square pieces in the dish. Using a pipet filler/dispenser and 10-mL pipet, aspirate the pieces of striata with the 1X PBS into a 15-mL screw cap centrifuge tube. Fill the tube with 1X PBS up to 10 mL.
2. Add 1 mL of 2.5% trypsin (which should be prewarmed to 37°C in a water bath) to the tube to achieve a final concentration of approx 0.25% for proteolytic digestion of striatal tissues (*see* **Note 3**). Place the tube in a 37°C water bath for 10 min, periodically and gently inverting.
3. Centrifuge for 5 min at 100*g*.
4. Remove supernatant (debris) from the tube with a sterile Pasteur pipet and add 10 mL of trituration solution. Trituration solution is used to inactivate trypsin and allow for mechanical dissociation of neural tissue to produce a final cell suspension. DNase in trituration solution is used to prevent cell clumping caused by release of DNA from cells damaged in the cell suspension procedures.
5. Triturate striata by gently aspirating and expelling striata in trituration solution 10–15 times with a sterile Pasteur pipet, until the tissue breaks into smaller pieces. Disperse the small clumps of tissue into single individual cells by repeating the trituration 8–10 times with a flame-narrowed (polished) Pasteur pipet, the bore of which is narrowed to about one third (~0.3–0.4 mm) of its original diameter (0.8–1.2 mm).
6. Centrifuged for 5 min at 100*g*.
7. Remove supernatant (debris) with a sterile Pasteur pipet carefully without disturbing precipitated cells.
8. Add 5 mL of DMEM culture medium and pipet up and down several times successively with smaller bore pipet tips (first with a 1-mL tip, then with a 200-µL tip) to resuspend cells.
9. Once thoroughly resuspended, cell concentration is counted on a Bright-line hemocytometer with trypan blue (shake or invert well before use). Combine 20 µL of cell suspension with 30 µL of DMEM culture medium and 50 µL of 0.4% trypan blue (1:5 dilution of original suspension). Add 20 µL to hemocytometer. Count on one central square (area = 1 mm^2; with 16 smaller squares) and each of four 1-mm^2 squares located diagonally from the central square. Multiply the total number of cells from the five squares by 10,000 (if depth = 0.1 mm) to give the concentration in the number of cells per milliliter. On an average, approx 600,000–1,000,000 cells can be yielded per neonate (from bilateral striata).
10. When counting cells, one can also determine the number of viable cells by dye exclusion. Trypan blue is visible when it leaks into cells that have damaged plasma membranes. By counting total cells and stained cells (damage cells) one can calculate the percent viability (*see* **Note 4**). In our practice, the cellular death was usually <10% of total cells *(2)*.
11. Dilute with DMEM medium to a final concentration of 300,000 cells/mL. Add 0.4 mL final suspension to each chamber of an eight-chamber glass

slide (120,000 cells/0.4 mL/0.7 cm^2 growth area/chamber). Add 2 mL to a 24-well plastic dish. Cell density can be adjusted to comply with objectives of experiment.
12. Place the slide or dish into a culture incubator at 37°C in 5% CO_2 and 100% humidity.

3.4. Cell Culturing

1. After overnight incubation, the culture medium is replaced with a fresh maintenance mixture of 70% DMEM and 30% NB.
2. On d 3–4 after plating, 5 μ*M* Ara-c, a mitotic inhibitor that inhibits the growth of dividing glial cells, is added to reduce the population of glial cells (*see* **Note 5**).
3. Cells will grow for 5–14 d, depending on objectives of experiment, before use. Exchange the medium every 5–7 d (*see* **Note 6**).

3.5. Culture Characteristics

Using the cell preparation procedures outlined in the preceeding, the following characteristics can be anticipated (for details, *see* **ref. 2**; *see* **Note 7**):

1. Newly dissociated cells immediately start to adhere to the surface of the slides after plating. The adherence usually finishes within the first hour (**Fig. 1A**).
2. The very first type of cells to grow during the first few days is the flat cell. These cells form a monolayer. Neuronal cells are seen in the cultures soon after the monolayer is formed. They extend their processes on the top of the monolayer *(5)* and build synapse-like connections among each other by the end of first week. During the next week, neurons increase slightly in size and greatly in the number, length, thickness, and complexity of their processes.
3. Soma and processes of individual cells are very distinctive throughout course of cell growth in culture. Glial cells, consisting of 5–10% of a whole population, include mainly astrocytes, microglia, and oligodendrocytes. Astrocytes are the major type of glial cells. They are large and flat with phase-dark, large pale nuclei (25–35 μm) and widely spread cytoplasm (monolayer) that covers the large surface. Microglia have large cell bodies (>25 μm) ranging from compact amoeboid cells to bipolar rod-shaped or occasionally ramified cells with variation in the length and complexity of branches. Under a light microscope, microglia show a large pale nucleus surrounded by a unique granular cytoplasm. Because they are phagocytic, phase-bright debris-filled microglia can sometimes be seen in the early culture. A few small oligodendrocyte-like cells are identified by their small, round, phase-bright soma (8–15 μm), which have a few thick processes emanating from it. These processes could branch extensively to form a dense, weblike halo around the cell body. Neuronal cells are a predominant cell type in cultures (>90% of the total cells), which are small (8–12 μm) or medium-sized (13–19 μm) cells with extended processes that are often long or branched or both

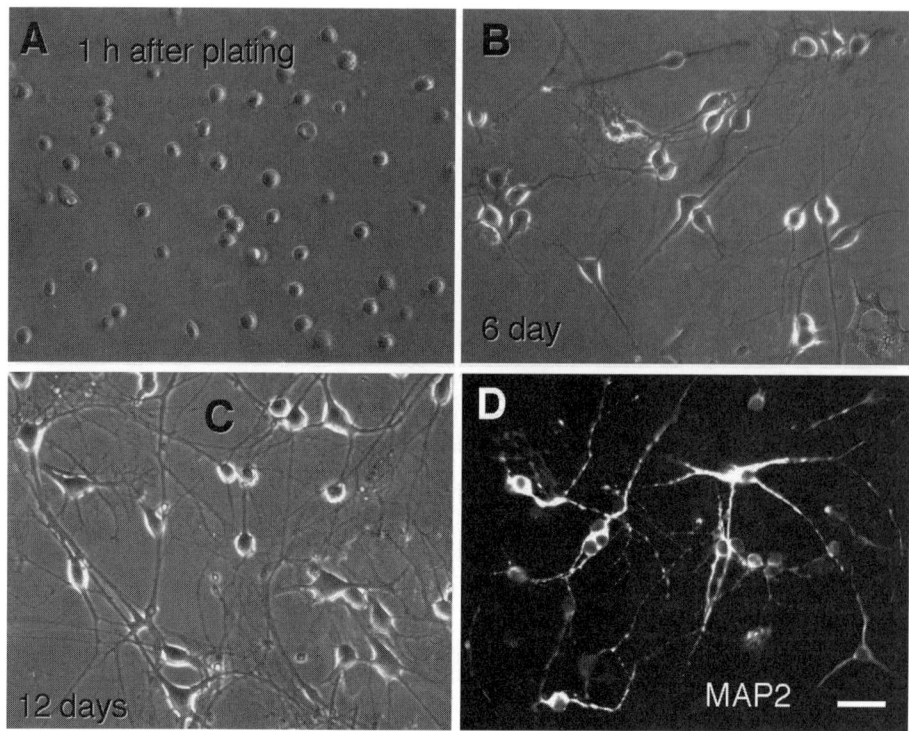

Fig. 1. Phase-contrast photomicrographs illustrating dynamic growth of primary rat striatal cells at 1 h (**A**), 6 d (**B**), and 12 d (**C**) in culture. A vast majority of neuronal cells with small (8–12 μm) to medium (13–19 μm) sized, phase-bright cell bodies were typically seen in 12-d-old cultures. These neuronal cells show round and bipolar, spindle and bipolar, triangular and tripolar or polygonal and multipolar shapes of cell bodies with extensive processes. All neuronal cells were immunoreactive to a neuronal specific marker microtubule-associated protein 2 (MAP2) as tested via fluorescein isothiocyanate (FITC) immunofluorescent labeling (**D**). Scale bar = 40 μm.

(**Fig. 1**). Neurons typically have a clear, centrally or eccentrically located nucleus with one or two prominent nucleoli. The body margins are clearly defined as round, spindle, triangular, or even polygonal. Bipolar or tripolar neurons are more numerous than multipolar neurons. Other contaminating cells, such as endothelial cells, ependymal cells, or fibroblasts, are rarely seen throughout culturing.

4. To identify chemically phenotypes of cultured cells, single or multiple immunofluorescent labeling can be employed with phenotype-specific markers 12–14 d after in culture (**Fig. 1D**). In our culture system, we were able to identify that the predominant number of cells were γ-aminobutyric acid-ergic (GABAergic) neurons. The rest is other types of neurons and non-neuronal cells, including

choline acetyltransferase-associated cholinergic cells and glial fibrillary acidic protein-associated astrocytes *(2)*.

4. Notes

1. Although we use E19–E20 fetuses in most cases, postnatal d 1 rat pups have also been used successfully with our cell preparation procedures *(2)*. However, our experience is that the tissues from younger animals can provide more viable cells and fewer non-neuronal cells. They seem to grow healthier and survive longer as well.
2. Coating substrates, poly-D- or poly-L-lysine, can be purchased from Sigma in sterile form over a range of molecular weights. The two isomers of polylysine have a comparable cell adhesion promoting effect. However, L-isomer amino acids seem to be biologically active; the D-isomer is usually used. We have had good outcomes using poly-D-lysine (molecular weight = 70,000–150,000).
3. More careful control of using trypsin is important. Trypsin is a pH-sensitive protease. It is almost completely inactive below pH 7. The length of time of using this enzyme should be determined empirically and adjusted proportionately according to the age of embryos. If substantial cell lysis occurs and the yield of cells is not near expectations during dissociation, incubation duration of trypsin should be shortened accordingly.
4. Caution is necessary when assessing cell viability by dye exclusion. Although dye uptake marks cells that have grossly disrupted membranes, it may not detect other forms of injury or an early stage of injury that affects cell adhesion or can progress to cell death afterwards.
5. We use Ara-c as an antimitotic agent to reduce the population of non-neuronal cells capable of DNA synthesis. To avoid possible toxic effects on neurons *(6)*, Ara-c is employed at its lowest effective concentration (5 μM). In addition, the timing of addition of this agent is critical. It should be added into the medium 3–4 d after culturing when a confluent monolayer of astroglial cells has formed, which serves as a needed bed for neurons to grow a network of perikarya and processes.
6. Cultured cells seem to be very sensitive to air. Therefore, we usually exchange three-fourths of the medium with a small amount of medium to be left to cover the bottom of cell layer. This prevents cells from exposure to air and minimizes the disturbance caused by feeding cultures with fresh medium. The optical schedule for feeding cultures is once every 5–7 d in our case. This infrequent feedings reflect an unnecessary need to maintain high rates of cell proliferation, and may benefit cells with less disturbance that occurs between feedings.
7. Primary striatal cultures provide the closest approximation of the situation in vivo, and it can be rationally employed for many purposes if one is aware of some of its disadvantages as fresh tissues. They are heterogeneous. Thus, growth of glial cells needs to be controlled by a mitotic inhibitor, and the phenotype of neurons or fraction of a given phenotype of neurons needs to be identified immunohistochemically. Cell preparation procedures vary from experiment to

experiment, and sometimes need to be adjusted empirically. It becomes apparent to us now that neural cell culture is an empirical process by large. There are many steps and reagents that should be checked if cultures do not grow as well as they should. Again, detailed records of an experiment can make an important difference.

Acknowledgments

This work was supported by the NIH Grants DA10355 and MH61469.

References

1. Barker, R. and Johnson, A. (1995) Nigral and striatal neurons, in *Neural Cell Culture* (Cohen, J. and Wilkin, G. P., eds.), Oxford University Press, New York, pp. 25–40.
2. Mao, L. and Wang, J. Q. (2001) Upregulation of preprodynorphin and preproenkephalin mRNA expression by selective activation of group I metabotropic glutamate receptors in characterized primary cultures of rat striatal neurons. *Mol. Brain Res.* **86,** 125–137.
3. Brewer, G. J., Torricelli, J. R., Evege, E. K., and Price, P. J. (1993) Optimized survival of hippocampal neurons in B-27-supplemented neurobasal, a new serum-free medium combination. *J. Neurosci. Res.* **35,** 567–576.
4. Misgeld, U. and Dietzel, I. (1989) Synaptic potentials in the rat neostriatum in dissociated embryonic cell culture. *Brain Res.* **492,** 149–157.
5. Banker, G. and Goslin, K., eds. (1996) *Culturing Nerve Cells.* The MIT Press, Cambridge, MA.
6. Wallace, T. L. and Johnson, E. M. (1989) Cytosine arabinoside kills postmitotic neurons: evidence that deoxycytidine may have a role in neuronal survival that is independent of DNA synthesis. *J. Neurosci.* **9,** 115–124.

26

Primary Rat Mesencephalic Neuron–Glia, Neuron-Enriched, Microglia-Enriched, and Astroglia-Enriched Cultures

Bin Liu and Jau-Shyong Hong

1. Introduction

Degeneration of dopaminergic neurons in the substantia nigra is a hallmark of Parkinson's disease (PD). However, despite decades of research, the cause of PD and the underlying mechanism of action responsible for the progressive degeneration of nigral dopaminergic neurons remain poorly understood (1). The creation of rodent and primate models for PD has provided a valuable tool in the study of the pathogenetic progression of the disease (2). On the other hand, primary neural cell cultures have become extremely valuable in the delineation of the molecular and cellular mechanisms of neuronal death. In light of the increasing appreciation of the role brain immune cells play in the neurodegenerative process (3), establishment of enriched primary cultures of neurons, microglia, and astroglia enables the dissection of the complex in vivo system in an in vitro setting. The utility of these cultures has helped gain critical information on the role each cell type plays and the potential factors produced by individual cell types that contribute to the neurodegeneration (4–10). Here we describe our laboratory's routinely used procedures for establishing primary mixed mesencephalic neuron–glia, mixed glia, neuron-enriched, microglia-enriched, and astroglia-enriched cultures.

2. Materials

2.1. Animals and Equipment

1. Timed-pregnant Fisher 344 rats are obtained from a commercial source (Charles River, Raleigh, NC). Pregnant rats are usually shipped to in-house animal facility

at around 9 days of gestation age. For mesencephalic neuron–glia cultures, embryonic d 13/14 (E13/14) fetuses were used (see **Note 1**). For glia cultures, mothers are kept until littering and 1-d-old pups are used.
2. Acrylic box connected to a pressurized CO_2 tank for anesthesia.
3. Spray bottle filled with 75% ethanol.
4. Sterile surgical tools including regular and microdissecting tweezers (Biomedical Research Instruments, cat. no. 10-1420), microdissecting scissors (BMI, cat. no. 11-1390), hemostats, and regular and fine forceps.
5. Sterile Petri dishes (60 × 15 mm round and 100 × 15 mm square).
6. Sterile 10-mL plastic pipets, and 1-mL and 200-µL pipet tips.
7. Sterile cell culture 50-mL conical tubes.
8. Sterile cell cultures 24-well plates and 75- and 150-cm^2 culture flasks.
9. Eppendorf repeat pipet and pipet-aid.
10. Hemocytometer.
11. Water bath, centrifuge, and dissecting microscope.
12. Cell culture incubator preset at 37°C and with humidified air and 5% CO_2.

2.2. Solutions and Reagents

1. Sterile water: De-ionized and distilled water is sterile filtered and stored at 4°C.
2. Sterile phosphate-buffered saline (PBS): 1X PBS is prepared from 10X stock solution (Gibco), sterile filtered, and stored at 4°C.
3. Poly-D-lysine (mol wt. 30,000–70,000, Sigma P-7280): A stock solution of 100 µg/mL is prepared with sterile water and aliquots are stored at –20°C.
4. Dissection medium: Base cell culture medium containing 50 U/mL of penicillin and 50 µg/mL of streptomycin. Sterile filter (0.2 µm) and store at 4°C.
5. Cytosine β-D-arabinofuranoside (ara-C): Prepare freshly a 1 mM solution in neuron–glia maintenance medium (see **item 7**).
6. Trypan blue solution (0.4% in PBS, Gibco).
7. Neuron–glia culture maintenance medium: Mix 380 mL of minimum essential medium (MEM) (Gibco cat. no. 11090-08) with 50 mL each of heat-inactivated fetal bovine serum (FBS, Gibco) and heat-inactivated horse serum (HS, Gibco), 5 mL each of nonessential amino acids (100X, Gibco), 100 mM sodium pyruvate (Gibco), 200 mM L-glutamine (Gibco), penicillin–streptomycin (5000 U/mL/5000 µg/mL, Gibco), and 0.5 g of D-glucose. Stir gently to ensure complete dissolution, sterile filter, and store in the dark at 4°C.
8. Mixed glia culture maintenance medium: Mix 430 mL of Dulbecco's MEM (DMEM)–F12 (Gibco cat. no. 11330-032) with 50 mL of FBS and 5 mL each of nonessential amino acids (100X), 100 mM sodium pyruvate, 200 mM L-glutamine, and 5000 U/mL/5000 µg/mL of penicillin–streptomycin. Sterile filer and store in the dark at 4°C.
9. Microglia/astroglia culture maintenance medium: Mix 440 mL of DMEM (Gibco cat. no. 12430-054) with 50 mL of FBS and 5 mL each of 100 mM sodium

pyruvate and 5000 U/mL/5000 μg/mL of penicillin–streptomycin. Sterile filter and store in the dark at 4°C.
10. Trypsin–EDTA solution (0.05%–0.53 mM in saline, Gibco).

3. Methods
3.1. Method I: Mesencephalic Neuron–Glia Cultures
3.1.1. Coating 24-Well Plates with Poly-D-Lysine

1. In a tissue culture hood, prepare a 20 μg/mL working solution of poly-D-lysine by diluting the stock solution with sterile water. Add 0.5 mL to each well of a 24-well-culture plate. Let stand in the hood for 2–3 h.
2. Before seeding the cells, rinse the wells twice with 1 mL/well of sterile water and once with sterile PBS.

3.1.2. Dissection of Embryonic Midbrain Tissues

1. Asphyxiate 2–4 E13/14 rats (*see* **Note 2**) with CO_2.
2. Lay the rats on their back and clean the surface of abdominal area with 70% ethanol. With a hemostat, lift the abdominal skin and, with scissors, make a vertical incision of about 5 cm in length to first expose the abdominal muscle layer. With a new pair of scissors, make a vertical incision of the muscle layer to expose the abdominal contents. Carefully remove the string of embryos and place it in a square Petri dish one third filled with cold dissection medium. With a pair of small scissors, carefully make an incision in the individual embryonic sac to free the embryo. Take extra caution not to damage the embryos. Carefully transfer embryos to round Petri dishes two thirds filled with cold dissection medium and keep the dishes on ice.
3. Put 5–10 embryos in a new round Petri dish two-thirds filled with cold dissection medium and place the dish on the floor of a dissection microscope.
4. Under the microscope, position the embryo with the head in the upright fashion. While holding the body of the embryo with a pair of microdissecting tweezers, make two perpendicular incisions with a pair of microdissecting scissors to cut out the midbrain portion of the brain (*see* **Note 3**). Carefully insert one blade of the scissors into the tubular structure and incise along the dorsal edge: now the tubular structure opens up to become a butterfly-shaped tissue with the ventral midbrain in the center. Use two pairs of microdissecting tweezers to separate the brain tissue carefully from the meninges and skin. With the microdissecting scissors, trim the dorsal one-third (i.e., outer edge of the butterfly) of the midbrain tissue to obtain the ventral midbrain tissue.
5. Carefully transfer and pool the ventral midbrain tissues in a round Petri dish two-thirds filled with cold dissection medium. Keep the dish on ice.

3.1.3. Tissue Trituration and Cell Seeding

1. In a tissue culture hood, using a 10-mL pipet, transfer the midbrain tissues and medium in the Petri dish to a 50-mL conical tube. If needed, bring the volume

to 8 mL with dissection medium. Slowly pass the tissues through the orifice of the 10-mL pipet three to five times without foaming. Fit a 1-mL pipet tip to the 10-mL pipet and again slowly pass the tissues through three to five times. Repeat the process with a 200-µL pipet tip fitted to the 10-mL pipet (*see* **Note 4**).
2. Cap the tube and centrifuge at 200*g* for 10 min at room temperature.
3. In a tissue culture hood, discard the supernatant, and resuspend the pellet in 5–10 mL of prewarmed neuron–glia maintenance medium (37°C). Transfer 30 µL of the cell suspension to a microtube.
4. Add 270 µL of Trypan blue to the microtube. Invert the tube to mix and load 10 µL of the mixture to a hemocytometer to estimate the density of cells.
5. In a culture hood, adjust the density of viable cells to 1×10^6/mL with warm maintenance medium. Gently invert the conical tube to ensure a good mix of the cell suspension. With a repeat pipet fitted with a 12.5-mL tip, gently but precisely add 0.5 mL of the cell suspension to each well of the poly-D-lysine-coated 24-well plates. Gently move the plate a few times in a left-to-right and back-and-forth fashion to ensure even plating of the cells in the wells.
6. Place the plate in the cell culture incubator.
7. Three days later, carefully add, along the side of the wells, 0.5 mL of the maintenance medium (prewarmed to 37°C) to each well of the 24-well plates.
8. For neuron–glia cultures, cultures are usually ready for treatment 7 d after the initial seeding.
9. The relative abundance of the major types of cells of interest, that is, astrocytes, microglia, neurons, and dopaminergic neurons in particular, can be determined following immunocytochemical staining with cell type specific markers (*see* **Note 5**). At the time of treatment, the cultures are usually made up of 50% astrocytes, 10% microglia, 40% neurons in general, and up to 1% dopaminergic neurons.

3.2. Method II: Mesencephalic Neuron-Enriched Cultures

The protocol that we routinely use to prepare neuron-enriched cultures involves the use of cytosine β-D-arabinofuranoside (ara-C) to suppress the growth of glial cells in a neuron–glia culture.

1. Neuron-enriched cultures are prepared by first establishing the neuron–glia cultures following the exact procedure described in the preceding.
2. At 48 h after initial seeding, an aliquot of ara-C stock solution was added to the culture medium to achieve a final concentration of 5–15 µ*M* in a final volume of 1 mL/well (*see* **Note 6**). Two to three days later, the ara-C-containing medium was removed and prewarmed midbrain neuron–glia medium was added back (1 mL/well). Cultures are ready for use 7 d after initial seeding.
3. The composition of the neuron-enriched cultures can again be determined following immunostaining for cell type-specific markers. In general the cultures contain 85–95% of neurons. The rest of the cells are primarily astrocytes with very few contaminating microglial cells.

3.3. Method III: Mixed Glia Cultures

Mixed-glia cultures are normally prepared in two formats. For the preparation of microglia- and astrocyte-enriched cultures, mixed-glia cultures are first grown in large culture flasks. For studying the characteristics of mixed-glial cells, cultures are prepared in 24-well culture plates.

3.3.1. Coating 24-Well Plates with Poly-D-Lysine

Follow the procedure described in **Subheading 3.1.** for coating the 24-well plates with poly-D-lysine.

1. For coating large culture flasks, in a cell culture hood, add 25 mL of the sterile working solution of 20 µg/mL of poly-D-lysine to each 150-cm^2 flask. Make sure the solution covers the entire surface area for cell culture.
2. Lay the flasks flat on the hood floor and let stand for at least 2 h. Before seeding cells, rinse the flasks twice with 25 mL/flask of sterile water and once with sterile PBS.

3.3.2. Dissection of Neonatal Whole Brain Tissues

1. Use gauze sponges soaked in 70% alcohol to lightly wipe clean the surface around the head of 1-d-old rat pups and quickly kill the pups by decapitation. Usually, brains from two pups are sufficient to seed one 150-cm flask or two or more 24-well plates.
2. To remove the whole brain, secure the head, with the dorsal side upward, by holding the anterior (nostril) end with a pair of forceps. Using a pair of small scissors, make a straight cut through the skull from the opening close to the neck all the way to a point close the eye. Make a similar cut on the opposing side. Use a pair of forceps to peel back the skull to expose the brain. With a pair of forceps, gently scoop up the entire brain and place it in a clean round Petri dish filled with enough dissection medium to immerse the brain completely. Keep the dish on ice.
3. Under a dissecting microscope, use microdissecting tweezers to remove the olfactory bulbs and cerebellum. Carefully remove as much meninges and blood vessels as possible. Transfer and pool all the brain tissues into a clean round Petri dish filled with a sufficient volume of dissecting medium and keep the dish on ice.

3.3.3. Tissue Trituration and Cell Seeding

1. In a tissue culture hood, transfer the brain tissues and medium to a 50-mL conical tube. If needed, bring the volume to 10 mL with dissection medium for 5–10 brains. Slowly pass the tissues first through the orifice of a 10-mL pipet 5–10 times without foaming, followed by passage (5–10 times) through a 1-mL pipet tip, and then a 200-µL pipet tip fitted to the 10-mL pipet (*see* **Note 4**).
2. Centrifuge at 200g for 10 min at room temperature.

3. Resuspend the pellet in 10 mL of prewarmed glia maintenance medium (37°C). For seeding into 150-cm² flasks, seed a volume of cell suspension equivalent to two brains to each poly-D-lysine-coated flask preloaded with glia maintenance medium for a total of 30 mL/flask.
4. For seeding into 24-well culture plates, first determine the density of cell suspension with trypan blue and a hemocytometer. Adjust the cell density to 1×10^6/mL with glia maintenance medium.
5. With a repeat pipet, add 0.5 mL of the cell suspension to each well of the 24-well plates.
6. Four days after initial seeding, completely remove the spent medium and add prewarmed fresh glia maintenance medium at 30 mL/flask or 0.5 mL/well of 24-well plates.
7. Cultures are usually ready for use 14 d after initial seeding.

3.4. Method IV: Microglia-Enriched Cultures

Microglial cells in the primary mixed glial cultures have the unique characteristic of growing loosely attached to a confluent monolayer of astroglial cells that grow very tightly adhered to the poly-D-lysine-coated plastic surface. Therefore, a nearly pure preparation of microglia can be obtained simply by dislodging the microglia off the astroglia layer by a gentle shaking the cultures for a sufficient period of time.

1. When the mixed glial cultures in the 150-cm² culture flasks grow to confluence (14 d after initial seeding), tightly cap the 150-cm² culture flasks and seal the cap area with Parafilm.
2. Stack the flasks on a flat platform shaker and shake the flasks at 180 rpm for 5 h at room temperature.
3. In a cell culture hood, collect the medium into 50-mL conical tubes and pellet the cells by centrifugation for 8 min at 200g at room temperature. Add 30 mL of the mixed glia culture maintenance medium to each 150-cm2 flask and put the flask back to the cell culture incubator. A second harvest of microglia may be possible after an additional 7 d in culture (*see* **Note 7**).
4. Resuspend the cell pellet in an appropriate volume of microglia/astroglia maintenance medium. Determine the cell density with trypan blue solution and a hemocytometer. Adjust the cell density to 1×10^6/mL.
5. Seed 0.1 mL, 0.5 mL, and 2 mL of the cell suspension to wells of 96-, 24-, and 6-well plates, respectively. Maintain the culture in 5% CO_2–95% air at 37°C in a humidified environment. The cultures are ready for treatment the next day. The purity of the enriched microglial cultures can be determined by immunostaining with microglia- and astroglia-specific markers. Usually, the cultures are at least 95% pure for microglia.

3.5. Method V: Astroglia-Enriched Cultures

Enriched astroglia are prepared by a repeated subculturing of the mixed glia cultures after the removal of the majority of microglia by shaking. After five consecutive passages in vitro, a near pure preparation of astroglia can be obtained (*see* **Note 8**).

1. After the removal of microglia by shaking, rinse the flask twice with 30 mL of prewarmed (37°C) sterile PBS in a cell culture hood.
2. Add 10 mL of the trypsin–EDTA solution (37°C) to each flask and incubate for 5 min in a 37°C incubator. The cells should be detached. If needed, gently tap the side of the flask to ensure a complete dislodging of the cells.
3. Transfer the cells to a 50-mL conical tube. Pellet the cells by centrifugation for 8 min at 200g at room temperature.
4. Reseed the cells at a ratio of one 150-cm^2 flask to two 75-cm^2 flasks. Maintain a total of 15 mL of microglia/astroglia maintenance medium for each 75-cm^2 flask. The cultures usually reach confluence in 5–7 d.
5. Subculture by first rinsing the flasks with sterile PBS, detaching the cells with trypsin–EDTA and then reseeding the cells to 75-cm flasks at a ratio of 1 to 1.5.
6. When the cells grow to confluence again (in 5–7 d), subculture at a ratio of 1 to 1.5 and repeat the process two more times.
7. After a total of five consecutive subculturings, the enriched astroglia are seeded ready for use.
8. The purity of the cultures can be determined by immunostaining with astroglia- and mciroglia-specific markers. Usually a >98% pure astrocyte enriched culture can be obtained (*see* **Note 8**).

4. Notes

1. We have been routinely using the E13/14 embryos of Fisher 344 rats supplied by Charles River for preparing mixed neuron–glia cultures. Differences in the assignment of exact embryonic age by different vendors and in the fetuses among different strains of rats may call for pilot studies to find the most suitable age to use.
2. Use of anesthetic agents should be avoided, and rapid asphyxiation with CO_2 gas appears to have a minimal impact on the embryonic brain cells.
3. Detailed illustrations are provided in the **ref. *11*** to help locate the ventral mesencephalic region for microdissection.
4. The number of passages of the tissue suspension through the various pipet and pipet tips is empirical. The actual number of passages may depend on the balance between achieving a single cell suspension and not damaging too many cells, which will be obvious when estimating the density of cell suspension with trypan

blue solution. Damaged cells will take up the dye and appear blue under the microscope.

5. For rat primary cultures, we have routinely identified neurons by immunocytochemical staining with an antibody against a nuclear neuron-specific protein (Neu-N, Chemicon, Temecular, CA). In addition, both the dendrites and cell bodies of neurons can be visualized by immunostaining with an anti-microtubule-associated protein-2 (MAP-2, BD PharMingen, San Diego, CA). Dopaminergic neurons are detected with an anti-tyrosine kinase (TH) antibody. Microglia are immunostained with the OX-42 antibody which recognizes the complement receptor CR3 (BD PharMingen) and astroglia with the anti-glial fibrillary acidic protein antibody (GFAP, Dako, Carpinteria, CA). Detailed protocols are described in **refs. 6–10**.

6. The concentration and/or time line for the inclusion of ara-C to inhibit glial proliferation may need to be determined by pilot studies. Toxicity of ara-C to neurons has been reported. Therefore, a balance needs to be reached between obtaining the purest possible neuron-enriched cultures and minimal damage to neurons in the cultures.

7. Sometimes the quality of the microglia from the second shaking is not as good as the first. Therefore, a second harvest may not always be desirable.

8. Microglia are generally more responsive than astroglia toward a variety of stimuli. Compared to microglia, the repertoire of proinflammatory factors produced by activated astroglia are limited and the quantities are in general small *(12)*. Therefore, in astroglia-enriched cultures, a minute amount of contaminating microglia may significantly skew the experimental results if one is interested in finding the absolute difference between microglia and astroglia.

References

1. Olanow, C. W. and Tatton, W. G. (1999) Etiology and pathogenesis of Parkinson's disease. *Annu. Rev. Neurosci.* **22**, 123–144.
2. Langston, J. W., Langston, E. B., and Irwin, I. (1984) MPTP-induced parkinsonism in human and non-human primates—clinical and experimental aspects. *Acta Neurol. Scand. Suppl. J. Immunol.* **100**, 49–54.
3. Liberatore, G. T., Jackson-Lewis, V., Vukosavic, S., et al. (1999) Inducible nitric oxide synthase stimulates dopaminergic neurodegeneration in the MPTP model of Parkinson disease. *Nat. Med.* **5**, 1403–1409.
4. Kong, L. Y., Maderdrut, J. L., Jeohn, G. H., and Hong, J. S. (1999) Reduction of lipopolysaccharide-induced neurotoxicity in mixed cortical neuron/glia cultures by femtomolar concentrations of pituitary adenylate cyclase-activating polypeptide. *Neuroscience* **91**, 493–500.
5. Jeohn, G. H., Wilson, B., Wetsel, W. C., and Hong, J. S. (2000) The indolocarbazole Go6976 protects neurons from lipopolysaccharide/interferon-gamma-induced cytotoxicity in murine neuron/glia co-cultures. *Brain Res. Mol. Brain Res.* **79**, 32–44.

6. Chang, R. C., Chen, W., Hudson, P., Wilson, B., Han, D. S., and Hong, J. S. (2001) Neurons reduce glial responses to lipopolysaccharide (LPS) and prevent injury of microglial cells from over-activation by LPS. *J. Neurochem.* **76,** 1042–1049.
7. Kim, W. G., Mohney, R. P., Wilson, B., Jeohn, G. H., Liu, B., and Hong, J. S. (2000) Regional difference in susceptibility to lipopolysaccharide-induced neurotoxicity in the rat brain: role of microglia. *J. Neurosci.* **20,** 6309–6316.
8. Liu, B., Du, L., and Hong, J. S. (2000) Naloxone protects rat dopaminergic neurons against inflammatory damage through inhibition of microglia activation and superoxide generation. *J. Pharmacol. Exp. Ther.* **293,** 607–617.
9. Qin, L., Liu, Y. X., Cooper, C. L., Liu, B., and Hong, J. S. (2001) The role of microglia in beta-amyloid (1-42) toxicity to cortical and mesencephalic neurons. *Abstr. Soc. Neurosci.* **27,** 548.13.
10. Gao, H. M., Hong, J. S., Zhang, W. Q., and Liu, B. (2002) Distinct role for microglia in rotenone-induced degeneration of dopaminergic neurons. *J. Neurosci.* **22,** 782–790.
11. Paxinos G., Tork, I., Tecott, L. H., and Valentino, K. L. (1991) *Atlas of the Developing Rat Brain.* Academic Press, San Diego.
12. Liu, B., Gao, H. M., Wang, J.-Y., Jeohn, G.-H., Cooper, C. L., and Hong, J. S. (2002) Role of nitric oxide in inflammation-mediated neurodegeneration. *Ann. NY Acad. Sci.* **962,** 318–331.

27

Primary Culture of Adult Neural Progenitors

Steffany A. L. Bennett and Lysanne Melanson-Drapeau

1. Introduction

In this chapter, we describe methodology for in vitro culture of adult neural stem and progenitor cells. The mammalian adult brain, once thought to be completely postmitotic, is now recognized to contain a finite number of neural stem cells, progenitor cells with the capacity for self-renewal and the ability to differentiate into functional neurons and glia *(1)*. The largest populations are found in the subventricular zone (SVZ) of the lateral ventricle and the subgranular zone (SGZ) of the dentate gyrus *(1)*. It is predicted that these populations, as well as smaller pools of stem and progenitor cells in other brain regions, will be affected by chronic exposure to drugs of abuse. In support of this hypothesis, psychomotor stimulants and opioids have been shown to influence activation of neural progenitors in adult tissue in vitro *(2)* and in vivo *(3)*. Toxic dopaminergic insult by repeated exposure to 1-methyl-4-phenyl-1,2,3,6-tetrahydropyridine (MPTP) increases progenitor proliferation and subsequent gliogenesis *(4)*. Chronic prenatal exposure to drugs of abuse such as cocaine impairs neurogenesis and migration of differentiating neurons *(5,6)*. One means of investigating the underlying molecular mechanisms responsible for these effects on neural precursors is through in vitro culture. In the protocol provided in the following subheadings, a methodology for the generation of primary stem and progenitor neurospheres from the subventricular zone of the C57Bl/6 mouse is described.

2. Materials

1. Male C57Bl/6 mice ($n = 6$) (Charles Rivers, St Foie, PQ), 8–10 wk of age.
2. Sodium pentobarbital.

3. Dissecting microscope (Leica).
4. Artificial cerebrospinal fluid (aCSF): 26 mM NaHCO$_3$, 124 mM NaCl, 5 mM KCl, 2 mM CaCl$_2$, 1.3 mM MgCl$_2$, 10 mM D-glucose, 100 U/mL of penicillin, 100 μg/mL of streptomycin, pH 7.3.
5. Surgical instruments for microdissection (scalpel, dissecting scissors, and needle probes).
6. Dissociation media: aCSF containing high Mg^{2+} and low Ca^{2+} (26 mM NaHCO$_3$, 124 mM NaCl, 5 mM KCl, 0.1 mM CaCl$_2$, 3.2 mM MgCl$_2$, 10 mM D-glucose, 100 U/mL of penicillin, 100 μg/mL of streptomycin, pH 7.3). This solution can be aliquoted and frozen at –20°C. Add 0.2 mg/mL of kynurenic acid (Sigma) before use. Add 1.33 mg/mL of trypsin (Sigma) and 0.67 mg/mL of hyaluronidase (Sigma) before use.
7. Hybridization oven.
8. Polypropylene 15-mL and 50-mL capped tubes (VWR, Mississauga, ON).
9. 35-mm, 100-mm diameter, and 12-well tissue culture dishes with transwel inserts (VWR).
10. Enzyme inactivation media: Neurobasal media, 2 mM glutamine, 10% fetal calf serum, 10% horse serum, 100 U/mL of penicillin, 100 μg/mL of streptomycin (Invitrogen, Burlington, ON).
11. 30-mL syringe and 18-gauge needles.
12. Maintenance media: Neurobasal media, 2 mM glutamine, 25 mg/mL of insulin (Sigma), 100 mg/mL of transferrin (Sigma), 20 nM progesterone (Sigma), 60 μM putrescine (Sigma), 30 mM sodium selenite (Sigma), 20 ng/mL of epidermal growth factor (Invitrogen), 10 ng/mL of basic fibroblast growth factor (Invitrogen).
13. Trypan blue stain (Invitrogen).
14. Orbital rotator/shaker.
15. Vortexer/mixer.
16. LabTek II Chamber Slide System (VWR).
17. Phosphate-buffered saline (PBS): 10 mM sodium phosphate, 154 mM NaCl.
18. Poly-D-lysine (1 mg/mL, Sigma) in double-distilled H$_2$O (ddH$_2$O). Dispense into 1-mL aliquots. Store at –20°C.
19. Poly-D-lysine/laminin coating solution (100 μg/mL of poly-D-lysine, 0.02 mg/mL of laminin). Add 1 mL poly-D-lysine stock (1 mg/mL) and 400 μL of 0.5 mg/mL of laminin (Invitrogen) to 8.6 mL of sterile ddH$_2$O. Use immediately.
20. Differentiation media with serum: Neurobasal media, 2 mM glutamine, 5% fetal calf serum, 5% horse serum.
21. Differentiation media without serum: Neurobasal media, 2 mM glutamine, 5 mg/mL of insulin (Sigma), 100 mg/mL of transferrin (Sigma), 20 nM progesterone (Sigma), 60 μM putrescine (Sigma), 30 mM sodium selenite (Sigma), 1X B27 Supplement (Invitrogen).
22. Brain-derived neurotrophic factor (Invitrogen).
23. Matrigel (Becton Diskinson, Bedford, MA).

3. Methods

Methodology is described for tissue preparation (**Subheading 3.1.**), single cell dissociation/neurosphere culture (**Subheading 3.2.**), and assessment of progenitor differentiation potential (**Subheading 3.3.**).

3.1. Dissection and Tissue Preparation

This protocol describes the euthanasia procedure (**Subheading 3.1.1.**) and dissection parameters (**Subheading 3.1.2.**) underlying neurosphere preparation.

3.1.1. Animal Euthanasia

Mice were deeply anesthetized with sodium pentobarbital and decapitated. Brains were rapidly removed and blocked at approx bregma 1.4 and 0 mm under a dissecting microscope (*see* **Note 1**). Tissue blocks were placed in 10 mL ice-cold aCSF in 10-cm tissue culture dishes with gentle shaking on an orbital rotator until all blocks could be collected.

3.1.2. Tissue Dissection

Individual tissue blocks were transferred to 35-mm dishes, placed on their anterior face, and covered in ice-cold aCSF. The subventricular zone (approx 1.5 mm × 0.2 mm, L × W), including the ependyma and subependyma, was dissected from the striatal side of the lateral ventricles (**Fig. 1**) under a dissecting microscope using surgical instruments (scalpel, dissecting scissors, and needle probes).

3.2. Single-Cell Dissociation

Single cells were dissociated enzymatically in stepwise fashion.

1. Dissected tissue was minced into fine strips with a scalpel.
2. Minced tissue was transferred and pooled in a single 15-mL polypropylene tube.
3. Two milliliters of dissociation media was added and the tube was lightly vortex-mixed at the lowest vortexer/mixer setting. Tubes were capped and incubated for 30 min at 37°C in an hybridization oven with rotation.
4. Five milliliters of enzyme inactivation media was added and the sample was centrifuged for 5 min at 200g.
5. The supernatant was removed. Ten milliliters of enzyme inactivation media was added to the cell pellet and the suspension was transferred to a 50-mL polypropylene tube. Tissue was dissociated by tituration through successive 10-mL, 5-mL, 1-mL, and pasteur pipets.
6. The suspension was transferred to a 30-mL syringe and forced through an 18-gauge needle into a 50-mL polypropylene tube.

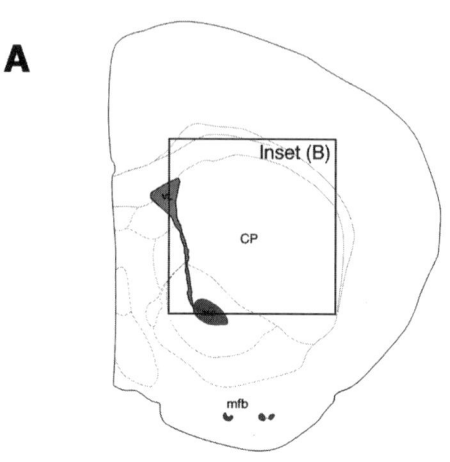

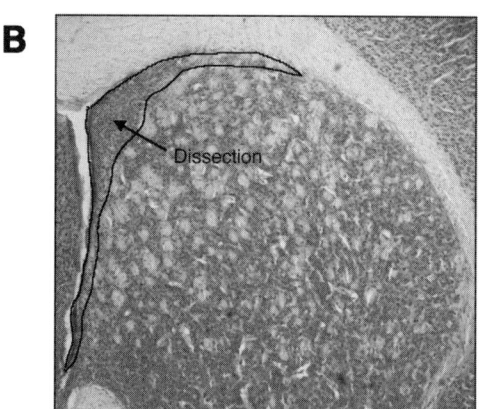

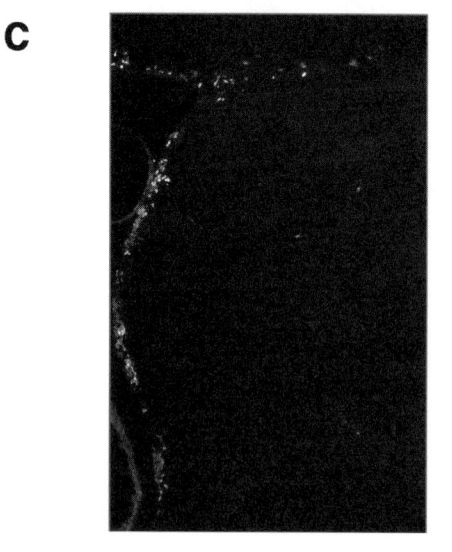

Neurosphere Culture

7. The tube was centrifuged for 5 min at 200g.
8. The supernatant was removed and cells were resuspended in 10 mL of maintenance media.
9. Cells were counted on a hemocytometer using trypan blue. Twenty microliters of cell suspension was added to 20 µL of trypan blue solution (Invitrogen). Ten microliters of this cell/trypan blue mixture was counted on a hemocytometer. To progagate neurospheres successfully, approx 95% of all cells should exclude the dye after the dissociation procedure.
10. A total of 1×10^5 cells were plated per transwell well in 12-well tissue culture plate containing 3-µm pore transwell inserts (*see* **Note 2**). Cells were cultured in maintenance media. Cells capable of neurosphere formation were allowed to expand for 7–15 d (**Fig. 2**). Cultures were fed every 5 d.

3.3. In Vitro Assessment of Lineage Potential

Cell differentiation can be examined by plating neurospheres on tissue culture plates covered with extracellular matrix or other compound promoting neural cell adherence. Spontaneous differentiation potential into cells of neuronal and/or glial lineages can be assessed in serum containing media followed by immunocytochemistry. Neuronal differentiation can be facilitated by the omission of serum and the inclusion of brain-deprived neurotrophic factor (BDNF) in the culture media.

3.3.1. Coating Sterile Tissue-Culture Treated Microscope Slides

1. Tissue-culture treated glass LabTek II 4-chamber slides were washed with 1 mL/well of PBS. PBS was removed by aspiration.

Fig. 1. (*see opposite page*) Dissection parameters. (**A**) Schematic diagram of a representative coronal section at bregma 0.9 in adult C57Bl/6 mice adapted from **ref. 10**. Inset B is boxed. (**B**) Cresyl violet-stained coronal section. The dissection area (subventricular, ventricular zone, and medial striatum) is outlined. (**C**) Increased magnification of the dissection area taken from an animal injected with bromodeoxyuridine (BrdU) to identify actively dividing cells. BrdU is a thymidine analog that readily passes the blood–brain barrier and is incorporated into the DNA of mitotic cells. BrdU (50 µg/Kg in sterile 10 m*M* PBS, pH 7.0) was administered intraperitoneally. Animals received two daily injections (4–5 h apart) over 2 consecutive days and a single injection on the third day. Mice were killed 24 h after the last injection and coronal sections processed for BrdU immunofluorescence. Note the presence of actively dividing cells in the dissection area. These cells have been shown to be a mixed population of neural stem and progenitor cells *(1,8,11)* that can be cultivated and expanded in vitro by neurosphere assay. VL, Lateral ventricle; CP, caudate putamen; mfb, medial forebrain bundle.

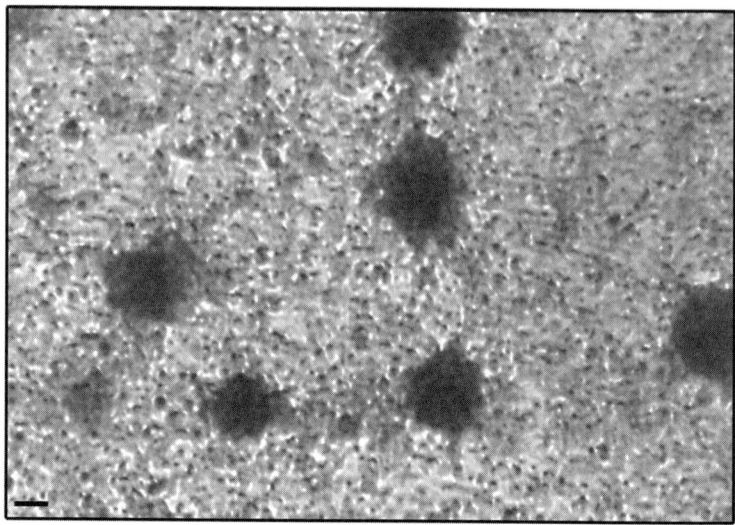

Fig. 2. Representative neurospheres. 7 DIV cultures photographed under bright-field illumination through the clear polyester transwell membrane on an inverted Leica microscope. Cellular debris and cells incapable of neurosphere formation can be seen among the neurosphere colonies. Scale bar, 50 μm.

2. Sufficient poly-D-lysine/laminin coating solution was added to cover each well (500 μL–1 mL) (see **Note 3**).
3. Wells were incubated at room temperature in a laminar flow hood for 1–2 h. Wells were covered with chamber lids to ensure that the coating solution did not dry. Slides were washed with PBS and used the same day (see **Note 4**).

3.3.2. Plating of Neurospheres

1. Neurospheres were gently removed from the transwell plates in a pasteur pipet.
2. Cells were gently dissociated by repeated pipetting through a 200-μL pipet tip.
3. Cells were plated in 1 mL of differentiation media with serum on poly-D-lysine/laminin-coated wells.
4. The following day, media was removed and cells were fed with 1 mL of differentiation media without serum. Cells were cultured for 20 d and fed every 5 d. Spontaneous differentiation to cells of neuronal or glial lineage can be assessed by monolayer culture in differentiation media containing 1% fetal calf serum followed by immunocytochemistry using antibodies raised against neuronal and glial lineage markers. Inclusion of 10 ng/mL of BDNF *(7)* in the differentiation media (in the absence of serum) can be used to promote selective differentiation of neurons.

4. Notes

1. Free-hand dissection can be imprecise. Precise anatomical alignment can be assured by sectioning whole brain into 500-μm slices on a vibratome as described in **ref. 8**. In this latter protocol, brains are removed, placed in 35-mm tissue culture dishes, and immediately covered with a thin layer of 1% Ultra-Pure low-melting point agarose (BRL) heated to between 38 and 40°C. Agarose-permeated tissue is placed at 4°C to allow the gel to harden rapidly. The agarose shell helps to maintain cerebral structure during sectioning *(9)*. Coronal sections (500 μm) are cut on vibratome in chilled aCSF, placed in 10-cm tissue culture dishes, covered in ice-cold aCSF, and processed as described in **Subheading 3.1.3**.
2. The number of cells plated per transwell will vary with the experimental design. To determine neurosphere yield per number of cells plated, total number of spheres per well are counted after 7–9 d in vitro (DIV). Uncoated 12- or 24-well culture plates can also be used but we have found that use make feeding of the neurosphere cultures easier for extended expansion. It has been observed by others that cells that do not divide and form neurospheres within 7 DIV are not subsequently activated by extended culture or by media replenishment *(8)*. Thus, if the experimental goal is to quantitate the number of colony forming units per cells plated, 7 DIV in uncoated tissue culture wells is sufficient. Larger neurospheres are obtained by longer culture times and repeated feeding. This expansion is necessary for experiments requiring large numbers of progenitors, although it should be cautioned that the effect of extended culture on neural stem and progenitor cell potential has yet to be established.
3. Plating efficiency and subsequent neuronal differentiation may be improved by coating cells with 1% fetal calf serum and Matrigel in lieu of poly-D-lysine/laminin coating solution *(8)*.
4. The coating solution can be reused immediately. To conserve costs, half of the total number of slides can be coated first. The coating solution can be removed and transferred to the remaining slides.

References

1. Gage, F. H. (2000) Mammalian neural stem cells. *Science* **267,** 1433–1438.
2. Hauser, K. F., Houdi, A. A., Turbek, C. S., Eide, R. P., and Maxson, W., 3rd (2000) Opiods intrinsically inhibit the genesis of mouse cerebellar granule neuron precursors in vitro: differential impact of mu and delta receptor activation on proliferation and neurite elongation. *Eur. J. Neurosci.* **13,** 1281–1293.
3. Mao, L. and Wang, J. Q. (2001) Gliogenesis in the striatum of the adult rat: alteration in neural progenitor populations after psychostimulant exposure. *Brain Res. Dev. Brain Res.* **130,** 41–51.
4. Mao, L., Lau, Y. S., Petroske, E., and Wang, J. Q. (2001) Profound astrogenesis in the striatum of adult mice following nigrostriatal dopaminergic lesion by repeated MPTP administration. *Brain Res. Dev. Brain Res.* **131,** 57–65.

5. Levitt, P., Reinoso, B., and Jones, L. (1998) The critical impact of early cellular environment on neuronal development. *Prev. Med.* **27,** 180–183.
6. Lidow, M. S. and Song, Z. M. (2001) Primates exposed to cocaine in utero display reduced density and number of cerebral cortical neurons. *J. Comp. Neurol.* **435,** 263–275.
7. Shetty, A. K. and Turner, D. A. (1999) Neurite outgrowth from progeny of epidermal growth factor-responsive hippocampal stem cells is significantly less robust than from fetal hippocampal cells following grafting onto organotypic hippocampal slice cultures: effect of brain-derived neurotrophic factor. *J. Neurobiol.* **38,** 391–413.
8. Seaberg, R. M. and van der Kooy, D. (2002) Adult rodent neurogenic regions: the ventribular subependyma contains neural stem cells, but the dentate gyrus contains restricted progenitors. *J. Neurosci.* **22,** 1784–1793.
9. Chiasson, B. J., Tropepe, V., Morshead, C. M., and van der Kooy, D. (1999) Adult mammalian forebrain ependymal and subependymal cells demonstrate proliferative potential, but only subependymal cells have neural stem cell characteristics. *J. Neurosci.* **19,** 4462–4471.
10. Hof, P. R., Young, W. G., Bloom, F. E., Belichenko, P. V., and Celio, M. R., eds. (2001) *Comparative Cytoarchitectonic Atlas of the C57BL/6 and 129/Sv Mouse Brains.* Elsevier Science, London, UK.
11. Doetsch, F., Garcia-Verduga, J. M., and Alvarez-Buylla, A. (1997) Cellualr composition and three-dimensional organization of the subventricular germinal zone in the adult mammalian brain. *J. Neurosci.* **17,** 5046–5061.

28

Organotypic Culture of Developing Striatum

Pharmacological Induction of Gene Expression

Fu-Chin Liu

1. Introduction

The major advantage of the organotypic culture of striatal slices is that the organotypic culture preserves much of the physiologically relevant environment of striatal neurons. The organotypic culture represents a system more anatomically and physiologically relevant than cultures of cell lines and dissociated primary striatal cells *(1,2)*. The maintenance of the infrastructure of cell–cell interactions in striatal slices allows analysis of neuronal development and cell biology of striatal neurons in an approximately *in situ* condition. This special property and the experimentally accessible condition of cells in the slices makes organotypic cultures of the striatum a value compromise paradigm between in vivo whole animal and in vitro dissociated cell culture approaches.

The striatum is organized into a neurochemical mosaic made up of two neuronal compartments, striosomes (or patches) and matrix *(3,4)*. Striosomes are embedded in the larger surrounding matrix to form an elaborate labyrinthe structure. These two neuronal populations in the striatum were first defined by differential expression of neurochemical molecules in the two compartments. Subsequent studies demonstrate that many neurochemical molecules including neuromodulators, receptors, and signal transduction molecules are differentially expressed in these two striatal compartments. Striosomes and matrix also differ in their neuronal connectivity as well as their neurogenesis *(3,5,6)*. Studies of gene regulation further demonstrate that neurons in these two compartments have different patterns of gene expression in response to a variety of pharmacological manipulations affecting dopamine receptors *(3)*.

A key feature of the organotypic culture system that we developed is that it makes the identification of the striatal compartment-specific cell types possible in vitro, because the maintenance of the structural architecture of the striatal slices permits the identification of compartment-specific cells in culture condition. Specifically, as we have demonstrated in our previous study, the striosomal and matrix compartments can be recognized, respectively, as dopamine and adenosine 3′:5′-monophosphate-regulated phosphoprotein (DARPP-32)-positive neuronal clusters and matrix of calbindin-D_{28KDa}-positive neurons in the cultured striatal slices *(7)*.

We have taken advantage of this system to study induction of different proteins with different kinetics of induction in the developing striatum. In our previous work, we have found that the phosphorylation of cAMP response element-binding protein (CREB), with induction times of minutes, and the expression of Fos-like protein, with induction times of hours, are compartmentally correlated *(7,8)*. They both occur differentially in striosomal cells of the cultured striatum at their respective peak induction times. The successful identification of compartment-specific cells in vitro itself is significant, as it opens a new avenue for cellular and molecular characterization of striatal compartmentation in vitro *(9,10)*.

To give one example of the value of this approach, I describe in the present chapter the experimental protocols for compartmental induction of Fos, the protein product of immediate-early gene c-*fos* by dopamine agonists in slice cultures of developing striatum (**Fig. 1**) *(7)*.

2. Materials

1. Postnatal day 0–1 rats (Sprague-Dawley, Taconic Farms, Germantown, NY).
2. 70% Alcohol.
3. D-(+)-Glucose (Sigma, St. Louis, MO).
4. Gey's Balanced Salt Solution (GIBCO BRL, Grand Island, NY) supplemented with 5.5 mg/mL of D-(+)-glucose (GBSS).
5. SF21 serum-free medium *(11)*.
6. SKF-81297 HCl (SmithKline Beecham Pharmaceuticals, King of Prussia, PA).
7. R(+)-SCH-23390 HCl (Research Biochemicals Inc., Natick, MA).
8. Low melting temperature agarose (FMC BioProducts, Rockland, ME).
9. 0.1 *M* Phosphate buffer (PB, pH 7.4).
10. 0.1 *M* Phosphate-buffered saline (PBS, pH 7.4).
11. 4% Paraformaldehyde in 0.1 *M* PB.
12. Rabbit polyclonal anti-c-Fos antiserum (Ab2, Oncogene Science Inc., Uniondale, NY).
13. 30% H_2O_2.

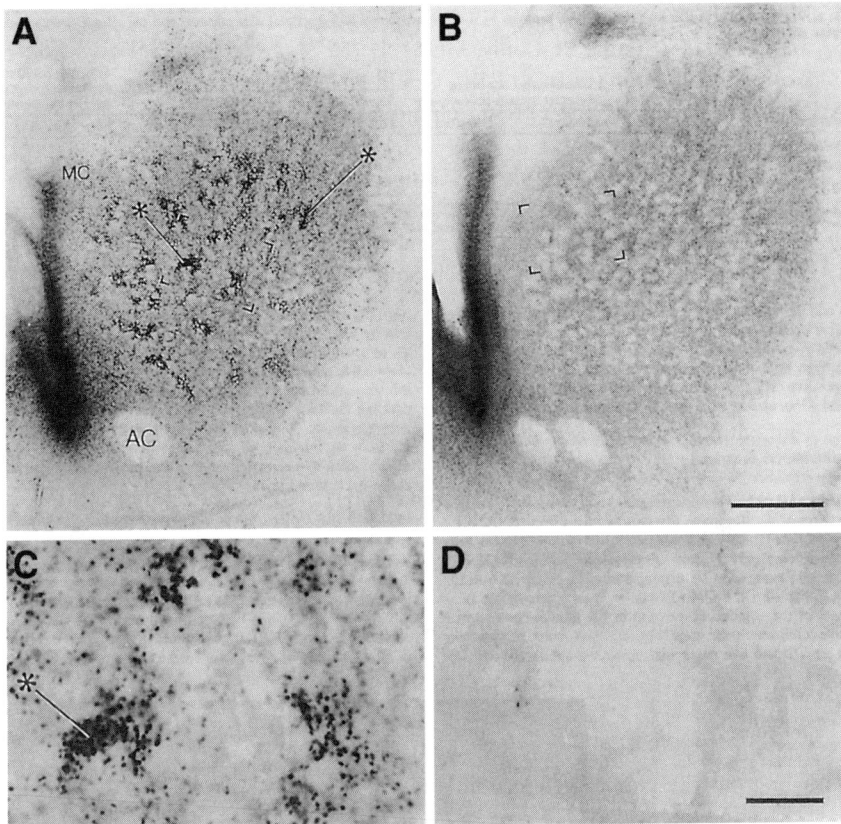

Fig. 1. Induction of Fos protein by dopamine D1 receptor-selective agonist SKF-81297 in striatal slice culture. (**A**) Immunostaining of Fos protein showing induction of Fos by SKF-81297 (100 nM) in striosomes of striatal slice culture derived from newborn rat brain. The slice was cultured for 3 d before treatment with SKF-81297. The bracketed region in **A** is shown at higher magnification in **C**. Clusters of Fos-positive nuclei (examples at *asterisks*) that represent striosomal loci are distributed through the striatum. Fos-positive nuclei are also present in the medial compartment (MC). (**B**) Photomicrograph of Fos immunostained striatal slice that was cultured for 3 d in parallel with the slice shown in **A**, but was pretreated with the D1-selective dopamine receptor antagonist, SCH-23390 (1 µM), before SKF-81297 treatment. The SCH-23390 pretreatment blocks the striatal Fos induction by SKF-81297. The bracketed region in **B** is shown at higher magnification in **D**. AC, Anterior commissure. Scale bar in **B** (for **A** and **B**) indicates 500 µm. Scale bar in D (for **C** and **D**) indicates 100 µm. (Reproduced from Liu et al. *[7]* with permission. © 1995 The Society for Neuroscience.)

14. Methanol.
15. Triton X-100 (Sigma, St. Louis, MO).
16. Normal goat serum (GIBCO BRL, Grand Island, NY).
17. Normal rat serum (GIBCO BRL, Grand Island, NY).
18. Azide.
19. Biotinylated goat anti-rabbit IgG antibody (Vector Laboratories, Inc., Burlingame, CA).
20. ABC kit (Vector Laboratories, Inc., Burlingame, CA).
21. 3,3'-Diaminobenzidine tetrahydrochloride (DAB, Sigma, St. Louis, MO).
22. Nickel ammonium sulfate.
23. Scissor.
24. Forceps.
25. Scalpel.
26. Dumont no. 5 forceps.
27. Spatula.
28. Krazy Glue Pen (Borden Inc., Columbus, OH).
29. Razor blades.
30. 35-mm tissue culture dish (Falcon, Becton-Dickinson and Company, Franklin Lakes, NJ).
31. Six-well culture plates (Falcon, Becton-Dickinson and Company, Franklin Lakes, NJ).
32. 30-mm culture plate inserts (Millicell-CM, PICM 03050, Millipore, Bedford, MA).
33. Water bath.
34. Vibratome (Ted Pella Inc., Redding, CA).
35. Dissecting microscope.
36. CO_2 incubator.

3. Methods

3.1. Preparation of Organotypic Culture of Striatum

The organotypic slice culture method of Stoppini et al. *(1)* was adapted for use with modifications.

1. Sterilize all surgical instrument by placing the instrument in a beaker containing 70% alcohol.
2. Prepare 4% low melting temperature agarose in Gey's Balanced Salt Solution (GBSS). Boil the solution to dissolve the agarose (*see* **Note 1**). Once agarose is dissolved, keep the bottle containing agarose in a 45°C water bath to maintain the agarose in liquid form.
3. Place rat pups on ice for 5–10 min for hypothermia anesthesia.
4. Clean the pup's head with 70% alcohol.
5. Decapitate the head with a scissor.

Organotypic Culture of Developing Striatum

6. Remove the skin and skull with scissor and forceps.
7. Use a scalpel to make a cut in the back at the hindbrain, and make another cut in the front before the olfactory bulbs.
8. Scoop out the brain from the ventral side with a spatula, and then gentle release the brain into a 35-mm culture dish containing GBSS.
9. Use a pair of Dumont no. 5 forceps to remove the pia mater of the brain tissue as much as possible.
10. Remove the bottle of the agarose solution from the water bath. Pour agarose solution into another 35-mm culture dish. When the agarose solution is cool down, pour the solution slowly into the dish where the brain is kept (*see* **Note 2**).
11. Use a spatula to reposition the brain to the center of the dish. The brain should now sit with its ventral side facing the bottom of culture dish. Make sure that the brain is completely immersed in agarose solution. Wait for the agarose to be completely solidified.
12. When the embedding is complete, use a scalpel to cut a rectangle agarose block containing the brain.
13. Put some Krazy Glue on the specimen block of the vibratome. Use a forceps to orient the agarose block in the dish so that the brain now stands up at its caudal end. Place the agarose block on the specimen block such that the caudal end of the agarose block would adhere to the specimen block with the aid of Krazy Glue (*see* **Note 3**).
14. When the agarose block is firmly stuck to the specimen block, fill the specimen tray with GBSS.
15. Set the blade angle at 20°, the vibration amplitude at 3, and the speed at 3 (*see* **Note 4**).
16. Cut the brain in the coronal plane at 300 µm. Typically, for one newborn rat brain, five to six coronal slices containing striatal tissue from rostral to caudal levels can be obtained.
17. Collect the brain slices into a 35-mm culture dish containing GBSS.
18. Remove the pia matter of the brain slices with a pair of Dumont no. 5 forceps under a dissecting microscope.
19. Separate two hemispheres of the slices by cutting through the midline of brain slices.
20. Place the brain slices of right and left hemispheres, respectively, into two 35-mm culture dishes (*see* **Note 5**).
21. In a laminar flow hood, transfer the slices into 30-mm Millipore culture inserts.
22. Prepare a six-well culture plate. For each well, add 1 mL of SF21 culture medium.
23. Place the culture inserts containing slices into the six-well plate. Remove GBSS from the inserts, and replenish 300 µL of SF21 medium from the underlying well. Cover the six-well plate with lid.
24. Place the six-well plate into a humidified incubator at 33°C with 5% CO_2/95% air. The slices are to be cultivated for 3 d.

3.2. Application of Drugs in Slice Culture

1. Prepare 100 μM stock of dopamine D1 receptor-selective agonist, SKF-81297 HCl, by dissolving it in 0.6% ethanol and 0.2% HCl in sterile 0.9% saline containing 1 mM EDTA and 0.01% ascorbic acid.
2. Prepare 1 mM stock of R(+)-SCH-23390 HCl by dissolving it in sterile 0.9% saline.
3. Remove the six-well culture plate from the CO_2 incubator, and place it in the laminar flow hood.
4. Add 1 μL of 1 mM stock of SCH-23390 into the culture medium of one well of slices. Add 1 μL of the vehicle of SCH-23390 to another well as control. The final concentration of SCH-23390 in the medium is 1 μM.
5. Thirty minutes after pretreatment with SCH-23390, add 1 μL of 100 μM stock of SKF-81297 per well into the wells containing SCH-23990 or its vehicle. The final concentration of SKF-81297 in the medium is 100 nM.
6. For SKF-81297 treatment alone, add 1 μL of 100 μM stock of SKF-81297 and 1 μL of its vehicle into two wells of slices without SCH-23390 pretreatment, respectively.
7. Place the six-well culture plate back to the incubator. The slices are further to be cultivated for 2.5 h before fixation.

3.3. Immunocytochemistry

1. After incubation, the slice tissues are fixed in ice-cold 4% paraformaldehyde in 0.1 M PB for 1 h.
2. 4 × 5 min rinses in 0.1 M PBS.
3. 1 × 5 min in 0.1 M PBS containing 0.2% Triton X-100.
4. 1 × 5 min in 1.5% H_2O_2 and 10% methanol in 0.1 M PBS containing 0.2% Triton X-100.
5. 3 × 5 min rinses in 0.1 M PBS.
6. 1 × 1 h in 5% normal goat serum in 0.1 M PBS.
7. 3 × 5 min rinses in 0.1 M PBS.
8. Incubate slice tissue for 48 h at 4°C in primary antibody solution containing 1 : 200 rabbit polyclonal anti-c-Fos antiserum, 0.4% Triton X-100, 1% normal rat serum, 1% normal goat serum, and 0.1% azide in 0.1 M PBS.
9. 3 × 5 min rinses in 0.1 M PBS.
10. 1 × 1 h in 1 : 500 biotinylated goat anti-rabbit IgG antibody containing 1% normal goat serum in 0.1 M PBS.
11. 3 × 5 min rinses in 0.1 M PBS.
12. 1 × 1 h in avidin–biotin complexes solution containing 6 μL of solution A and 6 μL of solution B in 1 mL of 0.1 M PBS (mixing A and B should be performed at least 30 min prior to tissue incubation).
13. 2 × 5 min rinses in 0.1 M PBS.
14. 1 × 5 min rinse in 0.1 M PB.

15. 1 × 5 min in 0.02% DAB solution containing 0.08% nickel ammonium sulfate (20 mg of DAB and 80 mg of nickel ammonium sulfate in 100 mL of 0.1 M PB) (*see* **Note 6**).
16. Prepare H_2O_2 stock solution by adding 2 µL of 30% H_2O_2 into 1 mL of distilled water.
17. For a 1-mL DAB solution, add 30 µL of H_2O_2 stock solution.
18. Observe the color development in brain slices under microscope (**Fig. 1**).
19. Stop the DAB reaction by replacing the DAB solution with 0.1 M PB.
20. 4 × 10 min rinses in 0.1 M PB.
21. Mount the slice tissue on microscopic slides.
22. Air-dry the slides overnight.
23. Dehydration and coverslip the slides.

4. Notes

1. Be sure to use a large bottle, for example, a 500-mL bottle for preparation of 100 mL of agarose solution to prevent the solution from overflowing from the bottle when boiling.
2. The control of the temperature of agarose solution is important. The agarose solution should not be too hot, as it may damage the viability of brain tissue. It should not be too cold either, as agarose will gel quickly before the embedding is complete. To determine the optimal temperature of agarose solution for embedding, place the dish containing the agarose solution against the skin of your wrist. When you feel that the dish is slightly warmer than your skin, the agarose solution is ready to be poured to mix with brain tissue.
3. It is better to have the ventral side of the brain facing the blade. The ventral part of the brain is flat, which has the advantage of taking the vibration pressure evenly when the blade cuts across the brain tissue.
4. The settings of the vibration amplitude and the speed of blade stage may vary with different machines. They may need to be empirically adjusted.
5. All striatal slices from a single hemisphere of a single brain (approx five or six slices) are cultured together in an individual culture insert.
6. DAB is a suspected carcinogen and should be handled accordingly.

Acknowledgments

The preparation of this work was supported by the National Health Research Institutes NHRI-EX91-9010NL, Taiwan.

References

1. Stoppini, L., Buchs, P. A., and Muller, D. (1991) A simple method for organotypic cultures of nervous tissue. *J. Neurosci. Methods* **37**, 173–182.
2. Gähwiler, B. H., Capogna, M., Debanne, D., McKinney, R. A., and Thompson, S. M. (1997) Organotypic slice cultures: a technique has come of age. *Trends Neurosci.* **20**, 471–477.

3. Graybiel, A. M. (1990) Neurotransmitters and neuromodulators in the basal ganglia. *Trends Neurosci.* **13,** 244–254.
4. Gerfen, C. R. (1992) The neostriatal mosaic: multiple levels of compartmental organization in the basal ganglia. *Annu. Rev. Neurosci.* **15,** 285–320.
5. Graybiel, A. M. and Hickey, T. L. (1982) Chemospecificity of ontogenetic units in the striatum: demonstration by combining [^{3}H] thymidine neuronography and histochemical staining. *Proc. Natl. Acad. Sci. USA* **79,** 198–202.
6. van der Kooy, D. and Fishell, G. (1987) Neuronal birthdate underlies the development of striatal compartments. *Brain Res.* **401,** 155–161.
7. Liu, F.-C., Takahashi, H., McKay, R. D. G., and Graybiel, A. M. (1995) Dopaminergic regulation of transcription factor expression in organotypic cultures of developing striatum. *J. Neurosci.* **15,** 2367–2384.
8. Liu, F.-C. and Graybiel, A. M. (1996) Spatiotemporal dynamics of CREB phosphorylation: transient versus sustained phosphorylation in the developing striatum. *Neuron* **17,** 1133–1144.
9. Liu, F.-C. and Graybiel, A. M. (1998) Region-dependent dynamics of cAMP response element-binding protein phosphorylation in the basal ganglia. *Proc. Natl. Acad. Sci. USA* **95,** 4708–4713.
10. Liu, F.-C. and Graybiel, A. M. (1998) Activity-regulated phosphorylation of cAMP response element binding protein in the developing striatum: implications for patterning the neurochemical phenotypes of striatal compartments. *Dev. Neurosci.* **20,** 229–236.
11. Segal, R. A., Takahashi, H., and McKay, R. D. G. (1992) Changes in neurotrophin responsiveness during the development of cerebellar granule neurons. *Neuron* **9,** 1041–1052.

VI

DETECTION OF TRANSMITTER RELEASE IN THE STRIATUM

29

Microdialysis Coupled with Electrochemical Detection

A Way to Investigate Brain Monoamine Role in Freely Moving Animals

Ezio Carboni

1. Introduction

Microdialysis technique coupled with electrochemical detection (ED) is a relatively new method that allows detection of neurotransmitters and other substances from brain and other tissues. It is based on the insertion of a dialysis probe in a specific area and perfusing it with artificial cerebrospinal fluid (CSF), which, passing in a chamber delimited by the dialysis fiber, becomes enriched with small molecular weight substances diffusing into the fiber because of their concentration gradient. Substances recovered can be assayed by high-performance liquid chromatography (HPLC) to evaluate their concentration in the dialysate, that is closely related to their extracellular concentration in the area investigated. After recovery from surgery, therefore, the effects of drugs or other treatments on the assayed substance can be evaluated in freely moving animals *(1)*.

1.1. Aim

The intention of this chapter is to provide basic information on assessing brain monoamines such as dopamine (DA), norepinephrine (NA), and 5-hydroxy-tryptamine (5-HT) by ED coupled to HPLC. Particular attention is given to a description of basic equipment and other material needed to set up a microdialysis laboratory that uses ED to detect monoamines in brain dialysates. This information could be used to decide either to start such a laboratory, taking into account both the costs and time required, or to learn quickly some useful insights that otherwise may require years of experience to gain. The last part of this chapter is devoted to prevention of some of the most common problems that may be encountered in microdialysis laboratories, limiting the discussion to points that are not discussed or only briefly mentioned in technical manuals.

The information regarding equipment should not be used as a "simple to read" substitute for operating manuals. Careful reading of operating manuals indeed has to be considered essential for understating the true role of each piece of equipment and to manage microdialysis and ED. Nevertheless it is recognized that they are often not very helpful to solve problems quickly when one is in trouble. Often in this case, a colleague with good experience may be more helpful even than the manufacturer's technical assistance. Moreover this chapter could allow a researcher entering the field to predict advantages and problems that can be encountered with ED, so that the decision whether to buy a microdialysis system can be better evaluated. Part of this chapter is dedicated to optimization of the system to increase productivity.

1.2. Fields of Application of ED

DA, NA, and 5-HT represent three major neurotransmitters in the mammalian central nervous system (CNS). Their wide distribution in brain, their involvement in CNS pathologies such as psychosis and depression, the relevant role they play in the effects of drugs of abuse, and in behavior in general, has promoted an endless series of research that has provided an enormous contribution to knowledge of CNS function. This extraordinary success is mainly the result of the relative ease of measuring these three neurotransmitters and their metabolites in biological fluids, extracts, and tissues *(2)*. The ease of estimating DA, NA, and 5-HT is due mainly to their chemical structure, whereas the growing facility with which their role in the brain has been ascertained is due to their unique function as neurotransmitters, in contrast to other neurotransmitters such as GABA or glutamate whose detection in vivo has been hampered by their presence in brain fluids with a function other than neurotransmission *(3)*.

The possibility of detecting DA, NE, and 5-HT by ED in extracts such as dialysate from freely moving animals has given an extraordinary advantage to researchers. No comparison can be made with old techniques in which a time course of a drug effect on a specific neurotransmitter could be made only by scarifying groups of animals at given time and assessing the total amount of a transmitter, without being able to distinguish the fraction that is actively released upon depolarization. The possibility of detecting an increasing number of neurotransmitters, by implanting concentric microdialysis probes almost everywhere in the brain, offers a wide spectrum of applications of ED. Moreover if one considers the large number of pharmacological tools that can be used to modify the level of a neurotransmitter or its metabolites in a specific brain area, the number of possible applications increases exponentially.

1.3. Basis of ED of Monoamines

The possibility of detecting biogenic amines such as DA, NE, and 5-HT and their metabolites is due to their readiness to be oxidized. Although oxidation of DA, NA, and 5-HT occurs spontaneously in solution, ED is based on application of an electrical potential to a carbon-based electrode on which DA, NA, and 5-HT flow after separation on a chromatographic column. The oxidation will produce a derivative with the giving up of two electrons per molecule to the electrode. Because the applied potential to the electrode immersed in an electrical conducting solution (mobile phase) results in an electrical current, the giving up of electrons to the electrode will produce alteration of the basal current which will be proportional to amount of substance that is oxidized. The electronic elaboration of this signal through a detector will allow the quantitative detection of oxidable substances present in the sample injected in the HPLC. The variation of current will produce a chromatogram in which each peak will correspond to a substance, and its height or area will be proportional to its concentration in the sample. By the use of a standard curve, the absolute amount of substance can be assessed. **Figure 1** illustrates the process of oxidation of DA, NE, and 5-HT.

1.4. Economic Considerations

1.4.1. Equipment

The basic equipment required to start a microdialysis laboratory equipped with ED includes a stock of dialysis fibers that can be obtained from a hospital supplier of artificial kidneys at modest cost and may last for 1–2 yr. To implant probes in the brain it is essential to purchase a stereotaxic apparatus, that although expensive (about $5000), if well kept, lasts for many years. A perfusion pump is necessary to complete the essential microdialysis equipment; those in the market range $2000–5000 but are precise and durable. To analyze samples a basic HPLC system is necessary. It includes a pump, a column, detectors with appropriate electrodes, and a handling data system that can cost up to $15,000–20,000 (*see* **Note 11**).

A good idea is to double the system to test five or six animals each time. This will allow to have a sufficiently elevated n to produce data statistically significant after each working day. (It needs to be considered that it is difficult to obtain usable data from each implanted animal.) One has to consider that

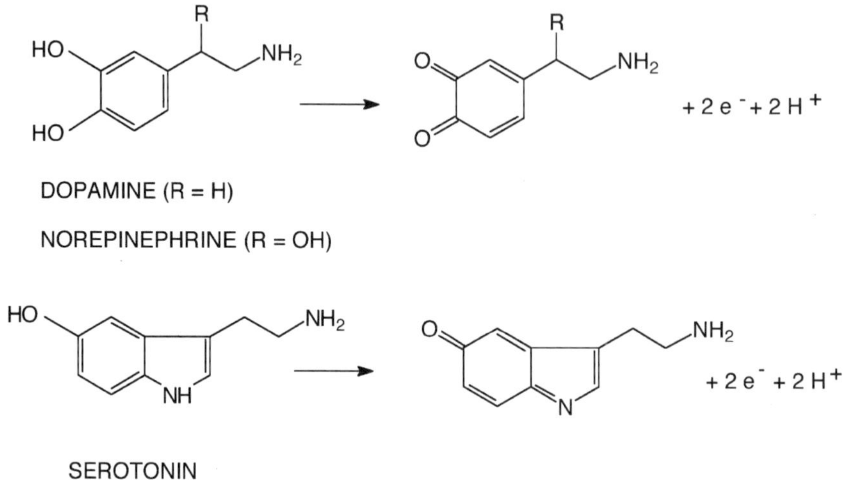

Fig. 1. The typical products of oxidization of DA, NE, and 5-HT.

once a researcher stays in the laboratory a full day to test two or three rats, it is worth working with five or six rats. In any case once the equipment has been acquired, the potential productivity is high, while expenses for keeping a working microdialysis laboratory are very limited. Indeed apart from money for animals, chemicals, and eventually for a HPLC column, or a new electrode, no other expenses have to be anticipated. It has also to be considered that if well kept, the equipment acquired can last up to 10 yr with little or no maintenance at all (*see* **Note 1**).

1.4.2. Time

Although productivity in microdialysis can be high, it is also very time consuming. Some simple considerations can reduce wasted time to a minimum. First, preparation of each fiber requires 30–45 min, considering preparation and assembling of parts (*see* fiber preparation section). Often even a skilled person will not reach more that 80% of success in assembling perfectly working fibers. Because ready to use fibers are sold by several companies, a researcher new in the field can start with those, and later can learn how to assemble fibers. Although a detailed description is provided, in this article and others, it is suggested that one obtain a sample of each part and an assembled probe to support self-instruction. Visiting the lab of course would be best. Self-assembled probes may be more adaptable to specific requirements and may be more efficient in terms of amount of substance recovered.

2. Materials
2.1. ED Equipment

1. Detectors and electrodes: There are basically two types of electrochemical detectors (*see* **Notes 2** and **3**): amperometric detectors (BAS that has been a leading company for years and Antec-Leyden market excellent detectors and electrodes) and the coulometric detectors (ESA is the leading company in this field). They both can oxidize or reduce the substance to be detected when a specific potential, that depends on the substance, is applied. The major difference between these two detectors is the type of electrode used.

   An amperometric electrode is capable of oxidizing or reducing the analyzed substance that flows in a thin layer onto its surface and can be made of carbon paste or glassy carbon. The first type can be made by filling the electrode site with carbon paste. This electrode offers a high sensitivity (much better than that of glassy carbon available 10 yr ago) but only a high skilled person can make them, and they have to be replaced often. Sensitivity for DA can be about 5–10 fmol/sample. Today excellent glassy carbon electrodes are available. The major advantage of the glassy carbon electrodes is the long life (more than 10 yr) while sensitivity is quite high and constant, with repolishing necessary not more than once a year.

   A Coulometric electrode, instead, is based on a porous graphite in which the solution containing the substance to analyze flows through. Because all the substances that pass trough the electrode become oxidized, a second electrode in series after the first can be used to reduce the substance previously oxidized. This increases selectivity, as a specific oxidizing and reducing potential can be selected. This feature permits cleaning of the chromatogram from disturbing peaks of substances that can become oxidized irreversibly. Coulometric electrodes have a great sensitivity when new, but they lose it even after a few months, and their average life (in my experience) is about 1 yr. When sensitivity decreases they can be still used for measuring samples with a high concentration of transmitter (50–100 fmol/sample). Their cost is rather high if compared with their average life (high unexplained price differences exist among coulometric electrodes from different countries).

   Reference electrodes have the function of producing a reference signal for the working electrode. They are generally Ag/AgCl electrodes in saturated KCl or NaCl solution. They are never problematic, but air bubbles that eventually stop in their vicinity have to be washed out (*see* **Note 10**). Their average life is generally long (many years).

2. HPLC equipment: The basic HPLC equipment includes an HPLC pump, a pulse damper, an injector, a chromatographic column, a guard column, and connections to the detectors.

   The HPLC pump delivers the solvent at high pressure through the system (column and electrode). Because detection of neurotransmitters in areas where they are present in a low concentration requires a stable baseline, a perfectly functioning HPLC coupled with ED is necessary to improve the signal-to-noise ratio. It can be obtained with a good pump (a two-piston pump offers better flow stability than a single-piston one). Here the choice in the market is very wide. Usually a low-cost pump is as good as a more expensive one only when new, but its efficiency declines more quickly. A colleague who has purchased the pump that one wants to buy may give useful advice.

A pulse damper is recommended to improve further flow stability (they are not expensive and have a long life). For measuring DA in the caudate (200–400 fmol/20 µL of sample) even an average system can be used, but if areas with a low concentration of DA such as the frontal cortex or the bed nucleus of stria terminalis (BNST) are investigated (10–30 fmol/20 µL of sample) a highly efficient system is recommended *(4)*.

Injectors permit introduction of the sample to be analyzed in the system, without interrupting flow. They are long lasting but care must be taken to avoid salts or organic material deposits. Water flushing or methanol respectively has to be used at the end of each working day to eliminate deposits.

Chromatographic columns and guard columns are the core of chromatograph separation and their choice depends on the substance to be detected. Their cost is not prohibitive, so several columns may be tested to obtain the best separation condition. Their efficiency, of course, depends also on the mobile phase used. It must be considered that injecting the sample without any previous purification will shorten the time of analysis and will allow detection of neurotransmitter levels almost in real time. We used reversed-phase or reversed-phase ion-pair chromatography with ODS columns to detect DA, NA, and 5-HT. Because the sample is injected without any previous purification, a guard column, inserted before the column, is recommended to increase column life. Choice depends on column type, and in general their cost is low. The columns should not be stored with buffer solutions or salts. Salts may precipitate and halogen salts may corrode metal. Wash columns with water and methanol–water (50:50) and close them before storage.

3. Mobile phase: The mobile phase delivers samples containing the substance to be detected to the chromatographic column and afterward to the detector. It can be prepared by diluting in double-filtered (or bidistilled) water stock solutions of each reagent (HPLC grade) kept in the refrigerator. Once the mobile phase has been brought to the definitive pH and volume, it must be filtered and degassed. Ultrasound can be used to shorten the degassing procedure. The composition of the mobile phase depends on the substance to be separated and on the column used. In general, in reversed-phase chromatography substances dissolved in a hydrophilic mobile phase are separated on the base of their affinity for the hydrophobic stationary phase. In this process, an increased concentration of methanol will shorten the retention time by reducing the hydrophilicity of the mobile phase. Reduction of pH will produce an increased polarity of weak bases, reducing their retention time (DA, NA, 5-HT retention time is not affected by

Monoamine Microdialysis

change of pH in the 3–6 range). On the contrary, in acidic metabolites retention time will be increased. If it is necessary to reduce the elution time reversed-phase ion-pair chromatography coupled with the addition of methanol can be used; it may allow injection of samples of three or four different animals in one HPLC. This outcome can be accomplished by adding sodium octyl sulfate to the mobile phase, which will retard positively charged amines. In particular catecholamines can be detected (within 4–5 min) because their retention time is increased, while most of the other components of the sample will be eluted immediately after the solvent front. The composition of the mobile phase used for detecting DA, NA, 5-HT with or without metabolites is discussed in **Subheading 3**. Increasing the temperature will produce a reduction in the retention time of both catecholamines and acidic metabolites.

2.2. Microdialysis Equipment

1. Microdialysis probe.
2. Perfusion pump: There are several commercially available perfusion pumps that are precise and durable. As 0.5–2 µL/min perfusion flows are generally included, a perfusion pump must deliver with precision liquids at flows of 0.1–10 µL/min and must hold 2.5–5 mL syringes (*see* **Note 4**). When a pump is chosen a future expansion of the system that can allow perfusion of up to six animals needs to be anticipated.
3. Perfusion Ringer (PR): Dialysis fibers are perfused with isotonic solution, the composition of which should resemble as much as possible the composition of brain extracellular fluid. Because fluid composition in each specific brain area is unknown and subject to time variations, PR composition has been validated on the base of its capacity to cause minimal alteration of brain function in that specific brain area, being compatible with voltage-responsive and calcium-dependent neurotransmitter release. The composition of PR used in our most recent investigations is: 147 mM NaCl, 2.2 mM CaCl$_2$, 4 mM KCl. It can be pumped through the dialysis probe at a constant flow of 1 µL/min. The rationale of ionic composition of PR has been debated in several articles *(5,6)*. The most critical ions in the PR composition are K$^+$ and Ca^{2+}. The former is critical because its presence in the PR may alter physiological depolarization of terminals, with a consequent change in neurotransmitter release. Experiments conducted by changing [K$^+$] in PR from 0 to 90 mM indicated that 3–5 mM should allow basal neurotransmitter release, whereas increasing the concentration above this will produce an artificial increase of neurotransmitter release mediated by sodium channels because it is blocked by tetrodotoxin as shown for DA *(7,8)* or 5-HT *(9)*. The large majority of microdialysis studies use [K$^+$] between 3 and 5 mM. The [Ca^{2+}] in PR has been debated more extensively. Because lowering [Ca^{2+}] to 0 causes an approx 80–90% reduction of neurotransmitter release, calcium dependency has been considered a main requirement for validation of neuronal vs non-neuronal origin of a neurotransmitter detected by microdialysis *(1,10)*. Many authors use 1.2 mM [Ca$^+$] because it is closer to extracellular [Ca^{2+}], whereas

others prefer to use 2.2 mM [Ca^{2+}] because this condition has permitted the validation of many pharmacological tools. A particular situation is encountered when a neurotransmitter is present at so low a concentration that can be barely detected. In this situation some authors prefer to increase [Ca^{2+}] to be able to detect the neurotransmitter; others prefer to add an inhibitor of reuptake or of degradation. We think the first option should be tested first because it should produce minor pharmacological effects. The influence of other ions such as Mg^{2+} has in a general minor influence but can be important in specific situations (*1*).

4. Swivels: The swivel is small piece of equipment with an inlet and an outlet for liquid. It is inserted between the probe and the perfusion syringe allowing the rat to move freely in the half bowl, without twisting the polyethylene (PE) tubing that delivers the Ringer. The use can be optional when testing drugs such as sedatives or drugs that do not stimulate locomotion in general, but becomes mandatory when drugs that stimulate locomotion such as psychostimulants are used. When buying swivels it is advisable to purchase two-channel swivels although they cost a little more than one-channel types, because they allow connecting two probes implanted in the same animal. Swivel use may be a source of leakage when the PE tubing does not fit tightly to the metal probe inlet or outlet. Leakage in turn can become a source of friction or complete free rotation block, when the salts, contained in the PR, solidify in the ball bearing. In this case, swivels can be washed with distilled water and dried with hot air.

3. Methods
3.1. Probe Construction and Implantation
3.1.1. Probe Geometry

Microdialysis probes can be inserted in brain tissue either transversally or vertically. The first method has the advantage of reduced tissue damage, because its flexible connection with the skull allows the fiber to follow the brain in its small movements in meningeal fluids. Major disadvantages are a complex and time-consuming surgery (*11*). The second method has the advantage of quick and simple surgery, but the presence of a rigid fiber support produces a tissue damage around the fiber caused by brain movements within the skull. We recently developed a vertical probe that combines some of the advantages of both methods (*see* **Subheading 3.1.2.**).

3.1.2. Probe Construction

Concentric dialysis probes can be prepared with a 7-mm piece of AN 69 (sodium methallyl sulfate copolymer) dialysis fiber (310 µm outer diameter [o.d.] and 220 µm inner diameter [i.d.]; Hospal, Dasco, Italy), sealed at one end with a drop of epoxy glue. The sealed end is gently sharpened with a thin grain abrasive disk to reduce tissue damage during implant. Two 4-cm long pieces of fused silica (Composite Metal Services, UK) tubing are introduced in the dialysis fiber, taking care to have the inlet reach the lower end and the outlet reach the higher end of the dialyzing portion (1.0 mm) of the fiber. The inlet and the outlet are then sealed to the fiber and to a 18-mm piece of stainless steel (obtained from a 24-gauge needle) that were then inserted into a piece of 200-µL micropipet tip 6 mm long and glued to it. The fiber is finally covered with a thin layer of epoxy glue except for the dialyzing part. The probe is left to dry for 24 h *(1,5)*. Among the wide spectrum of dialysis probes used, this type offers the distinction of being rigid enough to be implanted properly in brain, but at the same time is flexible enough to follow brain movements within the skull, preventing related tissue damage. We noticed that implanting probes with metal parts in contact with the brain resulted, at histological examination, in an increased space between the fiber and the tissue and in a reduced recovery of the probe. Probes have to be checked before use. At the beginning of surgery they can be connected to a perfusion pump delivering saline and left to dialyze until the end of surgery. There are three reasons for this: (1) checking that the fiber is not clogged; (2) the liquid pressure makes the fiber more rigid during implant, allowing a more precise positioning in deep areas; (3) removal of Ca^{2+} most likely reduces tissue damages due to calcium-dependent processes.

3.1.3. Time After Implant

The microdialysis technique was first developed in anesthetized rats *(7,12)* and later used in freely moving rats *(13)*. It was soon understood that a recovery period was necessary to clean up the excess of transmitter due to leakage from terminals damaged by the probe insertion. In the majority of studies a period of recovery of about 24 h is used although depending on the probe 6–8 h might be sufficient to get Ca^{2+} dependent tetrodotoxin (TTX)-sensitive neurotransmitter release. When the microdialysis probe is implanted in mice we noticed that 48-h recovery determines a more stable baseline of extracellular concentration of DA *(14)*.

3.1.4. Chronic Implants

Microdialysis can also be performed by implanting a guide cannula 1–2 mm above the brain area of interest, well in advance, to allow the animal to recover from the surgery. On the test day a fiber is implanted and the neurotransmitter is assayed. Drawbacks of this method are due to (1) the presence of a rigid metal guide that produces tissue damage above the area investigated and (2) probe insertion that produces an acute trauma in the area investigated. Inflammation of tissue at the site of repeated insertion may reduce the recovery of the probe *(1,15)*. The advantage of this method is that a different probe can be implanted in the same animal at long time interval to assess changes in the extracellular concentration of a neurotransmitter (e.g., due to a chronic treatment). Repeated neurotransmitter measurement also can be accomplished by implanting a probe and perfusing it each day for several days. We assayed DA in the nucleus accumbens from 9 A.M. to 7 P.M. for 8 d. We noticed that at d 1 and 2 basal levels of DA were stable, but from d 3 they kept decreasing to become undetectable at d 8, probably due to reactive gliosis clogging of the fiber *(1,7)*.

3.1.5. Two-Probe Experiment

Two probes can be implanted in an animal when the difference of the drug effect in two different regions of brain has to be ascertained. Recently this type of study has established a different response to drugs of abuse between the shell and the core region of the nucleus accumbens. No particular difficulties are encountered in implanting two probes but a two-channel swivel is mandatory to avoid twisting of perfusion tubing *(16)*.

3.2. General Considerations

3.2.1. Working Day

A typical microdialysis working day begins with preparation of perfusion Ringer. It can be quickly prepared, and diluted single salt stock solutions (100X) kept in the refrigerator. Stock solutions prepared and filtered under vacuum (0.22 μm) can last up to 3 mo. Filling syringes and connecting tubing may require about 1 h if four to six animals are used. Because samples recovered from the probe in the first hour of perfusion do not reflect a steady state of the neurotransmitter assayed, the samples collected during the first hour can be discarded. This hour can be used to prepare and inject some standards into the HPLC. The DA solution can be prepared daily and kept shielded from light under ice. To extend the life of standards, $0.1\ M$ perchloric acid and $0.5\ \text{m}M$ sodium metabisulfite can be added to 10 mM DA stock solution, which can to be kept as individual samples in a deep freezer ($-80°C$). At least four or five concentrations, in the range of that found in the sample recovered from the area investigated, must be injected. Thereafter samples can be assayed. Usually after a sharp decrease at beginning of perfusion, the concentration of neurotransmitter tends to remain constant, although it depends on the brain area under investigation. When the last three samples differ from each other <10% the animal can be tested and the mean of the last three samples can be considered as a reference basal. Stable levels of neurotransmitters can be obtained in about 2 or 3 h after start of dialysis (*see* **Notes 5–7**).

3.2.2. Collecting Samples

In the last decade we have collected samples using PE tubing (0.58 i.d., 0.96 o.d.; Portex, Hythe, UK) used to connect probes to the perfusion syringe. A 12-cm long coil is positioned in the probe outlet fitting tightly the 23-gauge needle. Every 10 or 20 min the collecting tubing (CT) is removed and its content is taken up by the HPLC syringe. Carefully rotating CT until the liquid reaches the CT edge will allow picking up the dialysate with an HPLC syringe, avoiding bubble formation (*see* **Notes 8** and **9**).

Systems are available that allow online microdialysis. It means that samples are loaded in the injector loop and at the established time, injected and detected without any interference of operator in the collecting procedure. These systems can be used for particular applications (e.g., sophisticated behavior studies, or long-term studies). Limiting operator influence is one of the reasons for using automatic injections. In this case intravenous injection of a drug using a previous implanted intravenous cannula *(17)* will allow administration of a drug without the need to hold a rat. It will, however, increase the time and complexity of surgery and risk of failure of the procedure.

3.2.3. Neurotransmitter Detection with or Without Metabolites

When metabolites are detected, the time required for elution of all detected substances may be as long as the collection time; it means that only samples from one animal can be injected in each HPLC with obvious reduction of capability of running multiple animal experiments. Because metabolite changes are often predictable, in many cases, detection of the neurotransmitter only is sufficient to characterize a drug effect pharmacologically. In this case modification of elution conditions (*see* **Subheading 2.1., item 3**) may allow reduction of elution time to 4–5 min, enabling injection in each HPLC of up to four samples in 20 min. In this way it is possible to assess a more detailed time course for a drug effect, or alternatively, to test more animals at the same time reducing the number of experiments necessary to obtain statistically significant results.

3.2.4. Results Expressed as % of Basal Rather than in Absolute Femtomoles

It has been debated often whether expression of microdialysis data should be in absolute amount of neurotransmitter detected or in % of basal. For unknown reasons, variability of basal neurotransmitter recovery is very high (out of six animals the lowest value recovered can be half of the highest) even when animals are implanted by the same person, on the same day, with probes prepared by the same person with the same material. It is difficult to obtain statistical significance using the raw data when the change of output induced by drug treatment is minimal. On the contrary, when the change of output is expressed as % of basal (considering the mean of the last three samples before treatment as a basal), reproducibility of results is very high and a drug effect can be statistically significant even when few animals are used.

3.3. Detecting Monoamines

In this subheading, the conditions we have recently used for detecting some neurotransmitters with or without metabolites are illustrated in detail.

3.3.1. Detecting Dopamine and Metabolites in the Nucleus Accumbens by Transcerebral Fiber

Twenty-minute samples were injected, without any purification, in a HPLC equipped with a reversed-phase ODS column (LC 18 DB Supelcosil, Supelco, Bellefonte, PA) and an electrochemical detector (BAS, Lafayette, IN) to quantitate DA, 3,4-dihydroxyphenylacetic acid (DOPAC), and homovanillic acid (HVA). The mobile phase was 0.23 M sodium acetate, 0.015 M citric acid, 100 mg/L EDTA pH 5.5. The electrode used was a silicon carbon paste TL-3 working electrode with potential set at +0.65 V in oxidation, coupled with an Ag/AgCl reference electrode. Experiments were always performed 24 h after probe implantation. Perfusion Ringer (147.2 mM NaCl, 3.4 mM Ca Cl$_2$, and 4 mM KCl) flow was 2 µL/min *(11)*.

3.3.2. Detecting Serotonin and HIAA in Frontal Cortex by Transcerebral Fiber

Twenty-minute samples were injected without any purification in HPLC equipped with a reversed-phase ODS column (Supelcosil LC18-DB, Supelco) and an electrochemical detector (BAS), to quantitate 5-HT and 5-hydroxyindole-3-acetic acid (5-HIAA). The mobile phase was a sodium citrate (30 mM)–acetate (78 mM) buffer, pH 4.7, containing 6.5% methanol. The flow cell was equipped with a silicon carbon paste Tl-3 working electrode (BAS) and an Ag/AgCl reference electrode. The potential was set at + 0.55 V (oxidation). Experiments were always performed 24 h after probe implant. Perfusion Ringer (147.2 mM NaCl, 3.4 mM Ca Cl$_2$, and 5 mM KCl) flow was 2 µL /min *(9)*.

3.3.3. Detecting Dopamine and Norepinephrine

Dialysate samples (10 or 20 µL) were injected without purification into an HPLC equipped with a reversed-phase column (LC-18 DB, 15 cm, 5 µm particle size; Supelco) and a coulometric detector with a 5014 A cell (ESA; Coulochem II, Bedford, MA) to quantitate DA and NA. The first electrode of the detector was set at +130 mV (oxidation) and the second at –175 mV (reduction). The mobile phase composition was 100 mM NaH$_2$PO$_4$, 0.1 mM Na$_2$–EDTA, 0.8 mM *n*-octyl sodium sulfate, and 18% methanol, pH adjusted to 5.50 with 5 mM Na$_2$HPO$_4$. Flow rate by a LKB 2150 pump was 1.0 mL/min. The sensitivity for DA and NA was 1 fmol/sample. Perfusion Ringer (147.2 mM NaCl, 2.2 mM Ca Cl$_2$, and 4 mM KCl) flow was 1 µL/min *(17)*.

3.3.4. Detecting Dopamine Without Metabolites in the Nucleus Accumbens by Vertical Probes

Ringer's solution (147 mM NaCl, 2.2 mM CaCl$_2$, and 4 mM KCl) was pumped at a constant flow of 1 µL/min. Samples were taken every 20 min and injected without any purification into an HPLC apparatus equipped with reversed-phase column (LC-18 DB Supelco) and an ANTEC Flowcell electrode coupled with a BAS detector. The oxidation potential was + 0.55 V. The mobile phase composition was 50 mM NaH$_2$PO$_4$–5 mM Na$_2$HPO$_4$, 0.1 mM Na$_2$EDTA, 0.5 mM octyl sodium sulfate, and 15% (v/v) methanol, pH 5.5. The mobile phase was pumped with an LKB 2150 pump at a flow rate of 1.0 mL/min. The sensitivity of the assay allowed detection of 5 fmol of DA *(18)*.

4. Notes

1. A great advantage in investigating the cause of noisy baseline is obtainable by having two similar HPLC systems. It will permit substitution of one piece of equipment at a time in the defective system, using parts (e.g., entire pump, single valve, pulse damper, column, reference electrode, etc.) of the system that function effectively; it can allow detection of the failure in the malfunctioning system. Furthermore, two similar systems could make it possible to have just one set of spare parts in the lab. It has to be considered that even though the microdialysis procedure is very economical, it is very time consuming, so every effort has to be made to avoid to wasting even one sample. One always needs to consider that, when a dialysate of three animals is injected into one HPLC, its failure means the experiment is wasted because if a total of six animals are used, the remaining three do not constitute a large enough n to obtain statistically significant data. In an experiment in which three animals are tested against three controls, the result is even worse and the experiment has to be repeated, to say nothing of testing animals chronically treated for several weeks. No one would like to waste an experiment like that because of equipment failure. The very best solution is to have a full spare system to use immediately when needed, while the less expensive solution is to have at least a spare working electrode and a spare HPLC pump.
2. Detector failure: When an amperometric detector is used, it is necessary to verify the reference electrode and the working electrode. Reference electrodes are usually very long lasting but substitution with a spare one (it can be borrowed from a well functioning second system) will ascertain if the reference electrode is defective. If the problem persists, the working electrode has to be checked and eventually changed, using the same procedure. In the case of failure of a coulometric electrode (ESA cell) again, the substituting procedure can be used to ascertain malfunctioning. Because malfunctioning cells, in most cases, cannot be repaired, a lower quality cell (noisy baseline) can be used for samples that require low sensitivity. Otherwise the cell has to be substituted.

3. It is very common to put the cabinet with the column and electrode above the detector cabinet, because it is presented in this way by the manufacturer and allows easy access to the column and the electrode. This organization represents a risk in case of flooding due to leakage. Damage to the detector can be prevented by placing a sheet of filter paper, such that when it gets wet it moves the liquid to the bench, saving the cabinet detector.
4. Pump failure: A malfunctioning pump is often a source of deteriorating baseline, but rarely does this condition prevent finishing the experiment. When failure occurs suddenly, the cause is very often the electronic control of the pump. If the operator's manual checks are not successful, the pump has to be delivered to the manufacturer. For this reason having a spare pump is mandatory; it will allow finishing experiments. We realized that when HPLC pumps fail, especially if old, they often continue to present problems even after being repaired. For this reason it is convenient to purchase a new pump, and keep the repaired one as an emergency spare pump. In case the noise increases slowly, over a 1-mo period, the cause is probably malfunctioning of the mechanical part of the pump (i.e., valves, piston, and seals). In this case substitution of the defective part will solve the problem. Purchasing a set of pistons and seals in advance will prevent interruption of research. A failure in one of the pistons is common when pumps are kept OFF for a while and then are turned ON. Failure may be due to a shrinkage of the piston seal. Injecting methanol in the pump will soften the seal and more than likely solve the problem.
5. Operator influence: While conducting a microdialysis experiment it can be observed an abrupt change in neurotransmitter output, in more than one rat at the same time. When it happens before the treatment, it causes only a delay, but if it happens once animals have been tested (either with a drug or other behavioral challenge), in most of the cases, the experiment has to be repeated. As the search for the cause is often unsuccessful, due to multiple factors change in the 10–20-min collecting period (people entering the lab, noise from outside, abrupt weather changes, etc.), a preventive action, on controllable factors, must be conducted. The ideal situation would be to conduct experiments in a soundproof room, having rats connected with an automatic injector to an HPLC that is in another room, with the possibility of administering drugs by intravenous catheter positioned previously. In reality, a microdialysis experiment is conducted in a noisy lab, often shared with other persons. This of course limits greatly the search for the cause that determines abrupt changes of basal neurotransmitter output. Nevertheless some tips may be useful to prevent problems of unknown origin. It is strongly recommended that the entire experiment from preliminary procedures (probe connection to the perfusion pump) to administration of drugs is performed by the same person. This may raise some problems when students are involved, as they often do not have good experience in administering animals (a stressful administration is the last thing one wants to see when a stable baseline is obtained after a few hours of perfusion). In this case it is advisable that the person who is in charge of administering animals collects two or three samples

before the administration of the test drug. Of course, if it causes a change of neurotransmitter baseline, administration of the test drug will have to be delayed. Stress may strongly influence neurotransmitter levels in specific areas (e.g., DA and NA in the prefrontal cortex) *(19)*. Inexperienced students should perform saline injections using the same route of the test drug, to check their expertise. If a drug effect is compared in two different groups of animals (e.g., a chronically treated group vs its control; a knockout group vs wild one), it is suggested to test in the same day by the same operator two groups of three animals rather than comparing results of one group of five or six animals with those obtained in the paired group, tested in a different day by a different operator.

6. Lunch time: This problem should not be ignored, although it seems of low interest if compared with technical problems. It requires the necessary attention to prevent an influence on drug effect and false results. Because the average microdialysis experiment may last about 6–8 h, it is predictable that the operator will have a break for lunch. The person who is collecting samples should have lunch early so he or she can collect at least three more samples before the test/drug treatment even in the case of stable baseline. It is advisable to wash the hands after lunch with a soap with a scent that is already familiar to animals. People should not visit the lab during the experiment (a "Do not enter" signal on the door is mandatory) because administrative employees and salespersons often wear strong perfumes.

7. Once the HPLC system has been properly tuned, it can be left "ON" overnight, taking care of reducing pump flow from about 1 mL/min to 0.2–0.4 mL/min at the end of the experiment. Therefore detectors have also to be left "ON." Even if lack of power is a rare event, it is advisable to connect EC detectors and pumps to an uninterruptible power supply. It will preserve electrode deterioration due to potential fluctuation and blackouts. Moreover, because some HPLC pumps do not restart when the electricity is restored, an uninterruptible power supply will prevent salt precipitation in both pump and electrode. If this happens, substitute the mobile phase with degassed water, disconnect columns and electrodes, and wash the pump by gravity first and at very low flow later (0.1 mL/min). Connect again all the components, one by one, keeping the flow low. If pressure does not rise abruptly, flow can be increased to 1 mL/min. After 30 more minutes at this flow, water can be substituted depending on the use with either water–methanol 50:50 or the original mobile phase.

8. Disappearance of a sample: Finding the collecting tubing empty is a frequent inconvenience. It is often due to leakage in the connection between the probe and the perfusion pump. When swivels are used, chances of leakage increase. A careful search by touching without gloves the entire line to look for wet spots is often successful. It can be prevented by cutting the PE tubing in contact with the metal tubing of probes, syringes, swivels, and connections (i.e., cut and smoothed 23-gauge needles used for joining two pieces of PE tubing) that become loose by prolonged use. Always buy at last three different sizes (below and above the size used) to have always a PE tubing that can fit tightly. Among others, a

loose positioning of collecting tubing in the probe outlet is the most frequent. Substitute the collecting tubing or cut the loose part frequently.
9. The fiber does not perfuse: When the search for leakage in unsuccessful, it has to be suspected that something went wrong in the probe. The probes can be checked by connecting directly a syringe to the probe outlet, pressing the barrel gently and slowly. Often an air bubble in the fused silica tubing can be eliminated in this way. If the liquid does not go in, or goes in but does not come out the probe, the animal has to be scarified.
10. Air in the system: Whenever the baseline looks noisy, sources of instability have to be found in the electrode or in the pump. The first thing to be checked is the presence of air bubbles in the proximity of the working or the reference electrode. Procedures to solve this problem are usually different and depend on the type of electrode used. Usually, following the manufacturer's instructions is sufficient to eliminate the problem. Because an abrupt deterioration of the baseline will often result in failure to assess a sample (it may be a crucial one, in the time course) it is mandatory to identify immediately the cause. Often, the cause is air injected with the syringe, an error that can be easily prevented. A second cause of deteriorated baseline could be the presence of air in the pump. It can be easily detected by verifying if the displayed pressure is unstable. In this case, the purge operation should be performed, according to the manufacturer's instructions. This operation consists of opening the HPLC circuit, either with a valve or by disconnecting the pump from the injector. The flow can be increased to eliminate the air trapped in pump. If this problem occurs often, the procedure of degassing the mobile phase has to be verified. If this procedure is not successful, an important failure in the electrode or in the pump has to be suspected.
11. It can be either an integrator or a computer with dedicated software or just a chart recorder. A failure in the integrator is definitive especially if all the tests indicated in the operator's manual have been performed without success. In this case having a spare recorder can save the experiment. Alternatively as last resource, one recorder/integrator can be used for two systems, connecting it in the proximity of the elution of the peak of interest.

Acknowledgment

This work was supported by Ministero Istruzione Università Ricerca (MIUR) project PRIN 2001-2002 and by Consiglio Nazionale delle Ricerche (CNR).

References

1. Di Chiara, G., Tanda, G., and Carboni, E. (1996) Estimation of in-vivo neurotransmitter release by brain microdialysis: the issue of validity. *Behav. Pharmacol.* **7,** 640–657.
2. Marsden, C. A. and Joseph, M. H. (1986) Biogenic amines, in *HPLC of Small Molecules: A Practical Approach.* (Lim, C. K., ed.), IRL Press, Oxford/Washington, DC.

3. Timmerman, W. and Westerink, B. H. C. (1997) Brain microdialysis of GABA and glutamate: what does it signify? *Synapse* **27,** 242–261.
4. Carboni, E., Silvagni, A., Rolando M. T. P., and Di Chiara, G. (2000) Stimulation of in vivo dopamine treansmission in the bed nucleus of stria terminalis by reinforcing drugs. *J. Neurosci.* **20,** RC 102.
5. Di Chiara, G. (1990) In vivo brain dialysis of neurotransmitters. *Trends Pharmacol. Sci.* **11,** 116–121.
6. Westerink, B. H. C. and De Vries, J. B. (1988) Characterization of in vivo dopamine release as determined by brain microdialysis after acute and subchronic implantations: methodological aspects. *J. Neurochem.* **51,** 683–687.
7. Imperato, A. and Di Chiara, G. (1984) Trans-striatal dialysis coupled to reverse phase high performance liquid chromaotography with electrochemical detection: a new method for the study of the "in vivo" release of endogenous dopamine and metabolites. *J. Neurosci.* **4,** 966–977.
8. Westerink, B. H. C., Tuntler, J., Damsma, G., Rollema, H., and de Vries, J. B. (1987) The use of tetrodotoxin for the characterization of drug-enhanced dopamine release in conscious rats studied by brain dialysis. *Naunyn Schmiedebergs Arch. Pharmacol.* **336,** 502–507.
9. Carboni, E. and Di Chiara, G. (1989) Serotonin release estimated by transcortical dialysis in freely-moving rats. *Neuroscience* **32,** 637–645.
10. Westerink, B. H. C., Hofsteede, H. M., Damsma, G., and de Vries, B. (1988) The significance of extracellular calcium for the release of dopamine, acetylcholine and amino acids in conscious rats, evaluated by brain microdialysis. *Naunyn Schmiedebergs Arch. Pharmacol.* **337,** 373–378.
11. Carboni, E., Imperato, A., Perezzani, L., and Di Chiara, G. (1989) Amphetamine, cocaine, phencyclidine and nomifensine increase extracellular dopamine concentrations preferentially in the nucleus accumbens of freely moving rats. *Neuroscience* **28,** 653–661.
12. Zetterstrom, T., Sharp, T., Marsden C. A., and Ungersted, U. (1983) "In vivo" measurement of dopamine and its metabolites by intracerebral dialysis: changes after *d*-amphetamine. *J. Neurochem.* **41,** 1769–1773.
13. Imperato, A. and Di Chiara, G. (1985) Dopamine release and metabolism in awake rats after systemic neuroleptics as studied by trans-striatal dialysis. *J. Neurosci.* **5,** 297–306.
14. Carboni, E., Spielewoy, C., Vacca, C., Nosten-Bertrand, M., Giros, B., and Di Chiara, G. (2001) Cocaine and amphetamine increase extracellular dopamine in the nucleus accumbens of nice lacking the dopamine transporter gene. *J. Neurosci.* **21,** RC 141.
15. Robinson, T. E. and Camp, D. M. (1991) The feasibility of repeated microdialysis for within-subjects design experiments: studies on the mesocortical dopamine system, in *Microdialys in Neuroscience* (Robinson, T. E. and Justice, J. B., Jr., eds.), Elsevier, Amsterdam-New York.
16. Di Chiara, G. (1998) A motivational learning hypothesis of the role of mesolimbic dopamine in compulsive drug use. *J. Psychopharmacol.* **12,** 54–67.

17. Tanda, G., Pontieri, F. E., Frau R., and Di Chiara, G. (1997) Contribution of blockade of the Noradrenaline carrier to the increase of extracellular dopamine in the rat prefrontal cortex by amphetamine and cocaine. *Eur. J. Pharmacol.* **9,** 2077–2085.
18. Carboni, E., Bortone, L., Giua, C., and Di Chiara, G. (2000) Dissociation of physical abstinence signs from changes in extracellular dopamine in the nucleus accumbens and in the prefrontal cortex of nicotine dependent rats. *Drug Alcohol Depend.* **58,** 93–102.
19. Abercrombie, E. D., Keefe, K. A., Di Frischia, D. S., and Zigmond, M. J. (1989) Differential effect of stress on in vivo dopamine release in the striatum, nucleus accumbens and medial frontal cortex. *J. Neurochem.* **52,** 1655–1658.

30

HPLC/EC Detection and Quantification of Acetylcholine in Dialysates

James A. Zackheim and Elizabeth D. Abercrombie

1. Introduction

Acetylcholine (ACh) is a neurotransmitter widely distributed in the central nervous system (CNS) and peripheral nervous system (PNS). Its role as a neurotransmitter was first elucidated by Dale, who noted that ACh mimicked the effects of parasympathetic nerve stimulation and by Otto Loewi, who demonstrated the vagal release of a substance that slowed heart rate *(1–3)*. More recently ACh in the CNS has been implicated in sensorimotor arousal, attention, sleep regulation, and learning and memory *(4–8)*. Its distribution in the CNS includes the entire cortical mantle and hippocampus innervated by cholinergic neurons of the basal forebrain and the interpeduncular nucleus innervated by the medial habenula. The striatum, nucleus accumbens, and olfactory tubercle each contain cholinergic interneurons *(9)*. The extraction of ACh from these areas in intact preparations historically has been largely via push–pull cannulae and now via microdialysis *(10,11)*. Once extracted from the brain, however, the separation (by high-performance liquid chromatography [HPLC]) and quantification (by electrochemical detection) of ACh presents an unusually difficult problem. Such a difficulty arises first from the extremely rapid hydrolysis of ACh in vivo by ACh esterase and second from the resistance of the ACh molecule to electrochemical oxidation. The first of these problems can be overcome by in vivo application of an esterase inhibitor such as neostigmine, although the use of too high a concentration can dramatically disturb the physiological regulation of the cholinergic system and introduce artifactual experimental results *(12–14)*. The second problem is overcome by a two-step enzymatic conversion of ACh to a more readily oxidizable species *(15)*. The use of enzymes strongly constrains all other HPLC parameters, however. These parameters and detailed methods are discussed in the following subheadings.

2. Materials

1. Mobile phase: 0.15 M K_2HPO_4, 2.5 mM KCl, 0.1 mM EDTA, 3 mM tetramethylammonium chloride (TMA), 0.08 mM sodium octyl sulfate (SOS), and antimicrobial agent MB (1:20,000; ESA, Bedford, MA). Adjust pH to 7.78 by adding NaH_2PO_4, then filter and thoroughly degas. SOS is added as an ion pairing agent and TMA is added to adjust ACh retention time (*see* **Note 1**). TMA is a quaternary amine that competes with ACh for column binding sites, thereby reducing retention time. MB is added to prevent microbial growth (*see* **Note 2**). EDTA is added to prevent chelation of oxidizable Fe^{3+} ions. Under the conditions described in **Subheading 3.**, this mobile phase results in retention times of 1.6 min for choline and 4.5 min for ACh.
2. Sodium lauryl sulfate (SLS) (Sigma).
3. Chromspher $5C_{18}$ packing material (Varian, Walnut Creek, CA).
4. Glutaraldehyde and CCl_4 (Fisher).
5. Lichrosorb-NH2 (10 µm) (Phenomenex, Torrance, CA).
6. Choline oxidase (EC 1.1.3.17) from *Alcaligenes* species and ACh esterase type VI-S from electric eel (EC 3.1.1.7) (Sigma).
7. Dual piston Shimadzu LC-10AD pump (Shimadzu Scientific Instruments, Columbia, MD).
8. Valve type injector (Rheodyne, Varian, Walnut Creek, CA) with a 10-µL sample loop.
9. Pulse dampener (SSI, Varian, Walnut Creek, CA).
10. Valco column (100 mm × 4.5 mm) (Varian).
11. Guard column and assembly (Chrompack type A2, 10 × 2 mm) (Varian).
12. VT-03 wall jet flow cell (Antec, Leiden, The Netherlands).
13. 0.25-µm diamond slurry (Pace-ATM, Northbrook, IL).
14. Macintosh-based chromatography hardware and software (Dynamax) (Varian).

3. Methods

3.1. Sample Preparation

Twelve-microliter dialysis samples are collected every 8 min. The sample volume is restricted to 10 µL by the sample loop. No sample dilution or purification is required.

Samples are typically immediately injected onto the HPLC; however, storage of dialysate at –80°C for several months is possible without significant degradation of ACh content.

Detection of Acetylcholine

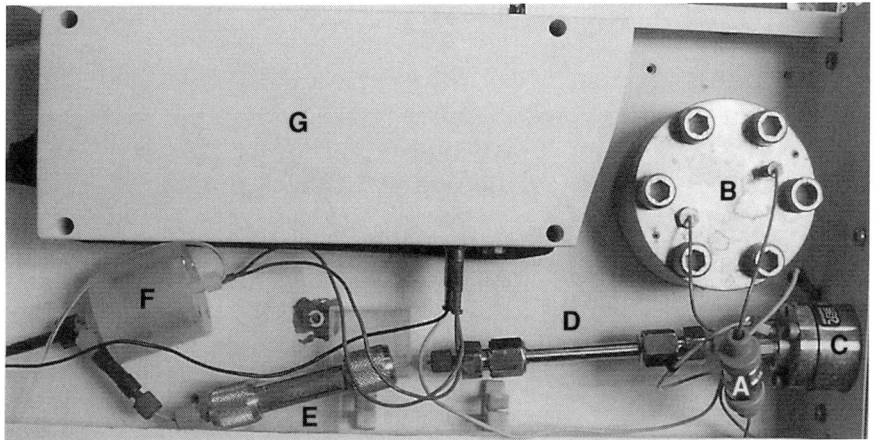

Fig. 1. Plan view of the ACh HPLC to show the interconnection of components. Mobile phase is pumped sequentially through a guard column (not shown), an in-line filter (A) and pulse dampener (B) toward a Rheodyne injector (C). Here, injected samples join the flow stream and pass onto the analytical column (D). Column effluent flows through the immobilized enzyme reactor (E) and then to the flow cell (F). The potentiostat (G) controls the potential of the electrode within the flow cell and measures induced oxidative current.

3.2. Arrangement of Components

An HPLC system in our laboratory is dedicated solely to the analysis of ACh. This system consists of a dual-piston Shimadzu LC-10AD pump connected to a valve type injector with a 10-µL sample loop. A guard column, in-line filter, and pulse dampener are placed between the pump and injector. All components are connected with polyetheretherketone (PEEK) tubing (0.18 mm inner diameter [i.d.]). The analytical column, enzyme reactor and flow cell are connected after the injector, in series, taking care to minimize tubing volume. All components excluding the pump are housed in a temperature-controlled Faraday case set at 30°C. The arrangement of system components is illustrated in **Fig. 1**. The flow rate of mobile phase through the system is maintained at 0.3 mL/min at all times.

3.3. Preparation of Column

1. Our laboratory uses a reversed-phase (nonpolar) column. To prepare the column, an empty Valco column (100 mm × 4.5 mm) serves as a reservoir and is connected to an empty analytical column (75 mm × 2 mm) fitted with a 0.22-µm filter inside the end cap. The arrangement of components is shown in **Fig. 2**. Both

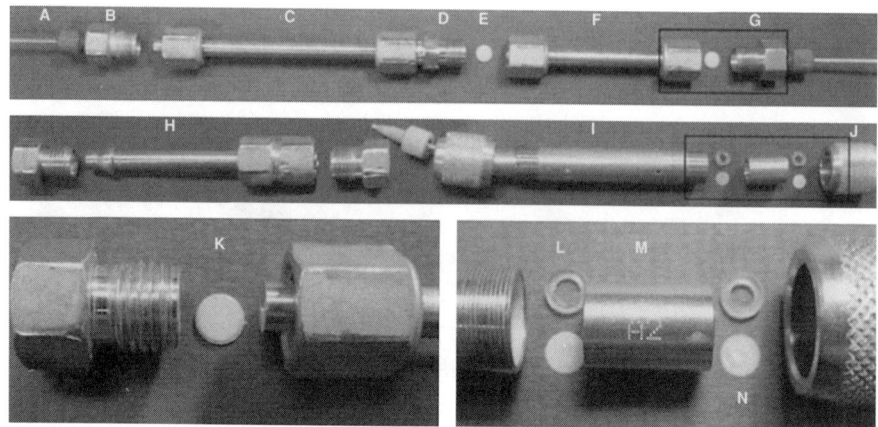

Fig. 2. The initial arrangement of components used to prepare a new analytical column (**upper panel**) and immobilized enzyme reactor (**middle panel**). The lower two panels show the boxed regions at higher magnification. **Upper panel:** An outlet line from the pump (A) is connected to the inlet cap (B) of an empty Valco column (C) serving as a reservoir. A union (D) connects the reservoir to the empty analytical column (F). A 0.22-µm filter (K, **lower left panel**) is placed inside the end cap (G) and connected to the column outlet. The column outlet flows to waste. Note that the filter in the column inlet (E) is not placed in-line until the initial column packing has been completed and the reservoir removed. **Middle panel:** In manner analogous to column packing, an empty Valco column (H) serves as a reservoir and is connected to the enzyme reactor cartridge housing (I). The packed enzyme cartridge (M, **lower right panel**) is capped with "O" rings (L) and 0.22-µm filters (N) and secured inside the housing with a screw type end cap (J).

    columns are aligned vertically and connected to a pump via a pulse dampener and in-line filter.

2. Chromspher 5C_{18} (0.4 g) is completely dissolved in 1 mL of CCl_4 and pipetted into the reservoir. The reservoir is then filled with 100% methanol and capped.
3. Methanol (100%) is pumped through the columns at an initial flow rate of 0.5 mL/min and the flow rate is then increased in 0.5 mL/min increments every 30 s until a flow rate of 4.0 mL/min is attained. This flow rate is maintained for 15 min and then gradually decreased to 0 mL/min.
4. The reservoir is removed, taking care to scrape away excess packing material evenly from the column inlet. The column inlet is quickly capped with a 0.22-µm filter to prevent expansion of the compressed packing material. The column is then packed further by pumping 100% MeOH and then HPLC grade water at 1.0 mL/min for 60 min each.

Detection of Acetylcholine

5. A cation-exchange column is prepared by the addition of the ion pairing agent SLS. SLS (0.5 g) dissolved in pure water and filtered is pumped through the column for 25 min at 1.0 mL/min.
6. Finally, the column is rinsed with mobile phase at 0.3 mL/min for at least 4 h, but preferably overnight. Columns typically last 3–5 mo (*see* **Note 3**). Once retired the analytical column is removed and used as a guard column. The use of a guard column prolongs the life of the analytical column by preventing physical disturbances and saturating the mobile phase with silica.

3.4. Preparation of Enzyme Reactor

ACh is not readily oxidizable and must therefore be converted to an oxidizable species to allow electrochemical detection. This conversion can be achieved using an enzyme reactor. Although enzyme reactors are commercially available we prepare a reactor in the laboratory. The enzyme reactor consists of ACh esterase and choline oxidase covalently immobilized in a silica matrix. These enzymes convert ACh to H_2O_2 as follows:

$$ACh + H_2O \xrightarrow{ACh\ esterase} Choline + acetate \quad (1)$$

$$Choline \xrightarrow{Choline\ oxidase} Betaine + 2\ H_2O_2 \quad (2)$$

1. The reactor is prepared within a commercially available guard column (Chrompack type A2, 10 × 2 mm). A 0.22-µm filter is attached to the outlet of the guard column. A solution of 75 mg of Lichrosorb-NH_2 in 1 mL of CCl_4 is pipetted into the column under mild vaccum and excess powder is removed until flush with the column inlet.
2. The column inlet is capped with a 0.22-µm filter and placed inside the guard column assembly. The assembly is connected to the outlet of an empty Valco column (80 mm × 4.5 mm) that serves as a reservoir (*see* **Fig. 2**).
3. The reservoir is filled with a filtered 25% glutaraldehyde solution and capped. The column reservoir and enzyme reactor are aligned vertically and connected to a pump via a pulse dampener and in-line filter. Mobile phase is pumped through the reservoir at 0.1 mL/min for 10 min.
4. The reservoir is removed, rinsed in deionized water, and reattached to the enzyme reactor. A freshly prepared solution of enzymes (80 U of ACh esterase and 40 U of choline oxidase in 400 µL mobile phase) is then pipetted into the reservoir. The mobile phase is pumped through the reservoir at 0.04 mL/min for 25 min.
5. The reservoir is removed and the reactor is rinsed with mobile phase overnight. Once prepared enzyme reactors last between 2 and 4 wk depending on usage. Maximal efficiency is obtained at a temperature of 30–35°C. When the HPLC is

not being used enzyme life can be prolonged by removing the reactor cartridge and storing under mobile phase at 4°C (*see* **Notes 4–6**). Cartridges stored in this manner maintain their efficacy for at least 4 wk.

3.5. Electrochemical Cell

Our laboratory employs a VT-03 wall jet flow cell (Antec) for the electrochemical oxidation of H_2O_2. The flow cell is fitted with a 25-µm gasket and a platinum electrode held at +0.5 V vs an Ag/AgCl reference electrode by an INTRO potentiostat (Antec) (*see* **Note 7**). The reference electrode must be filled with saturated KCl. NaCl is not compatible with the phosphate mobile phase.

The surface of the electrode is cleaned weekly by rinsing in deionized water and gentle wiping with a cotton swab soaked in 100% MeOH. If this treatment fails to restore analytical sensitivity, the surface is then gently polished with a silk cloth and 0.25-µm diamond slurry. The surface of the electrode is rinsed and cleaned with MeOH after polishing to remove residue. Polishing is typically required no more than once every 2 mo.

3.6. Data Analysis

Oxidative current output is integrated and analyzed using Macintosh based chromatography hardware and software (Dynamax). A detection limit (3:1 signal to noise ratio) of 5 fmol/sample is routinely obtained. The system is calibrated by injection of 20 and 200 n*M* standards (*see* **Note 8**).

4. Notes

1. Addition of TMA and SOS to the mobile phase also contributes to a gradual loss of enzymatic activity. Concentrations greater than 0.3 m*M* SOS rapidly denature the enzymes. TMA is an esterase inhibitor, although concentrations below 6 m*M* do not appear to noticeably compromise enzyme efficiency. Adjustment of retention time is preferably achieved by using oven-controlled temperature (for both choline and ACh), the addition of small amounts of TMA (ACh only) or a small alteration of buffer strength.
2. The loss of optimal enzymatic activity is the source of most ACh HPLC problems. Bacterial growth in the mobile phase is a common problem because of the pH of the solution. Some bacterial species can rapidly denature the enzymes. Other species produce the enzyme catalase that degrades hydrogen peroxide, thus diminishing the end product of the enzymatic conversion. The presence of such bacteria can lead to the erroneous conclusion that the enzyme reactor itself is compromised. These problems can be prevented primarily by using ultrapure water, clean glassware, never recycling mobile phase, changing the mobile phase at least weekly, and if necessary adding a low concentration of an antibacterial

agent to the mobile phase. Use of an excessive concentration of antibacterial agents, however, will also denature the enzymes.
3. One problem associated with HPLC analysis of ACh is specific to the analysis of ACh in neuronal dialysate: choline is present at concentrations 1000 times greater than ACh. The enzymatic conversion efficiency for choline to hydrogen peroxide is also twice that of ACh; a result of a one-step rather than two-step conversion. Thus the choline peak in a chromatogram can be several orders of magnitude greater than that of ACh. A 2-min peak-to-peak interval in choline and ACh retention times may thus still result in an ACh peak that coelutes with the tail of the choline peak. In this situation the peak height of ACh will be substantially underestimated. Alas, wide separation of ACh and choline is difficult owing to the constraints of the mobile phase composition. For this reason regular replacement of the analytical column is essential.
4. One further problem that is associated with the use of an enzyme reactor is the contribution of the reactor, especially when new, to the buildup of a residue on the surface of the working electrode. The detrimental effects of the residue can be diminished by turning off the applied potential whenever the HPLC is not in use. The cell may even be turned off each night if it is possible to allow enough time for the system to reequilibrate the following day.
5. There are several alternatives, both substantial and minor, to the procedures outlined in the preceding, although apparent optimization of one parameter often leads to a degradation of performance elsewhere in the system. This problem of constraint satisfaction is especially acute in the case of ACh HPLC. Nevertheless, our laboratory has had experience with some technical alternatives. These and other promising developments are outlined in this and in subsequent subheadings. Complete kits that include separation columns and enzyme reactors are commercially available, for example, Chrompack (Varian) or BAS. Our laboratory has only limited experience with these kits. However, preliminary investigation demonstrated that laboratory prepared hardware was more sensitive and significantly more durable. One advantage of one such commercially available kit (Chrompack), however, is the zero dead volume connection between guard column, column, and enzyme reactor which results in improved peak shape.
6. The lifetime of any enzyme reactor is proportional to the total converted amount of choline and ACh, rather than a function of time. Choline is often not a dependent measure of interest in neuroscientific experiments owing to its largely non-neuronal origin *(16)*. Thus the lifespan of the reactor can be prolonged by precolumn conversion of choline in a separate enzyme reactor. Moreover, the elimination of choline eliminates problems associated with the chromatographic separation of choline from ACh. Such a reactor can be prepared exactly as described in the preceding, but using the enzymes choline oxidase and either peroxidase or catalase to convert choline to peroxide and peroxide to water *(17)*.
7. The choice of electrochemical surface is limited for ACh HPLC. H_2O_2 will slowly damage the surface of glassy carbon electrodes and a high applied potential is

required for its oxidation at this surface. Platinum is an excellent surface, being inert and requiring a relatively low applied potential for the oxidation of peroxide. Moreover we have not found the reported drawback of a long equilibration time for platinum electrodes to be problematic. However, platinum is soft and easily scratched. The required applied potential of +0.5 V also contributes significantly to background noise, especially due to the continuous activity of the enzyme reactor. A newer electrode surface that is constructed from a glassy carbon electrode coated with a peroxidase-redox polymer is now commercially available from BAS. This surface requires a low applied potential and should thus reduce background noise. Maintenance of this surface may be significantly more problematic than for platinum surfaces.

8. The analytical methods described in the preceding provide a highly sensitive assay for ACh. However the detection of ACh in dialysate may still require the application of an ACh esterase inhibitor in vivo due to the very low concentration of ACh in certain CNS regions. Although this chapter discusses only the analysis of ACh and not its extraction from brain by microdialysis, the two techniques are highly interdependent. The need for esterase inhibitors can be minimized by careful surgical technique. Recovery of ACh is especially sensitive to local tissue damage, which can be minimized by extremely slow lowering of the microdialysis probe. In this way esterase inhibitor concentrations can be kept low and the interpretation of changes in ACh levels becomes more straightforward. The use of high concentrations of the esterase inhibitor neostigmine has been repeatedly shown not only to elevate basal ACh levels but also to alter the responsiveness of the striatal cholinergic system to drug challenge *(12–14)*. This result highlights the *variable* interdependence of basal ACh levels and the magnitude of changes in ACh level subsequent to experimental manipulation. For this reason our laboratory always reports basal ACh levels and the magnitude of changes in ACh levels in absolute terms.

Acknowledgments

The authors acknowledge the pioneering efforts of Dr. Peter DeBoer toward establishing and refining the methodology described in this chapter. The assistance of Dr. J. M. Tepper with the digital photography is also greatly appreciated. This work was supported USPHS Grants DA08086 and NS19608 and the Tourette's Syndrome Association.

References

1. Loewi, O. (1921) Ueber humorale Uebertragbarkeit der Herznervenwirkung. *Pflugers Arch.* **189**, 239–242.
2. Dale, H. H., Feldberg, W., and Vogt, M. (1936) Release of acetylcholine at voluntary motor nerve endings. *J. Physiol. (Lond.)* **86**, 353–380.
3. Dale, H. H. (1914) The action of certain esters and ethers of choline and their relation to muscarine. *J. Pharmacol.* **6**, 147–190.

4. McCormick, D. A. (1992) Neurotransmitter actions in the thalamus and cerebral cortex. *J. Clin. Neurophysiol.* **9,** 212–223.
5. McCormick, D. A. (1993) Actions of acetylcholine in the cerebral cortex and thalamus and implications for function. *Prog. Brain. Res.* **98,** 303–308.
6. Vanderwolf, C. H. (1992) The electrocorticogram in relation to physiology and behavior: a new analysis. *Electroencephalogr. Clin. Neurophysiol.* **82,** 165–175.
7. Sarter, M. and Bruno, J. P. (1997) Cognitive functions of cortical acetylcholine: toward a unifying hypothesis. *Brain Res. Brain Res. Rev.* **23,** 28–46.
8. Hasselmo, M. E. and Bower, J. M. (1993) Acetylcholine and memory. *Trends Neurosci.* **16,** 218–222.
9. Woolf, N. J. (1991) Cholinergic systems in mammalian brain and spinal cord. *Prog. Neurobiol.* **37,** 475–524.
10. Westerink, B. H. C. and Justice, J. B., Jr. (1991) in *Microdialysis in the Neurosciences* (Robinson, T. E. and Justice, J. B., Jr., eds.), Elsevier, Amsterdam, pp. 23–43.
11. Benveniste, H. (1989) Brain microdialysis. *J. Neurochem.* **52,** 1667–1679.
12. DeBoer, P., Westerink, B. H., and Horn, A. S. (1990) The effect of acetylcholinesterase inhibition on the release of acetylcholine from the striatum in vivo: interaction with autoreceptor responses. *Neurosci. Lett.* **116,** 357–360.
13. DeBoer, P. and Abercrombie, E. D. (1996) Physiological release of striatal acetylcholine in vivo: modulation by D1 and D2 dopamine receptor subtypes. *J. Pharmacol. Exp. Ther.* **277,** 775–783.
14. Acquas, E. and Fibiger, H. C. (1998) Dopaminergic regulation of striatal acetylcholine release: the critical role of acetylcholinesterase inhibition. *J. Neurochem.* **70,** 1088–1093.
15. Potter, P. E., Meek, J. L., and Neff, N. H. (1983) Acetylcholine and choline in neuronal tissue measured by HPLC with electrochemical detection. *J. Neurochem.* **41,** 188–194.
16. Tucek, S. (1985) Regulation of acetylcholine synthesis in the brain. *J. Neurochem.* **44,** 11–24.
17. Ichikawa, J., Dai, J., and Meltzer, H. (2000) Acetylcholinesterase inhibitors are neither necessary nor desirable for microdialysis studies of brain acetylcholine. *Curr. Separat.* **19,** 37–43.

31

Real-Time Measurements of Phasic Changes in Extracellular Dopamine Concentration in Freely Moving Rats by Fast-Scan Cyclic Voltammetry

Paul E. M. Phillips, Donita L. Robinson, Garret D. Stuber, Regina M. Carelli, and R. Mark Wightman

1. Introduction

Rapid, transient changes in extracellular dopamine concentrations following salient stimuli in freely moving rats have recently been detected using fast-scan cyclic voltammetry *(1,2)*. This type of neurotransmission had not been previously observed (for any neurotransmitter), but has been implicated by electrophysiological studies. Schultz et al. *(3)* reported synchronous burst firing of midbrain dopaminergic neurons following presentation of liquid reinforcers or associated cues. Such firing patterns would predictably produce transient (lasting no more than a few seconds), high concentrations (high nanomolar) of extracellular dopamine in terminal regions. This phasic dopaminergic neurotransmission has been heavily implicated in associative learning and reward processing, and therefore may prove essential in understanding the reinforcing actions of drugs of abuse.

Fast-scan cyclic voltammetry is a real-time electrochemical technique that can detect dopamine by its redox properties. It is capable of monitoring monoaminergic neurotransmission in the brain on a subsecond time scale while providing chemical information on the analyte. This chapter describes how the properties of fast-scan cyclic voltammetry make it uniquely suitable for making chemical measurements of phasic dopaminergic neurotransmission in freely moving animals. To emphasize the potential of this technique, three examples of its use are highlighted. First we describe experiments testing

From: *Methods in Molecular Medicine, vol. 79: Drugs of Abuse: Neurological Reviews and Protocols*
Edited by: J. Q. Wang © Humana Press Inc., Totowa, NJ

intracranial self-stimulation that provide information on dopamine's role in reward. Next we show transient dopamine signals evoked by natural stimuli, and finally, we present preliminary data demonstrating transient dopaminergic neurotransmission during cocaine self-administration behavior.

2. Materials
2.1. Electrochemistry

1. Multifunction input/output card: PCI-6052E (16 bit, 333 kHz) (National Instruments, Austin, TX).
2. Virtual instrumentation software: Locally written in LabVIEW (National Instruments).
3. Potentiostat: EI-400 (Cypress Systems, Lawrence, KS).
4. Headstage: Miniaturized current-to-voltage converter (UNC-CH Electronics Design Facility, Chapel Hill, NC).
5. Swivel: CAY-675-12 (12 electrical and 1 fluid channel) (Airflyte, Bayonne, NJ).

The software we use for fast-scan cyclic voltammetry and data analysis will be marketed with data acquisition hardware and a potentiostat for fast-scan cyclic voltammetry: FAST-FSCV16, Quanteon, Lexington, KY (http://quanteon.cc/).

2.2. Electrical Stimulation

1. Digital-to-analog card: PCI-6711 (National Instruments).
2. Constant-current stimulator: Neurolog NL800 (Harvard Apparatus, MA).

2.3. Electrode Fabrication

1. Carbon fiber: T-650 (Thornel, Greenville, SC).
2. Capillary glass: 6245 (0.6 mm outer, 0.4 mm inner diameter) (A-M Systems, Carlsborg, WA).
3. Vertical electrode puller: PF-2 (Narishige, Tokyo, Japan).
4. Micromanipulator (**Fig. 2**) (UNC-CH Machine shop).
5. Silver paint: 22-0202-0000 (Silver Print) (GC Electronics, Rockford, IL).
6. Heat shrink tubing: KYNAR 3/64-inch diameter, (3 M Electronics, Austin, TX).
7. Activated carbon: 102489 (NORIT A decolorizing carbon) (ICN Biomedicals, Aurora, OH).
8. Silver wire: 32,702-6 (0.5 mm diameter) (Sigma, St. Louis, MO).
9. Gold connector: 46N6790 (socket pin terminal) (Newark Electronics, Chicago, IL).

2.4. Surgery

1. Anesthetic: K-113 (ketamine HCl/xylazine HCl solution) (Sigma).
2. Stereotaxic frame: Model 900 Small Animal Stereotaxic (David Kopf Instruments, Tujunga, CA).

3. Guide cannula: MD-2251 (locking, with stylet) (Bioanalytical Systems, West Lafayette, IN).
4. Anchor screws: O-80 × 1/8 (stainless steel, 3.2 mm) (Plastics One, Roanoke, VA).
5. Cranioplastic cement: 675571/2 (Grip Cement) (DENTSPLY Caulk, Milford, DE).
6. Stimulating electrode: MS303/2 (stainless steel, 0.2 mm, bipolar) (Plastics One).

2.5. Experiment

1. Video character generator (UNC-CH Electronics Design Facility).

2.6. Electrode Calibration

1. Dopamine HCl (Sigma H6,025-5).
2. Flow injection apparatus (Rheodyne, Rohnert Park, CA).
3. Tris-HCl buffer solution: 12.0 mM Tris-HCl, 140 mM NaCl, 3.25 mM KCl, 1.20 mM CaCl$_2$, 1.25 mM NaH$_2$PO$_4$, 1.20 mM MgCl$_2$, 2.00 mM Na$_2$SO$_4$, pH 7.40. Store at room temperature.

2.7. Data Analysis

1. Software: Locally written in LabVIEW (National Instruments).

2.8. Verification of Electrode Placement

1. Stainless steel electrode: 5725 (0.5 mm, 8° taper, 12 MΩ) (A-M Systems).
2. Stimulator: A360R/A362/PRO4 (World Precision Instruments, Sarasota, FL).
3. Paraformaldehyde: 4% Paraformaldehyde in phosphate buffer (80 mM Na$_2$HPO$_4$, 20 mM NaH$_2$PO$_4$), pH 7.4. Refrigerate.
4. Thionin stain: 1.75 mM Thionin acetate, 450 mM sodium acetate in 3.5% acetic acid. Store at room temperature.
5. Potassium ferricyanide solution: 150 mM potassium ferricyanide in 10% HCl. Make up fresh.

3. Methods

3.1. Electrochemistry

Using real-time electrochemical techniques, dopamine can be monitored in brain tissue on the millisecond time scale (*see* **Note 1**). It can be electrochemically detected at carbon fiber microelectrodes because at physiological temperature and pH it is oxidized by application of a relatively modest potential to the electrode. This reaction converts dopamine to its *o*-quinone by oxidation of the two hydroxyl groups of the catechol, thus liberating two electrons from each dopamine molecule. These electrons are measured at the electrode as (faradaic) current.

With fast-scan cyclic voltammetry, the carbon fiber microelectrode (*see* **Note 2**) is held at a nonoxidizing potential and then periodically driven to an oxidizing potential and back by a triangular wave (**Fig. 1A**). Typically the electrode is held at −0.40 V vs Ag/AgCl between scans and then ramped to +1.00 V and back at a scan rate of 300 V/s, so that the entire scan takes a little under 10 ms. Scans are typically repeated every 100 ms, to give a dopamine sampling rate of 10 Hz.

When scanning the potential at such a high rate, there is a relatively large capacitance current produced due to charging of the electrode. In addition there is also faradaic current from redox processes of the surface groups of the carbon itself. These background currents can be more than 100 times greater than the faradaic current of the analyte (*see* **Fig. 1B,C**). Fortunately, carbon fiber microelectrodes are chemically and electrically stable and therefore this background remains very constant for each scan. This means background current prior to a stimulus or behavioral event can be subtracted from that during/afterwards to determine the chemical change in the solution around the electrode.

The background-subtracted current collected during the voltammetric scan (**Fig. 1C**) provides information on dopamine oxidation during the first phase (anodic scan) and on reduction of dopamine–*o*-quinone in the second phase (cathodic scan). For analyte identification, current during a voltammetric scan (~1000 data points at 100 kHz data acquisition) can be plotted against the applied input potential to yield a (background-subtracted) cyclic voltammogram (**Fig. 1D**). The cyclic voltammogram offers a wealth of chemical information: the positions of the peaks provide information on the redox potential of the analyte, the separation of the peaks provides information on the electron transfer kinetics of the redox reactions, and the relative amplitude of the peaks provides information on the diffusion or stability of the oxidation product (*see* **Note 1**). Because the combination of these chemical properties is fairly unique for each substance, this allows resolution of dopamine from other electroactive biological compounds (e.g., serotonin, 3,4-dihydroxyphenylacetic acid [DOPAC], ascorbic acid). One exception is norepinephrine, which has a cyclic voltammogram almost identical to dopamine. Interference from norepinephrine can be excluded in most cases because measurements are typically made in areas of the brain with high dopamine but very low norepinephrine tissue content and also the sensitivity to norepinephrine is only about half that to dopamine. There is some, albeit small, variation in the cyclic voltammogram for dopamine among individual electrodes. Therefore, to avoid erroneous identification, the template of dopamine provided by the in vitro calibration of each individual electrode is used for authentication of in vivo signals collected by that electrode.

Movement artifacts (due to bending of the wires in the tether or compromised electrical connections) cause DC shifts in the background current. These can

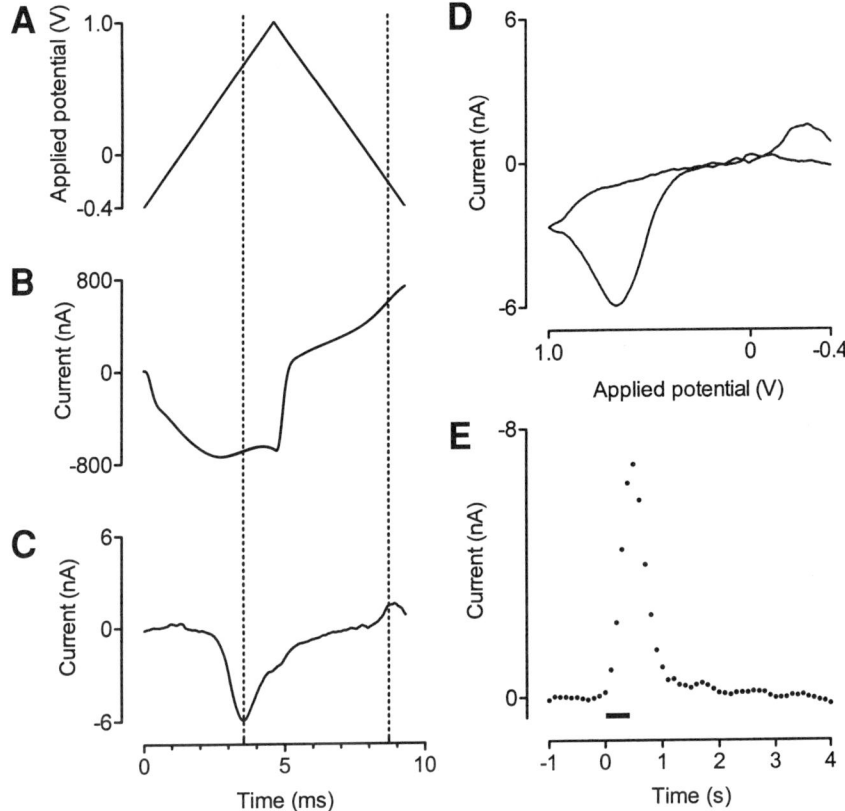

Fig. 1. Voltammetric signals. The input waveform (**A**) is applied to the carbon fiber microelectrode which produces a background output current (**B**). There are two background signals superimposed, one prior to and one after an electrical stimulation that evoked dopamine release. Because the background current is stable, the signals superimpose well. However, when there is oxidation of dopamine or reduction of dopamine–o-quinone (peak reaction times represented by the *dashed vertical lines*), there are slight deviations of the signals due to faradaic current. If the prestimulation background signal is subtracted from the post-stimulation signal, the faradaic current is revealed (**C**). Note from the scales of the y-axes that the faradaic current for dopamine oxidation is <1% of the background current despite this being a relatively large biologically signal (~1 μM). The background-subtracted signal can be plotted against the applied potential to give a cyclic voltammogram (**D**). This is the analytical tool for chemical resolution with fast-scan cyclic voltammetry. The current at the peak oxidation potential for dopamine from multiple scans can be plotted against time to reveal the temporal profile of dopamine (**E**). Each point represents current from one voltammetric scan, collected every 100 ms. Note that the current in this graph is plotted negative up so that an increase in dopamine is represented by an upward deflection. The signal shown is dopamine release in the caudate–putamen of a freely moving rat evoked by an electrical stimulation (24 pulses, 60 Hz, ±120 μA, 2 ms/phase) represented by the *bar*.

be distinguished from faradaic current by their cyclic voltammogram because the change in current for such artifacts is in the same direction for the anodic and cathodic scans (unlike for a redox couple). Ionic changes in the media can also produce background shifts. These give current in opposite directions for the anodic and cathodic scans (like genuine redox couples). Their cyclic voltammograms have broad features, particularly at inflection points of the background current, and so they can be distinguished from electroactive analytes which generate sharp peaks.

Once dopamine has been chemically verified, the current at its peak oxidation potential (approx +0.60 to 0.70 V vs Ag/AgCl) can be plotted against time to reveal the temporal profile of dopamine concentration changes (**Fig. 1E**). This current is directly proportional to the concentration of dopamine at the electrode and can be converted to concentration by a factor obtained by in vitro calibration of the electrode with a dopamine stock solution (*see* **Subheading 3.6.**).

Because of the requirement for background subtraction, fast-scan cyclic voltammetry is a differential technique and therefore best suited for monitoring chemical changes that take place over seconds rather than minutes or hours. For this reason, it is not good at monitoring slow, tonic changes in dopamine concentration (unlike microdialysis). However, owing to its exquisite temporal resolution, along with its chemical resolving power, it is the foremost technique for the measurement of fast, phasic changes in dopamine concentration.

Waveform generation and electrochemical data acquisition are carried out on a personal computer with a multifunction input/output card using software locally written in LabVIEW. The waveform is applied to the carbon fiber microelectrode via a potentiostat. Current generated at the carbon fiber microelectrode is converted to voltage (200 nA/V) by a head-mounted current-to-voltage converter (headstage; **Fig. 2**), sent to the potentiostat where it is amplified (2–10×) and low-pass filtered (2 kHz), and then acquired (100–333 kHz) on the personal computer. Connections between the potentiostat and the headstage are made through an electrical swivel.

3.2. Electrical Stimulation

By electrically stimulating dopaminergic neurons, a "snapshot" of the status of dopamine release and uptake can be taken. This strategy has been particularly useful for determining the neurochemical mechanisms involved in behavior following pharmacological manipulations (*4,5*). In experiments in which we wish to measure behaviorally evoked dopamine release, we employ electrical stimulations for optimization of the carbon fiber microelectrode placement into a dopamine-rich region. In addition, by electrically stimulating dopamine release before and after the experiment we not only get an in vivo template of dopamine at the carbon fiber microelectrode, but also verify that

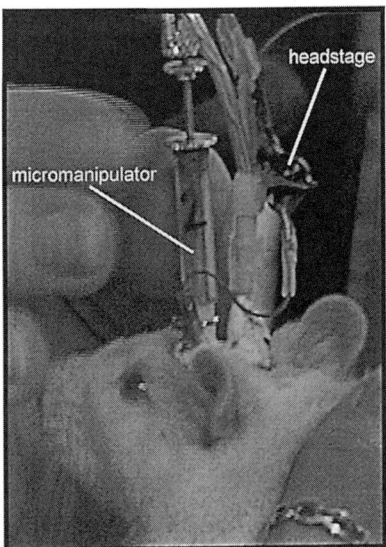 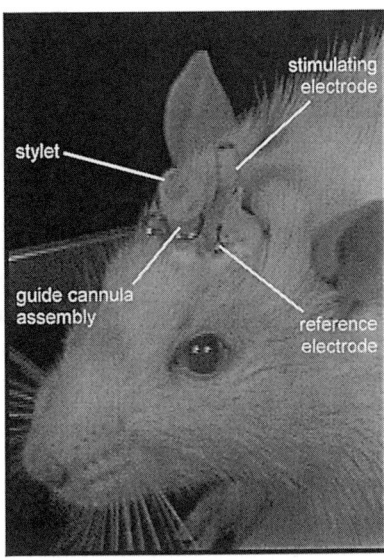

Fig. 2. Head-mounted apparatus for fast-scan cyclic voltammetry in freely moving rats. The **left panel** shows a rat that has been prepared for an experiment. A micromanipulator containing a carbon fiber microelectrode is coupled to the guide cannula, the stimulator cable is connected, and the carbon fiber microelectrode and reference electrodes are connected to the headstage. In the **right panel**, the experimental apparatus was removed from the rat and a stylet inserted into the guide cannula. This is the state in which the rat is housed after surgery.

the properties of the carbon fiber electrode and the recording site integrity have not changed during the course of the experiment. Electrical stimulation trains are generated on a personal computer with a digital-to-analog card using the data acquisition software. The stimulation is delivered to a stimulating electrode implanted in the rat brain (**Fig. 2**) using an optically isolated, constant-current stimulator (*see* **Note 3**). Connections between the stimulator and stimulating electrode are made through the swivel.

3.3. Electrode Fabrication

3.3.1. Carbon Fiber Microelectrodes

1. A single carbon fiber should be carefully inspected, and any noticeable debris removed (*see* **Notes 4** and **5**). The fiber is then aspirated into a borosilicate glass capillary, so that it extends from both ends.
2. The carbon fiber-filled capillary is pulled in an electrode puller. The magnet and heat settings should be optimized to ensure that the pulled glass forms a fluid-tight seal around the carbon fiber.

3. The exposed carbon fiber then needs to be cut with a sharp scalpel under a light microscope (~400× magnification) to a length of 75–150 μm. The electrode should also be inspected under the microscope for visible cracks or abnormalities in the fiber or glass seal, and discarded if any are present.
4. Trimmed carbon fiber microelectrodes are loaded into micromanipulators that can interface with the guide cannula implanted in the rat brain (**Fig. 2**). The electrode is inserted into the manipulator and a small diameter stainless steel wire, dipped in silver paint, is fed into the back of the capillary to make an electrical connection with the carbon fiber. The wire and capillary are then secured to the micromanipulator with heat shrink tubing.
5. The microelectrode is soaked in isopropanol for at least ten minutes, and then before use, the fidelity of the glass seal and the carbon fiber should be inspected once again under a light microscope.

3.3.2. Ag/AgCl Reference Electrodes

1. A piece of silver wire is cut to ~10 mm, inserted into the socket of a gold connector, and fixed with epoxy.
2. Once this has cured, the positive terminal of a 1.5-V battery is connected to the gold pin on the silver wire (anode) and the negative terminal to a stainless steel wire (cathode) immersed in a concentrated solution of hydrochloric acid (37%).
3. The silver wire is dipped into the hydrochloric acid for about 30 s until its surface becomes dark gray (bubbles should appear as gas is evolved during this process).

3.4. Surgery

3.4.1. Guide Cannula and Reference Electrode Implantations

1. The rat is anesthetized with intramuscular ketamine (80 mg/kg) and xylazine (12 mg/kg) and placed in a stereotaxic frame. The scalp is incised and retracted to reveal a 15–20 mm (longitudinal) and 10–15 mm (lateral) area of cranium.
2. Holes are drilled for the guide cannula, stimulating and reference electrodes and three anchor screws. The stimulating electrode can be positioned either in the dopaminergic cell bodies region (substantia nigra/ventral tegmental area; 1.0 mm lateral and 5.2 mm caudal from bregma, ~8.5 mm ventral from dura mater) or the ascending fibers in the medial forebrain bundle (1.4 mm lateral and 4.6 mm caudal from bregma, ~8.5 mm ventral from dura mater). The guide cannula and stylet are trimmed to 2.5 mm past the plastic hub and positioned above the target area (e.g., 1.3 mm lateral and 1.3 mm rostral from bregma would be above the caudate–putamen and the core of the nucleus accumbens). The reference electrode is placed in the contralateral hemisphere opposite the guide cannula.
3. Once the guide cannula, reference electrodes and anchor screws are in place, they are secured with cranioplastic cement, leaving the stimulating electrode hole exposed.

3.4.2. Stimulating Electrode Implantation

1. The stimulating electrode is prepared by separating the tips by 0.8–1.0 mm and cutting (or filing) them to a uniform length. Dura mater is thoroughly cleared from the stimulating electrode hole and the electrode is stereotaxically lowered into the tissue oriented so that the tips of the electrode are splayed in the coronal plane (*see* **Note 6**). The electrode is lowered to 1 mm above the target, taking care that it does not deflect against bone at the side of the hole. The stylet is now removed from the guide cannula and replaced with a micromanipulator containing a carbon fiber microelectrode. The carbon fiber microelectrode is then lowered to the target dopaminergic terminal region.
2. The stimulating electrode is connected to the stimulator and the carbon fiber and reference electrodes to the headstage. The voltammetric waveform is then switched on. After allowing the carbon fiber to stabilize for several minutes, an electrical stimulation (60 biphasic pulses, 60 Hz, ±120 µA, 2 ms/phase) is applied through the stimulating electrode while dopamine is monitored at the carbon fiber microelectrode. The stimulating electrode is then lowered in 0.2-mm increments until dopamine efflux is detected following a stimulation. It is then lowered further in 0.1-mm increments until dopamine release is maximal.
3. Finally, cranioplastic cement is applied to the part of the cranium that is still exposed, carefully covering the stimulating electrode and lower half of its plastic hub.

3.5. Experiment

3.5.1. Connecting the Tether

On the experiment day (*see* **Note 7**), all connectors are cleaned and the stylet is removed and replaced with a micromanipulator containing a carbon fiber microelectrode. The stimulator cable (on which the headstage is mounted) is connected and secured onto the stimulating electrode and the carbon fiber and reference electrodes are connected to the headstage (**Fig. 2**).

3.5.2. Positioning the Carbon Fiber Microelectrode

1. The carbon fiber microelectrode is slowly lowered into the brain and its output is observed on an oscilloscope. When the carbon fiber enters tissue the background current will be seen. The electrode is slowly lowered to just above the target tissue and allowed to stabilize for several minutes. If the electrode breaks, it will be apparent by a sudden change in the shape of the background current to a more resistive profile (approximating a triangular wave), and it should be removed and replaced with a fresh carbon fiber microelectrode.
2. An electrical stimulation (24 biphasic pulses, 60 Hz, ±120 µA, 2 ms/phase) is applied to the stimulating electrode while dopamine is monitored at the carbon fiber microelectrode. The behavioral reaction to the stimulation is usually a head turn to the ipsilateral side followed by sniffing. The carbon fiber microelectrode

is then lowered by small increments until electrically evoked dopamine release is optimized. The electrode can be secured in position by the locking device on the micromanipulator. At this time the rat is allowed to habituate to the environment (~30 min) before commencing the experiment (*see* **Note 8**).

3.5.3. Synchronization of Electrochemical Data with Behavior

The electrochemical and behavioral data must be synchronized for precise correlation of neurochemistry and behavior. A video character generator superimposes the file and scan number from the data acquisition computer onto the behavioral video recording in real time (*see* **Note 9**).

3.6. Electrode Calibration

Carbon fiber microelectrodes are calibrated in vitro with known concentrations of dopamine (*see* **Note 10**). Electrodes are lowered via the micromanipulator into the flowstream of a flow injection apparatus, perfused with Tris-HCl buffer. The flow injection apparatus is equipped with an injection loop, in which the analyte of interest (dopamine) is loaded. The carbon fiber microelectrode and a reference electrode (submerged in the buffer) are connected to the headstage and the voltammetric waveform is applied. The timing of the injection of dopamine into the flowstream is controlled by the data acquisition software. Because of the linearity of the response, one or two concentrations of dopamine (~1–2 μM) are sufficient to calibrate the electrode. Calibrations are done in triplicate and the average value for the current at the peak oxidation potential is used to normalize in vivo signals to dopamine concentration.

3.7. Data Analysis

Data analysis is carried out using software locally written in LabVIEW. The chemical integrity of each signal is checked from the cyclic voltammogram. Once a signal is confirmed to be dopamine, the concentration is determined from the current at the peak oxidation potential (*see* **Notes 11–13**).

3.8. Verification of Electrode Placement

To obtain unequivocal evidence of placement, histological methods must be used to verify the sampling site. Because the fiber is so small, a noticeable tract will not be seen using standard histological methods. However, a lesion can be made at the tip of the electrode that can be visualized using histological techniques (*see* **Note 14**).

1. The rat is anesthetized with 160 mg/kg of intramuscular ketamine and 24 mg/kg of xylazine, and an electrolytic lesion is made with a stimulator (100 µA, 5 s; anode connected to the carbon fiber microelectrode, cathode connected to a rectal probe).

2. The rat is immediately transcardially perfused with saline followed by 4% paraformaldehyde. The whole brain is then removed and stored in paraformaldehyde. Once the brain tissue is adequately fixed, it is rapidly frozen in an isopentane bath (~5 min).
3. The brain is then blocked, sliced on a cryostat (50-μm coronal sections, −20°C) and placed on slides. To aid visualization of anatomical structures, slices are stained with thionin.

4. Notes
4.1. Electrochemistry

1. The preeminent electrochemical techniques used to make real-time biological measurements *in situ* are constant-potential amperometry, high-speed chronoamperometry, and fast-scan cyclic voltammetry *(6)*. In constant-potential amperometry *(7)*, the electrode is continually held at an oxidizing potential (typically +0.30–0.65 V vs Ag/AgCl) so that any molecule at the electrode surface oxidized by this potential will produce faradaic current at the electrode. This technique provides measurements in absolute real time; the sampling rate is limited only by data acquisition. However, other than excluding molecules that are not electroactive, constant-potential amperometry has no chemical selectivity. It is also unable to resolve movement artifacts or ionic changes from faradaic current produced by analytes, and therefore is not ideally suited for making measurements in freely moving animals. High-speed chronoamperometry *(8)* is a pulsing technique, whereby the potential at the electrode is rapidly shifted to an oxidizing potential (typically +0.55 V vs Ag/AgCl), held for a period of time (typically 100 ms) and then returned to its resting potential (0.00 V). Faradaic current is provided by all molecules at the electrode surface that are oxidized by the change in potential in the first step, and current in the opposite direction is produced on the second step when the oxidized products are returned to their reduced state. Therefore, two data points are obtained for each pulse, one measurement of oxidative and one of reductive current. Together these data can provide a level of chemical selectivity. The ratio of reductive to oxidative current can provide information on the stability of the oxidized species or its diffusion from the electrode. Oxidized species generated at the electrode that rapidly diffuse away will not be available during the reductive phase and produce lower reductive-to-oxidative current ratios. Likewise, if the oxidized species is unstable (e.g., dehydroascorbic acid formed from oxidation of ascorbic acid), it will not be available during the reductive phase and will produce a reductive-to-oxidative current ratio of zero. High-speed chronoamperometry can exclude movement artifacts because they result in a change in current in the same direction for the oxidative and reductive phase. However, ionic changes cannot easily be distinguished from analytes with this technique. Fast-scan cyclic voltammetry *(9)* also utilizes reduction as well as oxidation reactions. However, rather than increasing the potential in a single step, the potential is more gradually ramped to an oxidizing potential and back which generates a whole array of data for each

voltammetric sweep (approx 1000 data points using 100-kHz data acquisition). This provides chemical information about the analyte, and allows selectivity of dopamine from other electroactive biological compounds. In addition, it allows analytes to be easily distinguished from movement artifacts or ionic changes (such as pH) that change the background signal to produce apparent faradaic current. The latter is most important because although many electroactive species can be excluded by other means (e.g., anatomical specificity, stimulation of specific pathway, use of ion-exchange polymers: *see* **Note 2**), the brain of the freely moving rodent is not a beaker and ionic changes occur regularly, particularly under conditions of changing metabolic demand. Thus, fast-scan cyclic voltammetry offers the most reliable real-time electrochemical detection of dopamine in freely moving animals.

2. Carbon fiber microelectrodes can be coated with the perfluorinated ion-exchange polymer Nafion *(10)* to improve their selectivity for dopamine. Because of its negative charge, Nafion allows dopamine and other positively charged molecules to diffuse to and preconcentrate at the electrode surface, while excluding negatively charged species (e.g., DOPAC, ascorbic acid). Electrochemical sensitivity to dopamine is enhanced by its adsorption to the carbon fiber microelectrode *(11)*. This occurs between voltammetric scans when the electrode is held at a negative potential (typically –0.40 V vs Ag/AgCl), probably by electrostatic attraction of dopamine to the carbon surface. The use of extended voltammetric waveforms (typically with a more positive anodic limit, up to +1.40 V vs Ag/AgCl) has been shown to improve the sensitivity to dopamine *(12)*. It is speculated that this augmentation is due to increased adsorption capacity of the electrode. However, this process may also slow the dynamic response to dopamine and lower the chemical selectivity.

4.2. Electrical Stimulation

3. To apply biphasic-current stimulations, we use two stimulators (each one provides current in one direction). Because they are not necessarily matched (i.e., the input–output characteristics may not be equal), we use a potentiometer to divide the input, so we can balance the outputs.

4.3. Electrode Fabrication

4. For easy handling of carbon fibers without damaging them, watchmakers' forceps with the tips encased in either heat-shrink or the insulator from an electrical wire can be used.
5. If carbon fiber microelectrodes are soaked in isopropanol for at least 10 min prior to use, their sensitivity can be improved by more than threefold. This is thought to clean the electrode and perhaps modify its surface functional groups. If the isopropanol is "cleaned" by loading it with activated carbon and then filtering before use, the sensitivity can be improved further by another twofold *(11)*.

4.4. Surgery

6. To provide adequate space between the stimulating electrode and the guide cannula, the stimulating electrode wires should be bent at a 90° angle from the plastic hub and then bent back down at another 90° angle, to give a horizontal distance of ~2 mm between the hub and the main axis of the wires. Following surgery, the animal should be monitored daily and gently handled to facilitate experimental procedures. While handling, the stylet should be removed from the guide cannula, cleaned with an alcohol wipe, and reinserted.

4.5. Experiment

7. Experiments should be conducted 2–10 d after surgery. The earliest post-surgical day that experiments can be carried out on is dependent on the rat's well being. Full recovery usually takes 2 d, but it can be as many as 5. The latest day that experiments can be carried out is dependent on a number of factors concerning the integrity of each of the electrodes. After a period of time in the brain, reference electrodes can drop their holding potential by approx 0.20 V. This is most likely due to dechlorination of the silver wire. After this occurs, all applied potentials will be 0.20 V lower than anticipated. This problem is easily identifiable to the trained eye by the shifted appearance of the background current on the oscilloscope. The occurrence of this problem potentially could be avoided with the use of removable reference electrodes. The stimulating electrode may also lose its functionality after several days implanted in the brain, resulting in both a lack of behavioral and neurochemical response to stimulation. Lastly, an accumulation of dried blood or perhaps gliosis at the tip of the guide cannula can cause carbon fiber microelectrodes to break as they are lowered into the brain. This problem can sometimes be resolved by inserting a stylet cut to extend past the tip of the guide cannula, but not as far as the target tissue to clear a path for the electrode. However, the problem becomes more prolific and the remedy less effective as post-surgical time increases.

8. During the initial period that voltammetric scans are applied to a fresh carbon fiber microelectrode, the background current tends to drift. This may be due to adsorption of brain material to the electrode surface (*see* notes on electrode calibration) and other surface changes. Therefore the electrode should be left to equilibrate (while applying voltammetric scans) for about 30 min before the start of the experiment. This time can also be used to habituate the rat to the environment.

9. Although fast-scan cyclic voltammetry has the best chemical selectivity of the real-time electrochemical techniques, it is not definitive. Therefore, in addition to confirming that cyclic voltammograms obtained in vivo correlate well with those from in vitro calibrations of dopamine, a set of identification criteria is used *(13)*. First, the anatomy should support the presence of the analyte, that is, the electrode must be in a dopamine-rich region. This is readily evaluated with common histological techniques. Next, the physiology should be consistent with

the release of the analyte, for example, stimulation of dopaminergic pathways should evoke release, and the known dopaminergic electrophysiology should be appropriate to sustain the release of the purported concentrations. In addition, a pharmacological approach should be used. This is generally in the form of a characterization rather than a routine experimental procedure. For instance, the electrochemical signal measured in the caudate–putamen following electrical stimulation of the medial forebrain bundle, was characterized many years ago *(14)*, and so we can be confident that under identical conditions (and where an appropriate cyclic voltammogram is obtained), a signal will always be dopamine. The most commonly used pharmacological agents for characterization are selective uptake blockers (e.g., GBR 12909), biosynthesis inhibitors (e.g., α-methyl-*p*-tyrosine), metabolism inhibitors (e.g., pargyline), and drugs that interfere with vesicular storage (e.g., Ro 4-1284). It is important to remember that the characterization is specific to the circumstances: for example, a pharmacological characterization for electrically evoked dopamine does not confirm the identity of an analyte evoked by a behavioral task. Systemic pharmacology may be a concern in freely moving animals, particularly when characterizing behaviorally evoked signals, because drugs that alter dopaminergic neurotransmission may modify the perception of a stimulus or the behavior itself. Therefore, the best way to carry out a pharmacological characterization in a behavioral context is by local application of drugs around the carbon fiber microelectrode so as to alter the immediate environment without grossly affecting the global behavior of the animal.

4.6. Electrode Calibration

10. In vitro calibration of the carbon fiber microelectrode may be performed before or after the experiment. The sensitivity of the electrode decreases when it is exposed to brain tissue *(15)*. We believe that this is due to adsorption of substances (including proteins) in the brain, which compete with dopamine for the surface of the electrode. Therefore we consider that post-experimental calibration gives a better estimate of the in vivo state. In addition, some pharmacological agents (dopamine uptake inhibitors) have also been shown to affect adversely the sensitivity of fresh carbon fiber microelectrodes *(16)*. In our hands, these did not alter the sensitivity if the electrode had previously been used in vivo. We assume that, like the endogenous material of the brain, these adsorb to the electrode to compete with dopamine. However, once the sensitivity has already been decreased by adsorption of contaminants in vivo, the pharmacological agents will have no further effect.

4.7. Data Analysis

11. To be confident that the signals purported to relate to specific behaviors are not simply ongoing changes in dopamine concentration, unrelated to that behavior, it is important to know the baseline level of occurrence of dopamine transients. As discussed previously, we cannot simply assume that all fluctuations in the current

at the peak oxidation potential for dopamine are due to changes in dopamine: pH and movement artifacts can also contribute to this signal. Therefore, the cyclic voltammogram of every voltammetric scan from the entire behavioral session (typically 72,000 scans in a 2-h experiment) must be checked for dopamine. To facilitate this process we developed a computerized scanning method with template matching. Systematically, each scan in the session has a background subtracted from it (typically 10 scans, 0.5–1.5 s prior to the foreground scan). This is then statistically compared to a template for dopamine. The template is a background-subtracted cyclic voltammogram from an in vitro calibration or electrically evoked dopamine release at the same electrode. The statistical comparison can be least squares analysis of normalized signals or linear regression. For either method, a confidence score of dopamine is obtained for every sample in the session. Signals exceeding a confidence threshold are then reexamined manually by an experienced investigator to verify their validity.

12. Plots of voltammetric current for a single applied potential (e.g., at the peak oxidation potential for dopamine) provide temporal information, but no chemical resolution (c.f., constant-potential amperometry). Examination of individual cyclic voltammograms provides chemical information for any point in time. However, it is much more desirable to be able to map temporal changes concurrently with chemical information. For this reason, it is useful to plot voltammetric data in three dimensions: time on the x-axis, applied potential on the y-axis, and current on the z-axis. This can be done either by plotting a "landscape" of consecutive cyclic voltammograms *(17)* or displaying the current in pseudo-color *(18)*. Color plots are generated in the data analysis software and can help an experimenter to identify quickly and reliably chemical changes of the species of interest. In addition they allow identification of any interference that may be altering the apparent temporal response (e.g., pH, *see* **Note 13**).

13. Local ionic changes, including pH, occur in the brain, particularly at times of changing metabolic demand (e.g., during neurotransmitter release or changes in postsynaptic neuronal firing rate). As discussed previously, these can alter the voltammetric signal by producing current over a large range of applied potentials. If changes in pH and dopamine occur at the same time, the pH can potentially mask the signal at the peak potential for dopamine. This scenario is readily identified from a color plot (*see* **Note 12**). To correct for this *(18)*, current from a potential that is affected by pH but not dopamine (e.g., 0.0–+0.1 V on the anodic scan; a measure of pH alone) is subtracted from current at the peak oxidation potential for dopamine (a measure of dopamine plus pH). This process can be automated in the data analysis software.

4.8. Verification of Electrode Placement

14. The passage of sufficient current across the carbon fiber microelectrode to generate a lesion will render it unsuitable for calibration. Because most experiments require post-experimental calibration, we have adopted an alternative approach. The carbon fiber microelectrode is removed at the end of the experiment and

replaced with a stainless steel electrode. This is lowered to the depth of the carbon fiber microelectrode during the experiment (using an identical micromanipulator), and the lesion is made by passing current (100 µA, 5 s) through it. It is possible to make several lesions in a single animal to demonstrate electrode placement from multiple experiments. During histological preparation, slices are stained with potassium ferricyanide (in addition to thionin) to aid visualization of the electrode-induced lesion.

5. Applications
5.1. Intracranial Self-Stimulation

Intracranial self-stimulation is a behavioral paradigm whereby subjects carry out an operant task to receive direct electrical stimulation of specific regions of their brains *(19)*. Stimulation of the regions including the mesencephalic dopaminergic cell bodies or of the lateral hypothalamus that includes the ascending dopaminergic pathways are both reinforcing in this paradigm. It is believed that dopamine is involved in the reinforcing properties of this behavior, as it can be abolished by selective lesion of dopaminergic neurons *(20)*. To investigate further the role of dopamine, fast-scan cyclic voltammetry was used to monitor dopamine in terminal regions during intracranial self-stimulation of the ventral tegmental area/substantia nigra.

Rats were allowed free access to a lever, depression of which caused an electrical stimulation (24 pulses, 60 Hz, ~120 µA) to be applied across the stimulating electrode on a fixed-ratio 1 schedule. Within 1 d, rats usually learned to freely lever-press for intracranial self-stimulation. Dopamine was monitored using fast-scan cyclic voltammetry throughout the training and test sessions.

In animals in which the stimulating electrode was positioned such that stimulation produced detectable dopamine release, intracranial self-stimulation behavior was acquired. However, in those in which the stimulating electrode did not evoke dopamine release, the animals did not acquire the behavior, suggesting that activation of dopaminergic neurons is required for intracranial self-stimulation to be reinforcing.

In animals that did acquire intracranial self-stimulation behavior, the normally robust evoked dopamine signals disappeared as the animal acquired the lever-pressing for intracranial self-stimulation. This attenuation occurred despite the fact that experimenter-delivered electrical stimulation evoked dopamine release prior to the intracranial self-stimulation session. Furthermore, detectable dopamine release could be evoked, once again, by experimenter-delivered stimulations at the end of the session. Finally, if a lever-press record during an intracranial self-stimulation session from one animal was "played back" to another animal, as a yoked control, dopamine release could be detected for a significant part of the session. These data are highly indicative that there is higher, executive attenuation of dopamine release during contingent responses. This phenomenon has been observed in both the core and shell of the nucleus accumbens *(21)* and in the caudate–putamen *(22)*.

These results seem to be consistent with the proposed role of dopamine in the prediction error theory for reward *(23)* in which only unanticipated responses produce dopaminergic responses. Evidence for the source of the attenuation may be provided by the work of Henriksen and co-workers, who have recently shown that during intracranial self-stimulation session in trained rats, a group of γ-aminobutyric acid-ergic (GABAergic) neurons in the ventral tegmental area increase their firing rates just prior to the lever-press *(24)*. These neurons may interact with dopaminergic neurons to inhibit propagation of the stimulus and/or dopamine release.

5.2. Natural Stimuli

As well as being suitable for the measurement of transient dopamine concentrations produced by electrical stimulation of dopaminergic neurons, fast-scan cyclic voltammetry can also be used to measure changes in dopamine levels resulting from endogenous neuronal activity. Specifically, the burst firing of dopaminergic neurons, synchronized by gap junctions *(25)*, is likely to produce extracellular dopamine transients that achieve high concentrations (>100 nM). Electrophysiological studies demonstrate that such burst firing occurs during the presentation of rewards and associated stimuli *(3)* and various alerting stimuli *(26)*.

The first chemical measurement of a rapid dopaminergic response to a natural stimulus was by Rebec et al. *(1)* using fast-scan cyclic voltammetry. An increase in extracellular dopamine concentration (indicated by the cyclic voltammogram), lasting <10 s, was measured in the shell of the nucleus accumbens of rats at free-choice entry into a novel environment. Neither subsequent exploratory behavior nor reentry to the environment was associated with further increases in dopamine concentrations, emphasizing the role of novelty as opposed to locomotion in eliciting the neurochemical response. The response was also regionally specific, as it was not observed in the core of the nucleus accumbens or the caudate–putamen.

We recently measured faster dopamine transients in the core of the nucleus accumbens of male rats during sexual interaction *(2)*. In the first experiment, we monitored dopamine during introduction of a receptive female into the test cage. Dopamine transients occurred during the initial interaction with the female, that is, approach and sniffing behaviors, but not during copulatory behaviors such as mounting, intromission, or ejaculation. Additional experiments addressed whether this phasic dopamine response is specific for recogni-

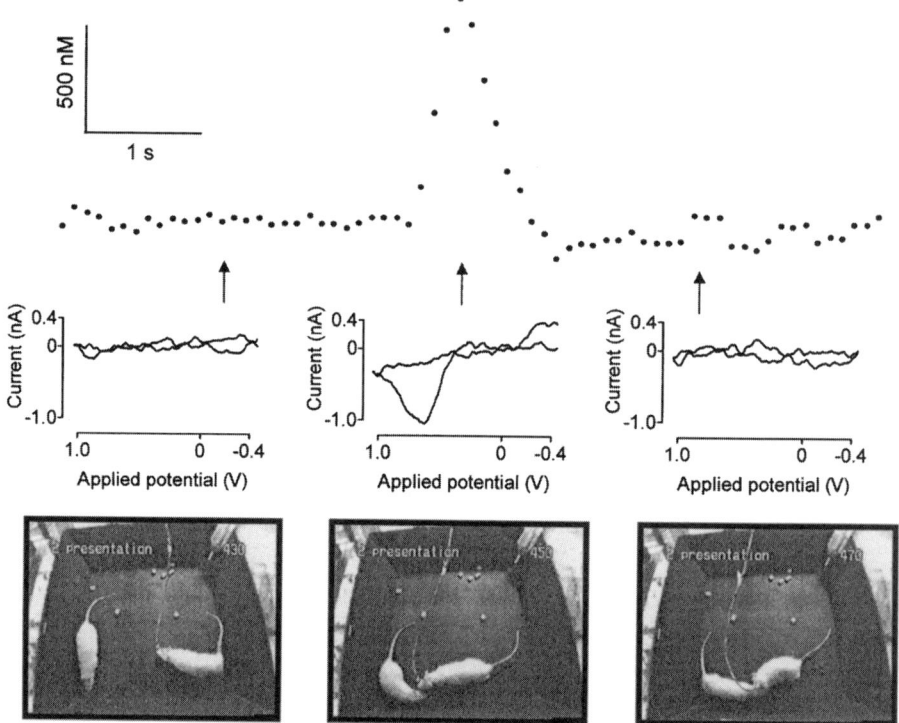

Fig. 3. Extracellular dopamine transient in the nucleus accumbens core of a male rat at investigation of another male. The voltammetric trace at the peak oxidation potential for dopamine is shown at the **top**, and the *arrows* indicate the time points of the cyclic voltammograms and video stills. Each point represents current from one voltammetric scan, collected every 100 ms. The test rat (on the *right* in each video frame with a tether) approaches the target rat (**left panel**) and at the point he makes contact (**middle panel**), there is a rapid but transient rise in the voltammetric signal. On further interaction (**right panel**), there is no further change in the voltammetric signal. The background-subtracted voltammograms indicate that the rise in the voltammetric signal was due to an increase in dopamine concentration (**middle**), and that there was no change in dopamine concentration either 2 s before (**left**; before initial contact) or 2 s after (**right**; during further interaction) the transient. The concentration of dopamine was estimated from in vitro post-experimental calibration of the carbon fiber microelectrode, and is shown in the scale bar.

tion of receptive females. Preliminary data suggest that similar responses occur with presentation of males (**Fig. 3**), suggesting a nonsexual role for this signal. The transients tended to occur at the initial interaction with the target rat and are followed by intense sniffing and often locomotion. These results and those of Rebec et al. *(1)* are consistent with a role for phasic dopamine in alerting and attention processes.

5.3. Cocaine Self-Administration

Extracellular dopamine has previously been measured during cocaine self-administration with both electrochemical *(27)* and microdialysis techniques *(28)*. The preliminary data presented here were obtained by monitoring extracellular dopamine using fast-scan cyclic voltammetry with much better temporal resolution (25–12,000+ times greater) than previous studies.

Rats ($n = 3$) were surgically implanted with indwelling jugular catheters. Following 1 wk of recovery, they were trained to self-administer cocaine over a 2-h behavioral session. Each active lever-press response (fixed-ratio 1 schedule with 20 s time out) resulted in a 0.33 mg of intravenous infusion of cocaine (6 s) paired with a tone/houselight stimulus (20 s). Once stable self-administration behavior was seen for at least 3 consecutive days, voltammetry surgery was performed. Following recovery, rats were allowed to self-administer cocaine until stable behavior was reestablished (1–2 d). On the experimental day, a carbon fiber microelectrode was lowered into the core of the nucleus accumbens and dopamine was monitored throughout the session. Electrically stimulated (24 pulses, 60 Hz) release of dopamine was monitored before and after rats were allowed to self-administer cocaine to verify that the working electrode was in a dopamine-rich area. At the beginning of the cocaine self-administration session rats typically lever-pressed two to four times in quick succession (termed load-up behavior) and then approx every 5 min for the remainder of the session.

Transient (~5 s) increases in dopamine concentration (75–150 n*M*) were seen immediately following lever-presses for cocaine (**Fig. 4**). These were concurrent with the onset of the tone and houselight (paired with cocaine delivery). Because the observed dopamine transients ended within 1–5 s of the onset of cocaine delivery, they are probably not related solely to a direct pharmacological effect of cocaine. Indeed, preliminary findings indicate that they are still present for a lever-press when the cocaine pump is switched off. Instead, these transient changes in dopamine are likely related to the goal-directed behaviors for intravenous cocaine, as well as associative factors operating within the self-administration context.

5.4. Conclusions

These data establish that fast-scan cyclic voltammetry can measure rapid, transient dopaminergic events in behaving rats on a physiological time scale, and invite research on phasic dopaminergic transmission during a variety of experimental paradigms, including motivation and learning. The time resolution

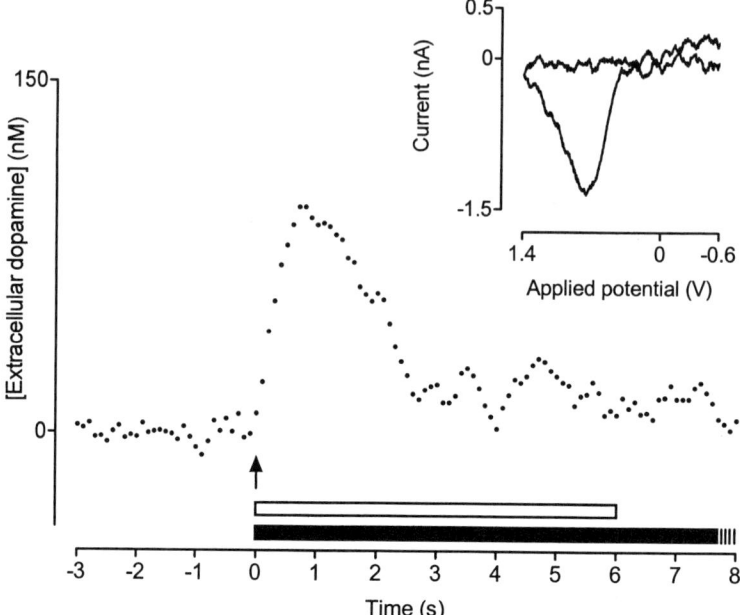

Fig. 4. Transient increase in extracellular dopamine concentration in the core of the nucleus accumbens following a lever-press for cocaine. The voltammetric trace at the peak oxidation potential for dopamine is shown. Each point represents current from one voltammetric scan, collected every 100 ms. Immediately following the lever-press (*vertical arrow*) there is a rapid rise in the voltammetric signal. The cyclic voltammogram (*inset*) is indicative that this is an increase in dopamine concentration. The concentration of dopamine was estimated from in vitro post-experimental calibration of the carbon fiber microelectrode (peak response is ~100 nM). The *open bar* represents the time of the cocaine infusion and the *filled bar* shows the period of the tone/houselight stimulus (note that this lasts for 20 s and so continues beyond the time window shown).

of the technique, along with the ability to identify signals chemically, provide an unprecedented ability to monitor dopamine in real time within a behavioral context. In particular, investigation of phasic dopamine release with fast-scan cyclic voltammetry in freely moving rats may prove important in aiding the understanding of the reinforcing properties of drugs of abuse.

References

1. Rebec, G. V., Christensen, J. R., Guerra, C., and Bardo, M. T. (1997) Regional and temporal differences in real-time dopamine efflux in the nucleus accumbens during free-choice novelty. *Brain Res.* **776,** 61–67.

2. Robinson, D. L., Phillips, P. E. M., Budygin, E. A., Trafton, B. J., Garris, P. A., and Wightman, R. M. (2001) Sub-second changes in accumbal dopamine during sexual behavior in male rats. *NeuroReport* **12,** 2549–2552.
3. Schultz, W. (1992) Activity of dopamine neurons in the behaving primate. *Semin. Neurosci.* **4,** 129–138.
4. Budygin, E. A., Kilpatrick, M. R., Gainetdinov, R. R., and Wightman, R. M. (2000) Correlation between behavior and extracellular dopamine levels in rat striatum: comparison of microdialysis and fast-scan cyclic voltammetry. *Neurosci. Lett.* **281,** 9–12.
5. Budygin, E. A., Phillips, P. E. M., Robinson, D. L., Kennedy, A. P., Gainetdinov, R. R., and Wightman, R. M. (2001) Effect of acute ethanol on striatal dopamine neurotransmission in ambulatory rats. *J. Pharmacol. Exp. Ther.* **297,** 27–34.
6. Michael, D. J. and Wightman, R. M. (1999) Electrochemical monitoring of biogenic amine neurotransmission in real time. *J. Pharm. Biomed. Anal.* **19,** 33–46.
7. Kawagoe, K. T. and Wightman, R. M. (1994) Characterization of amperometry for in vivo measurement of dopamine dynamics in the rat brain. *Talanta* **41,** 865–874.
8. Gerhardt, G. A. and Hoffman, A. F. (2001) Effects of recording media composition on the responses of Nafion-coated carbon fiber microelectrodes measured using high-speed chronoamperometry. *J. Neurosci. Methods* **109,** 13–21.
9. Kawagoe, K. T., Zimmerman, J. B., and Wightman, R. M. (1993) Principles of voltammetry and microelectrode surface states. *J. Neurosci. Methods* **48,** 225–240.
10. Gerhardt, G. A., Oke, A. F., Nagy, G., Moghaddam, B., and Adams, R. N. (1984) Nafion-coated electrodes with high selectivity for CNS electrochemistry. *Brain Res.* **290,** 390–395.
11. Bath, B. D., Michael, D. J., Trafton, B. J., Joseph, J. D., Runnels, P. L., and Wightman, R. M. (2000) Subsecond adsorption and desorption of dopamine at carbon-fiber microelectrodes. *Analyt. Chem.* **72,** 5994–6002.
12. Hafizi, S., Kruk, Z. L., and Stamford, J. A. (1990) Fast cyclic voltammetry: improved sensitivity to dopamine with extended oxidation scan limits. *J. Neurosci. Methods* **33,** 41–49.
13. Marsden, C. A., Joseph, M. H., Kruk, Z. L., et al. (1988) In vivo voltammetry—present electrodes and methods. *Neuroscience* **25,** 389–400.
14. Millar, J., Stamford, J. A., Kruk, Z. L., and Wightman, R. M. (1985) Electrochemical, pharmacological and electrophysiological evidence of rapid dopamine release and removal in the rat caudate nucleus following electrical stimulation of the median forebrain bundle. *Eur. J. Pharmacol.* **109,** 341–348.
15. Logman, M. J., Budygin, E. A., Gainetdinov, R. R., and Wightman, R. M. (2000) Quantitation of in vivo measurements with carbon fiber microelectrodes. *J. Neurosci. Methods* **95,** 95–102.
16. Davidson, C., Ellinwood, E. H., Douglas, S. B., and Lee, T. H. (2000) Effect of cocaine, nomifensine, GBR 12909 and WIN 35428 on carbon fiber microelectrode

sensitivity for voltammetric recording of dopamine. *J. Neurosci. Methods* **101**, 75–83.
17. Phillips, P. E. M. and Stamford, J. A. (1999) Voltammogram "landscapes" aid detection and identification of in vivo electrochemical signals. *Electroanalysis* **11**, 301–307.
18. Michael, D., Travis, E. R., and Wightman, R. M. (1998) Color images for fast-scan CV. *Analyt. Chem.* **70**, 586A–592A.
19. Olds, J. and Milner, P. M. (1954) Positive reinforcement produced by electrical stimulation of septal area and other regions of the rat brain. *J. Comp. Physiol. Psychol.* **47**, 419–427.
20. Fibiger, H. C., LePiane, F. G., Jakubovic, A., and Phillips, A. G. (1987) The role of dopamine in intracranial self-stimulation of the ventral tegmental area. *J. Neurosci.* **7**, 3888–3896.
21. Garris, P. A., Kilpatrick, M., Bunin, M. A., Michael, D., Walker, Q. D., and Wightman, R. M. (1999) Dissociation of dopamine release in the nucleus accumbens from intracranial self-stimulation. *Nature* **398**, 67–69.
22. Kilpatrick, M. R., Rooney, M. B., Michael, D. J., and Wightman, R. M. (2000) Extracellular dopamine dynamics in rat caudate–putamen during experimenter-delivered and intracranial self-stimulation. *Neuroscience* **96**, 697–706.
23. Schultz, W., Dayan, P., and Montague, P. R. (1997) A neural substrate of prediction and reward. *Science* **275**, 1593–1599.
24. Steffensen, S. C., Lee, R. S., Stobbs, S. H., and Henriksen, S. J. (2001) Responses of ventral tegmental area GABA neurons to brain stimulation reward. *Brain Res.* **906**, 190–197.
25. Grace, A. A. (1991) Phasic versus tonic dopamine release and the modulation of dopamine system responsivity: a hypothesis for the etiology of schizophrenia. *Neuroscience* **41**, 1–24.
26. Overton, P. G. and Clark, D. (1997) Burst firing in midbrain dopaminergic neurons. *Brain Res. Rev.* **25**, 312–334.
27. Gratton, A. and Wise, R. A. (1994) Drug- and behavior-associated changes in dopamine-related electrochemical signals during intravenous cocaine self-administration in rats. *J. Neurosci.* **14**, 4130–4146.
28. Wise, R. A., Newton, P., Leeb, K., Burnette, B., Pocock, D., and Justice, J. B., Jr. (1995) Fluctuations in nucleus accumbens dopamine concentration during intravenous cocaine self-administration in rats. *Psychopharmacology* **120**, 10–20.

32

Measurement of Dopamine Uptake in Neuronal Cells and Tissues

Yuen-Sum Lau

1. Introduction

Biogenic amines such as dopamine (DA), norepinephrine, and serotonin (5-hydroxytryptamine) are key neurotransmitters found in the central nervous system (CNS). Major DA containing cells, nuclei, and neurons are located in retina, olfactory bulb, hypothalamic–pituitary, nigrostriatal, mesolimbic, and mesocortical areas. Stimulation of DA containing neurons releases DA from storage vesicles at the nerve terminals into the synaptic region. Binding and activation of the high-affinity postsynaptic receptors by DA or DA agonists within the tissue will manifest changes in physiological and behavioral responses. Central DA receptors are known target sites for typical antipsychotic drugs as well as dopaminomimetic anti-Parkinson's drugs. Therefore, the amount and duration that DA accumulates at the synaptic region will determine the intensity of DA receptor responses.

Following transmitter release, central DA action is terminated mainly by an active reuptake process into the presynaptic nerve terminals, in addition to some amount being diffused away or subject to metabolic degradation. The uptake of DA at nerve terminals is a saturable, temperature-sensitive, and energy-dependent process occurring against a concentration gradient. The uptake system involves a high-affinity (with low K_m between 0.1 and 1 μM), Na^+-dependent, and carrier-mediated transporter *(1,2)*. The DA uptake transporter has been molecularly cloned and its functional expression characterized *(3–7)*. The structure and function of the DA transporter have been reviewed further more recently *(8)*.

Many pharmacological agents exert profound effects on the neuronal uptake of DA. Certain antidepressant (e.g., clomipramine) and anticholinergic drugs (e.g., benztropine) are shown to inhibit DA uptake *(9–11)*. Drugs of abuse (e.g., cocaine and amphetamine) are also potent neuronal DA reuptake inhibitors *(12,13)*. Because the inhibition of neuronal DA uptake by cocaine will enhance synaptic DA activity, manifest acute behavioral and reinforcing effects, and significantly contribute to abuse liability, preclinical studies indicate that DA transporter inhibitors (e.g., GBR 12909, WIN 35065-2) may be potentially effective in reducing cocaine use *(14)*. Several DA transporter inhibitors have been employed as ligands for labeling and quantitating the transporter within innervated neurons *(15)*. Nevertheless, the neuronal DA uptake assay is still considered a very useful experimental approach for determining the action of neuroactive drugs on neurotransmission. In addition, detecting the active DA uptake can also serve as a functional indication for neuronal integrity, as degenerated neurons will lose their ability to take up the neurotransmitter.

A wide variety of tissue preparations have been used in vitro to study the neuronal uptake transport of DA including cultured cells, tissue slices, homogenates, crude P_2 synaptosomal fraction, purified synaptosomes, or synaptic vesicles. The selection of an appropriate preparation for study depends on the tissue source and availability and time and equipment limitations. To investigate the effect of a potential new drug on DA uptake, a purified bovine synaptosomal preparation may be used, as it provides a sufficient source of tissue such that a broad range of [^{3}H]DA concentrations can be applied for the determination of uptake kinetics. On the other hand, to study in vivo effects of drug treatment on DA uptake in small rodents (such as in the mouse), where tissue is very limited and performing further purification is not feasible, a crude homogenate would be the best choice. In this chapter, DA uptake assays in cultured primary mesencephalic cells and in rat and mouse striatal tissue preparations are described.

2. Materials

2.1. Solutions for Cellular [^{3}H]DA Uptake

1. Phosphate-buffered saline (PBS): 140 mM NaCl, 2.6 mM KCl, 1 mM KH$_2$PO$_4$, 1 mM Na$_2$HPO$_4$, pH 7.4. Dissolve 8.18 g of NaCl (mol wt 58.44), 0.19 g of KCl (mol wt 74.56), 0.14 g of Na$_2$HPO$_4$ (mol wt 141.96), and 0.14 g of KH$_2$PO$_4$ (mol wt 136.09) in 900 mL of H$_2$O (*see* **Note 1**). After adjusting the pH to 7.4 with HCl, the volume of the solution is adjusted to 1000 mL with H$_2$O. Store the solution in a closely capped container in refrigerator.
2. Cell preincubation buffer: PBS supplemented with 0.75 mM MgCl$_2$, 0.75 mM CaCl$_2$, and 33 mM D-glucose. To prepare 100 mL of this buffer, dissolve

DA Uptake Assay

7.14 mg of $MgCl_2$ (mol wt 95.21), 8.32 mg of $CaCl_2$ (mol wt 110.99), and 0.6 g of D-glucose (mol wt 180.2) in PBS to a final volume of 100 mL.
3. Acidified ethanol: 25% 1 M HCl and 25% ethanol in 50% H_2O.

2.2. Solutions for Tissue [³H]DA Uptake

1. Tris-HCl stock solution: 1 M Tris-HCl, pH 7.4. Dissolve 121.14 g of Tris base (mol wt 121.14) in 850 mL of H_2O. The chemical is not readily soluble in H_2O, but the solubility speeds up when the solution is acidified by HCl. Adjust the pH to 7.4 with concentrated HCl (it takes approx 70 mL). Allow the solution to cool to room temperature. Make a final adjustment of the pH, if necessary, and then the volume of the solution to 1000 mL with H_2O. The Tris-HCl stock can be stored in a closely capped container in refrigerator.
2. Tissue homogenization buffer: 50 mM Tris-HCl, pH 7.4, and 0.32 M sucrose. Add 5 mL of 1 M Tris-HCl stock to 75 mL of H_2O. Mix and recheck for pH; make sure it is still at 7.4 or make necessary adjustment. Add and dissolve 10.95 g of sucrose (mol wt 342.3) to the mixture by constant stirring. Make a final adjustment of volume to 100 mL with H_2O. This solution should be made fresh each time and used on the day of experiment. The leftover solution should be discarded.
3. 10X DA uptake assay stock buffer: 500 mM Tris-HCl, pH 7.4; 1.2 M NaCl; and 50 mM KCl. Dissolve 30.3 g of Tris base (or use 250 mL of 1 M Tris-HCl stock), 35.1 g of NaCl, 1.86 g of KCl in 400 mL (150 mL if 1 M Tris-HCl stock is used) of H_2O. Adjust the pH to 7.4 with concentrated HCl, and then adjust the volume of the solution to 500 mL with H_2O. The 10X stock buffer can be stored in a closely capped container in refrigerator.
4. Tissue DA uptake assay buffer: 50 mM Tris-HCl, pH 7.4; 120 mM NaCl; 5 mM KCl; and 0.32 M sucrose. Add 50 mL of 10X DA uptake assay stock buffer to 350 mL of H_2O. Mix and recheck for pH; make sure it is still at 7.4 or otherwise make necessary adjustment. Add and dissolve 54.8 g of sucrose to the mixture by constant stirring. Make a final adjustment of volume to 500 mL with H_2O. This solution should be made fresh each time and used on the day of experiment. The leftover solution should be discarded.

2.3. Ligand and Drugs

1. [³H]DA: Ligand concentration before assay: 0.5 µM. After the ligand is received from a commercial vendor, it is diluted to 20 µM with H_2O. The ligand is stored in 50-µL aliquots at –20°C. Before each assay, the frozen 20 µM [³H]DA aliquot is thawed and diluted 40-fold with either cell preincubation buffer or tissue DA uptake assay buffer to yield a concentration of 0.5 µM. This ligand solution in the amount of one tenth (50 µL) of the final assay volume (500 µL) will be injected; therefore, the final assay ligand concentration is 0.05 µM. The remaining diluted ligand solution should be discarded after each use.
2. Mazindol: Mazindol concentration before assay: 100 µM. To prepare 1 mM mazindol solution, dissolve 2.85 mg of mazindol (mol wt 284.75) first in 50 µL

of 1 N HCl, then adjust the final volume to 10 mL with H_2O. Store the 1 mM mazindol solution in 100-μL aliquots at –20°C. Before each assay the frozen 1 mM mazindol aliquot is thawed and diluted 10-fold with either cell preincubation buffer or tissue DA uptake assay buffer to yield a concentration of 100 μM. This drug solution in the amount of one tenth (50 μL) of the final assay volume (500 μL) will be injected; therefore, the final assay drug concentration is 10 μM. The remaining diluted drug solution should be discarded after each use.

3. Pargyline: Pargyline concentration before assay: 100 μM. To prepare 1 mM pargyline solution, dissolve 1.96 mg of pargyline hydrochloride (mol wt 195.7) in 10 mL of H_2O. Store the 1 mM pargyline solution in 100-μL aliquots at –20°C. Before each assay the frozen 1 mM pargyline aliquot is thawed and diluted 10-fold with either cell preincubation buffer or tissue DA uptake assay buffer to yield a concentration of 100 μM. This drug solution in the amount of one tenth (50 μL) of the final assay volume (500 μL) will be injected; therefore the final assay drug concentration is 10 μM. The remaining diluted drug solution should be discarded after each use.

3. Methods
3.1. [³H]DA Uptake in Cultured Neuronal Cells

[³H]DA uptake in primary cultures of rat mesencephalic cells is measured according to the method previously described by Prochiantz et al. *(16)* with modifications *(17)*.

1. Wash cultured cells (1×10^{-5} cells per 16-mm well) twice with cell preincubation buffer.
2. Incubate the cells in 500 μL of cell preincubation buffer, containing 10 μM pargyline, 0.05 μM [³H]DA with or without 10 μM mazindol at 37°C for 15 min (*see* **Note 2**). A typical DA uptake assay can be set up as follows (volumes in μL):

Cell culture well	Cell preincubation buffer	Pargyline (100 μM)	Mazindol (100 μM)	[³H]DA (0.5 μM)
Total uptake	400	50	0	50
Nonspecific uptake	350	50	50	50

3. Terminate the uptake assay by placing the plates on ice and carefully aspirate the incubation medium into a radioactive waste container.
4. Rapidly wash the cells three times with 1 mL of ice-cold PBS. Aspirate the washes into a radioactive waste container.
5. Lyse the cells with 300 μL of 0.5 N NaOH for 1 h at room temperature. Collect the lysate into a scintillation vial.
6. Wash the well with 500 μL of distilled water, followed by 500 μL of acidified ethanol. Collect these washes and combine them with the lysate in the scintillation vial.

3.2. [³H]DA Uptake in Striatal Tissues

3.2.1. Preparation of Crude P_2 Synaptosomal Fraction

Measuring [³H]DA uptake in striatal tissues depends on the amount of tissue collected (*see* **Note 3**). To prepare a crude P_2 synaptosomal fraction, striata obtained from one rat (or striata from five mice) would be required *(18)*.

1. Homogenize the tissue in 1.0 mL of tissue homogenization buffer with an electronic tissue homogenizer (Tekmar Tissumizer or Brinkmann Polytron), five times for 5-s bursts with ice cooling in between each burst.
2. Centrifuge at 1000*g* for 20 min at 4°C.
3. Discard pellet and centrifuge the supernatant at 27,000*g* for 20 min at 4°C.
4. Resuspend the P_2 pellet in 1.0 mL of tissue homogenization buffer for assay use.

3.2.2. Preparation of Striatal Homogenate

If tissue is limited, for example, to measure [³H]DA uptake in a treated mouse where tissue cannot be pooled, striatal homogenate is therefore used. In this situation, the mouse striatum is homogenized in 500 µL of tissue homogenization buffer and directly used for the assay.

3.2.3. [³H]DA Uptake Assay Procedure

1. Set up the assay tubes (16 × 100 mm) on ice as suggested below. Pipet all the necessary ingredients, except the P_2 synaptosomes or tissue homogenate. All ingredients should be made in the assay buffer and kept on ice. A typical DA uptake assay can be set up as follows (volumes in µL):

Tube	Uptake assay buffer	Pargyline (100 µM)	Mazindol (100 µM)	[³H]DA (0.5 µM)	Membrane protein
Total uptake	300	50	0	50	100
Nonspecific uptake	250	50	50	50	100

2. Initiate the assay by adding 100 µL of P_2 membrane protein (or homogenate) to each tube, vortex-mix, and incubate at 37°C for 6 min.
3. Terminate the reaction by adding 5 mL of ice-cold uptake assay buffer to the assay tube and immediately pour the sample over a buffer-saturated Whatman GF/B (2.4 cm) filter on a filter manifold apparatus under vacuum.
4. Rinse the filter three times with 5 mL of the assay buffer. Dry filter in oven (50°C) overnight before counting for activity with 10 mL of scintillation fluid.

3.3. Data Treatment

1. Subtract the nonspecific uptake count (with mazindol) from the total uptake count (without mazindol) to obtain the level of specific DA uptake in cell or striatal tissue.
2. Use the information of ligand specific radioactivity, calculate the amount of DA uptake in pmol. This value can be expressed either in terms of number of cells or protein content contained in each assay.

4. Notes

1. In our laboratory, we routinely use good quality, distilled-deionized water in the preparation of all solutions and reagents. If that is not available in a research facility, high-performance liquid chromatography (HPLC) grade H_2O may be purchased commercially.
2. The assay incubation time should be kept as short as possible. In some laboratories, ascorbic acid is added to prevent [^{3}H]DA oxidation. If the assay time is kept short, we found there is no significant difference with or without ascorbic acid. It is critical, however, to select a concentration of [^{3}H]DA very close to the K_m for the uptake assay.
3. It is important to use freshly obtained and prepared tissues for the uptake assay. The uptake in preparations made from frozen tissues is found to be very low and unsatisfactory.

References

1. Horn, A. S. (1978) Characteristics of neuronal dopamine uptake. *Adv. Biochem. Psychopharmacol.* **19**, 25–34.
2. Paton, D. M. (1980) Neuronal transport of noradrenaline and dopamine. *Pharmacology* **21**, 85–92.
3. Giros, B., El Mestikawy, S., Bertrand, L., and Caron, M. G. (1991) Cloning and functional characterization of a cocaine-sensitive dopamine transporter. *FEBS Lett.* **295**, 149–154.
4. Kilty, J. E., Lorang, D., and Amara, S. G. (1991) Cloning and expression of a cocaine-sensitive rat dopamine transporter. *Science* **254**, 578–579.
5. Shimada, S., Kitayama, S., Lin, C. L., et al. (1991) Cloning and expression of a cocaine-sensitive dopamine transporter complementary DNA. *Science* **254**, 576–578.
6. Usdin, T. B., Mezey, E., Chen, C., Brownstein, M. J., and Hoffman, B. J. (1991) Cloning of the cocaine-sensitive bovine dopamine transporter. *Proc. Natl. Acad. Sci. USA* **88**, 11,168–11,171.
7. Wu, X. and Gu, H. H. (1999) Molecular cloning of the mouse dopamine transporter and pharmacological comparison with the human homologue. *Gene* **233**, 163–170.
8. Chen, N. and Reith, M. E. (2000) Structure and function of the dopamine transporter. *Eur. J. Pharmacol.* **405**, 329–339.

9. Friedman, E., Fung, F., and Gershon, S. (1977) Antidepressant drugs and dopamine uptake in different brain regions. *Eur. J. Pharmacol.* **42,** 47–51.
10. Randrup, A. and Braestrup, C. (1977) Uptake inhibition of biogenic amines by newer antidepressant drugs: relevance to the dopamine hypothesis of depression. *Psychopharmacology (Berl.)* **53,** 309–314.
11. Petrali, E. H. (1980) Differential effects of benztropine and desipramine on the high affinity uptake of paratyramine in slices of the caudate nucleus and hypothalamus. *Neurochem. Res.* **5,** 297–300.
12. Holmes, J. C. and Rutledge, C. O. (1976) Effects of the *d*- and *l*-isomers of amphetamine on uptake, release and catabolism of norepinephrine, dopamine and 5-hydroxytryptamine in several regions of rat brain. *Biochem. Pharmacol.* **25,** 447–451.
13. Hadfield, M. G. and Nugent, E. A. (1983) Cocaine: comparative effect on dopamine uptake in extrapyramidal and limbic systems. *Biochem. Pharmacol.* **32,** 744–746.
14. Howell, L. L. and Wilcox, K. M. (2001) The dopamine transporter and cocaine medication development: drug self-administration in nonhuman primates. *J. Pharmacol. Exp. Ther.* **298,** 1–6.
15. Laakso, A. and Hietala, J. (2000) PET studies of brain monoamine transporters. *Curr. Pharm. Des.* **6,** 1611–1623.
16. Prochiantz, A., Di Porzio, U., Kato, A., Berger, B., and Glowinski, J. (1979) In vitro maturation of mesencephalic dopaminergic neurons from mouse embryos is enhanced in presence of their striatal target cells. *Proc. Natl. Acad. Sci. USA* **76,** 5387–5391.
17. Lau, Y. S., Hao, R., Fung, Y. K., et al. (1998) Modulation of nigrostriatal dopaminergic transmission by antisense oligodeoxynucleotide against brain-derived neurotrophic factor. *Neurochem. Res.* **23,** 525–532.
18. Lau, Y. S., Fung, Y. K., Trobough, K. L., Cashman, J. R., and Wilson, J. A. (1991) Depletion of striatal dopamine by the N-oxide of 1-methyl-4-phenyl-1,2,3, 6-tetrahydropyridine (MPTP). *Neurotoxicology* **12,** 189–199.

VII

BEHAVIORAL ASSESSMENT AND OTHERS

33

Laboratory Analysis of Behavioral Effects of Drugs of Abuse in Rodents

Neil M. Richtand

1. Introduction

Because the neurological effects of drugs of abuse are ultimately reflected in behavioral alteration, rodent behavioral assays are often an important component of drug abuse research. Unfortunately, although there are numerous excellent sources of information for equipping a laboratory for cellular and molecular analysis, far fewer resources are available to guide the researcher in building and equipping a behavioral lab. In our experience this problem is more difficult to navigate both in terms of information available from academic, as well as commercial, resources. Here we describe information we found very useful in designing a laboratory to measure the behavioral effects of drugs of abuse in rodents.

2. Materials

A clear determination of available funding and space, the two major variables determining the parameters for the planned behavioral lab, should precede all development. Prices for individual components of equipment vary widely, depending on source. Comparison shopping is highly recommended. Residential Activity Chambers may be stacked two-high on shelving constructed to hold the boxes in order to conserve space (*see* **Note 1**).

1. Residential activity chambers: The residential activity chamber serves to limit extraneous visual and auditory stimuli during behavioral testing. In addition, the activity chambers may be used to house rodents throughout the course of long-term behavioral studies, limiting the contribution of conditioning to observed behavioral effects. Residential activity chambers are available from several

Fig. 1. Residential activity chamber, *front view*. Removable Plexiglas top is added to interior Plexiglas enclosure to prevent rodents from jumping out of interior enclosure during long-term studies.

scientific commercial sources. Alternatively, a local cabinetmaker (such as Cline Builders, Covington, KY) will be able to manufacture chambers, in our experience for a fraction of the cost of obtaining them through a scientific vendor. Our residential activity chambers were constructed as follows (*see* **Fig. 1**): The chamber is a box with exterior dimensions of 26-inches wide × 25-inches tall × 24-1/2-inches deep. Interior dimensions are 22-inches wide × 21-inches tall × 24-inches deep. The sides are constructed of an outer layer of 3/4-inch melamine, middle layer of 1/2-inch acoustic barrier, and inner layer of 3/4-inch melamine. The back is 1/2-inch melamine. The door is constructed of melamine, and held with a continuous piano hinge and latch. A "camera port" hole in the chamber top is sealed with Plexiglas on the inside roof (**Fig. 2**). A smaller hole in the chamber roof accommodates both microdialysis tubing, as well as electrical wire to a light fixture in the chamber roof (**Fig. 2**). A blower fan (Dayton Model 2C782 blower, Dayton Electric Manufacturing Co., Chicago) affixed to the outside back of the chamber ensures air circulation and provides constant background noise (**Fig. 3**). Experimental data are automatically collected into computers equipped with software to run the system (**Fig. 4**).
2. Photobeam monitors: These are available from several commercial sources, including San Diego Instruments (San Diego, CA) and Columbus Instruments (Columbus, OH). We employ a Flex-Field Activity System (San Diego Instru-

Laboratory Analysis of Rodent Behavior

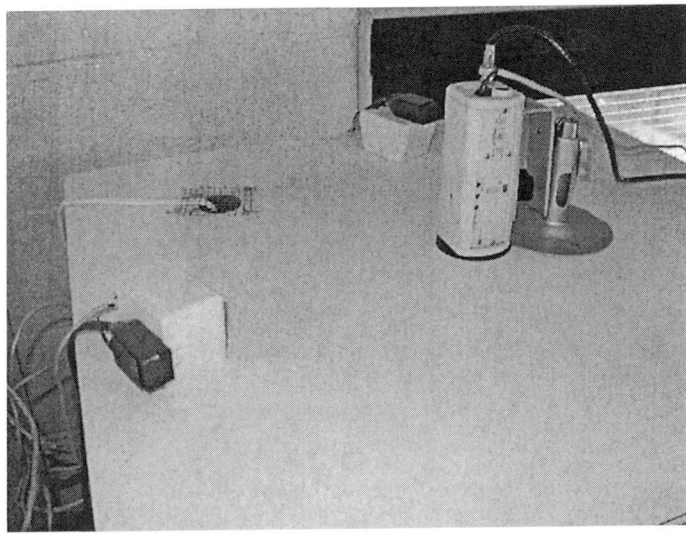

Fig. 2. Residential activity chamber, *top view*. Surveillance camera is attached by camera mount above camera port in center of roof. A hole is available for light fixture wire and microdialysis tubing.

Fig. 3. Residential activity chamber, *rear view*. Blower fans are mounted at center rear of chamber. Electrical wire from interior light fixture is connected to timer for synchronization of light cycle.

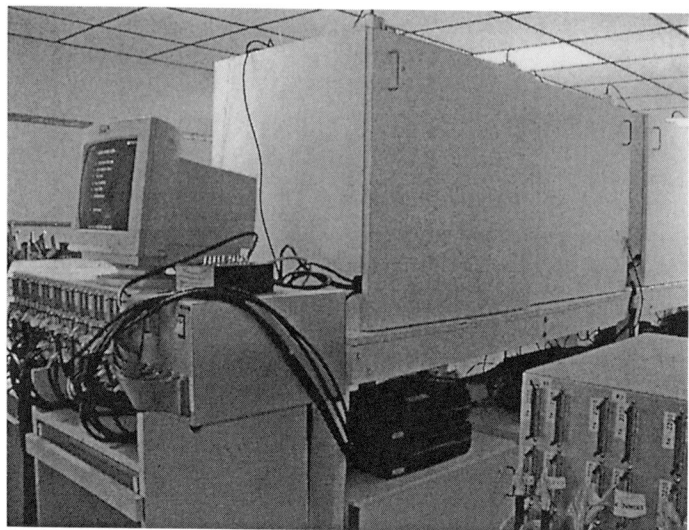

Fig. 4. Experimental data are automatically collected into computers equipped with software to run the system. One computer is needed for each 10 residential activity chambers.

ments), which includes a 16 × 16 photobeam array, surrounding a 16-inch × 16-inch × 15-inch Plexiglas enclosure. The locomotion photobeams are placed at a height of 1.5 inches above the enclosure floor for rats, and 0.5 inches above the enclosure floor for mice. A second photobeam array may be placed above the locomotor photobeam array to monitor animal rearings. The 16-inch × 16-inch chamber size is sufficiently large to accommodate zone maps for either rat or mouse distinguishing between time spent in proximity to chamber walls vs central zones.
3. Video camera: We have equipped each chamber with a Burle 1/3-inch BW high-resolution video surveillance camera, equipped with Computar 4–8 mm lens. Each camera is directly interfaced to an individual videocassette recorder.

3. Methods

3.1. Locomotion

Measures of locomotion are particularly well suited to research studies because data collection may be automated, once use of the specific measure is validated. Validation of the method of analysis by direct observation of the behaving animal, however, must precede automation. The two most commonly employed measures of locomotion are "distance traveled," and "crossovers." "Distance traveled" measurements are calculated from the pattern of photobeam breaks using software supplied by the photobeam manufacturer. "Crossovers" are calculated by dividing the chamber into quadrants, and tabulating a "crossover" each time the rodent enters one of the predefined quadrants subdividing the chamber. Again, this measure may generally be calculated from the pattern of photobeam breaks using software supplied by the manufacturer.

The calculated locomotion measure may be validated by videotaping a session, and replaying the session on a video monitor with masking tape placed vertically and horizontally bisecting the video image of the behavioral chamber. The locomotor measure calculated by computer is compared to the locomotor measure determined manually in the test session. We have employed manual observation of rodents treated with increasing amphetamine doses (0.1, 2.5, 5, and 10 mg/kg) to validate computer-generated locomotion measures. Because "distance traveled" and "crossovers" are not identical measures, choice of a particular locomotion measure will depend on the given study and specific drug of abuse studied.

3.2. Stereotypical Behaviors

Stereotypy refers to repetitive, perseverative patterns of behavior. Measurement of stereotypy can be accurately determined only by direct observation of the animal, and cannot be calculated from analysis of photobeam breaks. This translates into a labor intensive undertaking for experiments analyzing the effect of drugs of abuse on stereotypy.

A variety of rating scales to estimate stereotyped behaviors quantitatively have been described *(1–4)*. In practice, we have found measurement of the percentage of time engaged in stereotyped behavior to be the simplest method of analyzing stereotyped behavior *(5)*. Observers rate behavior for 30 s of each 3-min interval (*see* **Note 2**). The percent of each 30-s interval engaged in stereotyped behaviors including focused sniffing, repetitive chewing, licking, or biting, or repetitive head and limb movements is determined *(5)*.

4. Notes

1. The specific plans for equipment housing animals will in almost all cases require approval of the local Institution Animal Care and Use Committee. It is highly recommended to review plans with a representative of the committee at an early stage of the planning process.
2. Measures of stereotyped behavior are subjective. It is important that raters be blind to animal treatment. For best results, two raters should score animals independently and a measure of inter-rater reliability established. In addition, stereotypy scores should be compared to automated measures of data collection (such as locomotor measurement) to ensure that the independent measurements are consistent (i.e., locomotion goes down as stereotypy goes up).

Acknowledgments

This work was supported by the Department of Veterans Affairs Medical Research Service, and by a NARSAD Essel Investigator Award.

References

1. Creese, I. and Iversen, S. D. (1973) Blockage of amphetamine induced motor stimulation and stereotypy in the adult rat following neonatal treatment with 6-hydroxydopamine. *Brain Res.* **55,** 369–382.
2. Fray, P. J., Sahakian, B. J., Robbins, T. W., Koob, G. F., and Iversen, S. D. (1980) An observational method for quantifying the behavioural effects of dopamine agonists: contrasting effects of *d*-amphetamine and apomorphine. *Psychopharmacology (Berl.)* **69,** 253–259.
3. Rebec, G. V. and Segal, D. S. (1980) Apparent tolerance to some aspects of amphetamine stereotypy with long-term treatment. *Pharmacol. Biochem. Behav.* **13,** 793–797.
4. MacLennan, A. J. and Maier, S. F. (1983) Coping and the stress-induced potentiation of stimulant stereotypy in the rat. *Science* **219,** 1091–1093.
5. Segal, D. S. and Kuczenski, R. (1987) Individual differences in responsiveness to single and repeated amphetamine administration: behavioral characteristics and neurochemical correlates. *J. Pharmacol. Exp. Ther.* **242,** 917–926.

34

Conditioned Place Preference

A Simple Method for Investigating Reinforcing Properties in Laboratory Animals

Ezio Carboni and Cinzia Vacca

1. Introduction
1.1. Aim

A large number of studies (more than 800) have been conducted with the conditioned place preference (CPP) method. The relative popularity of CPP is due to its wide variety of applications. It indeed allows assessment of the rewarding property of drugs or other natural reinforcers and to identify the neuronal basis of reinforcement expression. Destruction of drug-induced CPP, obtained either by local or systemic drug administration, or also by lesioning discrete brain areas, can prove the involvement of specific neuronal circuits in the expression of CPP (*see* **refs. *1–3*** for recent reviews). The aim of this chapter is to provide methodological details together with a theoretical background that justifies the choice of a particular application.

1.2. Basis of Place Preference and Place Aversion
1.2.1. Place Preference

The CPP protocol is generally based on three phases: pretest, conditioning, and test. In the pretest investigators estimate the preference of the experimental animal, for each of two different environments of CPP apparatus, that can be recognized for visual, tactile, or olfactory cues. This estimation is expressed as the time spent in each environment while the animal is moving freely between the two. In the conditioning phase, the animal is paired alternately, in one of the two environments (usually the nonpreferred one), with the drug under investigation for its potential motivational effects or other unconditioned stimulus (US), and in the other environment, without any specific stimulus. Number and length of conditioning periods may vary. After the conditioning, the animal without any treatment, is tested by placing it in the apparatus where can freely move between the two environments. An increase in the time spent in the environment in which the animal has experienced the rewarding stimulus is considered CPP *(4)*. In particular, drugs, such as drugs of abuse, induce CPP because their administration produces an intense pleasurable sensation (as reported by humans) that acts as US. The association between the rewarding effect of the drug and the affective state experienced in the drug-paired environment allows the acquisition of incentive properties by the environment, that from neutral stimulus (NS) becomes conditioned stimulus (CS). These properties are expressed during the test, by an impulse toward the place where animals have experienced the rewarding affective state and consequently resulted in an increase in the time spent in that environment (with respect to the time spent in the same environment before conditioning).

1.2.2. Place Aversion

Conditioned place aversion (CPA) is based on the same principles of CPP. Because the US can be also aversive, the administration of a drug with "aversive" properties/effects will have the potential, by establishing a negative association with the drug-paired environment, of expressing repulsion toward that environment. In general, the aversive stimulus will be paired with the preferred environment during conditioning. During the test the animal will tend to avoid the environment in which it has experienced the aversive affective state.

1.3. Advantages and Disadvantages of CPP

One of the main advantages of CPP is that the animal is tested in a drug free condition, so confounding effects due to drug administration can be excluded. One criticism to CPP is that novelty-seeking behavior can confound the estimation of the preference, and consequently the reinforcing effect of a drug. Indeed the compartment paired with the drug investigated will look new to the animal on the day of the test because it will be explored in a drug-free state. This new sensation may induce the animal to increase the time spent in that environment because the experienced novelty. It has been reported *(5)* that rats show CPP for a novel compartment when tested in a two-compartment apparatus, of which only one has been previously explored. An apparatus equipped with a compartment that has never been explored by the animal can be used to monitor the importance of novelty in the preference of the animal. It has been reported that, in this condition, the time spent in the novel environment is significantly less than that spent in the drug (morphine or amphetamine)-paired environment, although more than that spent in the saline-paired environment *(6,7)*.

1.4. Do Drugs of Abuse Elicit CPP?

The property of drugs of abuse to induce CPP has received much attention since the early years of development of CPP. Because many drugs of abuse induce CPP, this method has assumed now the potential of being discriminatory for those substances for which abuse potential is unknown. Discrimination of rewarding or aversive properties of drugs can be obtained also by self-administration. In the large majority of cases a drug that produces CPP is also self-administered. On the contrary, drugs that produce place aversion are not. Here are briefly summarized the characteristics of the most studied drugs of abuse in CPP procedures (for a comprehensive review *see* **ref.** *1*).

1.4.1. Psychostimulants

The ability of psychostimulants, such as amphetamine and cocaine, to induce CPP is supported by a broad range of data. Other drugs that share the characteristic of cocaine and amphetamine of increasing synaptic dopamine (DA) concentration, such as nomifensine, GBR 12783, and methylphenidate *(8,9)*, have also been reported to induce CPP.

1.4.2. Opiates

A large number of studies reveal that morphine *(10)*, heroin *(11)*, methadone *(12)*, morphine-6-glucuronide *(13)*, and nonpeptide δ opiate agonists *(14)* induce CPP. Conversely, opiate antagonists such as naloxone and naltrexone can block CPP induced by agonists *(15,16)*. Further, naloxone and naltrexone have been also shown to produce CPA in naive rats *(15,17)* as well as in animals chronically treated with morphine *(18,19)*. In particular, these studies open the question on the role of endogenous opiates in reinforcement. Indeed, if it is clear that morphine-induced CPP can be easily prevented by the antagonist pretreatment, CPA induced by naloxone in naive animals (although negative results have been reported on this issue) can be explained only by admitting the existence of endogenous μ-receptor-mediated tonic role in the expression of reward/reinforcing mechanisms. Alternatively, or concurrently, endogenous opiate transmission may mediate incentive learning of behavior tasks that can be disrupted by μ-antagonists. A support of this hypothesis is provided by the antagonist action of naloxone or related compounds on ethanol- and nicotine-induced CPP *(20)*.

1.4.3. Nicotine

Several studies reported that nicotine induced CPP *(21,22)* whereas other studies were not able to confirm this finding *(23,24)*. The incongruity of available data may be due to crucial points of the procedure used, such as the dose. Lack of studies using chronically exposed rats to nicotine does not support clarification of the matter although previous exposure to nicotine allows the expression of nicotine-induced CPP *(24)*.

1.4.4. Ethanol

CPP induced by ethanol has been reported to be strongly influenced by the species, strain, and methodology used. In particular, ethanol has been reported to produce both CPP in rats *(25)* as well as in mice *(26)* and CPA *(27,28)*. These opposite effects can be related to the time-lag administration-conditioning in which the metabolite acetaldehyde can possibly play a role *(29,30)*. A further controversy is due to the report that ethanol produced dose-dependent CPA when self-injected *(31)*. Studies in rats that have a genetic preference for ethanol have not clarified this issue, although they have suggested that ethanol preferring rats have an innate reduced aversion for ethanol *(28)*. The aversive effect of ethanol may be due to administration route *(32)*, taurine levels *(33)*, or genetic preference for ethanol *(32,33)* (*see* **Note 1**).

1.4.5. Cannabinoids

Although the high human consumption should prove without any doubt the reinforcing effects of cannabinoids, it has been reported that they induce both CPP *(34)* and CPA *(35)*. On the other hand, cannabinoids do stimulate DA transmission in the shell of nucleus accumbens (NAc) similar to other drugs of abuse, such as morphine, cocaine, and nicotine *(36)*. As seen for nicotine and ethanol, administration and procedural characteristics as well as sensitization process may influence evaluation of CPP properties.

1.4.6. Caffeine

Caffeine is the most widely utilized central nervous system (CNS) stimulant by humans, although it cannot be considered as a drug of abuse. Nevertheless indications on its ability to induce CPP are not univocal. Low doses of caffeine as well as theophylline seem to produce CPP whereas high doses induce CPA *(37)*. Lack of stimulant action on DA output in the NAc differentiates this class of compounds from psychostimulants and opiates *(38)*.

1.5. Behavioral Reinforcers

Natural reinforcers stimulate DA transmission and produce CPP-like drugs of abuse. In particular food (only when was eaten during conditioning, but not when was only seen or smelled) produced CPP *(39)*. Exposition to food can also activate DA transmission with a particular pattern that involves novelty and habituation *(40)*. Sexual behavior (i.e., intercourse with receptive female, and ejaculation), maternal behavior, and social interaction can also produce CPP *(1)*. Some studies pointed to interactions between drugs of abuse and behavior, in particular, isolation rearing decreases the reinforcing properties of morphine and amphetamine *(41,42)*.

1.6. Dopamine Role in Place Preference

A large number of studies confirm that drugs of abuse as well as natural reinforcers not only produce CPP, also stimulate DA transmission in areas of the extended amygdala such as NAc shell *(38,43)*, and the bed nucleus of stria terminals (BNST) *(44)*. The role of DA transmission in the expression of reward/reinforcement has been confirmed by studies with DA antagonists or by lesioning dopaminergic areas. Indeed it has been reported that the specific D_1 receptor blocker SCH 23390 impairs the acquisition of rewarding properties of drugs, such as amphetamine, morphine, and nicotine *(45,46)*. The finding that SHC 23390 prevents also CPA due to aversive drugs that do not stimulate DA transmission, such as naloxone, phencyclidine, and lithium *(46,47)* has been used to propose an alternative role of DA in CPP and CPA (*see* **Note 2**). Indeed if the D_1 antagonist blockade, by preventing DA receptor interaction, prevents reward induced by amphetamine it is rather puzzling understanding how D_1 blockade can block CPA. A possible explanation has been provided *(37,42)* hypothesizing that DA transmission plays a role in the acquisition of the association between the primary unconditioned stimulus (reward or punishment) and the secondary (conditioned stimulus) environment. According to this hypothesis DA antagonists would act by interrupting the process of associative learning that in turn allows the expression of CPP or CPA. A further support of DA role in reinforcement is provided by findings showing that rats self administer nomifensine (DA reuptake inhibitor) when injected in the NAc shell but not in the caudate *(48)*. A mixture of dopamine D_1/D_2 agonists such as quinpirole and SKF 38393 is also self-administered when injected in the NAc *(49)*.

1.7. Interactions Among Drugs

A way to prove the specificity of a drug effect is the administration of an antagonist prior the agonist. Examples of antagonist action in CPP are provided by mecamylamine antagonism of nicotine-induced CPP *(21)* or by naloxone antagonism of morphine-induced CPP as discussed in the preceding. Mecamylamine can also produce CPA when administered in nicotine-dependent rats *(50)*. Antagonism can be also indirect as suggested by the action of 5-HT$_3$ antagonist in blocking nicotine CPP *(22)*. These types of studies may allow to investigate whether the interaction with CPP is at cue perception level, or at learning process level (*see* **Note 3**).

1.8. Local Injections and Lesions

The use of local drug injections in CPP procedures offers a wide spectrum of pharmacological applications. First, the local injection of lesioning agents can be used to investigate the role of definite brain areas in the expression of CPP. When brain lesions are performed, area specificity can be ascertained whereas no direct information on the transmitter involved in the disrupted process is available unless a neuronal specific drug is used. Indeed the fact that 6-hydroxydopamine (6-OHDA) lesions of NAc blocked novelty-induced CPP *(51)* may support the involvement of DA transmission in that process. Moreover, lesions of striatum and prefrontal cortex did not affect morphine-induced CPP *(52)*, indicating that these areas are not essential in that process. Lesions with ibotenic acid, kainic acid, or quinolinic acid or electrolytic lesions are also used to produce lesions in specific brain areas (*see* **Note 4**). Second, drugs known to induce CPP may be locally injected to investigate the role of definitive brain areas in their reinforcing/rewarding mechanism. Specificity of this effect can be obtained by locally injecting agonists or antagonists. The evidence that injection of morphine into the ventral tegmental area (VTA) induced CPP *(53)*, an effect that was antagonized by naloxone, offers an example of how area and receptor specificity can be ascertained in CPP studies. Amphetamine also produced CPP when injected into the NAc but not in the medial prefrontal cortex, the amygdala, and the striatum *(53–56)*. Nevertheless, the precise role of an area in CPP process may remain partially obscure. Indeed, because expression of CPP involves a complex learning process that includes experimentation, acquisition, association, and recalling, the identification of the disrupted subprocess may be difficult (*see* **Note 5**).

2. Materials

One of the reasons for the large diffusion of CPP is the relative inexpensiveness apart for animals that represent the main cost. It is possible to start with a pilot apparatus and manual time measurement. Later an apparatus with an automatic estimation of time (e.g., photobeam measurement), video-tracking, and eventually locomotion estimation can be bought from specialized companies (e.g., Columbus, Columbus, OH, USA).

2.1. Animals

1. Different strains of rats: Effectiveness of drugs inducing CPP can be influenced by the strain of rats used. Lewis rats have shown higher sensitivity than Fisher

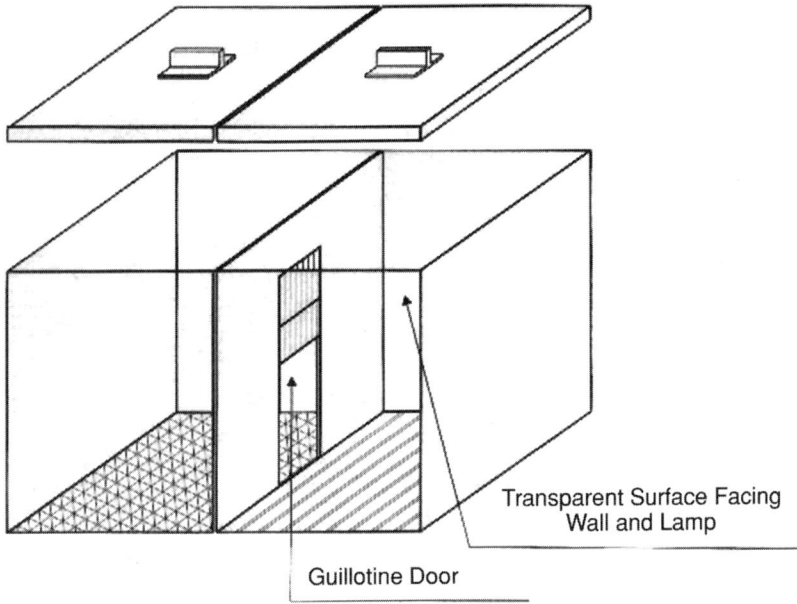

Fig. 1. Schematic drawing of a two-compartment apparatus for conditioned place preference.

344 rats to nicotine *(57)*- and cocaine *(58)*-induced CPP. Sprague–Dawley rats, on the other hand, are more sensitive than Wistar rats to the CPP induced by morphine *(59)*. The different sensitivity of rats may generate doubts on the efficacy of methodology. Again, differences may be found when tested drugs do not produce a strong CPP. In the studies performed in our laboratory *(22,46,47)* we have used male Sprague–Dawley rats (Charles River, Calco, Italy) weighing 200–250 g at the start of the experiment. All animals have been individually housed at the start of the experiment with food and water freely available. Animals have been housed under an artificial 12 h light–dark cycle (light at 0700 h) and with constant temperature (23°C) and relative humidity (60%) (*see* **Note 6**).
2. Control group: A control group is absolutely mandatory in CPP procedure. A control group has to be sorted, housed, administered (with the solvent used), and exposed to the paired environment exactly in the same manner as the test group. If an intracerebroventricular injection or a local injection is performed, control rats must be implanted with the same type of cannula and injected as the treated group at the same time interval prior to exposition (*see* **Note 7**).

2.2. Apparatus

1. Two-compartment apparatus: A two-compartment apparatus (illustrated in **Fig. 1**) may consist of two square base compartments (height 38 × 30 × 30 cm): one with

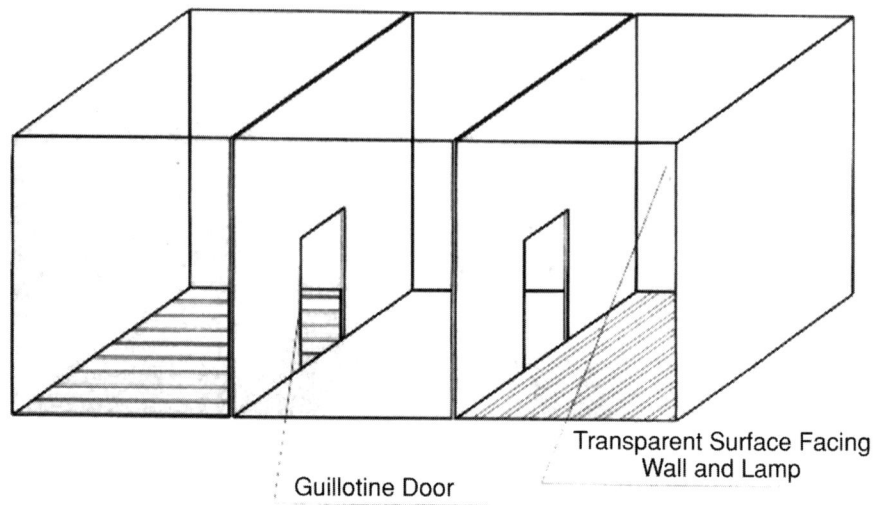

Fig. 2. Schematic drawing of a three-compartment apparatus for conditioned place preference.

white and the other with gray walls (except for the front wall facing the lamp) separated by a guillotine door to match the respective wall. The door has to be kept closed during the conditioning period while it is open during the pretest and the test. A transparent cover is placed above each compartment.

2. Three-compartment apparatus: A three-compartment apparatus (illustrated in **Fig. 2**) may consist of two square base compartments (height 38 × 30 × 30 cm), one with white and the other with gray walls. Between two compartments, there is a smaller compartment (height 38 × 15 × 30 cm) with walls and floor of clear gray. Covers and doors were similar to the two compartment apparatus. The apparatus can be made of 4 mm thick Plexiglas. Covers can be made of transparent Plexiglas which is sheltered with a semitransparent film to allow the experimenter to see the rat but not vice versa. Six apparatus can be placed 50 cm far from each other and 50 cm far from a wall where one fluorescent lamp (40 W) per apparatus is placed. To avoid dark corners in the apparatus, each compartment has a transparent wall facing the lamp to provide uniform illumination. It is advisable to keep apparatus in a soundproof room with 60-dB white noise. When the video camera is used to record the sessions, it is placed above two apparatuses. It is suggested to have the experiment room attached to a small room provided with washing facilities. In some tests, tactile cues can be included to allow distinction among compartments. One floor can be made by 2-mm diameter rods placed in a frame at 1.5-cm distance. The second floor can be made by 4 mm high squares (1 × 1 cm) placed on a base at 1 cm distance from each other. Alternatively a wire mesh floor can be used. Ease of cleaning has to be considered in choosing the type of floor. When a third compartment is used,

the floor is left smooth. It is possible to include in the CPP procedure olfactory cues by using different bedding or other particular odors (*see* **Note 8**).

3. Methods

3.1. Experimental Procedures

Procedures may vary. A standard experiment may last 8 d and consists of three phases: preconditioning (pretest), conditioning, and post-conditioning (test).

3.1.1. Preconditioning

During the first phase (d 1, 2, and 3; preconditioning) the guillotine door is kept lifted and each rat is given access to both compartments of the apparatus for 15 min. On d 3 the time spent by the rat in each compartment is recorded. Rats showing "freezing" postures during the test (about 10% of the total) are discarded. Measure of the time spent in a compartment starts when an animal has both forepaws in that compartment. The time recorded indicates the "unconditioned preference" of each rat for each compartment.

When animals do not show a strong preference for one compartment (unbiased design) animals can be assigned randomly to a group, then tested in the preconditioning phase, treated, and tested in the post-conditioning phase, no matter for their preference in the pretest (*see* **Note 9**). A second possibility is that animals show a clear preference for one environment. In this case a drug with potential of abuse can be paired with the less preferred compartment while the control group will receive saline in the nonpreferred compartment. This type of study (biased design) is the most used although it may produce false-positive results. Indeed whatever is the action of a drug, it is difficult to see an increase of time spent in preferred compartment (*see* **Note 10**). A third possibility occurs when, even in the presence of some animals that show a strong preference, animals are chosen in a way that the mean of each group is not significantly different from the others.

3.1.2. Conditioning

During conditioning, environment cues are established because the effect of the reinforcing drug in a manner that they become secondary reinforces. In this phase (d 4–7) the rats are administered drugs and are placed for 30 or 60 min in the "preferred" or "nonpreferred" compartment, depending on whether CPA or CPP is expected. After an interval of about 4 h, rats are administered with saline and placed in the other compartment. Care has to be taken to balance the daily order of exposure to each compartment. Alternatively only one conditioning session can be performed each day so that the conditioning phase has a duration of 8 d instead of 4 d. As a result of this conditioning phase, drugs or saline are paired four times to a specific compartment.

3.1.3. Post-Conditioning Test

During the post-conditioning phase (d 8, post-conditioning test), 24 h after the last injection, the guillotine door is removed and the time spent by each rat in the drug-paired compartment is recorded during 15 min of observation. The difference in seconds between the time spent in the drug-paired compartment during the post-conditioning test and that spent in the preconditioning test is taken as a measure of the degree of conditioning induced by the drug (*see* **Note 11**). A stopwatch that is started when the animal enters one context and is stopped as soon as the animal leaves the context is sufficient to estimate CPP. When more than two animals are tested at the same time, an automatic estimation of time (by infrared photobeam) has to be preferred.

3.2. Experimental Designs

3.2.1. Multiple vs Single-Trial Conditioning

It is known that drug of abuse vulnerability depends on the effect of both first experience reward and/or repeated treatment exposures *(60)*. Also in animals, drug-induced CPP may involve tolerance and sensitization processes *(61)*. To clarify this issue some authors have tested capability of morphine *(62)* and amphetamine *(63)* to induce CPP on a single exposition to the non preferred compartment. These studies show that intravenous injection of 1–3 mg/kg of amphetamine induces CPP or stimulates locomotor activity through a DA mechanism as it can be blocked by both the D_1 (SCH 23390) and the D_2 (Eticlopride) antagonists *(63)*. Injection of morphine can also produce CCP on a single treatment *(62)*.

3.2.2. Drug Administration

3.2.2.1. Systemic Administration

Administration route has received much attention in CPP procedures because of its influence on CPP expression. Cocaine, when injected intravenously *(64)* or intracerebroventricularly *(65)*, produced CPP that could be disrupted by haloperidol or pimozide, respectively. Systemic administration of cocaine instead was not blocked by DA antagonists *(65,66)*. Among administration-related factors that could influence CPP procedure, injection characteristics (time of the day, skills of experimenter, location of the apparatus) may become inhibitory when a nonrewarding stimulus is administered. At the same time, factors related to pharmacokinetics may become relevant because the association between the US (drug or other stimulus) and the context must be close in time and therefore depends on the availability of drug at site of action at time of exposition to the context.

3.2.2.2. Intracranial Administration

A drug that produces its reinforcing or CPP disrupting properties by acting in a specific area may be administered locally. Moreover an intracerebroventricular administration can be performed when a drug is not available in sufficient amount to perform systemic injection. In both cases a cannula has to be implanted in the brain. Guide cannulae can be aseptically implanted in the lateral ventricle (coordinates from *71*): A = 0.9; L = 1.3; V = 3.3 mm from the dura or in the area of interest under ketamine (80 mg/kg i.p.) and xylazine (10 mg/kg i.p.) anesthesia. Cannulas can be fixed to the skull with three screws and acrylic cement. Rats can be allowed to recover for 1 wk before testing. It must be considered that the cannula has to stay in place for 10–14 d to allow to complete a CPP study.

3.2.2.3. Time After Injection

To have the maximal expression of the coupling between the US and the context the animal must be exposed when the drug is producing the peak effect, which in turn depends on the administration route (*see* **Note 12**).

3.2.3. Organizing the Experiment

The number of groups and number of animals per group is the first item to be defined when planning an experiment. Considering that the procedure is rather time consuming, a well planned experiments is necessary to optimize time use. Conditioning represents the most time-consuming phase. Considering that each rat may be exposed for 20–30 min, and about 20–30 min are necessary for cleaning the apparatus, even with five or six apparatuses available, conditioning each group of five or six rats will take about 1 h. On the other hand, reducing the number of rats per group owing to high animal variability may produce inconsistent results. In conclusion, no more than four groups can be run, especially if two exposition per day are programmed. In any case during conditioning, rotation of exposition has to be done to avoid rats to be exposed always in the early morning or at the lunch time. Cleaning the apparatus is necessary to avoid that odors of previously exposed animals become a noncontrollable variable. Washing the apparatus with water and the same soap used for the housing cages is the best.

4. Notes
1. Understanding the expression of CPP by both ethanol and nicotine is more complex than that of other drugs of abuse, such as cocaine (DA reuptake blocker) or morphine (µ-opiate agonists). On the other hand, responses to ethanol and nicotine in humans may not be rewarding or may even be aversive at the beginning of use. Only later, after chronic use, a strong dependence is developed by

a process that involves complex phenomena such as tolerance and dependence. The rewarding effects of nicotine and ethanol depend strongly on a continuous exposure. Indeed although the first experience with nicotine is frequently aversive, the smoker driven by a strong psychological motivation soon develops tolerance to the aversive effect of nicotine, learning to appreciate the stimulating and rewarding effects that drive to dependence. Alcohol dependence also develops slowly only on a long period of recreational use in a limited portion of population, often in predisposed individuals who learn to control impairing effects of ethanol. In experimental animals, both nicotine *(67)* and ethanol *(68)* are not easily self-administered, but on continuous exposure both drugs can be self-administered. Sensitization and tolerance appear to play an important role in the expression of CPP by both ethanol and nicotine.
2. This hypothesis is supported by recent evidence showing that conditioned taste aversion is also blocked by the D1 antagonist SCH 391160. In particular this study has shown that a rewarding stimulus such as saccharin drinking can become aversive if followed by intraperitoneal injection of lithium chloride, which produces a visceral malaise. SCH 23390 given 5 min after saccharin prevents the association between the taste of saccharin and the malaise induced by lithium. In other words the D_1 antagonist would block the formation of the short-term memory of the gustatory taste to be later associated with the negative state produced by lithium *(69)*.
3. In any case, this type of experiment has to be conducted absolutely at the same time with a saline control group and a group treated with the agonist. Results showing a blockade of CPP by an antagonist without testing a group in which the agonist produces CPP can be misleading.
4. When local injections are performed, extreme care must be taken to avoid diffusion of a volume too large that can cause nonspecific tissue damage. To prove specificity of site, the drug can be injected in a close vicinity.
5. CPP was produced by bilateral injection either by the D_1 agonist SKF 38393 or the D_2 agonist quinpirole *(70)*. To prove specificity of drug, the nonactive stereoisomer can be injected. These and other findings recently have been summarized by McBride *(2)*.
6. Because this apparently simple behavior is rather complex and is affected by many variables, it is important to sort randomly animals into each group. It is necessary to avoid picking up rats from the cage just as they let themselves be captured.
7. Because CPP is related to motivational status (a water-deprived rat will show a preference for the environment paired with water), the effect of a rewarding drug many vary depending on the stress level of the rat. The stressful effect of keeping rats in a single cage for several days also must be considered.
8. Plexiglas is preferred to wood based material because it can be cleaned more easily and does not become imbedded with odors. Panels (3 mm thick) may offer the advantage of lightness, but may be less resistant to continuous handling for cleaning procedures.

9. Such a design can be obtained by choosing cues in a way that animals do not show a strong preference for one compartment. A white compartment will be avoided in favor of a gray one, allowing one to perform a biased experiment. A solution to a strong bias in preference for one context can be obtained by changing the conditions of the apparatus. For example, if the white compartment is strongly avoided, adding bedding in the nonpreferred compartment will reduce the difference in time spent between compartments. Tactile cues in two compartments made of an identical material may allow one to obtain an unbiased protocol.
10. On the other hand, an increase in time spent may be considered as a reduction of aversion to the nonpreferred compartment. This last possibility should not be considered as a negative aspect of CPP because reduction of aversion is undoubtedly and fully rewarding. (Morphine is rewarding when is administered in a painful status [aversive status] to alleviate pain.)
11. In a manual assessment a strict observance of blindness of the experimenter is mandatory to avoid false-positive results. Moreover, all the trials have to be conducted by the same person, who must be able to administer animals with precise avoidance of any stress to them. Unfortunately this can be a problem when students are involved in the trial, and in this case, a crucial variable in a study that lasts several months can be represented by differences in administering animals between the beginning and the end of the study. Obtaining a false-positive result means that, in different conditions, the results cannot be reproduced by others, and often also by the group who first obtained it, generating waste of time, money, and, above all, confusion in the literature. For the same reason, animals should be used only once.
12. It is possible to obtain information on time course of drug effects in the literature, if drugs have been used in other behavioral or biochemical tests (e.g., evaluation of neurotransmitter or metabolite release by in vivo microdialysis studies). In the absence of any information, a pilot study can be conducted by injecting the animal and observing appearance and disappearance of a particular behavior.

Acknowledgments

The authors thank Mrs. Adelaide Marchioni for typing assistance and Mrs. Emma Zanda for technical assistance. This work was supported by Ministero Istruzione Università e Ricerca (MIUR) project PRIN 2001-2002 and by Consiglio Nazionale delle Ricerche (CNR).

References

1. Tzschentke, T. M. (1998) Measuring reward with the conditioned place preference paradigm: a comprehensive review of drug effects, recent progress and new issues. *Prog. Neurobiol.* **56,** 613–672.
2. McBride, W. J., Murphy, J. M., and Ikemoto, S. (1999) Localization of brain reinforcement mechanisms: intracranial self-administration and intracranial place-conditioning studies. *Behav. Brain Res.* **101,** 129–152.

3. Bardo, M. T. and Bevins, R. A. (2000) Conditioned place preference: what does it add to our preclinical understanding or drug reward? *Psychopharmacology* **153,** 31–43.
4. Bozarth, M. A. and Wise, R. A. (1981) Intracranial self-administration of morphine into the ventral tegmental area in rats. *Life Sci.* **28,** 551–555.
5. Mucha, R. F. and Iversen, S. D. (1984) Reinforcing properties of morphine and naloxone revealed by conditioned place preferences: a procedural examination. *Psychopharmacology* **82,** 241–247.
6. Bardo, M. T., Neisewander, J. L., and Pierce, R. C. (1989) Novelty-induced place preference behavior in rats: effects of opiate and dopaminergic drugs. *Pharmacol. Biochem. Behav.* **32,** 683–689.
7. Parker, L. A. (1992) Place conditioning in a three- or four-choice apparatus: role of stimulus novelty in drug-induced place conditioning. *Behav. Neurosci.* **106,** 294–306.
8. Martin-Iverson, M. T., Ortmann, R., and Fibiger, H. C. (1985) Place preference conditioning with methylphenidate and nomifensine. *Brain Res.* **332,** 59–67.
9. Le Pen, G., Duterte-Boucher, D., and Costentin, J. (1996) Place conditioning with cocaine and the dopamine uptake inhibitor GBR12783. *NeuroReport* **25,** 2839–2842.
10. Mucha, R. and Herz, A. (1986) Preference conditioning produced by opioid active and inactive isomers of levorphanol and morphine in rat. *Life Sci.* **38,** 241–249.
11. Schenk, S., Ellison, F., Hunt, T., and Amit, Z. (1985) An examination of heroin conditioning in preferred and nonpreferred environments and in differentially housed mature and immature rats. *Pharmacol. Biochem. Behav.* **22,** 215–220.
12. Steinpresis, R. E., Rutell, A. L., and Parrett, F. A. (1996) Methadone produces conditioned place preference in the rat. *Pharmacol. Biochem. Behav.* **54,** 339–341.
13. Abbott, F. V. and Franklin, K. B. (1991) Morphine-6-glucuronide contributes to rewarding effects of opiates. *Life Sci.* **48,** 1157–1163.
14. Longoni, R., Cadoni, C., Mulas, A., and Di Chiara, G. (1998) Dopamine-dependent behavioural stimulation by non-peptide delta opioids BW373U86 and SNC 80: 2. Place preference and brain microdialysis studies in rats. *Behav. Pharmacol.* **9,** 9–14.
15. Shippenberg, T. S. and Herz, A. (1986) Differential effects of mu and kappa opioid systems on motivational processes. *NIDA Res. Monogr.* **75,** 563–566.
16. Bardo, M. T. and Neisewander, J. L. (1987) Chronic naltrexone supersensitizes the reinforcing and locomotor-activating effects of morphine. *Pharmacol. Biochem. Behav.* **28,** 267–273.
17. Acquas, E., Carboni, E., Garau, L., and Di Chiara, G. (1990) Blockade of acquisition of drug-conditioned place aversion by 5HT3 antagonists. *Psychopharmacology* **100,** 459–463.
18. Kesley, J. E. and Arnold, S. R. (1994) Lesions of the dorsomedial amygdala, but not the nucleus accumbens, reduce the aversiveness of morphine withdrawal in rats. *Behav. Neurosci.* **108,** 1119–1127.

19. Funada, M., Schutz, C. G., and Shippenberg, T. S. (1996) Role of delta-opioid receptors in mediating the aversive stimulus effects of morphine withdrawal in the rat. *Eur. J. Pharmacol.* **300,** 17–24.
20. Biala, G. and Langwinski, R. (1996) Rewarding properties of some drugs studied by place preference conditioning. *Pol. J. Pharmacol.* **48,** 425–430.
21. Fudala, P. J, Teoh, K. W., and Iwamoto, E. T. (1985) Pharmacologic characterization of nicotine-induced conditioned place preference. *Pharmacol. Biochem. Behav.* **22,** 237–241.
22. Carboni, E., Acquas, E., Leone, P., and Di Chiara, G. (1989) 5HT$_3$ receptor antagonists block morphine- and nicotine- but not amphetamine-induced reward. *Psychopharmacology* **97,** 175–178.
23. Clarke, P. B. and Fibiger, H. C. (1987) Apparent absence of nicotine-induced conditioned place preference in rats. *Psychopharmacology* **92,** 84–88.
24. Shoaib, M., Stolerman, I. P., and Kumar, R. C. (1994) Nicotine-induced place preferences following prior nicotine exposure in rats. *Psychopharmacology* **113,** 445–452.
25. Bozarth, M. A. (1990) Evidence for the rewarding effects of ethanol using the conditioned place preference method. *Pharmacol. Biochem. Behav.* **35,** 485–487.
26. Cunningham, C. L., Niehus, J. S., and Noble, D. (1993) Species difference in sensitivity to ethanol's hedonic effects. *Alcohol* **10,** 97–102.
27. van der Kooy, D., O'Shaughnessy, M., Mucha, R. F., and Kalant, H. (1983) Motivational properties of ethanol in naive rats as studied by place preference. *Pharmacol. Biochem. Behav.* **19,** 441–445.
28. Stewart, R. B., Murphy, J. M., McBride, W. J., Lumeng, L., and Li, T. K. (1996) Place conditioning with alcohol in alcohol-preferring and -nonpreferring rats. *Pharmacol. Biochem. Behav.* **53,** 487–491.
29. Amit, Z. and Smith, B. R. (1985) A multi-dimensional examination of the positive reinforcing properties of acetaldehyde. *Alcohol* **2,** 367–370.
30. Suzuki, T., Shiozaki, Y., Moriizumi, T., and Misawa, M. (1992) Establishment of the ethanol-induced place preference in rats. *Arukoru Kenkyuto Yakubutsu Ison* **27,** 111–123.
31. Stewart, R. B. and Grupp, L. A. (1986) Conditioned place aversion mediated by orally self-administered ethanol in the rat. *Pharmacol. Biochem. Behav.* **24,** 1369–1375.
32. Ciccocioppo, R., Panocka, I., Froldi, R., Quitadamo, E., and Massi, M. (1999) Ethanol induces conditioned place preference in genetically selected alcohol-preferring rats. *Psychopharmacology* **141,** 235–241.
33. Quertemont, E., Lallemand, F., Colombo, G., and De Witte, P. (2000) Taurine and ethanol preference: a microdialysis study using Sardinian alcohol-preferring and nonpreferring rats. *Eur. Neuropsychopharmacol.* **10,** 377–383.
34. Braida, D., Pozzi, M., Cavallini, R., and Sala, M. (2001) Conditioned place preference induced by the cannabinoid agonist CP 55,940: interaction with the opioid system. *Neuroscience* **104,** 923–926.

35. Cheer, J. F., Kendall, D. A., and Marsden, C. A. (2000) Cannabinoid receptors and reward in the rat: a conditioned place preference study. *Psychopharmacology* **151,** 25–30.
36. Tanda, G., Pontieri, F. E., and Di Chiara, G. (1997) Cannabinoid and heroin activation of mesolimbic dopamine transmission by a common mu1 opioid receptor mechanism. *Science* **276,** 2048–2050.
37. Brockwell, N. T., Eikelboom, R., and Beninger, R. J. (1991) Caffeine-induced place and taste conditioning: production of dose-dependent preference and aversion. *Pharmacol. Biochem. Behav.* **38,** 513–517.
38. Di Chiara, G., Tanda, G., Bassareo, V., et al. (1999) Drug addiction as a disorder of associative learning. Role of nucleus accumbens shell/extended amygdala dopamine. *Ann. NY Acad. Sci.* **877,** 461–485.
39. Maes, J. H. and Vossen, J. M. (1993) Context conditioning: positive reinforcing effects of various food-related stimuli. *Physiol. Behav.* **53,** 1227–1229.
40. Bassareo, V. and Di Chiara, G. (1997) Differential influence of associative and nonassociative learning mechanisms on the responsiveness of prefrontal and accumbal dopamine transmission to food stimuli in rats fed ad libitum. *J. Neurosci.* **17,** 851–861.
41. Wongwitdecha, N. and Marsden, C. A. (1995) Isolation rearing prevents the reinforcing properties of amphetamine in a conditioned place preference paradigm. *Eur. J. Pharmacol.* **279,** 99–103.
42. Wongwitdecha, N. and Marsden, C. A. (1996) Effect of social isolation on the reinforcing properties of morphine in the conditioned place preference test. *Pharmacol. Biochem. Behav.* **53,** 531–534.
43. Di Chiara, G. (1995) The role of dopamine in drug abuse viewed from the perspective of its role in motivation. *Drug Alcohol Depend.* **38,** 95–137
44. Carboni, E., Silvagni, A., Rolando, M. P. T., and Di Chiara, G. (2000) Stimulation of in vivo dopamine transmission in the bed nuclesus of stria terminalis by reinforcing drugs. *J. Neurosci.* **20,** RC102 1–5.
45. Leone, P. and Di Chiara, G. (1987) Blockade of D-1 receptors by SCH 23390 antagonizes morphine- and amphetamine-induced place preference conditioning. *Eur. J. Pharmacol.* **135,** 251–254.
46. Acquas, E., Carboni, E., Leone, P., and Di Chiara, G. (1989) SCH 23390 blocks drug-conditioned place-preference and place-aversion: anhedonia (lack of reward) or apathy (lack of motivation) after dopamine-receptor blockade? *Psychopharmacology* **99,** 151–155.
47. Acquas, E. and Di Chiara, G. (1994) D1 receptor blockade stereospecifically impairs the acquisition of drug-conditioned place preference and place aversion. *Behav. Pharmacol.* **5,** 559–565.
48. Carlezon, W. A. and Wise, R. A. (1996) Rewarding actions of phencyclidine and related drugs in nucleus accumbens shell and frontal cortex. *J. Neurosci.* **16,** 3112–3122.

49. Ikemoto, S., Glazier, B. S., Murphy, J. M., and McBride, W. J. (1997) Role of dopamine D1 and D2 receptors in the nucleus accumbens in mediating reward. *J. Neurosci.* **17,** 8580–8587.
50. Suzuki, T., Ise, Y., Maeda, J., and Misawa, M. (1999) Mecamylamine-precipitated nicotine-withdrawal aversion in Lewis and Fischer 344 inbred rat strains. *Eur. J. Pharmacol.* **369,** 159–162.
51. Pierce, R. C., Crawford, C. A., Nonneman, A. J., Mattingly, B. A., and Bardo, M. T. (1990) Effect of forebrain dopamine depletion on novelty-induced place preference behavior in rats. *Pharmacol. Biochem. Behav.* **36,** 321–325.
52. Bals-Kubik, R., Ableitner, A., Herz, A., and Shippenberg, T. S. (1993) Neuroanatomical sites mediating the motivational effects of opioids as mapped by the conditioned place preference paradigm in rats. *J. Pharmacol. Exp. Ther.* **264,** 489–495
53. Phillips, A. G. and Le Piane, F. G. (1980) Reinforcing effects of morphine microinjection into the ventral tegmental area. *Pharmacol. Biochem. Behav.* **12,** 965–968.
54. Carr, G. D. and White, N. M. (1986) Anatomical dissociation of amphetamine's rewarding and aversive effects: an intracranial micro-injection study. *Psychopharmacology* **89,** 340–346.
55. Josselyn, S. A. and Beninger, R. J. (1993) Neuropeptide Y: intraaccumbens injections produce a place preference that is blocked by cis-flupenthixol. *Pharmacol. Biochem. Behav.* **46,** 543–552.
56. Schildein, S., Agmo, A., Huston, J. P., and Schwarting, R. K. (1998) Intraaccumbens injections of substance P, morphine and amphetamine: effects on conditioned place preference and behavioral activity. *Brain Res.* **790,** 185–194.
57. Horan, B., Smith, M., Gardner, E. L., Lepore, M., and Ashby, C. R., Jr. (1997) (–)-Nicotine produces conditioned place preference in Lewis, but not Fischer 344 rats. *Synapse* **26,** 93–94.
58. Kosten, T. A., Miserendino, M. J., Chi, S., and Nestler, E. J. (1994) Fischer and Lewis rat strains show differential cocaine effects in conditioned place preference and behavioral sensitization but not in locomotor activity or conditioned taste aversion. *J. Pharmacol. Exp. Ther.* **269,** 137–144.
59. Shoaib, M., Spanagel, R., Stohr, T., and Shippenberg, T. S. (1995) Strain differences in the rewarding and dopamine-releasing effects of morphine in rats. *Psychopharmacology* **117,** 240–247.
60. Haertzen, C. A., Kocher, T. R., and Miyasato, K. (1986) Reinforcements from the first drug experience can predict later drug habits and/or addiction: results with coffee, cigarettes, alcohol, barbiturates, minor and major tranquilizers, stimulants, marijuana, hallucinogens, heroin, opiates and cocaine. *Drug Alcohol. Depend.* **11,** 147–165.
61. Schenk, S. and Partridge, B. (1997) Sensitization and tolerance in psychostimulant self-administration. *Pharmacol. Biochem. Behav.* **57,** 543–550.

62. Bardo, M. T. and Neisewander, J. L. (1986) Single-trial conditioned place preference using intravenous morphine. *Pharmacol. Biochem. Behav.* **25,** 1101–1105.
63. Bardo, M. T., Valone, J. M., and Bevins, R. A. (1999) Locomotion and conditioned place preference produced by acute intravenous amphetamine: role of dopamine receptors and individual differences in amphetamine self-administration. *Psychopharmacology* **143,** 39–46.
64. Spyraki, C., Fibiger, H. C., and Phillips, A. G. (1982) Dopaminergic substrates of amphetamine-induced place preference conditioning. *Brain Res.* **253,** 185–193.
65. Morency, M. A. and Beninger, R. J. (1986) Dopaminergic substrates of cocaine-induced place conditioning. *Brain Res.* **399,** 33–41.
66. Mackey, W. B. and van der Kooy, D. (1985) Neuroleptics block the positive reinforcing effects of amphetamine but notof morphine as measured by place conditioning. *Pharmacol. Biochem. Behav.* **22,** 101–105.
67. Caggiula, A. R., Donny, E. C., White, A. R., et al. (2001) Cue dependency of nicotine self-administration and smoking. *Pharmacol. Biochem. Behav.* **70,** 515–530.
68. Meisch, R. A. (2001) Oral drug self-administration: an overview of laboratory animal studies. *Alcohol* **24,** 117–128.
69. Fenu, S., Bassareo, V., and Di Chiara, G. (2001) A role for dopamine D1 receptors of the nucleus accumbens shell in conditioned taste aversion learning. *J. Neurosci.* **17,** 6897–6904.
70. White, N. M., Packard, M. G., and Hiroi, N. (1991) Place conditioning with dopamine-D1 and D2 agonists injected peripherally or into nucleus accumbens. *Psychopharmacology* **103,** 271–276.
71. Paxinos, G. and Watson, C. (1987) *The Rat Brain Stereotaxic Coordinates.* Accademic press, Sydney.

35

Detection of Cell Proliferation and Cell Fate in Adult CNS Using BrdU Double-Label Immunohistochemistry

Zin Z. Khaing and Mariann Blum

1. Introduction

Traditionally, qualitative and quantitative detection of cell proliferation in the central nervous system (CNS) has been performed using tritiated thymidine autoradiography. In more recent years immunohistochemical detection of the thymidine analog 5-bromo-2′-deoxyuridine (BrdU), which gets incorporated into DNA during the S-phase, has been used to detect proliferating cells in many tissues including brain (*1*). Single-stranded DNA with incorporated BrdU can then be detected using specific monoclonal antibodies raised against BrdU. The antibody staining can be readily identified in the nucleus, and the distribution can vary from a punctate pattern to homogeneous staining within the nucleus (*2*) (*see* **Fig. 1**).

Immunohistochemical detection methods have advantages over traditional tritiated thymidine labeling. First, the detection of tritiated nucleotides requires exposure to emulsion coating and it can take from 3 to 12 wk or more to develop whereas immunological techniques take between 1 and 3 d. Second, because the emulsion coating is only on the surface of the tissue sections and the β particles emitted can penetrate only about 1 µm (*3*), the signal seen is only from the most superficial part of the specimen. Therefore it is not possible to estimate proliferating cells using unbiased stereological methods within a brain structure. However, by using immunohistochemical detection of BrdU labeling, stereological methods can be utilized to estimate the number of

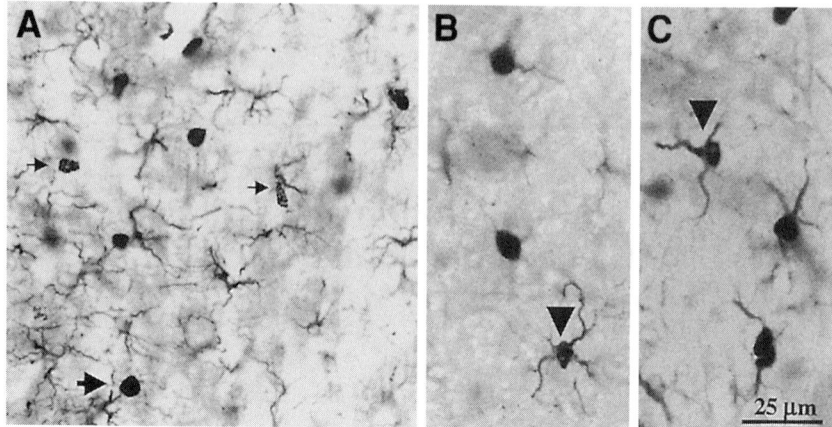

Fig. 1. BrdU and GFAP double immunohistochemistry. Immunolabeled sections from animals collected 10 d post-MPTP/7d post-BrdU injection *(4)*. BrdU-positive cells can be seen in both substantia nigra (**A**) and striatum (**B, C**). Examples of punctate (*small arrow*) and smooth (*large arrow*) nuclear profiles are shown. Cells double labeled with BrdU, in the nucleus, and GFAP, in the cytoplasm (*arrowhead*, **B, C**) are indicated.

proliferating cells within a structure. Finally, using immunohistochemistry, double labeling with cell-type specific markers may be performed to follow the fate of BrdU labeled cells (*see* **Notes** for technical considerations).

Because BrdU is integrated into replicating DNA only during the S-phase, a single pulse of BrdU labels only a fraction of all proliferating cells. The labeling index then is the proportion of labeled cells to the total number of cells in an area or structure. Multiple injections should be used to label a large portion of dividing cells. The number of BrdU injections needed to estimate the population of dividing cells within a structure will depend on cell cycle time and the length of the S-phase. In addition, the amount of BrdU within a particular nucleus can decrease if the cell divides after incorporation. Hence it is of importance to determine the appropriate post injection time to collect the tissue samples. Thus, one pulse BrdU labeling (with short post injection time, ~2 h) is used if one simply wants to know if any proliferation has occurred. Repeated injections of BrdU are done to estimate the population of dividing cells within a structure. To follow the fate of divided cells, however, longer survival time after BrdU injections may be required. By using double immunohistochemical staining, time of birth and morphological and neurochemical phenotypes of cells can be determined. Detailed procedures to label diving cells with BrdU and the detection of BrdU as well as other cell markers using immunological methods are presented.

2. Materials

1. BrdU: 20 mg/mL dissolved in 0.1 M Tris-HCl, pH 7.5. Store at –20°C.
2. Cryostat or vibrotome to cut thin (10–40 μm) sections.
3. 1 M phosphate buffer, pH 7.4.
4. 4% Paraformaldehyde in 0.1 M phosphate buffer, pH 7.4. To make 1 L, heat 900 mL of dH$_2$O to 55–60°C (never over 60°C); add 40 g of paraformaldehyde, 10 N NaOH dropwise until the solution becomes clear (~20 drops), and 100 mL of 1 M phosphate buffer. After the solution has cooled, the pH of the paraformaldehyde should be 7.2. Filter and store at 4°C (good for 14 d).
5. 30% Sucrose with 0.03% sodium azide (Sigma).
6. 30% H$_2$O$_2$.
7. 0.1 M Tris-HCl, pH 7.8.
8. Formamide.
9. 20X Saline sodium citrate (SSC): 0.3 M sodium chloride in 0.3 M sodium citrate buffer.
10. 0.1 M phosphate-buffer saline (PBS), pH 7.4.
11. 0.1 M PBS with 0.03% sodium azide (PBS-AZ).
12. 0.1 M PBS with 2 mM CaCl$_2$ and 6 mM MgCl$_2$ (PBS-Ca^{2+} and Mg^{2+}).
13. 2 N HCl.
14. 0.1 M Borate buffer, pH 8.5: 0.15 M sodium borate and 0.15 M boric acid. To make this buffer, sodium borate and boric acid are made up separately and then the sodium borate is titrated with boric acid to obtain the desired pH. Approximately four times the amount of boric acid is needed, so add generously until the solution gets near the desired pH value.
15. DNase I (Sigma).
16. Triton X-100.
17. Normal serum.
18. Bovine serum albumin (BSA, Sigma).
19. BrdU antibody (may be obtained from Sigma, Amersham, Harlan-Sera Labs as well as others).
20. Biotinylated secondary antibodies (Vector Laboratories, Inc. or Jackson Labs).
21. Fluorescent conjugated secondary antibodies (Vector Laboratories, Inc.).
22. Extra-Avidin Peroxidase (Sigma).
23. Peroxidase enzyme based substrates, such as 3, 3′-diaminobenzidine (DAB) or SG substrate (Vector Labs).
24. 4′,6-Diamidino-2′phenylindole (DAPI, Sigma).
25. Nonaqueous (Permount) and aqueous mounting media (Gel/Mount from Biomeda, Foster city, CA).

3. Methods

3.1. BrdU Injection

To label a large portion of dividing cells, for example, in response to brain injury, subcutaneous injections of BrdU (50–100 mg/kg) every 2 h for 8 h (five injections total) can be done *(4)*. Brains can be collected 2–15 h after the last BrdU injection. Alternatively, multiple survival time after BrdU injection may be done to follow the fate of cells divided (*see* **Note 1**).

3.2. Tissue Collection

1. Fixation: Perfuse the animals with ice-cold PBS for about 10 min until the perfusate becomes clear of blood, followed by 4% paraformaldehyde for 10 min. Remove brains and post-fix overnight in 4% paraformaldehyde at 4°C. The next day place brains in 30% sucrose solution containing 0.03% sodium azide for cryoprotection and store at 4°C until ready for use.
2. Freezing: If planning to cryosection, snap freeze brain tissue with methyl butane that is cooled on dry ice and store the frozen tissues at –20°C. No freezing is necessary if vibrotome sections are to be collected.
3. Sections: Mount and section brains to desired thickness using a vibrotome or cryostat. Collect freely floating sections into PBS-AZ. The sections can then be stored at 4°C.

3.3. BrdU Immunohistochemistry

3.3.1. DNA Denaturing

Because most monoclonal antibodies for BrdU are directed against single-stranded DNA, a DNA denaturing step is required before BrdU can be detected immunologically. There are three methods to denature DNA that have been commonly utilized: treatment with acid, a combination of formamide followed by acid, and nuclease digestion. Protocols to denature DNA with formamide followed by acid (2 N HCl) and nuclease (DNase I) are described in the following subheadings.

3.3.1.1. OPTION A: PROTOCOL ADAPTED FROM KUHN ET AL. *(5)*

1. Wash the sections with PBS, 0.6% H_2O_2 for 30 min at room temperature (RT) to block endogenous peroxidase activity (this step can be omitted if peroxidase detection method is not being used).
2. Denature DNA in 50% formamide/2X SSC for 2 h at 65°C.
3. Wash for 5 min in 2X SSC.
4. Incubate sections in 2 N HCl at 37°C for 30 min (incubation time can be varied).
5. Rinse for 10 min in 0.1 M borate buffer, pH 8.5, to neutralize the acid.
6. Wash several times in PBS.
7. Block with PBS/0.3% Triton X-100/3% normal serum for 30 min at RT.
8. Incubate with primary antibody anti-BrdU mouse monoclonal (Sigma 1:1500) at 4°C overnight.
9. Wash in PBS for three times, 10 min each.
10. Use desired detection method described in **Subheading 3.3.2**.

3.3.1.2. OPTION B

1. Wash the sections with PBS, 0.6% H_2O_2 for 30 min at RT to block endogenous peroxidase activity (this step can be omitted if not using peroxidase detection method).
2. Wash three times in PBS, 3 min each.
3. Block with PBS/0.1% Triton X-100/3% normal serum or 3% BSA for 30 min–1 h at RT.
4. Dilute primary antibody in PBS-Ca^{2+} and Mg^{2+}/0.3% Triton X-100/3% normal serum or 3% BSA/50–100 µg/mL of DNase I, and incubate at 37°C for 1 h.
5. Wash three times in PBS, 5 min each.
6. Use desired detection method.

3.3.2. Detection of Primary Antibody

3.3.2.1. Enzyme-Based Method

1. Incubate with secondary antibody (biotinylated anti-mouse, Vector Labs, 1:200) for 1 h at RT.
2. Wash three times in PBS, 10 min each.
3. Incubate sections with Extra-Avidin Peroxidase (1:200) for 1 h at RT.
4. Wash three times in PBS, 10 min each.
5. Perform peroxidase detection using DAB (0.25 mg/mL of DAB/0.01% H_2O_2) for 5 min (many other enzyme based substrates are available and can be used in this step).
6. Counter stain, dehydrate, and mount sections.

3.3.2.2. FLUORESCENT IMMUNOHISTOCHEMISTRY

1. Incubate with secondary antibody (e.g., anti-mouse fluorescein, Vector Labs, 1:200) for 1 h at RT.
2. Wash three times in PBS, 10 min each.
3. Counterstain using DAPI and mount sections.
4. Coverslip using aqueous mounting media.

3.4. Double Labeling with BrdU

If double labeling with BrdU is desired, option B using fluorescent immunohistochemistry is highly recommended (*see* **Note 2**). One can include another primary antibody with the BrdU antibody (given that the antibodies are from two different species). Then two species-specific secondary antibodies conjugated with different fluorophores can be used to detect the primary antibodies. However, if an enzyme based method for double-label immunohistochemistry is desired there are two different options: sequential detection of antibodies using the same enzyme (i.e., peroxidase) with two different substrates (e.g., DAB, SG) or a combination of two enzymes (e.g., peroxidase and alkaline phosphatase, both available from Vector Labs) can be used. Refer to manufacturer recommended protocols for various substrates.

The following protocol may be used if option A is desired.

1. Wash the sections with PBS, 0.6% H_2O_2 for 30 min at RT to block endogenous peroxidase activity (this step can be omitted if not using peroxidase detection method).
2. Wash three times in PBS, 10 min each.
3. Block with PBS/0.1% Triton X-100/3% normal serum or 3% BSA for 30 min–1 h at RT.
4. Incubate with primary antibody (not BrdU antibody) diluted in blocking buffer at 4°C overnight.
5. Wash three times in PBS, 10 min each
6. Fix sections lightly for 10–30 min in 4% paraformaldehyde at RT.
7. Then continue with BrdU pretreatment (**steps 2–7** from Option A, **Subheading 3.3.1.1.**).
8. Incubate section with BrdU antibody for 1 h at 37°C.
9. Wash three times in PBS, 10 min each.
10. Use desired detection method.

4. Notes

1. To examine the fate of cells that were going through S-phase at the time of BrdU injection, long survival times after BrdU injections can be performed. It is important to predetermine the length of time needed for cell specific markers to be expressed to establish appropriate post-BrdU injection time to kill the animals.
2. There are a few things to keep in mind when performing this protocol. The DNA denaturing procedure (option A) to detect incorporated BrdU molecules in single-stranded DNA is a rather harsh treatment. Although some antigens are very stable even after treatment in 2 N HCl (e.g., glial fibrillary acidic protein), others are not. Therefore, depending on the cell specific marker that is being detected by double labeling, the primary antibody can be incubated with the tissue section either before or after the denaturing steps (**steps 2–5**). Alternatively option B can be employed. Historically, the monoclonal antibody BU1 was characterized by the presence of nuclease in the supernatant (*6*) which allowed immunological detection of BrdU without harsh acid or alkali treatment. Thus by including DNase I nuclease in the primary antibody incubation step, BrdU antibody from any source can be used. Standard immunohistochemical techniques can then be used to detect the second antigen.

Acknowledgments

We thank Jeremy N. Kay for supplying the immunohistochemical material illustrated in this chapter and Dr. Joanne Cousins for helpful comments. This work was supported by NIH Grant AG08538.

References

1. Miller, M. W. and Nowakowski, R. S. (1998) Use of bromodeoxyuridine-immunohistochemistry to examine the proliferation, migration and time of origin of cells in the central nervous system. *Brain Res.* **457,** 44–52.
2. Takahashi, T., Nowakowski, R. S., and Caviness, V. S., Jr. (1992) BUdR as an S-phase marker for quantitative studies of cytokinetic behaviour in the murine cerebral ventricular zone. *J. Neurocytol.* **21,** 185–197.
3. Rogers, A. W. (1973) *Techniques of Autoradiography.* Elsevier, Amsterdam.
4. Kay, J. N. and Blum, M. (2000) Differential response of ventral midbrain and striatal progenitor cells to lesions of the nigrostriatal dopaminergic projection. *Dev. Neurosci.* **22,** 56–67.
5. Kuhn, H. G., Winkler, J., Kempermann, G., Thal, L. J., and Gage, F. H. (1997) Epidermal growth factor and fibroblast growth factor-2 have different effects on neural progenitors in the adult rat brain. *J. Neurosci.* **17,** 5820–5829.
6. Risio, M. (1994) Methodological aspects of using immunohistochemical cell proliferation biomarkers in colorectal carcinoma chemoprevention. *J. Cell. Biochem. Suppl.* **19,** 61–67.

Index

A

Acetylcholine, 56, 433
Adult neural progenitor cells, 33, 397, 499
Adult neural stem cells, 33, 499
Affymetrix, 246
Agarose gel, see gel
Alkaline phosphatase, 154, 333
Alzheimer's disease, 46
Ammonium persulfate, 264, 316
AMPA, 14
Amphetamine, 14, 34, 37, 82, 175, 322, 483
Amygdala, 82
Antisense approach
    efficacy and selectivity, 371-373
    implantation of guide cannula, 369-370
    infusion of oligonucleotide, 370
Aprotinin, 285
Ara-c, 381
β-D-arabino-furanoside, see Ara-c
Aseptic technique, 369
Autoradiography
    gel shift assay, 315-327
    in situ hybridization, 119-135, 137-151, 153-159
    receptor binding assay, 297-313
Azide, 408

B

B-27, 380
Basic fibroblast growth factor (bFGF), 35, 398
Behavioral neuroscience, 14, 75, 331, 475
Behavioral sensitization, 14, 18
Biting, 479
Borate buffer, 501
Brain-derived neurotrophic factor (BDNF), 36, 398
BrdU immunohistochemistry
    BrdU injection, 501
    detection with secondary antibody, 503
    DNA denaturing, 502
    primary antibody incubation, 503
    tissue collection, 502
5-Bromo-4-chloro-3-indolyl phosphate (BCIP), 154, 333
5-Bromo-2'-deoxyuridine (BrdU), 34, 499
Bromophenol blue, 212, 263, 317
Butanol, 264

C

Ca^{2+}/calmodulin-dependent protein kinase (CaMK), 279
Caffeine, 484
Calbindin, 406
cAMP response element-binding protein (CREB), 7, 87, 270, 279, 322, 331, 406
Cannabinoids, 484
CdK5, 7
Cell proliferation, 33, 397, 499
Cell viability, 385

c-fos
   gene expression, 4, 406
   Northern blot, 175
Chewing, 479
Choline oxidase, 434
Cocaine, 4, 14, 34, 294, 461, 483
Conditioned place preference
   conditioning, 489
   postconditioning, 490
   preconditioning, 489
   unbiased place conditioning apparatus, 345, 487-489
Connexin32 (Cx32), 283
Cultures
   primary striatal neuronal culture, 379-386
   mesencephalic neuron-glia culture, 387-395
   mesencephalic neuron-enriched culture, 387-395
   mesencephalic microglia-enriched culture, 387-395
   mesencephalic astroglia-enriched culture, 387-395
   adult neural progenitor culture, 397-404
   organotypic culture, 405-412
Cyclophilin, 175

D

DARPP-32, 406
Denhardt's reagent, 164
Dextran sulfate, 165
3,3'-Diaminobenzidine tetrahydrochloride (DAB), 276, 334, 408
Differential display PCR
   confirming, 204
   electrophoresis, 202
   introduction, 193-198
   PCR, 201
   reamplification, 204
   reverse transcription, 199
   RNA extraction, 199
Digoxygenin, 155
3,4-Dihydroxyphenylacetic acid (DOPAC), 446
Distance traveled, 478
Dithiothreitol (DTT), 164, 182, 263, 316
DNA
   array, 243
   cDNA macroarray, 243
   cDNA plasmid, 182
   cDNA primer, 212, 217
   cDNA probe, 156, 172, 201
   DNA-binding activity, *see* electrophoretic mobility shift assay
   DNA incorporation by BrdU, 499
   DNA probe for gel shift assay, 317
DNase, 198, 212
Dopamine and adenosine monophosphate regulated phosphoprotein (DARPP-32), 406
Dopamine and dopamine receptor
   D1 receptor mRNA, 142
   D2 receptor mRNA, 142
   receptor binding, see radioligand-binding assay
   receptor knockout, 351
   release, 55, 57, 81, 415-432, 462
Dopamine transporter, 465
Dopamine uptake
   data analysis, 470
   preparation of neural tissue, 468-469
   uptake assay, 469

Index

Dopaminergic neurons, 43
Double labeling
  in situ hybridization with immunohistochemistry, 141-143
  in situ hybridization with retrograde tracing, 140-141
  isotopic and nonisotopic, 153-159
DPX mounting medium, 155, 276
Dulbecco's minimum essential medium (DMEM), 380, 388
Dynorphin, 43

E

Electrochemical detection (ED), *see* HPLC
Electrochemical detector, 419
Electrode
  amperometric electrode, 419
  carbon fiber, 444
  coulometric electrode, 419
  reference, 419
  stimulating electrode, 445
Electrophoresis
  differential display PCR, 202-204
  gel-shift assay, 315-327
  Northern blot, 168-170
  Western blot, 263-271
  Electrophoretic mobility-shift assay (EMSA)
  DNA probe annealing, 317
  DNA probe design, 317
  DNA probe labeling, 318-319
  EMSA, 321-323
  film autoradiography, 323-324
  gel preparation, 321
  protein extraction, 319-321
Elk-1, 279
Embryonic cells, 351

Emulsion, 126
β-Endorphin, 43
Enkephalin, 43
Enzyme-linked immunosorbent assay (ELISA)
  data analysis, 290-293
  detection of antigen-antibody complex, 288
  protein capture onto microtiter plate, 287
  protein extraction, 286-287
Epidermal growth factor (EGF), 35, 398
Ethidium bromide, 198, 212, 247
Extracellular signal-regulated protein kinase (ERK), 279

F

Fast-scan cyclic voltammetry
  application, 458-462
  data analysis, 452
  electrical fabrication, 449-450
  electrical stimulation, 448-449
  electrochemistry, 445-448
  electrode calibration, 452
  electrode placement verification, 452
  place microelectrode, 451
  surgery, 450-451
Fluorogenic TaqMan probe, 231
FluoroGold, 140
Forebrain, 33

G

GABA, 384
Gap junctional intercellular communication (GJIC), 285
Gel
  agarose gel, 163, 198, 212, 247

polyacrylamide gel, 198, 264
Gel electrophoresis, *see*
electrophoresis
Gel shift assay, *see* electrophoretic
mobility-shift assay
Gene chip, 245
Glial fibrillary acidic protein
(GFAP), 35, 500
Gliogenesis, 35, 397
Glutamate receptor, 13-25
Glutaraldehyde, 434

H

Heroin, 483
Herpes simplex virus, for gene
infection, 331-333
High-density oligonucleotide arrays,
246
High-performance liquid
chromatography (HPLC)
acetylcholine, 433-441
dopamine, 415-432
5-HT, 426
norepinephrine, 426
Hippocampus, 33, 56, 77
Hybridization histochemistry
brain tissue preparation, 123, 139
emulsion slide dipping and
developing, 126-127
film autoradiography, 126
film quantification, 127
hybridization, 124, 139, 156
hybridization buffer, 154, 165,
183
isotopic riboprobe, 119-135
microscopic autoradiography,
126
pretreatment, 123, 155

post-hybridization washing, 125,
140
probe labeling, 124, 139, 155
sectioning, 123
silver grain quantification, 129
Hydrogen peroxide (H_2O_2), 275
5-Hydroxytryptamine (5-HT), 36,
55, 415

I

Immediate early gene, 4
Immunocytochemistry, *see*
immunohistochemistry
Immunohistochemistry
blockade of nonspecific staining,
277
DAB staining, 278
secondary antibody, 278
transcardial fixation, 276-277
primary antibody, 277
Inflammation, 43
Interleukin-1β (IL-1β), 46
In situ hybridization, *see*
Hybridization
histochemistry
In vivo microdialysis
collecting sample, 424-425
detection, 425-427
probe construction, 422-423
probe implantation, 423-424
Isotope, *see* radioligands

K

Knockout mice
blastocysts and ES cell
preparation, 358
blastocyst injection, 359
establishing a mouse colon, 358
gene cloning, 353-354

Index

homologous recombinant, 355-357
hybridization probe, 355
identification of mice carrying desired gene, 360
targeting construct, 354-355
transfer of blastocyst to foster mother, 359

Kodak
BioMax MR film, 198
hypoclearing agent, 155

Kynurenic acid, 398

L

Laemmli buffer, 263
Learning and memory, 75
Licking, 479
Limbic regions, 23
Lipopolysaccharide (LPS), 47
Locomotion, 478-479
Long-term depression (LTD), 13
Long-term potentiation (LTP), 13, 86

M

Macroarray
data analysis, 255-257
hybridization, 254
prehybridization, 253
probe purification and equilibration, 252
reverse transcription, 252
RNA extraction, 249-252
washing, 255

Matrigel, 398
Matrix, 405
Mazindol, 467
Membrane
nylon, 168
polyvinylidene difluoride (PVDF), 264

Methamphetamine, 36
1-Methyl-4-phenyl-1,2,3,6-tetrahydropyridine (MPTP), 37, 397
Metabotropic glutamate receptor (mGluR), 7, 22, 157
Methadone, 483
Methylenedioxymethamphetamine (MDMA), 55
Microarray, see macroarray
Microdialysis, see in vivo microdialysis
Mitogen-activated protein kinase (MAPK), 7
Microglia, 47
Microtome, 275
Microtubule-associated protein-2 (MAP2), 35, 384
MMLV reverse transcriptase, 198
Mobile phase
acetylcholine, 434
dopamine, 420
5-HT, 420
norepinephrine, 420
Modified Eagle medium (MEM), 388
Morphine, 37, 483
Morpholinopropanesulfonic acid (MOPS), 163, 212

N

Naloxone, 43, 483
Naltrexone, 483
Neurobasal (NB) medium, 381, 398
Neurogenesis, 35
Neuron specific enolase (NSE), 35

Neuronal nuclear antigen (NeuN), 35
Neurosphere, 402
Nichel chloride (NiCl$_2$), 276
Nicotine, 484
Nitric oxide (NO), 46
Nitric oxide synthase (nNOS), 25, 60
Nitroblue tetrazolium chloride (NBT), 154, 333
N-methyl-D-aspartate (NMDA), 14, 36, 86
Norepinephrine, 415
Northern blot
  cDNA probe labeling, 172-174
  film autoradiography, 175
  hybridization, 174
  nylon membrane transfer, 168-170
  RNA electrophoresis, 168-170
  RNA extraction, 166-168
  RNA probe labeling, 170-172
  RNA workplace setting, 165
Nucleus accumbens, 7, 18, 292, 304, 462

O

Okadaic acid, 316
Olfactory bulb, 34
Oligonucleotide
  antisense oligonucleotide, 369
  mismatched oligonucleotide, 369
  scrambled oligonucleotide, 369
  sense oligonucleotide, 369
Organotypic culture
  application of drug in slice culture, 410
  immunocytochemistry, 410
  preparation, 408-409

P

Pargyline, 468
Parkinson's disease
  acute MPTP model, 106
  chronic MPTP/Probenecid model, 108
  MPTP model, 103-116
  subacute MPTP model, 106
PCR
  differential display PCR, 193-209
  PCR primer, 185, 199
Phenol-chloroform-isoamyl alcohol (PCI), 162
Phenylmethanesulfonyl fluoride (PMSF), 285, 317
Phosphorylation
  CREB, 270
  pCaMK, 273
  pERK, 273
  pElk-1, 279
  phosphoprotein, 273
Photobeam, 476
Place preference, *see* conditioned place preference
Polyacrylamide gel, *see* gel
Poly-D-lysine, 333, 380, 388, 398
Polyvinylidene difluoride (PVDF), 264
Prefrontal cortex (PFC), 7, 23, 80
Preproenkephalin mRNA, 157
Primary striatal neuronal culture
  cell culturing, 383
  coating slides, 381
  dissecting striata, 381
  plating, 382
  suspension, 382
  trituration, 382
Protein
  electrophoresis, 265

ELISA, 283-295
gel-shift assay, 319-321
immunocytochemistry, 273-281
immunohistochemistry, 273-281
protein extraction, 264-265, 286
Western blot, 263-271
Protein kinase A (PKA), 7
Proteinase K, 155, 183
Pump, 275, 419

R

Radioligands
[^{3}H]dopamine, 467
[^{3}H]PD 128907, 299
[^{3}H]propylnorapomorphine, 297
[^{3}H]quinpirole, 297
[^{3}H]SCH 23390, 299
[^{3}H]Spiperone, 299
[^{3}H]YM 09151-2, 297
[^{125}I]iodosulpiride, 297
Radioligand-binding assays
assay protocol, 301
data analysis, 303
membrane homogenate preparation, 301
receptor autoradiography, 305-309
Residential activity chambers, 475-476
Reverse transcriptase, 212
Reverse transcription
differential display, 199
macroarray, 252
RT-PCR, 218
Reward Circuitry, 8
RIPA buffer, 285
RNA
in situ hybridization, 119-135, 137-151, 153-159

Northern blot, 161-179
RNAase protection assay 181-192
RNA electrophoresis, 168
RNA extraction, 166, 187, 199, 213, 231-232, 249-252
RNA polymerase, 183
RNA probe, 164, 183
RNase, 161-162, 164, 198
RNase H, 367
RNase protection assay
data analysis, 190
mRNA preparation, 187
plasmid construction, 184
plasmid linearization, 186
probe preparation, 187
protection assay, 188-190
RT-PCR analysis of mRNA
cDNA primer, 217, 232-233
data analysis, 220-223, 235-238
real-time PCR, 219
reverse transcription, 218
RNA extraction, 213-216, 231-232
TaqMan, 229-241
TaqMan PCR, 234-235

S

S100β, 35
Salmon sperm DNA, 165, 248
Slide
Nunc removable chamber coverslip, 380
Nunc removable chamber glass slide, 380
Sniffing, 479
Subgranular zone (SGZ), 397
Sodium chloride and sodium citrate (SSC), 154, 165, 248

Sodium dodecyl sulfate (SDS), 162, 165, 183, 263, 285
Sodium lauryl sulfate (SLS), 434
Sodium octyl sulfate (SOS), 434
Sodium orthovanadate, 285
Software
   Agilent Bioanalyzer, 248
   Array analysis, 248
   Array image capture, 248
   TINA, 248
   voltammetry, 444-445
Sonication buffer
   EMSA, 316
   Western blot, 264
Spectrophotometer, 213, 248
Stereotypy, 479
Striatum, 6, 18, 35, 47, 56, 78, 292, 407
Striosome, 405
Substantia nigra, 15, 37, 47
Subventricular zone (SVZ), 33
SYBR Green PCR kit, 212

T

Taq DNA polymerase, 198
Terminal deoxynucleotidyle transferase (TdT), 156
Tetramethylammonium chloride (TMA), 434
3,3',5,5'-Tetramethylbenzidine (TMB), 285
Thalamus, 294
Thermocycler, 213, 248
Thermus aquaticus (Taq), 229
Thionin, 445

Trituration, 382
Trypan blue, 381, 388, 398
Trypsin, 380, 389, 398
Tumor necrosis factor-α (TNF-α), 46

V

Ventral tegmental area, 6, 14, 486
Virus, 331
Virus-mediated gene transfer
   behavioral assessment, 344
   cell culture, 339-340
   histology, 342—344
   immunohistochemistry, 344
   surgery, 340-342
   titering the vector, 338-339
   viral packaging, 335-338
Voltammetry, see fast-scan cyclic voltammetry

W

Water-saturated phenol, 163
Western blot
   antibody detection, 267-269
   gel electrophoresis, 265
   protein extraction, 264-265
   protein quantification, 265
   transfer to PVDF membrane, 266-267
Withdrawal, 7

Y

Yeast tRNA, 155

Z

Zif/268, 4